The Application of Computational Methods in Industrial Fan Design

Edited by Geoff Sheard

Sigel Press

Sigel Press
51A Victoria Road
Cambridge CB4 3BW
England

4403 Belmont Court
Medina, Ohio 44256
USA

Visit us on the World Wide Web at: www.sigelpress.com

Internal Design: Professional Book Compositors, Lorain, Ohio, USA
Cover Design: Harp Mando
Cover Images: Alessandro Corsini, *Sapienza* University of Rome

ISBN 10: 1-905941-24-2
ISBN 13: 978-1-90541-24-7

A catalogue record for this book is available from the British Library.
Typeset in Times Roman

The publisher's policy is to use paper manufactured from sustainable forests.

Permissions Acknowledgements

Chapter 1, *A Critical Review of Computational Methods and Their Application in Industrial Fan Design*, was originally published in *International Scholarly Research Network, Mechanical Engineering*, Volume 2013, an Open Access publication. Reprinted under the terms of the Hindawi Publishing Corporation's Creative Commons Attribution license.

Chapter 2, *Computational Analysis of Noise Reduction Devices in Axial Fans with Stabilised Finite Element Formulations*, was originally published in *Computational Mechanics*, Volume 50. Reprinted with the permission of Springer. Permission to publish received 28 May, 2014.

Chapter 3, *Large-eddy Simulation of a Tunnel Ventilation Fan*, was published in *Proceedings of the 57th American Society of Mechanical Engineers Turbine and Aeroengine Congress*, paper number GT2012-69046 and subsequently in *Transactions of the American Society of Mechanical Engineers Journal of Fluids Engineering*, Volume 135. Reprinted with the permission of The American Society of Mechanical Engineers. Permission to publish received 2 December, 2013.

Chapter 4, *Large-eddy Simulation of the Aerodynamic and Aeroacoustic Performance of a Ventilation Fan*, was originally published in *Proceedings of the Fan 2012 Conference* and subsequently in *Advances in Acoustics and Vibration*, Volume 2013, an Open Access publication. Reprinted under the terms of the Hindawi Publishing Corporation's Creative Commons Attribution license.

Chapter 5, *A Numerical Investigation into the Aerodynamic Effect of Pressure Pulses on a Tunnel Ventilation Fan*, was originally published in *Proceedings of the Institute of Mechanical Engineering Part A, Journal of Power and Energy*, Volume 228, doi:10.1177/0957650913517881. Reprinted with the permission of SAGE Publications Limited. Permission to publish received 10 December, 2013.

Chapter 6, *A Parametric Study of Reversible Jet-fan Blades' Aerodynamic Performance*, was originally published in *Proceedings of the 57th American Society of Mechanical Engineers Turbine and Aeroengine Congress*, paper number GT2012-69920 and subsequently in *Transactions of the American Society of Mechanical Engineers Journal of Engineering for Gas Turbines and Power*, Volume 135. Reprinted with the permission of The American Society of Mechanical Engineers. Permission to publish received 2 December, 2013.

Chapter 7, *The Effects of Inlet Box Aerodynamics on the Mechanical Performance of a Variable Pitch in Motion Fan*, was originally published in *Advances in Acoustics and Vibration*, Volume 2012, an Open Access publication. Reprinted under the terms of the Hindawi Publishing Corporation's Creative Commons Attribution license.

Chapter 8, *The Conceptual Design of High-pressure Reversible Axial Tunnel Ventilation Fans*, was originally published in *Advances in Acoustics and Vibration*, Volume 2012, an Open Access publication. Reprinted under the terms of the Hindawi Publishing Corporation's Creative Commons Attribution license.

Chapter 9, *Predicting Blade Leading Edge Erosion in an Axial Induced Draft Fan*, was originally published in *Proceedings of the 56th American Society of Mechanical Engineers Gas Turbine and Aeroengine Congress*, paper number GT2011-45719 and subsequently in *Transactions of the American Society of Mechanical Engineers Journal of Engineering for Gas Turbines and Power*, Volume 134. Reprinted with the permission of The American Society of Mechanical Engineers. Permission to publish received 2 December, 2013.

Chapter 10, *Numerical Simulation of Coal Fly ash Erosion in an Induced Draft Fan*, was originally published in *Proceedings of the 56th American Society of Mechanical Engineers Gas Turbine and Aeroengine Congress*, paper number GT2012-69048 and subsequently in *Transactions of the American Society of Mechanical Engineers Journal of Fluids Engineering*, Volume 135. Reprinted with the permission of The American Society of Mechanical Engineers. Permission to publish received 2 December, 2013.

Chapter 11, *Predicting the Performance of an Industrial Centrifugal Fan Incorporating Cambered Plate Impeller Blades*, was originally published in *Budapest University of Technology and Economics Periodica Polytechnica, Mechanical Engineering*, Volume 58, an Open Access publication. Reprinted under the terms of the Budapest University of Technology and Economics Creative Commons Attribution license.

Chapter 12, *Simulation of Particle-laden Flows in a Large Centrifugal Fan for Erosion Prediction*, was originally published in *Proceedings of the 59th American Society of Mechanical Engineers Gas Turbine and Aeroengine Congress*, paper number GT2014-25865. Reprinted with the permission of The American Society of Mechanical Engineers. Permission to publish received 14 July, 2014.

Appendix 1, *A Comparison of U.S. and European Approaches to Regulating Fan Efficiency*, was originally published in *Proceedings of the CIBSE ASHRAE Technical Symposium 2014*, www.cibse.org/symposium. Reprinted with permission of the Chartered Institution of Building Services Engineers. Permission to publish received 28 May, 2014.

Appendix 2, *Variable Pitch Fans*, was originally published as Patent Number GB 2,485,634 B. Reprinted in accordance with the requirements of the Copyright, Designs and Patents Act, Section 30 (1) and Section 47 (1).

Contents

List of Figures

List of Tables

Foreword

Fans and air control products are different than most products because their performance varies widely depending on conditions which are unique to each air system design. Because every air system is different, how can a system designer accurately predict performance in the field? There are codes and standards that we can use to characterise performance. They enable us to compare different products on a like-for-like basis. Developing and adapting codes and standards by the air movement and control community is a form of self-regulation which has served the industry well for almost 100 years, until now.

Motor driven systems consume around one half of the electricity in the European Union. In 2005 the European Union identified these systems as a priority product group to analyse, with the objective of defining minimum efficiency requirements. The European Union set minimum efficiency requirements for fans in 2011 which became legally binding on 1 January 2013. As a result, the average product in the European market now has increased efficiency. The European Union imposed more stringent requirements on 1 January 2015, which will result in estimated energy savings of 34 TWh per year by 2020.

Given the importance of these products, the European Union is not the only jurisdiction that has regulated them. The United States was the pioneer in setting minimum requirements for electric motors. On 1 February 2013, the government published a framework document in the Federal Register outlining the intended approach to fan regulation that aims to eliminate inefficient fans within the United States by 2019.

With today's regulatory framework it is reasonable to assume that governments will require increased minimum efficiency over time, making an empirical approach to fan design increasingly less capable of dealing with the requirements. The resulting fan efficiencies will not meet the minimum allowable targets as traditional design techniques reach their limit. Therefore, there is increasing pressure for the air movement and control community to apply the same computational methods that are commonplace within the aerospace community. These methods enable engineers to optimise fan performance within the mechanical constraints imposed by each specific application. They enable engineers to design products that meet or exceed current and forthcoming minimum allowable fan and motor efficiency targets.

Unlike the aerospace industries' products, intrinsically non-aerodynamic features characterise air movement and control industry products. This makes it difficult to apply computational methods. Despite the challenges, the collection of papers in this volume illustrates how one can use an open-source computational code to both accurately and economically predict and improve fan performance. In collecting and editing these papers, Geoff Sheard has defined how the air movement and control community can apply aerospace developed computational methods, thus providing that community with a road-map. This road-map shows how to improve fan performance, and respond to both current and forthcoming fan efficiency requirements.

Marcos González Álvarez
Policy Officer
European Commission — Directorate General for Energy
1st March 2015

Opinions expressed in this foreword are those of the author and do not reflect necessarily the positions of the European Commission.

Acknowledgements

Many individuals have helped in writing the academic papers upon which this volume is based and it is my pleasure to acknowledge this. Foremost among these are the authors, Stefano Bianchi, Domenico Borello, Lucio Cardillo, Alessandro Corsini, Kasra Daneshkhah, Giovanni Delibra, Mario Fiorito, Michael Ivanovich Neil Jones, Andrea Marchegiani, Andreas Rhoden, Franco Rispoli, Tayfun Tezduyar and Paolo Venturini.

I also wish to thank Thomas Sigel, first for his tremendous contribution in turning around extensive scripts into succinct and easily readable text, and second for his assistance with the process of converting the final text into this published volume.

I would also like to acknowledge the contribution of the Institution of Mechanical Engineers (IMechE) library staff. They helped ensure that the references for the academic papers upon which this volume is based are both complete and technically correct. The resulting edited volume is a better reference text as a consequence.

I would like to acknowledge the contribution of Kirsten Greenzweig, who converted the original academic papers upon which this volume is based into electronic files that could then be edited and typeset. Kirsten also recreated the original figures in electronic form, an undertaking that proved to be far more time consuming then either of us imagined.

Geoff Sheard
Fläkt Woods Group

About the Editor

Wearing many hats, Geoff Sheard is the Fläkt Woods Group Vice President of Fan Technology, a leading global supplier of energy-efficient solutions operating in both the air climate for buildings and air movement for the infrastructure and industry markets. He is a director of Fläkt Woods Limited, Solyvent Fläkt Woods Limited and Fläkt Woods Fans (Australia) Pty Limited.

Geoff is an Honorary Professor at the Aston University Department of Engineering and Applied Science, a Visiting Professor at Sapienza University of Rome Dipartimento di Ingegneria Meccanica e Aerospaziale and a Visiting Professor at the University of Northampton Business School.

Geoff is President of the Air Movement & Control Association (AMCA), member of the AMCA Executive Board and Chairman of the European Air Movement & Control Association Board of Directors. He is also a member of the International Gas Turbine Institute (IGTI) Board of Directors and Vice Chairman of the Institution of Mechanical Engineers (IMechE) Fluid Machinery Group.

Geoff has doctorate degrees from the University of Oxford in turbomachinery aerodynamics and from the University of Northampton in leadership and team development. He also holds a master's degree in business administration from Cranfield University and a bachelor's degree in mechanical engineering from Liverpool University.

A chartered engineer, a Liveryman of the Worshipful Company of Engineers, a fellow of the Institution of Mechanical Engineers, a fellow of the Royal Aeronautical Society, a fellow of the American Society of Mechanical Engineers, and a fellow of the Chartered Institute of Building Service Engineers, Geoff has published widely in both technical and management areas. His publication list extends to 236 publications including 10 books, two monographs plus over 60 journal articles and patents. Geoff is editor of the *Journal of Management Development*, a member of the *Journal of Power & Energy* and *African Journal of Engineering* editorial boards, and the *Leadership and Organisational Development Journal* editorial review board.

Introduction: Computational Methods

Geoff Sheard

This volume of collected papers documents research that addresses the need for the air movement and control community to improve fan efficiency. The volume arises from discussions around thirteen papers and one patent that collectively comprise the chapters and appendixes. Each makes a contribution that collectively clarifies how the air movement and control community can apply computational methods developed by the aerospace community. They provide fan designers with the necessary knowledge to predict performance, and then go on to optimise that performance within the constraints imposed by each application.

The chapters relate to one of three separate, but complementary aspects of the reported research that collectively comprised a single fundamental research and applied development effort:

- Modelling turbulent flows in fans;
- Predicting fan system performance; and,
- Modelling erosion.

When an object moves through a fluid, forces arise due to the fluid's viscosity, inertia, elasticity and gravity. These forces are represented directly by the various terms in the Navier–Stokes equations, named after Claude-Louis Navier and George Gabriel Stokes, which describe the motion of substances that can flow. These equations arise from applying Newton's second law to fluid motion, together with the assumption that the fluid stress is the sum of a diffusing viscous term proportional to the gradient of velocity, plus a pressure term. Solving the Navier–Stokes equations is made challenging when regions of turbulent flow characterise the blade-to-blade flow-field. Turbulent flow regions inevitably characterise fans intended for application in air systems as a consequence of separated flow from electric motors, flat-plate motor supports and conduit. Therefore, developing robust modelling techniques has been a priority for scholars working with the air movement and control community, as they are a pre-requisite when solving the Navier–Stokes equations.

Developing robust modelling techniques enables designers to predict fan performance. If that predicted performance matches experimentally measured performance, we may reasonably assume that predicted blade-to-blade flow-field features are also a fair representation of those actually at play within the blade-to-blade passage. A characterisation of blade-to-blade flow-field features enables the impact of changing inlet and exhaust conditions on those features requiring evaluation. This then facilitates a prediction of the system's impact on a fan's performance. The air movement and control community typically tests fans in standardised airways, to facilitate a like-for-like comparison of fan performance with that of other fans. In practice, the community applies those self-same fans into a wide variety of air system designs. It is not practical to test every fan in every system configuration, and therefore it is helpful to use computational methods to characterise fan system performance. Insight into the impact of changing fan system configuration on a fan's blade-to-blade flow-field enables the designer to optimise a fan's design for a specific application.

In combination, the developed modelling techniques and techniques for modelling the system's impact within which a fan is installed result in an accurate application specific prediction of a fan's blade-to-blade flow-field. One may use this as the basis for prediction erosion. Some of the most demanding air movement and control applications are associated with the presence of erosive particles in the airstream. These particles in the airstream destroy the fan blade's aerodynamic profile, reducing both a fan's pressure developing capability and efficiency. When a fan is part of a wider process, erosion of both fan static and rotating components can limit the performance of the wider process. Consequently, the air movement and control community has an interest in predicting particles' erosive consequences on fan performance. Insight into the location and extent of erosion provides designers with a basis for optimising the positioning of wear-protection to extend fan in-service life. It also provides them with a basis for optimising fan geometry, producing designs that are inherently less susceptible to the effects of erosive particles.

Chapter 1: Literature Review

The adopted research philosophy was first, to characterise the turbulent flows associated with fans intended for application in air systems and second, to complement the characterisation by applying it to both the prediction of fan system performance and the modelling of erosion. However, researchers did not undertake characterisation of turbulent flows in a vacuum. Other scholars' findings working in both the aerospace and air movement and control communities informed the research.

Chapter 1 presents a review of computational methods and their application in air system fan design. The authors wrote this paper to place the work undertaken and reported in the other papers that form the body of this volume into context. The review identifies the key challenges those within the aerospace fan, aerospace compressor and air movement and control communities are addressing. The review's specific focus is on identifying how the air movement and control community can apply computational methods developed by, and for the aerospace community. The review identifies how one may use computational methods to elucidate aspects of the flow-field physics that one cannot easily study in a laboratory environment. Thus, the review provides a road map for researchers who are both developing computational methods and attempting to apply them to predict air system fan performance.

Chapters 2, 3 and 4: Modelling Turbulent Flows in Fans

Scholars within the air movement and control community have studied the application of computational methods to industrial fan design for two decades. Researchers developed commercial codes to a point where they modelled accurately the flow-field physics around the turn of the century. Modelling accurately the flow-field physics made the computational codes potentially useful analysis tools. In practice, the limited capability of the hardware upon which the computational codes ran limited their ability to accurately predict fan performance. By 2010, the cost of computer hardware had fallen far enough to enable all those within the air movement and control community to use computational grids fine enough to resolve the blade-to-blade flow-field features of interest.

Predicting the blade-to-blade flow-field features of interest is an essential precursor to predicting fan performance. However, a computational method that modelled accurately the flow-field physics and low cost computational hardware are not alone enough. The application of computational methods to predict the blade-to-blade flow-field is made challenging by the inevitability of separated flow from electric motors, flat-plate motor supports and conduit. Obtaining converged computational solutions is challenging. Consequently, the three chapters that follow focus on the modelling approaches that facilitate predicting the blade-to-blade flow-field. Collectively, they provide the necessary insight to facilitate predicting both fan aerodynamic and acoustic performance.

Chapter 2 addresses the fundamental issues associated with applying a computational method to the low-speed fans typical of those used in air systems. An issue when applying a computational method into any turbomachinery application is that even in the most favourable cases, Reynolds numbers are high resulting in complex flow-fields. A direct numerical simulation (DNS) of the Navier–Stokes equations would require fine computational grids and therefore a computational effort beyond that available within the air movement and control community. The only viable approach available when attempting to solve the Navier–Stokes equations is a Reynolds-averaged Navier–Stokes (RANS) simulation.

A key aspect of a Reynolds-averaged Navier–Stokes simulation is the use of turbulence models that constitute a range of simplifying computational approaches for modelling aspects of the blade-to-blade flow-field fluid-dynamics. These simplifying assumptions allow Reynolds-averaging to reduce the computational effort that occurs with a direct numerical simulation of the Navier–Stokes equations and therefore facilitate simulation of high Reynolds number flows. The authors based their Reynolds-averaged Navier–Stokes modelling on the streamline-upwind/Petrov– Galerkin (SUPG) and pressure-stabilising/Petrov–Galerkin (PSPG) methods. The authors present the rationale for the chosen stabilisation approach for the SUPG and PSPG formulations. The authors developed a stabilisation approach based on a diffusion for reaction dominated with jump (DRDJ) operator that facilitates the management of discontinuities. They then analyse the resultant formulations performance and characterise its interaction with the chosen turbulence model.

The characterisation of a Reynolds-averaged Navier–Stokes modelling technique presented in Chapter 2 provides fan designers with a way to predict fan performance. However, when modelling the physics of all turbomachinery flow, the limitations of a Reynolds-averaged Navier–Stokes approach hinder the accuracy with which one can model the blade-to-blade flow-field. Consequently, a Reynolds-averaged Navier–Stokes approach limits the accuracy with which one may calculate fan performance characteristics. This approach is not able to model the turbulence's spectral content and is not sensitive to the flow-field's multi-scale features. Scholars who have studied the flow-field physics within all turbomachinery classes are increasingly turning to large eddy simulations (LES) to overcome Reynolds-averaged Navier–Stokes approach limitations.

Chapter 3 presents a large eddy simulation modelling technique that reduces the required computational effort classically associated with a large eddy simulation. The principle issue with all large eddy simulations is the formidable required computational effort. In Chapter 3 the authors present a hybrid approach, using a computational grid too course for a classical large eddy simulation, coupled with a sub-grid scale model sensitive to each grid cell's size. Using this approach, the authors were able to recover part of the turbulence spectrum that is lost with a Reynolds-averaged Navier–Stokes approach and therefore to analyse the turbulent structures within a fan's blade-to-blade passage, reproducing the most relevant flow features.

Chapter 4 extends the large eddy simulation modelling technique presented in Chapter 3 by incorporating an acoustic model to predict fan noise. The authors use computed unsteady parameters as input to Ffowcs Williams and Hawkings' acoustic model. Thus, they were able to develop a tool that predicted both fan broadband noise and spectrum. The prediction was able to resolve the spectrum up to the fifth harmonic of blade passing frequency. Accuracy of the prediction was limited by the limitations inherent in the acoustic model that neglects duct modes. As a consequence, the acoustic model did not model the modal amplification of dipole-like noise sources. These characterise the inter-tonal and broadband noise when a fan is fitted in a ducted system. Despite the limitations of the acoustic model, agreement with experimentally measured broadband noise and spectrum was reasonable. A reasonable prediction of fan broadband noise and spectrum provides designers with an insight into the acoustic consequences of a change in fan geometry. It is therefore a useful tool, facilitating the extension of an aerodynamic and mechanical optimisation process into one that also considers the acoustic consequences of any change in fan geometry.

Chapters 5, 6, 7 and 8: Prediction of Fan System Performance

The research reported in Chapters 2, 3 and 4 collectively characterise the techniques needed to model turbulent flows in the low-speed fan typically used in air systems. In Chapters 5, 6, 7 and 8, the authors apply those techniques to fans associated with a range of systems. In Chapter 5 the authors study the impact of a pressure pulse on the performance of a central tunnel ventilation fan. They concluded that a pressure pulse too small to drive the studied fan into stall can still result in a doubling of the unsteady forces induced in fan blades. In Chapter 6 the authors optimised a jet fan's performance. Jet fans are classically used in longitudinal tunnel ventilation systems where they are required to operate in both forward and reverse directions. A need for reversibility constrains the aerodynamic design, making optimisation challenging. Chapter 7 presents an analysis of the effect of an inlet box on the mechanical performance of a variable pitch in motion fan. The inlet box turns the incoming flow by 90 degrees, resulting in periodic forces on fan blades. The authors minimised these forces with an asymmetric inlet design. Chapter 8 is an example of how one may use the application of computational methods to design a higher pressure reversible fan. As previously mentioned, a need for reversibility constrains the aerodynamic design of reversible fans. An ability to predict the aerodynamic performance for a range of blade designs facilitated the conceptual design of the reported range of high pressure reversible fans.

Chapter 5 presents the results of a research programme undertaken by the authors in response to in-service central tunnel ventilation fan failures. Although tunnel ventilation fan in-service failures are rare, engineers increasingly associate those failures with new or renovated tunnel systems with larger pressure pulses. Consequently, we require insight into the aerodynamic and mechanical consequences that occur with subjecting a tunnel ventilation fan to a pressure pulse. In the research reported in this chapter, the authors model the pressure pulse induced by a train in a tunnel system as a rapidly changing fan volume flow rate. The authors utilised the large eddy simulation modelling technique presented in Chapter 3. An analysis of the computational results provides an insight into the effects of turbulent structures that develop within the fan blade passage. The analysis indicates that a pressure pulse too brief to drive a tunnel ventilation fan into stall still results in a doubling of the unsteady aerodynamic and mechanical forces to which the blade is subjected. This is an increase significant enough to contribute to the tunnel ventilation fan's in-service mechanical failure.

Chapter 6 constitutes a complementary research programme to that presented in Chapter 5, focusing on the optimisation of reversible jet fans' used for both routing ventilation and smoke control in longitudinal tunnel ventilation systems. The authors utilised the Reynolds-averaged Navier–Stokes modelling technique presented in Chapter 2. They based their design methodology on a sensitivity analysis which they obtained from a response surface approximation. The sensitivity analysis calculates a rank and a weight for each design variable that affects the jet fan thrust and efficiency. It therefore facilitates insight into each design's relative performance. Having developed their design methodology, the authors apply it to a redesign of a baseline jet fan blade. The authors study the redesigned blade's aerodynamic and structural characteristics and compare it to that of the baseline design. They then manufactured the redesigned blade and measured the thrust and efficiency of a jet fan fitted with the baseline blade and then the redesigned blade. The redesigned blade's measured performance confirmed the predicted improvement in jet fan performance.

As with the research presented in Chapter 5, the author initiated the research reported in Chapter 7 following an in-service failure. In this case the failure was of the blade bearings of a variable pitch in motion fan intended for application in forced- and induced-draft boiler applications. Variable pitch in motion fans rotate at a constant speed, with the changing blade angle varying the load. A pitch-change mechanism facilitates the change in blade angle. A blade bearing supports each blade enabling it to rotate. The author observed that as the fan aerodynamic stage-loading progressively increased, so did the rate of blade bearing wear. The reported research addressed two separate, but linked needs. First, the on-going need to increase fan pressure development capability required an increase in fan loading. This increase was within the context of an erosive operating regime which systematically reduced fan pressure development capability. The second need was to identify the root cause of blade bearing failures. The author addressed the linked needs using the Reynolds-averaged Navier–Stokes modelling technique presented in Chapter 2. The rotor inflow aerodynamic characteristics were improved through an analysis of the inlet box and design of inlet guide vanes to control flow non-uniformities at the fan inlet. The results of the improvement facilitated both an increase in fan pressure developing capability and identifying the root cause of the blade bearing failures.

Chapter 8 builds on the work presented in Chapter 5, 6 and 7, presenting an evaluation of the available design concepts for a high pressure reversible fan. The reported research is underpinned by the ability to apply the Reynolds-averaged Navier–Stokes modelling technique presented in Chapter 2. The authors present three high pressure reversible fan concepts. These comprise a two-stage counter rotating fan, a single-stage high-speed fan and a two-stage fan with a single motor and impeller on each end of the motor shaft. The authors consider the relative merits of each concept. The third concept offers the most compact fan, transform, silencer and damper package size. The authors discuss the mechanical design challenges that occur with a two-stage fan with a single motor and impeller on each end of the motor shaft. They present and consider a selected motor bearing arrangement and casing design for maintainability. Finally, the authors present both prototype fan and full scale package aerodynamic and acoustic performance, before discussing the challenges presented by high temperature certification in accordance with Euro Norm EN 12101-3 requirements.

Chapters 9, 10, 11 and 12: Modelling Erosion

The research presented in Chapters 5, 6, 7 and 8 reports the results of four separate research programmes. Collectively, they provide an insight into how one may apply an ability to model the turbulent flows that characterise the low-speed fans typically used in air system to model the fan performance in a range of air systems. In contrast, the research reported in Chapters 9, 10, 11 and 12 constitute elements of a single research programme. This research was initiated as a consequence of the need to predict the erosion rate for both axial and centrifugal fans operating in erosive environments. Some of the most challenging fan applications are associated with erosive environments. The four chapters present the development of a numerical method for particle tracking, and the application of that method to predict particle impact location and frequency. A prediction of particle impact location and frequency facilitates the prediction of erosion rate, and hence in-service life.

Chapter 9 presents the development and first air movement fan application of the particle cloud tracking technique. The authors clarify the rationale for choosing a cloud particle, as opposed to a single particle tracking technique. A cloud particle tracking technique predicts the trajectory of cloud centres, rather than each individual particle in a cloud. The authors first describe the physics of particle laden flow and then move on to study the performance of the developed cloud particle tracking algorithm. The resultant particle cloud tracking technique requires at least an order of magnitude less computational effort than a single particle tracking technique, whilst still predicting particle trajectory well.

The authors applied the cloud tracking technique to an axial fan, predicting the blade-to-blade flow-field using the Reynolds-averaged Navier–Stokes approach presented in Chapter 2. An effect of erosive particles is the erosion of the fan blades' leading edge. Erosion results in the loss of the blade leading edge aerodynamic profile and a reduction of blade chord and effective camber that together degrade fan aerodynamic performance. The authors studied fan performance when operating in an erosive environment. They concluded that whilst the degradation of aerodynamic performance due to blade erosion begins gradually, it collapses as erosion reaches a critical limit. The authors then used a particle cloud tracking technique to clarify the influence of flow-field features, blade geometry, particle size and concentration on erosion patterns. A qualitative comparison indicated reasonable agreement between computed particle impact frequency and in-service blade erosion patterns. The agreement was good enough for the authors to consider the cloud particle tracking technique to have the potential to form the basis of an erosion prediction method. An erosion prediction method has the potential to predict fan in-service life when subjected to different particle sizes and concentrations. It would therefore be a valuable tool for designers attempting to establish if a specific fan design will have a service life long enough to enable operation from one scheduled maintenance outage to the next.

Chapter 10 continues the research presented in Chapter 9, presenting a quantitative prediction of erosion rate for a range of particle sizes derived from in-service measurements of particle size. The resultant prediction of erosion rate correlated well with in-service erosion rates, clarifying the role of particle size and mass distribution. The prediction of blade erosion that the authors made using the particle cloud tracking technique slightly over-predicted in-service blade life when compared with the actual blade in-service life. The authors concluded the discrepancy was a consequence of the discrepancy between the actual in-service conditions and those assumed in the numerical modelling. In practice, particle size and mass distribution and the actual fan operating point vary. Additionally, as the blade's leading edge erodes the blade-to-blade flow-field progressively departs from that associated with the as-manufactured blades used to predict particle trajectory. Despite these reservations, the authors concluded that the predicted fan in-service life was good enough to provide designers with useful input into the design process.

Chapter 11 also applies the particle tracking technique presented in Chapter 9, this time not in an axial fan, but in a centrifugal fan. The centrifugal fan that the authors used in the reported research incorporated cambered plate impeller blades that are inevitably associated with separated flow regions. Additionally, the modelling of centrifugal fans is not possible using periodic boundary conditions which simulate the flow-field in just one blade passage. When modelling a centrifugal fan it is necessary to model the entire impeller in order to account for the coupling between the impeller and the asymmetric volute casing within which it is housed. The authors adapted the Reynolds-averaged Navier–Stokes approach presented in Chapter 2. They predicted the studied fan's pressure rise and efficiency reasonably accurately at the peak pressure condition when stall is incipient. This success occurred despite the modelling assumption that neglected unsteady interaction between the impeller and its volute casing. The success of the developed modelling approach provides designers with a cost effective method for predicting centrifugal fan performance when the fan incorporates an impeller with cambered plate blades. The authors went on to use the numerical prediction of the blade-to-blade flow-field to predict impeller erosion rate. The resultant predictions were not in good agreement with those they saw in impellers that have operated in service. The authors concluded that although the developed modelling technique was able to provide a reasonable prediction of fan performance, they did not predict the blade-to-blade flow-field with sufficient accuracy to facilitate a useful prediction of erosion rate.

The research presented in Chapter 11 was encouraging, as predicting the performance of a centrifugal fan with cambered plate blades is challenging. However, the chosen modelling technique was unable to predict blade-to-blade flow-field features accurately enough to facilitate a prediction of erosion rate. The research presented in Chapter 12 extends that presented in Chapter 11, using the same computational mesh in conjunction with an unsteady computational method based on that presented in Chapter 3. The unsteady computational technique enables a more complete simulation of the flow-field physics by considering the unsteady coupling between the impeller and its housing. This increases the necessary computational effort by approximately an order of magnitude. However, modelling the unsteady coupling between the impeller and its housing is necessary to reproduce the evolution of flow-field features in the blade-to-blade passages, and the interaction of those flow-field features with the fans discharge volute. The authors concluded that this interaction was critical when predicting the trajectory of erosive particles, and therefore on the resultant prediction of erosion rate. They went on to study the effect of particle size on erosion rate, identifying critical regions of the impeller most prone to erosion for each combination of particle sizes. This provided a basis for the optimisation of impeller geometry in an effort to reduce its susceptibility to erosion, minimising erosion rate and consequently maximising the time between scheduled maintenance.

Summary of Chapters

Geoff Sheard

This chapter-by-chapter summary of technical contribution provides the reader with a detailed description of the work in each chapter. It clarifies each chapter's content and summarises its contribution to knowledge. This summary augments the description of the work itself with discussion as to why the authors undertook the work. It also lays out the rationale for undertaking the research, and the logic underpinning the move from one reported research programme to the next. Therefore, this summary links each chapter and makes explicit their collective contribution to knowledge.

The papers comprising each chapter constitute selected publications from the published scholarly work relating to the integration of computational methods into the design process of low-speed fans intended for application in air systems. The Editor selected them for inclusion based on the degree to which the content contributed towards the creation of a coherent body of knowledge.

Chapter 1 A Critical Review of Computational Methods and Their Application in Industrial Fan Design

The authors specifically wrote the paper that forms the basis of Chapter 1 to be the first chapter of this edited volume. It presents a review of computational methods and their application in air system fan design. The review identifies the key challenges those within the aerospace fan, aerospace compressor and air movement and control communities are addressing. The review's focus is on identifying how the air movement and control community can apply computational methods developed by, and for the aerospace community.

The aerospace fan community has systematically developed computational methods over the last five decades. The complexity of the developed methods, and the difficulty associated with their practical application, ensured that although commercial computational codes date back to the 1980s, they were not fully exploited by designers of low-speed fans intended for application in air systems until the beginning of the 2000s. The integration of commercial codes into the design process proved problematic as unlike aerospace fans, low-speed fans include electrical motors and other components from which the flow will invariably separate. Consequently, designers within the air movement and control community found the application of commercial codes challenging. The decade from 2000 to 2010 was focused on developing techniques that would facilitate converged solutions that predicted the fans' performance characteristics over the stable part of their operating range with reasonable accuracy, using a practical computational effort.

By 2010, scholars working with the air movement and control community had overcome the basic issues associated with commercial code application. Designers were able to apply computational methods incorporating a Reynolds-averaged Navier–Stokes (RANS) methodology to the fan design processes. For the first time they were able to simulate the overall fan performance. Over the stable part of the fan's characteristic these predictions are accurate enough to be of practical use to fan designers. Further, they achieve the predictions with a computational effort that is realistic for those working within the air movement and control community.

Computational codes also offer fan designers the possibility of simulating in detail the blade-to-blade flow-field. Codes incorporating a Reynolds-averaged Navier–Stokes methodology predict a fan's overall performance characteristics well, but the limitations inherent in their formulation make them less suitable for studying the blade-to-blade flow-field. Real insight requires a code that incorporates a more accurate model of the flow-field physics, with the favoured approach being a large eddy simulation (LES). The formidable computational effort required to undertake a large eddy simulation results in the air movement and control community favouring less accurate hybrid simulations on those occasions when a Reynolds-averaged Navier–Stokes methodology does not adequately model the flow-field physics.

An impediment to the widespread adoption of computational methods within the air movement and control community is the cost of licences for commercial codes. In response, scholars working with the air movement and control community are increasingly turning to open-source codes. These codes are licence free and can run on clusters of computers. At the time of writing, open-source codes have user interfaces that are less user friendly than commercial codes. Despite this caveat, the open source movement generally has made dramatic progress over the last five years. Therefore, it is likely that over the next five years open-source computational codes will become established within the air movement and control community.

Chapter 2 Computational Analysis of Noise Reduction Devices in Axial Fans with Stabilised Finite Element Formulations

Chapter 2 addresses the fundamental issues associated with application of a computational method to the low-speed fans typical of those used in air systems. The authors develop a modelling technique that is robust enough to be applied by designers to predict both the overall axial fan performance and blade-to-blade flow-field features. A reliable prediction of overall fan performance that matches experimentally measured fan performance is a fundamental requirement when incorporating a computational method into the fan design process. It enables a designer to evaluate the impact of changes in blade geometry more rapidly than would be possible by building and testing physical prototypes. A prediction of the evolution of blade-to-blade flow-field features through the blade passage provides the designer with an insight into the flow-field physics. This insight is what drives a designer's decision making process when deciding what changes in geometry are needed.

The authors observed that an issue when applying a computational method into any turbomachinery application is that even in the most favourable cases, Reynolds numbers are high resulting in complex flow-fields. A direct numerical simulation (DNS) of the Navier–Stokes equations would require fine computational grids and therefore a computational effort beyond that available within the air movement and control community. The authors concluded that the only viable approach available when attempting to solve the Navier–Stokes equations is a Reynolds-averaged Navier–Stokes (RANS) simulation. They based their Reynolds-averaged Navier–Stokes modelling technique on the streamline-upwind / Petrov–Galerkin (SUPG) and pressure-stabilising / Petrov–Galerkin (PSPG) methods. The application of computational methods to predict the blade-to-blade flow-field is made challenging by the inevitability of separated flow from electric motors, flat-plate motor supports and conduit. Therefore, obtaining converged computational solutions is challenging. The authors addressed this challenge by developing a novel stabilisation approach based on a 'diffusion for reaction dominated with jump' (DRDJ) operator. This stabilisation approach facilitated the management of discontinuities and therefore converged computational solutions.

The authors analysed the resultant formulations performance, and characterised its interaction with the chosen turbulence model. They concluded that the developed Reynolds-averaged Navier–Stokes modelling technique was able to predict reliably both overall axial fan performance and blade-to-blade flow-field features. The modelling technique represented a significant advance for the air movement and control community. The inevitability of separated flow from electric motors, flat-plate motor supports and conduit had constrained the accuracy of previous modelling techniques. The simplifying assumption when modelling the flow-field physics did facilitate converged computational solutions. However, the accuracy of the resultant prediction of fan performance was compromised and blade-to-blade flow-field features were not characterised accurately.

Chapter 3 Large-eddy Simulation of a Tunnel Ventilation Fan

The modelling technique presented in Chapter 2 provides designers with a computational method robust enough to be routinely incorporated into a fan design process. An advantage of the modelling technique is that as it is based on the Reynolds-averaged Navier–Stokes approach, the required computational effort is within the reach of designers working in the air movement and control community. However, the Reynolds-averaged Navier–Stokes approach is not able to model the turbulence's spectral content and is not sensitive to the flow-field's multi-scale features. Therefore, simplifying assumptions inherent in a Reynolds-averaged Navier–Stokes approach ultimately limit the accuracy with which one can predict fan performance. Scholars who have studied the flow-field physics within all turbomachinery classes are increasingly turning to large eddy simulations (LES) to overcome Reynolds-averaged Navier–Stokes approach limitations.

Chapter 3 presents a large eddy simulation modelling technique that reduces the required computational effort classically associated with a large eddy simulation.

The principle issue with all large eddy simulations is the formidable required computational effort that at the time of writing is simply beyond the reach of all those working within the air movement and control community. However, the developed modelling technique requires less computational effort than a classical large eddy simulation. It still requires an effort approximately one to two orders of magnitude greater than that associated with the Reynolds-averaged Navier–Stokes modelling technique presented in Chapter 2. Therefore, it is still beyond the reach of fan designers within the air movement and control community. However, it is not beyond the reach of those scholars working with the air movement and control community. Those scholars have access to high performance computing clusters. Therefore, they are able to utilise modelling techniques that demand more computational effort. Thus, they are able to develop the next generation of modelling techniques that designers working within the air movement and control community may apply when computer hardware costs fall.

The authors' modelling technique utilised a computational grid too course for a classical large eddy simulation. They coupled the computational grid with sub-grid scale (SGS) closure that relied on a one-equation model which required the authors to solve a differential transport equation to calculate the sub-grid scale turbulent kinetic energy. The use of this closure technique was a remedial strategy that partially recovered the turbulence spectrum when using a low resolution computational grid that alone was unable resolve the turbulence spectrum. The authors evaluated the resulting modelling technique and concluded that it was able to predict fan performance well. It was also capable of modelling separation and secondary flow features in the blade-to-blade flow-field. The authors concluded that the developed large eddy simulation modelling technique was better able to model the flow-field physics than a Reynold-averages Navier–Stokes approach. When insight into the flow-field physics is needed, the authors concluded that the developed technique was preferable to a modelling technique based on a Reynolds-averages Navier–Stokes approach.

Chapter 4 Large-eddy Simulation of the Aerodynamic and Aeroacoustic Performance of a Ventilation Fan

The air system fans developed by the air movement and control community have historically been sold on first cost. An intensely competitive market place has resulted in the community being manufacturing led, with specific competence in manufacturing minimising production costs and therefore being a source of competitive advantage. Fan efficiency has typically been a secondary consideration. The recent European and forthcoming US regulation of fans intended for application in air systems has resulted in a change of focus. It is now necessary to develop air system fans that at least meet current regulatory requirements and ideally forthcoming regulatory requirements as well. Consequently, designers within the air movement and control community are redesigning historic products.

A consequence of redesigning a historic product may be a change in its acoustic emissions. Although the air movement and control community has historically been focused on first cost, the market place into which it sells has expectations regarding the acceptability of fan broadband noise and spectrum. Some market segments are more sensitive than others to fan noise. However, the trend is towards a requirement for lower broadband noise, with some market segments going further and specifying acceptable noise levels in each octave band of the spectrum. Any redesign of a historic product, or new product design must aim to at least maintain acoustic performance and preferably minimise it.

In this chapter the authors use the large eddy simulation modelling technique presented in Chapter 3 to compute unsteady flow-field parameters. They then use these as input to Ffowcs Williams and Hawkings' acoustic model. Thus, they were able to develop a tool that predicted both fan broadband noise and spectrum. The prediction was able to resolve the spectrum up to the fifth harmonic of blade passing frequency. The authors compared their prediction of broadband noise and spectrum with experimental measurement, and found agreement within the uncertainty of the measurements. This agreement enabled the authors to reach two conclusions. First, the large eddy simulation modelling technique and Ffowcs Williams and Hawkings' acoustic model were able to predict broadband noise and spectrum. Second, the agreement between predicted and measured broadband noise and spectrum gave confidence in the accuracy with which the large eddy simulation modelling technique modelled the evolution of blade-to-blade flow-field features.

The accuracy of the prediction was restricted by the limitations inherent in the Ffowcs Williams and Hawkings' acoustic model that neglects duct modes. As a consequence, the acoustic model did not model the modal amplification of dipole-like noise sources. Dipole-like noise sources characterise the inter-tonal and broadband noise when a fan is fitted in a ducted system. Despite the limitations of the acoustic model, the agreement between predicted and measured broadband noise and spectrum was good enough to provide designers with an insight into the acoustic consequences of a change in fan geometry. Therefore, it is a useful tool, facilitating the extension of an aerodynamic and mechanical design optimisation process into one that also considers the acoustic consequences of any change in fan geometry.

Chapter 5 A Numerical Investigation into the Aerodynamic Effect of Pressure Pulses on a Tunnel Ventilation Fan

Chapter 5 presents the results of the authors' research programme in response to in-service central tunnel ventilation fan failures. Although tunnel ventilation fan in-service failures are rare, engineers increasingly associate those failures with new or renovated tunnel systems with larger pressure pulses. Consequently, we require insight into the aerodynamic and mechanical consequences that occur with subjecting a tunnel ventilation fan to a pressure pulse.

Tunnel ventilation fans are subjected to pressure pulses as a consequence of trains passing the ventilation shafts within which they are installed. These pressure pulses alter the volume flow rate through a ventilation fan, and consequently, the static pressure field around fan blades. Today's trains typically travel through railway tunnels faster than has been the historic norm. Additionally, platform screen doors are now a standard feature of metropolitan metro systems. Both result in the trains involved inducing larger pressure pulses than tunnel ventilation fan designers have traditionally assumed.

The authors modelled the pressure pulse induced by a train in a tunnel system as a rapidly changing fan volume flow rate. They computed the fan operating point using the large eddy simulation modelling technique presented in Chapter 3. An analysis of the computational results provides an insight into the effects of turbulent structures that develop within the fan blade passage. Large-scale vortical structures released from the blade are not affected by either the negative or positive pressure pulse. However, both positive and negative pressure pulses do affect small-scale structures in the blade-to-blade passage that are characterised by a time-scale comparable with those of the incoming pressure pulse. The negative pressure pulse leads to choking the flow, with evidence that the fan mid-span section stalls transiently. The positive pressure pulse occurs with an increase in blade loading, but no span-wise redistribution of the blade static pressure field. A consequence of the blade loading increase is that the blade tip section stalls transiently.

The analysis indicates choking the flow followed by transiently stalling the blade tip section results in a doubling of the unsteady aerodynamic and mechanical forces to which the blade is subjected. This doubling of induced mechanical stress each time a train passes the ventilation shaft within which a tunnel ventilation fan is situated may, over time, induce a fatigue crack in one or more fan blades. Once a fatigue crack is induced in at least one blade, it will propagate under the influence of the unsteady mechanical stress induced in a blade by stable operation, ultimately resulting in an in-service fatigue failure.

The authors observe that the introduction of higher speed trains into railway tunnels and the addition of platform screen doors during the renovation of metropolitan metro systems results in pressure pulses larger than the historic norm. They go on to conclude that larger pressure pulses are probably responsible for some recent in-service tunnel ventilation fan failures. Therefore, tunnel ventilation fan designers must respond by modelling pressure pulses' aerodynamic effect, calculating the mechanical consequences and adapting the mechanical design of blades to withstand the actual imposed loads.

Chapter 6 A Parametric Study of Reversible Jet-Fan Blades' Aerodynamic Performance

Chapter 6 constitutes a complementary research programme to that presented in Chapter 5. Transverse tunnel ventilation systems are classically designed around central tunnel fans, with a typical example being the subject of the research presented in Chapter 5. Longitudinal tunnel ventilation systems are classically designed around jet fans, with a typical example being the subject of the research presented in this chapter. In practice, the majority of tunnel systems are complex enough to require a combination of both central tunnel ventilation fans and jet fans. Consequently, overall ventilation system efficiency is a function of both central tunnel fan and jet fan efficiency.

A need to improve jet fan efficiency motivated the research presented in this chapter. The authors report the results of a parametric design study of the impact of design variables on the performance of fully reversible jet fan blades. The selected blade design variables are tip solidity, twist and camber distribution. The authors base their design methodology on a sensitivity analysis which they obtained from a response surface approximation. They analysed four combinations of the selected design variables, with the combinations of variables generated using an experimental design approach.

The authors used the Reynolds-averaged Navier–Stokes modelling technique presented in Chapter 2 to calculate the efficiency and thrust of the jet fan blades generated using each of the four combinations of design variables. They studied the computational results, using the sensitivity analysis to provide insight into the flow-field physics. The results indicated that a requirement for both high thrust and efficiency are contrasting objectives. Therefore, an optimisation of design variables is needed to maximise efficiency and over the stable region of the jet-fan's operating range and also to extend the stable operating range as far as possible to maximise thrust. Having developed their design methodology, the authors identify each design variables' effects on thrust and efficiency. Insight into the relative importance of each design variable enabled them to select optimum values, and propose a redesigned blade with both higher predicted thrust and efficiency.

The authors manufactured the redesigned blade and measured the jet fan's thrust and efficiency fitted with first the baseline blade and then the redesigned blade. The measured performance of the redesigned blade confirmed that the redesign resulted in a jet fan with higher thrust and efficiency at a pitch angle above that at which a jet fan fitted with the baseline blade would have stalled. Acoustic measurements indicated that this increase in thrust and efficiency was associated with an increase in jet fan noise. The authors expected this increase in noise as fan acoustic emissions correlate with blade loading. The redesigned blade camber was higher than that of the baseline blade, resulting in an increase in blade loading and the associated thicker boundary layers and wakes. Therefore, the authors concluded that the developed parametric design method captured jet fan blade design rules, facilitating the optimisation of blade aerodynamic design within a given set of mechanical constraints.

Chapter 7 The Effect of Inlet Box Aerodynamics on the Mechanical Performance of a Variable Pitch in Motion Fan

Chapter 7 presents research that the author initiated following an in-service failure. In this case, the failure was of the blade bearings of a variable pitch in motion fan intended for use in forced- and induced-draft boiler applications. Variable pitch in motion fans rotate at a constant speed, with the changing blade angle varying the load. A pitch-change mechanism facilitates the change in blade angle. A blade bearing supports each blade enabling it to rotate. The author observed that as the fan aerodynamic stage-loading progressively increased, so did the rate of blade bearing wear.

The author conceived the reported research to both increase fan aerodynamic stage-loading and identify the root cause of the blade bearing failures. The increase in stage-loading was required to increase the fan pressure developing capability that systematically reduced as a consequence of the fan's erosive operating environment. The author addressed the linked needs by studying the flow-field through the fan's inlet box. He analysed the flow-field through the inlet box in its five stays original form and an improved configuration with eight stays plus eight inlet guide vanes.

The author designed the improved configuration to pre-swirl the flow before discharge onto a fan inlet, thus increasing blade loading and fan pressure developing capability. The author was able to verify that the studied fan's stall margin was adequate to accommodate the additional stage-loading, with fan pressure developing capability increasing by up to 40 per cent at the fan's design operating point. However, the improved inlet box configuration did not reduce the fan's susceptibility to the effects of erosion.

The two numerical simulations enabled the author to identify handed-flow effects in the original inlet box configuration, resulting in a difference in pressure from one side of the fan to the other. An effect of this pressure difference was that each blade experienced a once per revolution increase and decrease in both lift and drag which induced a once per revolution orbital motion of each blade. Although the peripheral and axial movement of each blade associated with this orbital motion was small, it was sufficient to result in bearing ball procession which caused the bearing balls to migrate towards a single point around the bearing race. A result of this migration was that bearing ball-to-ball forces became large, resulting in the bearing ball separator's rapid wear and ultimately premature bearing failure.

Analysis of the improved eight stays plus eight inlet guide vanes inlet box computational results indicated only a two per cent cyclic variation in pressure associated with the asymmetric flow-field entering the fan from the inlet box. Although the aerodynamic impact of a two per cent cyclic variation in pressure is small, the resultant bearing ball-to-ball forces remain high. Unable to reduce the cyclic variation in pressure further, the author chose to mitigate their effect using a novel bearing separator design that is better able to withstand the induced bearing ball-to-ball forces.

The author fitted the novel bearing separator to an induced-draft fan with the original five stay inlet box configuration. After 22,000 hours' in-service operation, the author removed and inspected the bearings, estimating that the blade bearings fitted with the novel bearing separator had approximately 50 per cent bearing life remaining. As such, the novel bearing separator design facilitates an estimated five year bearing life, self-consistent with the major service intervals of variable pitch in motion fans in induced-draft application.

Chapter 8 The Conceptual Design of High Pressure Reversible Axial Tunnel Ventilation Fans

Chapter 8 presents an evaluation of the available design concepts for a high pressure reversible fan. Central tunnel ventilation fans classically must have the ability to both supply and extract air from a tunnel system, with the operator's choice dependent on the tunnel ventilation system's operating mode most appropriate at any given point in time. Consequently, tunnel ventilation fans must incorporate a reversible aerodynamic design which limits the maximum fan pressure rise.

The current state of the art in reversible fan design limits the maximum pressure rise that an operator can practically achieve in a single reversible fan stage to 2,000 Pa. The pressure rise is limited as the reversibility requirement results in the necessity for symmetrical fan blade aerofoil sections. Also, one must reduce guide vanes to flow-straighteners if the fan is to produce the same flow and pressure in both directions.

As tunnels become longer and deeper, the pressure drop that occurs with the flow necessary to ventilate the tunnel can rise above the 2,000 Pa limit of a current state of the art single reversible fan stage. In such cases, the traditional solution is to offer a two-stage fan with counter-rotating stages. A disadvantage of the two-stage counter-rotating fan is that it produces 15 dB more noise than one of the fans operating in isolation. Consequently, silencers must be larger for the same overall acoustic emissions at the silencer inlet and outlet.

In a situation where minimum overall fan, transform, silencer and damper package size is desirable, the two-stage counter rotating fan may not be the optimum fan concept as a consequence of the requirement for large silencers. In such circumstances, tunnel ventilation fan designers have two further fan concepts available. These concepts are a high-speed single-stage fan or a lower speed two-stage double ended motor fan. Each represents a valid technical solution, and has the potential to achieve a specified duty point.

The two-stage double ended motor fan concept resulted in the minimum package size, thus the authors selected it as the preferred concept. During the practical realisation of this concept, the authors encountered two mechanical design challenges. First, there was the issue of whether the motor bearing system design could accommodate the imposed aerodynamic and gravitational loads in both horizontal and vertical configurations. Second, there was the issue of designing a casing that operators could dismantle within a plant room's confines where engineers installed the fans and their associated package.

Having successfully addressed the mechanical design challenges, the authors measured the prototype fan's aerodynamic and acoustic performance. They then measured the full size package aerodynamic performance and full size fan acoustic performance. Both aerodynamic and acoustic results were in good agreement with the original performance prediction scaled from single-stage fan data. The authors concluded that the two-stage double ended motor fan concept represented an attractive option for tunnel ventilation system designers. The concept is most attractive when the required pressure rise exceeds that which one can achieve in a single fan stage and when plant room costs are high enough to mandate minimum fan, transform, silencer and damper package size.

Chapter 9 Predicting Blade Leading Edge Erosion in an Axial Induced Draft Fan

Chapter 9 presents the development of a particle cloud tracking technique that predicts the trajectory of cloud centres, rather than each individual particle in a cloud. The reported research was motivated by a need to predict the erosion rate of induced-draft fans. These fans remove combustion products from coal-fired boilers, which invariably contain unburnt carbon and silica particles. These particles can be highly erosive, resulting in a degradation of the fan blades' profile, a reduction in the blade chord and an associated loss of blade chamber. The net result of blade erosion is to reduce fan pressure developing capability and efficiency. For a variable pitch in motion axial fan, the loss of pressure developing capability can result in the fan's stall margin reducing to zero at the operating point. The fan then transitions from stable to stalled operation, with the resultant risk of mechanical failure.

The size and concentration of erosive particles is dependent on the quality of the coal burnt in the boiler. The quality of coal is unpredictable, and therefore the size and concentration of the resultant erosive particles is also unpredictable. This compounds the challenges associated with predicting the erosion rate of induced-draft fan blades. Despite the challenges, a prediction of erosion rate is important as one may use it to predict in-service life. Coal-fired boilers and their associated equipment is typically overhauled once every five years. Therefore, fan designers require a method of evaluating the effect of changing particle size and concentration on service life, and of evaluating the impact of any change in blade geometry on that blade's susceptibility to erosion.

The authors studied the in-service performance of an induced-draft fan in coal-fired boiler application. They observed that whilst the degradation of aerodynamic performance due to blade erosion begins gradually, it collapses as erosion reaches a critical limit. The authors present a numerical prediction of the evolution of induced-draft fan blade erosion patterns. They then present a qualitative assessment of the agreement between numerically predicted erosion patterns and those they observed on blades removed from service. The numerical predictions of erosion patterns utilised the Reynolds-averaged Navier–Stokes modelling technique presented in Chapter 2 to predict the blade-to-blade flow-field. The peak efficiency operating point prediction of the studied fan's blade-to-blade flow-field indicated that the blade passage was not adversely affected by the presence of large secondary flows. The only exception was on the blade pressure surface's hub region that featured a separation that distorted particle trajectories radially outward. The authors then used this blade-to-blade flow-field prediction as input to the cloud particle tracking technique presented in Chapter 9. The resulting particle trajectory prediction clarifies the influence of secondary flow structures in the blade-to-blade flow-field, initial blade geometry, particle size and concentration on erosion pattern.

The authors concluded that three factors dominate the erosion process: inflow incidence angle, flow deviation correlated to the blade profile and the blade tip speed. They were able to correlate predicted erosion patterns with those observed over blades that had been removed from service, making the general observation that erosion was concentrated at the blade tip due to a combination of higher peripheral speed and particle migration from the blade pressure to suction side through the blade tip-to-casing gap. The agreement between predicted erosion patterns and those they observed over the pressure and suction surface of blades that had been removed from service was good. This gave the authors confidence in both the Reynolds-averaged Navier–Stokes modelling technique they used to predict the blade-to-blade flow-field and the cloud particle tracking technique.

The authors concluded that the prediction of erosion patterns was good enough for the cloud particle tracking technique to form the basis of an erosion prediction method which has the potential to predict the change in fan service life when subjected to different particle sizes and concentrations. Therefore, it is a valuable tool for designers attempting to establish if a change in particle size or concentration will reduce a fan's service life to the point where one can no longer expect it to operate until the next scheduled maintenance outage. It also has the potential for application during the fan design process, helping designers to establish if a proposed change in fan geometry will increase or reduce the susceptibility of fan blades to erosion.

Chapter 10 Numerical Simulation of Coal Fly-ash Erosion in an Induced Draft Fan

The qualitative comparison of erosion patterns presented in Chapter 9 indicated that the cloud particle tracking technique could predict erosion patterns with reasonable accuracy. This constituted a proof of concept for the cloud particle tracking technique. Chapter 10 continues the research reported in Chapter 9, presenting a quantitative prediction of erosion rate for a range of particle sizes derived from in-service measurements of particle size.

The reported research first characterises the size and concentration of erosive particles collected from an in-service induced draft fan. The authors then used a numerical study to identify the impact of particle size and mass distribution on the resultant erosion patterns. The authors concluded that different size and mass particles resulted in distinctly different erosion patterns, indicating that particle size and mass affected particle trajectory. However, irrespective of the particle size and mass, the blade leading edge suffered the most severe erosion.

The authors focused on elucidating the erosion mechanisms responsible for blade leading edge erosion. They concluded that erosion was a function of particle impact velocity, angle of impact and the eroded surface's physical properties. They concluded that the most erosive particles were those between 32.5–67.5 μm diameter. Having characterised the factors driving the erosive process and the influence of particle size, the authors were able to predict erosion rate. They used the cloud particle tracking technique to predict impact frequency patterns that they then transformed into predictions of erosion rate. They could then use the erosion rate to predict the loss of blade mass over time, which the authors were then able to compare with differences in mass between a new blade entering service and an eroded blade after removal from service.

The predicted erosion rate was lower than observed when they removed the blades from service. The authors concluded the discrepancy was most likely a consequence of the fan operating at reduced load at night. Despite the discrepancy between the predicted erosion rate and the erosion rate they observed when they removed blades from service, the agreement between the two was good. The predicted erosion patterns were in good agreement with those of the blades they removed from service. This agreement indicated that the authors had characterised well the effect of particle size and mass distribution, and the factors that affected the erosive process. As such, although it is never easy to predict how operators will vary the fan operating point, the prediction of erosion rate was accurate enough to allow the authors to assess the effect of changing particle size, mass and concentration. Further, for a fixed combination of particle size, mass and concentration, one may use the erosion rate prediction to establish if a proposed change in blade geometry will increase or reduce its susceptibility to erosive effects.

Chapter 11 Predicting the Performance of an Industrial Centrifugal Fan Incorporating Cambered Plate Impeller Blades

Chapter 11 also applies the particle tracking technique presented in Chapter 9, this time not in an axial fan, but in a centrifugal fan. The centrifugal fan that the authors used in the reported research incorporated cambered plate impeller blades. Centrifugal fans with cambered plate blades are in widespread use across the air movement and control industry. They have a higher pressure developing capability than single-stage axial fans and are inherently low cost. At their design point they are also able to achieve an efficiency of over 80 per cent. Therefore, they have the potential to be both efficient in operation with an economical first cost.

A disadvantage of centrifugal fans with cambered plate blades is that the blade-to-blade flow-field is inevitably characterised by regions of separated flow. This makes predicting fan performance challenging. Additionally, modelling centrifugal fans is not possible using periodic boundary conditions which simulate the flow-field in just one blade passage. When modelling a centrifugal fan, it is necessary to model the entire impeller in order to account for the coupling between impeller and the asymmetric volute casing within which it is housed. Consequently, the numerical prediction of centrifugal fan performance typically requires approximately one order of magnitude more computational effort than an axial fan.

In the research reported in this chapter, the authors utilised the Reynolds-averaged Navier–Stokes modelling technique presented in Chapter 2 to predict the studied fan's pressure rise and efficiency. The prediction was in reasonable agreement with experimentally measured performance at the peak pressure condition when stall is incipient. This success occurred despite the modelling assumption that neglected unsteady interaction between the impeller and its volute casing. The authors concluded that the developed modelling approach provides designers with a cost effective method for predicting centrifugal fan performance when the fan incorporates an impeller with cambered plate blades.

The authors went on to use the numerical prediction of the blade-to-blade flow-field to predict impeller erosion rate. The authors used a blade-to-blade flow-field prediction as input to the cloud particle tracking technique presented in Chapter 9. The authors used resulting predictions of particle trajectory to evaluate the impact of particle size on erosion patterns. A qualitative assessment of the variation in erosion patterns with particle size indicated that 5 μm particles resulted in a prediction of erosion patterns that most closely matched those the authors saw over impellers removed from service.

Despite the qualitative agreement between predicted erosion patterns and those they saw on impellers removed from service, the resultant predictions of erosion rate were not in good agreement with those of impellers that had operated in service. The authors concluded that although the developed modelling technique was able to provide a reasonable prediction of fan performance, they could not predict blade-to-blade flow-field with sufficient accuracy to facilitate a useful prediction of erosion rate.

Chapter 12 Simulation of Particle-laden Flows in a Large Centrifugal fan for Erosion Prediction

Chapter 11 reported a research programme that resulted in an adapted form of the Reynolds-averaged Navier–Stokes modelling technique presented in Chapter 2. This modelling technique enabled the authors to predict the performance of a centrifugal fan with cambered plate blade. However, the developed modelling technique was unable to predict blade-to-blade flow-field features with sufficient accuracy to facilitate an accurate prediction of erosion rate. The authors concluded that a more accurate prediction of erosion rate required the blade-to-blade flow-field prediction to account for the unsteady interaction between the impeller and fan discharge volute. Accounting for this unsteady interaction would enable them to predict the trajectory of particle clouds as they transitioned from the stationary to rotating frame of reference.

In the research presented in Chapter 12, the authors used the computational mesh developed in Chapter 11. They used this mesh in conjunction with an adapted form of the large eddy simulation modelling technique presented in Chapter 3. This modelling technique required approximately one order of magnitude more computational effort than the Reynolds-averaged Navier–Stokes modelling technique. Although at the limit of the computational effort designers within the air movement and control community have available, the authors, nevertheless, felt the approach was valid. Hardware costs have historically fallen steadily and are likely to continue to do so. As they fall, a computational effort that one might consider excessive today is likely to be more acceptable tomorrow.

The authors used the modelling technique to predict fan performance, which matched experimentally measured performance within the uncertainty of the measurements. The good agreement between predicted and measured performance gave the authors confidence in the modelling technique. They then analysed evolution of the blade-to-blade flow-field as the impeller rotated inside its scroll, highlighting a series of unsteady flow features. They concluded that these unsteady flow features were responsible for the motion of secondary flow features in the blade-to-blade passage. Secondary flow feature motion then drives the trajectory of particles within the impeller blade-to-blade passage, thus having a primary effect on the predicted erosion rate.

The authors used the cloud particle tracking technique presented in Chapter 9 to predict particle trajectory. They then used the method presented in Chapter 10 to predict erosion rate. They concluded that modelling the unsteady coupling between the impeller and fan discharge volute had resulted in a prediction of erosion rate that was in reasonable agreement with the erosion rates they observed when studying impellers that were removed from service. They concluded that this interaction was critical when predicting the trajectory of erosive particles, and therefore on the resultant prediction of erosion rate. The authors concluded by characterising the effect of particle size on erosion rate and identifying critical regions of the impeller most prone to erosion for each combination of particle sizes. This provided a basis for the optimisation of impeller geometry in an effort to reduce its susceptibility to erosion, minimising erosion rate and consequently, maximising the time between scheduled maintenance.

Appendix 1 A Comparison of U.S. and European Approaches to Regulating Fan Efficiency

This appendix is a revised and extended version of a technical paper that presents a comparison of U.S. and European approaches to the regulation of fan efficiency. The appendix is not directly concerned with the application of computational methods to the design of low-speed fans intended for air system application. However, it does clarify why there is increasing interest in the application of computational methods across the air movement and control community. The use of these computational methods is the only viable method by which fan efficiency may be significantly increased. If fan efficiency is not significantly increased, fan manufacturers will be regulated out of business.

The air movement and control community has historically developed its own codes and standards that they use to characterise fan performance. These codes and standards enable them to compare different fans on a like-for-like basis and their adoption constitutes a form of self-regulation. Self-regulation has prevented manufacturers from making exaggerated claims about fan performance, thus creating a 'level playing field' where all compete on equal terms. However, self-regulation influences consumer choice only partially. A consumer may be inclined to choose from a range of fans that all comply with applicable codes and standards, but they are free to choose between compliant low efficiency fans.

To those unfamiliar with the air movement and control industry, it may seem counterintuitive to choose a fan with a low efficiency. Low efficiency implies that the fan will consume more power than necessary. As a rule of thumb, the initial cost of a fan is approximately equal to three per cent of the electricity it consumes over its life. Therefore, if consumers were economically rational, they would prefer to pay for a high efficiency fan rather than accept free of charge a low efficiency fan as the difference in efficiency between the two is typically far greater than three per cent.

The fact of the matter is that the air movement and control community is not economically rational. Consumers buy fans with little or no thought given to through-life cost. Consequently, consumers typically buy small, noisy and low efficiency fans as they have the lowest first cost. The only practical way to disrupt this market dynamic is for governments to step in and regulate the industry, making it illegal to sell fans below a specified efficiency.

However, improbable as it may seem that governments would take such draconian regulatory action, the European Union has decided to regulate the air movement and control industry, setting minimum fan and motor efficiency targets. The 2013 targets have resulted in approximately a third of all fans sold before 1 January, 2013 now illegal within Europe as a consequence of their low efficiency. The European Union raised again minimum fan and motor efficiency targets on 1 January, 2015. In the U.S., the Department of Energy has been monitoring the regulatory activity within Europe. On 1 February, 2013, the U.S. federal government published a framework document in the Federal Register outlining the intended approach to fan regulation that aims to eliminate inefficient fans within the U.S. by 2020. The air movement and control community widely anticipates that the Department of Energy will adopt the same approach as the European Union, increasing the minimum allowable fan and motor efficiency within three years of introducing the initial 2020 targets.

This regulatory environment has disrupted the air movement and control industry in Europe, but not negatively. More efficient fans tend to be more expensive fans. Therefore, a consequence of regulation has been that many fan manufacturers have seen the value of sales actually increase. A second consequence of regulation is that European fan companies are being forced to invest in research and development to ensure products comply with the 2015 minimum efficiency targets. This investment will help ensure that products remain competitive in a global market. Lastly, the consumer benefits even though first cost increases as through-life cost reduces.

Although regulation has its critics, the fact of the matter is that it is here to stay. The air movement and control community must pick up and apply computational methods that researchers have developed within the aerospace community over the last 50 years. The alternative, persisting with empirical design techniques first developed in the 1950s, will not enable fan efficiency to increase far enough or fast enough. Therefore, not integrating computational methods into the fan design process is an exit strategy. Some will undoubtedly choose to bury their head in the sand and deny the reality of what is happening. An inconvenient truth is that those manufacturers will be regulated out of existence, and the markets they leave behind will be taken over by others who have risen to the challenge.

Appendix 2 Variable Pitch Fans

The paper that forms the basis of this appendix is a developed form of a patent. Patents are written in a formal style that is difficult for those unfamiliar to penetrate. As such, we tend not to read or reference patents. To overlook patents, however, is to overlook a significant source of intellectual property. The Editor has adapted the patent into a more readable form, in an effort to make its content more accessible to the reader.

The research reported in the papers that constitute the chapters of this edited volume document a development effort that underpinned the application of computational methods to the design of the low-speed fans typically used in air systems. An assumption underpinning the application of computational methods is that if one can predict overall fan performance with reasonable accuracy, the predicted blade-to-blade flow-field features will also be a fair representation of those at play within the blade passage. If we accept this assumption, then it follows that one outcome associated with the application of computational methods is an insight into the flow-field physics. Although theoretically possible to study the flow-field physics using an experimental approach, the air movement and control community does not have the necessary resources. Therefore, a numerical prediction of the blade-to-blade flow field facilitates the study of flow-field features that would not have been possible otherwise.

The patent that the Editor has converted into this appendix's text is an outcome of the research reported in Chapter 7. The author initiated this research following the in-service failure of a variable pitch in motion fan blade bearing. The studied fan incorporates an inlet box that turns the flow 90 degrees into the fan inlet. A numerical prediction of the flow through the inlet box and into the fan indicated that the 90 degree bend, in combination with the fan blades' rotation, resulted in a significant difference in pressure on each side of the fan. An effect of this pressure difference was that each blade experienced a once per revolution increase and decrease in both lift and drag. This variation in lift and drag induced a once per revolution orbital motion of each blade. Although the peripheral and axial movement of each blade associated with this orbital motion was small, it was sufficient to result in bearing ball procession which resulted in the bearing balls migrating towards a single point around the bearing race, with the resultant bearing ball-to-ball forces becoming large.

Knowing that ball-to-ball forces were large resulted in a realisation that bearing lubricant would squeeze out of the contact zones between balls and separators. However, there was a secondary effect of high ball-to-ball forces within the bearing. The curvature of the bearing resulted in contact points between the balls on each side of each separator being offset. In combination with high ball-to-ball forces, this offset caused the separators to spin. The inventors realised that by adapting the design of the separators, they could exploit the rotation, using it to transport lubricant back to the ball and separator contact points where necessary.

The intellectual property that forms the basis of this patent is a consequence of an insight into the flow-field physics facilitated by a numerical simulation. The inventors went on to use their chosen computational method to develop an asymmetric inlet box vane arrangement that resulted in a more uniform flow-field entering the fan. However, as the inlet flow-field was not perfectly uniform, the procession of bearing balls persisted, and hence there was a need to transport lubricant back into the contact zone.

This appendix is an example of how a computational based design methodology is capable of providing insight into the flow-field physics that result in the discovery of new intellectual property. Further, it was a purely aerodynamic numerical prediction of the flow-field that facilitated an understanding of the root cause of a mechanical failure. An analysis of failed bearings alone would have been unlikely to identify the aerodynamic root cause of the bearing failure mechanism.

Chapter 1

A Critical Review of Computational Methods and Their Application in Industrial Fan Design

A. Corsini, G. Delibra and A.G. Sheard

ABSTRACT

Members of the aerospace fan community have systematically developed computational methods over the last five decades. The complexity of the developed methods, and the difficulty associated with their practical application, ensured that although commercial computational codes date back to the 1980s, they were not fully exploited by the air movement and control community until the beginning of the 2000s. The application of commercial codes proved problematic as unlike aerospace fans, the low-speed fans intended for application in air systems include electrical motors and other components from which the flow will invariably separate. Consequently, designers found the application of commercial codes challenging. The decade from 2000 to 2010 was focused on developing techniques that would facilitate converged solutions that predicted the fans' performance characteristics over the stable part of their operating range with reasonable accuracy, using a practical computational effort.

By 2010, the air movement and control community had overcome the basic issues associated with commercial code application, with the result that designers moved on to consider applying computational methods to the wider fan design processes. The specific areas of the process that can benefit from computational methods are:

- research: elucidating aspects of the flow physics that one cannot easily study in a laboratory environment;
- development: the use of computational methods to reduce or eliminate the need for traditional laboratory characterisation of fans intended for application in air systems, and;

This chapter is a revised and extended version of Corsini, A., Delibra, G. and Sheard, A.G. (2013), 'A Critical Review of Computational Methods and Their Application in Industrial Fan Design'. *International Scholarly Research Network, Mechanical Engineering*, vol. 2013, article ID 625175, pp. 1–20.

- application: formalising the fan design process such that relatively inexperienced engineers can use computational codes routinely.

In this chapter we focus on the first of these three areas, discussing the challenges involved and the relative merits of the available modelling techniques. The chapter ends with a discussion of the practical problems associated with the use of commercial codes in a development environment, and finally the legislative requirements in the European Community and the U.S. that are driving the need for widespread application of aerospace style computation methods by the air movement and control community.

NOMENCLATURE

Latin letters

c	[mm]	Blade chord
CFL	[-]	Courant-Friedrichs-Lewy Number or Courant number; CFL = U×Δt/Δx
c_μ, c_1,..., c_7	[-]	Numerical coefficients
C_p		Pressure coefficient
f_t	[1/s]	Turbulence frequency
f_t^+		Normalised turbulent frequency
F	[1/s]	Elliptic relaxation function
I_f		Impact factor
k	[m²/s²]	Turbulent kinetic energy
M^+		Normalised moment component
M_a		Axial moment component
M_p		Peripheral moment component
p	[Pa]	Static pressure
p_t	[Pa]	Total pressure
Δp_t	[kPa]	Total pressure rise
P		Power
P_φ		Production of φ
Pr		Prandtl number
Q	[1/s²]	Q criterion for vortical structures deduction; $Q = 0.5 \times (\Omega_{ij}\,\Omega_{ij} - S_{ij}S_{ij})$
Re		Reynolds number based on reference velocity and length
Re_τ		Reynolds number based on friction velocity
S_{ij}	[1/s]	Rate of strain tensor (symmetric component of the velocity gradient)
t	[s]	Time
u^+	[-]	Normalised velocity
U_{rel}	[m/s]	Relative velocity
U	[m/s]	Velocity
$\overline{u'_i u'_k}$	[m²/s²]	Reynolds stresses

u_τ	[m/s]	Friction velocity; $u_\tau = \tau_w^{0.5}$
V	[m^3/s]	Volume flow rate
v^2	[m^2/s^2]	Normal-to-the-wall component of the Reynolds stresses
x,y,z		Cartesian coordinates
y^+	[-]	Normalised wall-distance; $y^+ = y_n \times u_\tau / \nu$
y_n	[m]	Distance from the wall
y_{n1}	[m]	Distance from the wall of the first cell centre

Greek letters

α_b	[deg]	Blade angle
δ_{ij}	[-]	Kronecker delta
ε	[m2/s3]	Turbulent kinetic energy dissipation rate
ε^+		Normalised turbulent kinetic energy dissipation rate
ε_φ		Dissipation of j
ζ	[-]	Zeta; $\zeta = v_R^2/k$
η	[%]	Efficiency
κ	[-]	Von Karman constant; $\kappa = 0.41$
ν	[m^2/s]	Kinematic viscosity
ν_t	[m^2/s]	Turbulent kinematic viscosity or eddy viscosity
ρ	[kg/m^3]	Air density
τ_w	[m^2/s^2]	Wall shear stress
φ		Generic quantity
Ω_{ij}	[1/s]	Rate of rotation tensor (anti-symmetric component of the velocity gradient)

Subscripts

i,j,k	[-]	Indices

Acronyms

BEM	Boundary element method
CFD	Computational fluid dynamics
DES	Detached eddy simulations
DNS	Direct numerical simulation
HPC	High-performance computing
LE	Leading edge
LES	Large eddy simulations
PS	Pressure side
RANS	Reynolds-averaged Navier–Stokes
RSM	Reynolds stress models (same as SMC)
SAS	Scale adaptive simulations
SMC	Second moment closure (same as RSM)
SS	Suction side
TE	Trailing edge
U-RANS	Unsteady Reynolds-averaged Navier–Stokes
SST	Shear stress transport

INTRODUCTION

The air movement and control community has historically relied on empirical design methodologies based upon an Euler analysis of velocity triangles (Lakshminarayana, 1996), empirical correlations (Cumpsty, 2004), experimental fluid-dynamics (Tropea *et al.*, 2007) and fan noise measurements (Blake, 1986a, 1986b). Engineers have developed these empirical design methodologies over many decades, with each manufacturer developing its own proprietary empirical correlations that aid in applying the basic methodology in specific applications. Unlike aerospace fan designers, designers of low-speed fans have to apply their design methodologies into a wide range of different applications. At one extreme are the cooling fans required for electronic equipment that can be no more than a few centimetres in diameter (Huang and Wang, 2005), and at the other are fans absorbing up to 25 MW in induced-draft power plant applications (Bianchi *et al.*, 2013). It is this breadth of application that has resulted in the different empirical approaches adopted by competing air system fan manufacturers, each treating their proprietary empirical approaches as a source of competitive advantage.

The historic view that the empirical approaches to fan design constitute a form of competitive advantage has resulted in a lack of cooperation and collaboration between members of the air movement and control community. In contrast, the aerospace community has cooperated and collaborated, with the result that the available computational methods have progressed steadily over the least five decades. A result has been steadily improving aerospace fan efficiencies, with the drive for improved fan efficiency originating from efficiency as a source of competitive advantage. In the air movement and control community, low cost has historically been the primary source of competitive advantage. However, this focus on lowest cost has recently changed because of new regulations in the European Community and planned regulation in the USA.

Within the European Union (EU), Commission Regulation No. 327/2011 became legally binding on 1 January 2013, setting minimum Fan and Motor Efficiency Grades (FMEGs) for commercial and industrial fans (Commission Regulation (EU) No. 327/2011). For applicable fans the 2013 minimum fan and motor efficiency grades have resulted in approximately 33 per cent of those fans sold before 1 January 2013 now illegal within Europe (Hauer and Brooks, 2012). Moreover, these minimum fan and motor efficiency grades became more stringent on 1 January 2015 (Hauer and Brooks, 2012).

In the USA, the US Department of Energy has been monitoring activity within the European Union. On 1 February 2013, the US federal government published a framework document in the Federal Register, outlining the approach to fan efficiency regulation within the USA (US Department of Energy, 2013). The framework reflected a desire to be consistent with many elements of the European approach in EU Commission Regulation No. 327/2011 (US Department of Energy, 2013).

With both Europe and the USA now regulating or declaring intent to do so, it is likely that Asian countries will introduce regulations setting minimum commercial and industrial fan or fan and motor efficiencies. Currently, Malaysia, Korea, and Taiwan have considered adopting fan efficiency requirements based on the Air Movement and Control Association (AMCA) Standard 205 *Energy Efficiency Classification for Fans,* as a mandatory requirement for government and private-sector projects (Gho, 2013).

Given today's regulatory environment it is reasonable to assume that in all global regions, minimum fan or fan and motors efficiencies will become mandatory and then increase over time. As a direct result, pressure on the air movement and control community will increase to both develop fans with high peak efficiency, and also specify them such that they operate closer to their peak efficiency point when installed. In response to this pressure, designers have started to adapt aerospace fan design methodologies for application in commercial and industrial fan design processes (Horlock and Denton, 2005), with a particular emphasis on the use of computational fluid dynamics (CFD). Unlike aerospace fan designers who have the resources to develop their own CFD codes, designers working within the air movement and control community must rely on commercial CFD codes. The community is fragmented, with no individual manufacturers having the resources to develop in-house CFD codes.

Commercial CFD codes first become available in the early 1980s. However, modelling the physical flow equations involved significantly simplifying assumptions. The mesh generation techniques were rudimentary and the available hardware to run the codes lacked computational power with the result that meshes were inevitably coarse. Over the last three decades, engineers have systematically addressed these issues, and today commercially available CFD codes model the flow equations accurately. Engineers can construct well-conditioned meshes and the available hardware is adequate to achieve mesh-independent results. Consequently, it is now possible to predict a fan's performance characteristics using commercially available CFD codes with reasonable accuracy.

In this chapter we first provide an overview of the computational methods utilised by the air movement and control community. We start with a description of Reynolds-averaged Navier–Stokes (RANS) turbulence modelling. We then move on to consider how one may apply the RANS modelling technique in practice, followed by a consideration of the application of unsteady RANS, hybrid large eddy simulation (LES) / RANS and large eddy simulation as part of the design process for low-speed fans intended for air system application. We conclude with a consideration of the potential of open-source CFD codes.

RANS TURBULENCE MODELLING

An issue when considering using CFD in any turbomachinery application is that even in the most favourable cases, Reynolds numbers are high resulting in complex flow-fields. A direct numerical simulation (DNS) of the Navier–Stokes equations would require fine computational grids and therefore a computational effort beyond that available within the air movement and control community.

Consequently, the only viable approach available to industrial fan designers when attempting to solve the Navier–Stokes equations is the RANS modelling technique. A key aspect of the RANS modelling technique is the use of turbulence models. These turbulence models constitute a range of simplifying computational approaches for modelling aspects of the blade-to-blade flow-field fluid-dynamics (Durbin and Pettersson Reif, 2001; Davidson, 2004; Hirsch, 2007; Hanjalić *et al.*, 2009; Kundu *et al.*, 2012). These assumptions allow Reynolds-averaging to reduce the computational effort associated with a direct numerical simulation of the Navier–Stokes equations and therefore facilitate simulation of high-Reynolds number flows.

Reynolds-averaging the Navier–Stokes equations refers to a process of Reynolds-averaging the quantities within the equations. We express the generic φ quantity as the sum of its mean value $\bar{\varphi}$ and the fluctuating value φ':

$$\varphi = \bar{\varphi} + \varphi' \tag{1}$$

Successive averaging of the equations themselves results in the creation of a series of additional unknowns called Reynolds stresses and in the case of heat transfer, Reynolds fluxes. To solve the RANS equations requires a closed system of equations modelling these additional unknowns.

Moving from a direct numerical simulation to a RANS simulation reduces the required computational effort for a converged solution to a level practical within the air movement and control community. However, the trade-off is a need to develop methods of modelling the turbulent flow's underlying physics. This requirement is responsible for the significance of turbulence models in RANS codes. The appropriate choice of turbulence model for the application has a primary influence on the accuracy of the fan blade-to-blade flow-field's resulting prediction.

RANS models

For incompressible flows (without heat transfer) the RANS equations read:

$$\frac{\partial u_k}{\partial x_k} = 0 \tag{2}$$

$$\frac{\partial u_i}{\partial t} + u_k \frac{\partial u_i}{\partial x_k} = -\frac{1}{\rho}\frac{\partial P}{\partial x_i} + \nu \frac{\partial u_i}{\partial x_j \partial x_j} - \frac{\partial}{\partial x_k}\left(\overline{u_i' u_k'}\right) \tag{3}$$

In Equation (3) the term in parentheses is the additional unknown Reynolds stress that requires modelling. Most models assume that one can model Reynolds stresses following a Bossiness approximation that utilises a single gradient diffusion hypothesis (Pope, 2000):

$$\overline{u_i'u_j'} = \frac{2}{3}k\delta_{ij} - \nu_t\left(\frac{\partial u_i}{\partial x_j} + \frac{\partial u_j}{\partial x_i}\right) \tag{4}$$

which reduces the closure problem and requires an expression for eddy viscosity. One may model eddy viscosity using Prandtl's mixing length approximation (Pope, 2000):

$$\nu_t = \left|\frac{du}{dy}\right|(\kappa y_n)^2 \tag{5}$$

which results in an accurate model for wall-bounded, attached flows with small pressure gradients. Although potentially useful in some applications, a modelling approach that is accurate with attached flows and small pressure gradients is not applicable in low-speed fans intended for air system applications where separated flow is inevitable. This lack of applicability is the primary drawback of a RANS approach. In practice, the models are derived from a series of simplifying assumptions that one can use to generate and calibrate coefficients using a series of benchmark cases. The benchmarking cases are not representative of the separated flow and large pressure gradients in low-speed fans. This results in a very real risk that computational solutions do not adequately model the flow-field physics. This negative aspect that occurs with RANS solutions is unknown to many of those using commercial codes, and consequently they do not take the necessary care when setting up the solver.

The limitations that occur with using Prandtl's mixing length approximation is a reason why the majority of eddy viscosity models rely on one or more transport equations to derive a value for eddy viscosity. Spalart-Allmaras' one-equation model (Spalart and Allmaras, 1992) has become the established eddy viscosity model, used in the majority of commercial codes. The air movement and control community typically utilises the Spalart-Allmaras model in two ways:

- computation of lift and drag coefficients of isolated aerofoils; and,
- forming a baseline RANS closure for detached eddy simulations (DES).

When studying the modelling of lift and drag coefficients, Rábai and Vad (2005) reported that the Spalart-Allmaras model was capable of correctly predicting an isolated aerofoil's pressure distribution, lift and drag coefficients, Figure 1.1. Despite its success with isolated aerofoils, Šarić *et al.* (2008) concluded that the model did not lend itself to application with the complex geometries that occur with internal flow calculations. In this context, we define an internal flow as a flow through turbomachinery blading, as opposed to the external flow that occurs with isolated aerofoil or aircraft wing flow-field predictions.

When dealing with internal flows, one can overcome the limitations of a one-equation model with a two-equation model. The currently favoured two-equation

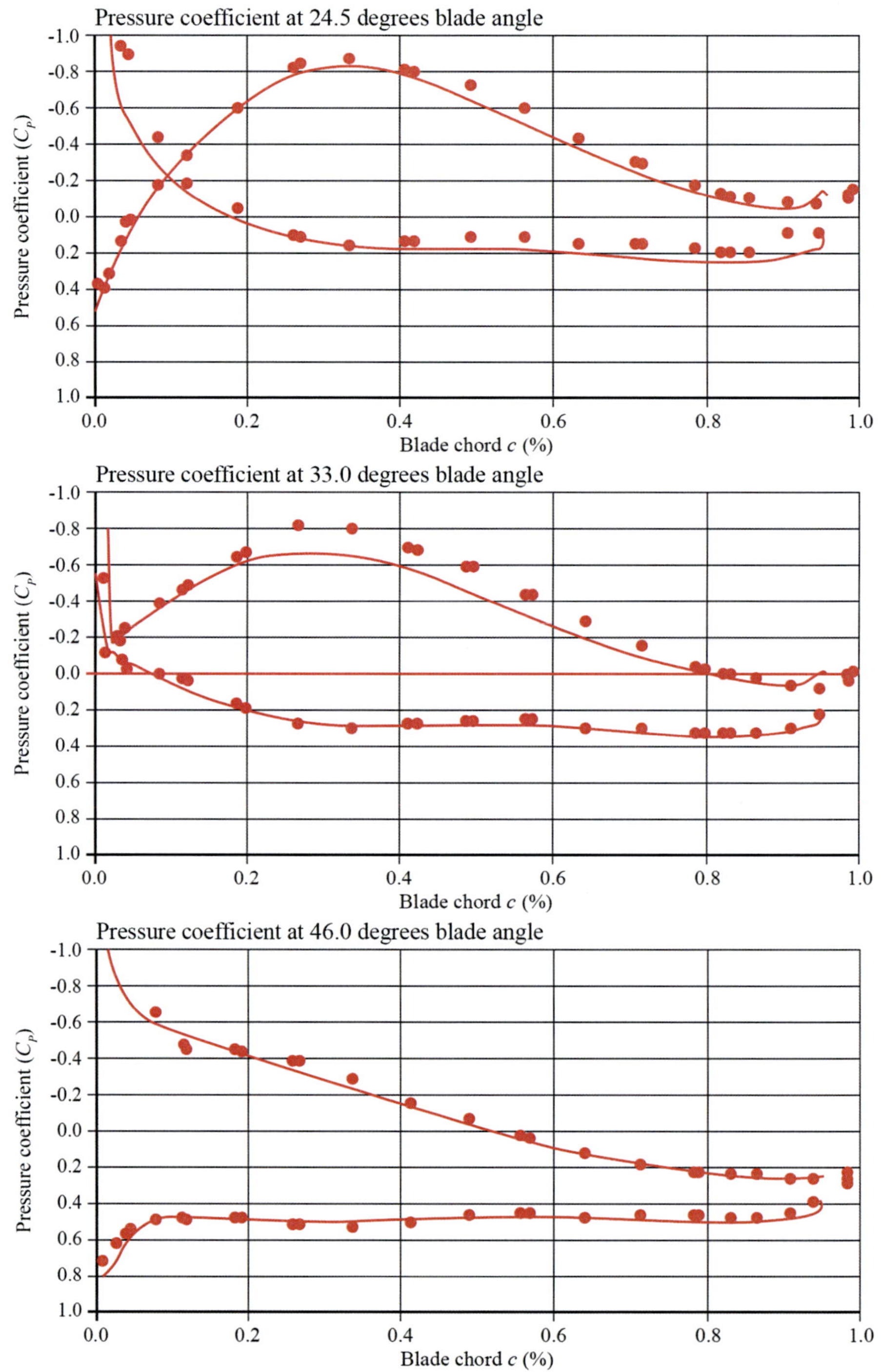

FIGURE 1.1. Two-dimensional Spalart-Allmaras computations of pressure coefficient for an isolated compressor blade (line) and three-dimensional measurements of pressure coefficient made in a compressor cascade at mid-span (dots), from Rábai and Vad (2005).

models are either the k-ε model (Jones and Launder, 1972) or the k-ω model (Wilcox, 1988, 1993). Both models require one to solve two transport equations to compute eddy viscosity as an algebraic expression of turbulent kinetic energy (k) and a dissipation rate of turbulent kinetic energy (ε) or turbulence frequency (f_t). Turbulence frequency is denoted by ω when referred to within the context of the k-ω turbulence model. Both the k-ε model and the k-ω model rely on an assumption that one can link eddy viscosity to a time- and length-scale that characterises turbulence that then links to the computed flow-field's characteristics. A feature of the k-ε and k-ω models is that the additional transport equations for k, ε and f_t share the same form and therefore for a generic φ quantity it reads:

$$\frac{D\varphi}{Dt} = P_\varphi - \varepsilon_\varphi + \frac{\partial}{\partial x_j}\left[\left(\nu + \frac{\nu_t}{\sigma_\varphi}\right)\frac{\partial \varphi}{\partial x_j}\right] \tag{6}$$

On the left-hand side of the Equation (6) is the quantity's material derivative. On the right-hand side are one (or more) production terms, a dissipation term, a diffusion term dependant on molecular viscosity and another given as the turbulent viscosity's function, corrected using the Prandtl number P_r. The primary difference between the k-ε and k-ω models is the different trend of ε and f_t at the wall and the definition of the wall boundary conditions for the same variables. When one studies normalised values of ε and f_t for an attached flow, it is evident that f_t is less dependent on the Reynolds number then ε in the wall's near vicinity, Figure 1.2.

There is a general consensus within the computational fluid dynamics community that the k-ε model better reproduces the energy cascade of large scale structures in the main flow core, whilst the k-ω model performs better near the wall (Menter, 1993). A realisation that k-ε models perform better in the main flow whilst the k-ω models perform better near the wall leads to the natural conclusion that ideally, one would use the two models in combination. Hanjalić *et al.* (2009) observed that it is possible to combine k-ε and k-ω models as one can reformulate every two-equation model into every other by changing model coefficients. This realisation has enabled engineers to formulate the k-ω shear stress transport (k-ω SST) model (Menter, 1994) that solves the equation for ω near the wall and ε elsewhere.

The use of two-equation models has become established within the air movement and control community. Pinelli *et al.* (2012) reported a series of simulations with the objective of optimising a double inlet centrifugal fan's design by utilising a k-ω SST model. Pinelli *et al.* (2012) were able to select the best geometry from three competing alternatives and correctly predict the pressure rise, volume flow rate characteristic curve, Figure 1.3. Corsini and Rispoli (2005) and Sheard *et al.* (2009) reported a series of computations to predict the characteristic curve for an axial fan using a RANS simulation with a two equation k-ε model. Other researchers (Corsini and Rispoli, 2004; Corsini and Sheard, 2007; Corsini *et al.*, 2007, 2009, 2010, 2013b; Vad and Horváth, 2013) report similar studies that use two-equation models.

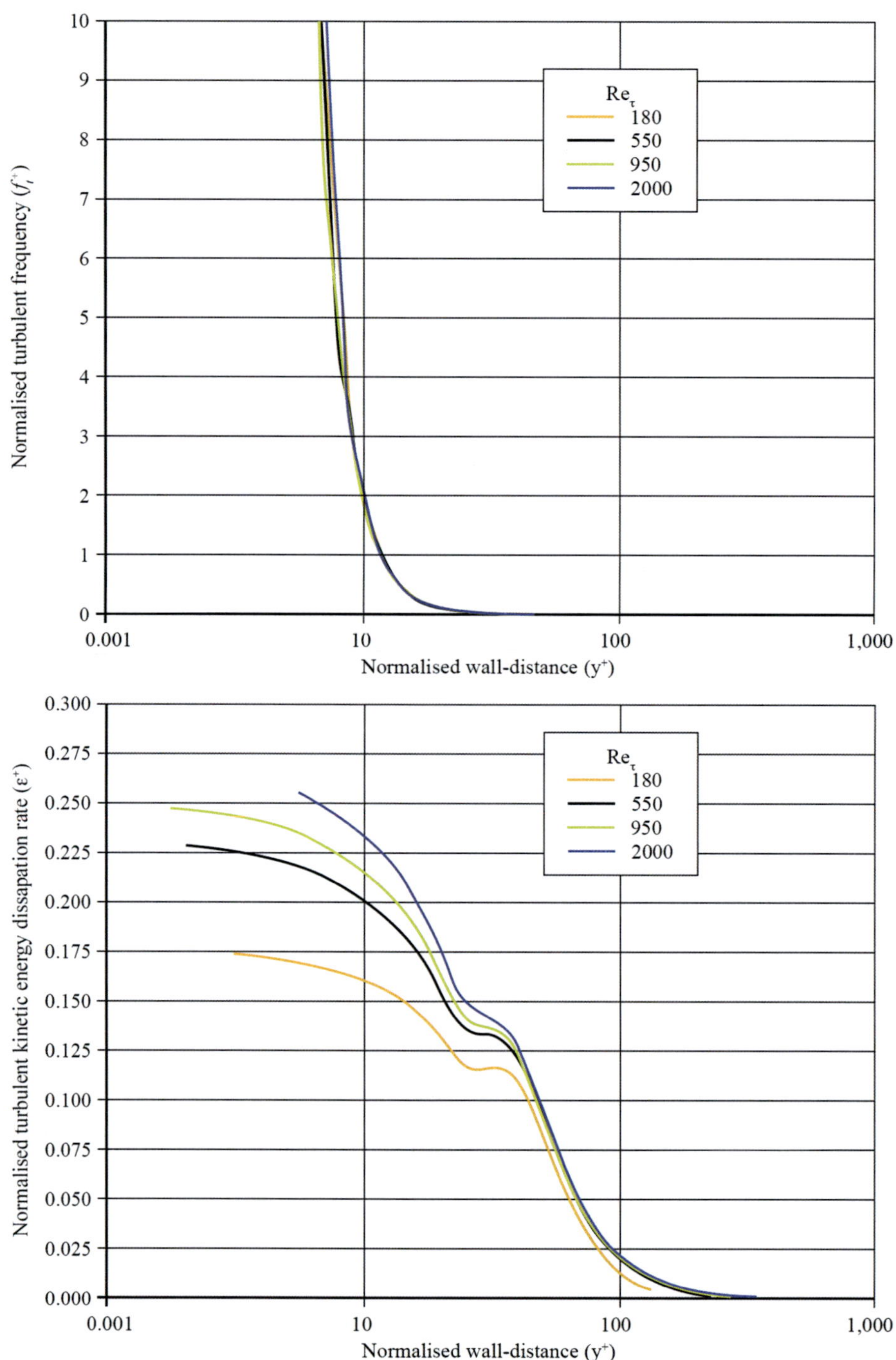

FIGURE 1.2. Channel flow computed using a direct numerical simulation (DNS). Normalised turbulent frequency (f_t^+) and normalised turbulent kinetic energy dissipation rate (ε^+) against normalised wall-distance (y^+) calculated over a range of Reynolds number based on friction velocity (Re_τ).

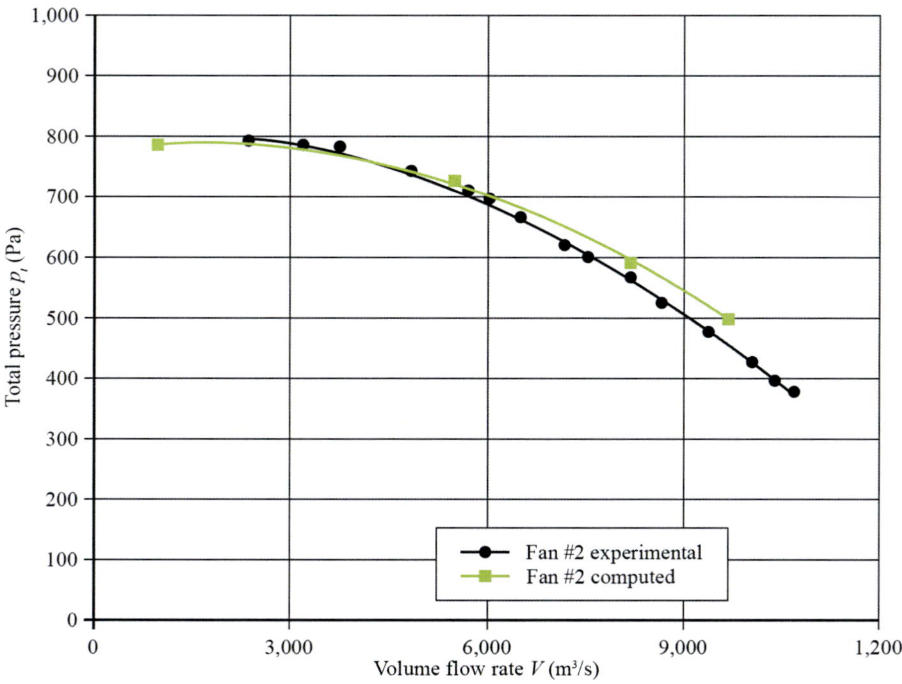

FIGURE 1.3. Characteristics computed using a RANS simulation with a k-ω SST turbulence model and experimentally measured characteristics of a double-inlet impeller centrifugal fan, from Pinelli *et al.* (2012).

Wall treatment

When studying and modelling the fluid mechanics of low-speed fans intended for air system application, researchers have concluded that the key phenomena of interest are induced by large velocity gradients that occur with the presence of solid walls. When using a RANS approach to wall treatment, one can formulate every model into what we may characterise as either a high- or low- Reynolds formulation. The difference between the two is integrating transport equations at the walls into the computational method.

When one studies the normalised velocity profile for channel flow at increasing Reynolds numbers, it is apparent that the profile is independent of Reynolds number. Researchers have studied the impact of Reynolds number on near-wall flow (Pope, 2000; Davidson, 2004; Hirsch, 2007; Kundu *et al.*, 2012) and have characterised the profile into three zones, a viscous sub-layer for $y^+<5$, a logarithmic layer for $y^+>30$ and a buffer layer for the range of $5<y^+<30$. It is the viscous sub-layer that is the most critical zone when modelling the flow-field, Figure 1.4.

The high-Reynolds formulation integrates transport equations into the computational method using wall functions that model fluid behaviour through the viscous sub-layer to the wall. The first node of the computational grid is placed in the flow's fully turbulent region ($y^+ \approx 40$), reducing the grid density and therefore the required computational effort for a simulation. A drawback of this approach is that one must define the wall function in a way that accounts for the wall's presence.

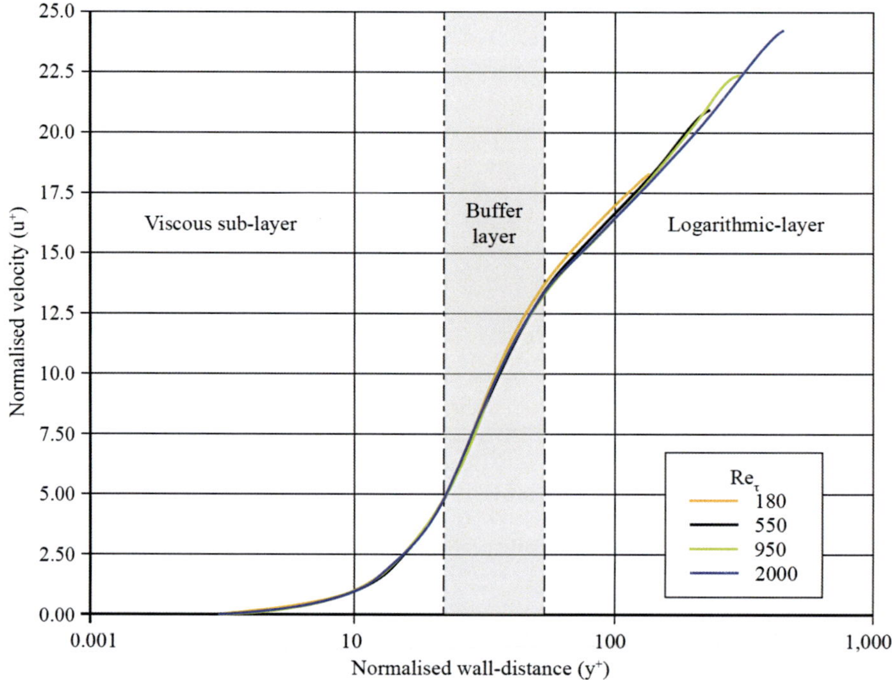

FIGURE 1.4. Channel flow computed using a direct numerical simulation (DNS). Normalised velocity (u^+) against normalised wall-distance (y^+) calculated over a range of Reynolds number based on friction velocity (Re_τ) illustrating the range of normalised wall-distance associated with the viscous sub-layer, buffer layer and logarithmic-layer.

Whilst this is practical, one typically derives wall functions from a series of assumptions:

- the boundary layer is attached; and,
- the local turbulent energy is in equilibrium.

The above assumptions are not valid when the flow is separated, there is an adverse pressure gradient or there is a wall curvature or impinging flow onto the wall. Those in the field inevitably associate low-speed fans intended for air system application with separated flow. Their geometry induces adverse pressure gradients in the flow-field as a consequence of the fan blades' curvature. Impinging flow onto walls is inevitable because of blade tip-to-casing flow and blade wake impingement on static components. Consequently, the use of wall functions in a high-Reynolds formulation facilitates a reduction in the computational effort required for the simulation, but the wall functions themselves have their deficiencies.

A RANS model may address these deficiencies by integrating up to the wall. However, in order to integrate up to the wall, we must reformulate the RANS equations to account for viscous effects in the near-wall region. We may reformulate the equations by incorporating a logarithmic function of either Reynolds number based on friction velocity (Re_τ) and normalised velocity (u^+)or normalised wall-distance (y^+) and normalised velocity (u^+). This makes it possible to account for the viscous effect in the viscous sub-layer, Figure 1.4. Davidson (2004) provides a comprehensive description of the use of wall functions for high-Reynolds representations of Equation (6) and their reformulation into a low-Reynolds form of Equation (6).

Grid refinement and grid independency

We may reformulate the RANS equations to at least partly account for viscous effects in the near-wall region by using a low-Reynolds formulation. When using a low-Reynolds formulation the mesh needs to be refined in the wall-normal direction. The objective is to ensure that the first grid node lies in the viscous sub-layer ($1 < y^+ < 5$). Both low- and high-Reynolds formulations place requirements on the computational grid. The low-Reynolds formulation requires the first grid node to lie in the viscous sub-layer. The high-Reynolds formulation requires placing the first node of the computational grid in the flow's fully turbulent region. However, it is not possible to know a priori if the first node lies in either the viscous sub-layer or the flow's fully turbulent region. In practice, designers must rely on experience, using validated results from similar fan geometries operating over a range of similar mass flow rate. This experience informs two key choices:

- grid spacing near the solid boundaries (blades, hub and casing) in order to fulfil the requirements of either low- or high- Reynolds formulations; and,
- estimating the number of cells required in order to achieve a grid independent solution.

When considering the required number of cells to achieve a grid independent solution, it is noteworthy that not all the computational domain parts are equally important. For example, we associate the blade tip-to-casing region with a tip-leakage vortex, and in order to model this flow feature, it requires a relatively high grid density. However, we must associate this cell clustering in regions of known flow-field features with a smooth transition from regions of low to high cell density. We must avoid cells with a large aspect ratio or those that are heavily distorted. Despite this caveat, we can minimise the required number of cells to achieve a grid independent solution through a grid refinement process, and therefore designers routinely employ grid refinement.

One may also use grid refinement to cluster cells near to every wall. This maximises the probability that there will be a node in the viscous sub-layer and hence that a low-Reynolds formulation will accurately model the flow-field physics. A way to check that grid refinement has successfully resulted in a node in the viscous sub-layer is to adapt the RANS formulation, with the formulation now checking the y^+ value during a simulation. If the y^+ value is greater than the first node's distance from the wall, the RANS formulation dynamically switches from a low- to high-Reynolds formulation. Popovac and Hanjalić (2007) characterised this dynamic switching from a low- to high-Reynolds formulation as a compound wall treatment. Popovac and Hanjalić (2007) concluded that a compound wall treatment resulted in significantly more accurate computational simulation results.

A low-Reynolds formulation typically requires a grid with twice the number of required cells for a grid-independent high-Reynolds formulation. Despite the increase in computational effort that occurs with low-Reynolds formulations, designers working in the air movement and control community have judged the improvement in accuracy that occurs with low-Reynolds formulations to be more important than the increase in computational effort. Therefore, the majority of designers use a RANS formulation with low-Reynolds formulation to predict overall fan performance over the stable part of the fan's characteristic.

Anisotropic modelling of eddy viscosity

A limiting factor in a RANS formulation occurs with the definition of eddy viscosity, Equation (5). Modelling eddy viscosity using Prandtl's mixing length approximation results in the Reynolds stress tensor aligning with the velocity's gradient. The Reynolds stress tensor is not aligned with the velocity's gradient in regions of separated flow or impingement, and therefore, the modelling approach for eddy viscosity does not accurately model the flow-field physics. A method to improve the accuracy with which one models the flow-field physics without increasing the computational effort required is to use a non-linear eddy viscosity model. Non-linear eddy viscosity models utilise a cubic definition of eddy viscosity (Craft *et al.*, 1996; Lien *et al.*, 1996). A cubic definition of eddy viscosity results in the definition of Reynolds stresses in Equation (4) assuming a more complex form:

$$
\begin{aligned}
\overline{u_i u_j} &= \frac{2}{3} k\delta_{ij} - \nu_t S_{ij} \\
&+ c_1 \nu_t \frac{k}{\varepsilon}\left(S_{ik}S_{kj} - S_{kl}S_{kl}\frac{1}{3}\delta_{ij}\right) + c_2 \nu_t \frac{k}{\varepsilon}\left(\Omega_{ik}S_{kj} - \Omega_{jk}S_{kl}\right) \\
&+ c_3 \nu_t \frac{k}{\varepsilon}\left(\Omega_{ik}\Omega_{kj} - \Omega_{kl}\Omega_{kl}\frac{1}{3}\delta_{ij}\right) \\
&+ c_4 c_\mu \nu_t \left(\frac{k}{\varepsilon}\right)^2 \left(S_{ki}\Omega_{lj} + S_{kj}\Omega_{li} - S_{km}\Omega_{lm}\frac{2}{3}\delta_{ij}\right) S_{kl} \\
&+ c_5 c_\mu \nu_t \left(\frac{k}{\varepsilon}\right)^2 \left(S_{ik}S_{jl} - S_{km}S_{lm}\frac{1}{3}\delta_{ij}\right) S_{kl} + c_6 c_\mu \nu_t \left(\frac{k}{\varepsilon}\right)^2 S_{ij}S_{kl}S_{kl} \\
&+ c_7 c_\mu \nu_t \left(\frac{k}{\varepsilon}\right)^2 S_{ij}\Omega_{kl}\Omega_{kl} \qquad (7)
\end{aligned}
$$

The eddy viscosity's third order tensorial formulation in Equation (7) partially accounts for surface curvature. The formulation is therefore better able to account for producing turbulent kinetic energy in impingement regions as it constitutes a more realistic anisotropic reproduction of Reynolds stresses with respect to linear eddy viscosity. Non-linear eddy viscosity models are a key feature of the formulation that enables a RANS simulation to simulate the flow-field in both turbomachinery (Chen *et al.*, 1998) and low-speed fan applications (Corsini and Rispoli, 2005). Corsini *et al.* (2013b) illustrated the effectiveness of an advanced cubic k-ε turbulence model. They were able to accurately predict the axial flow fan's characteristics within the uncertainty with which they could experimentally measure the fan's characteristics, Figure 1.5.

An alternative to a non-linear eddy viscosity model when accounting for the anisotropy of Reynolds stresses are the elliptic relaxation models v^2-f (Chen *et al.*, 1998) and ζ-f (Hanjalić *et al.*, 2004). By solving two additional transport equations these elliptic relaxation models are able to account for the effect of wall blockage on Reynolds stresses, and increase the RANS simulation's accuracy without a significant increase in computational effort (Durbin, 1995).

Incorporating either a non-linear eddy viscosity model or an elliptic relaxation model constitutes an increase in complexity over that of a basic RANS formulation. The next level of complexity is the addition of second moment closure (SMC), also known as Reynolds stress models (RSM). The addition of SMC requires the RANS formulation to solve additional transport equations for the Reynolds stresses (Hanjalić and Jakirlić, 1993, 1998; Launder and Sandham, 2002; Borello *et al.*, 2005; Hanjalić and Launder, 2011) whilst correctly reproducing anisotropy. Durbin (1993) observed that the addition of second moment closure significantly increases a RANS simulation's accuracy when modelling the complex flow-fields that occur with turbomachinery applications (Borello *et al.*, 2005, 2007). A drawback of the addition of second moment closure into a RANS formulation is the increase in computational effort as one must solve at least six additional transport equations. A further disadvantage is that the increase in numerical stiffness can lead to the solver's (Hanjalić and Jakirlić, 1993) non-convergence, and consequently, designers working within the air movement and control community do not favour RANS formulations with second moment closure.

Rotation, swirl and transition

As we discussed at the beginning of this chapter, a drawback of a RANS approach is the assumptions on which the model relies. When studying physical transport phenomenon, Hanjalić *et al.* (2009) observed that the equation for *k* in RANS formulations is not sensitive to rotation, and that this is a primary deficiency of any RANS formulation. A way to improve a RANS formulation is to include terms into the formulation that account for the effect of rotation (Sander and Lilley, 1983; Kobayashi and Yoda, 1987). Although potentially helpful, in practice these attempts to account for the effect of rotation have not improved the accuracy of

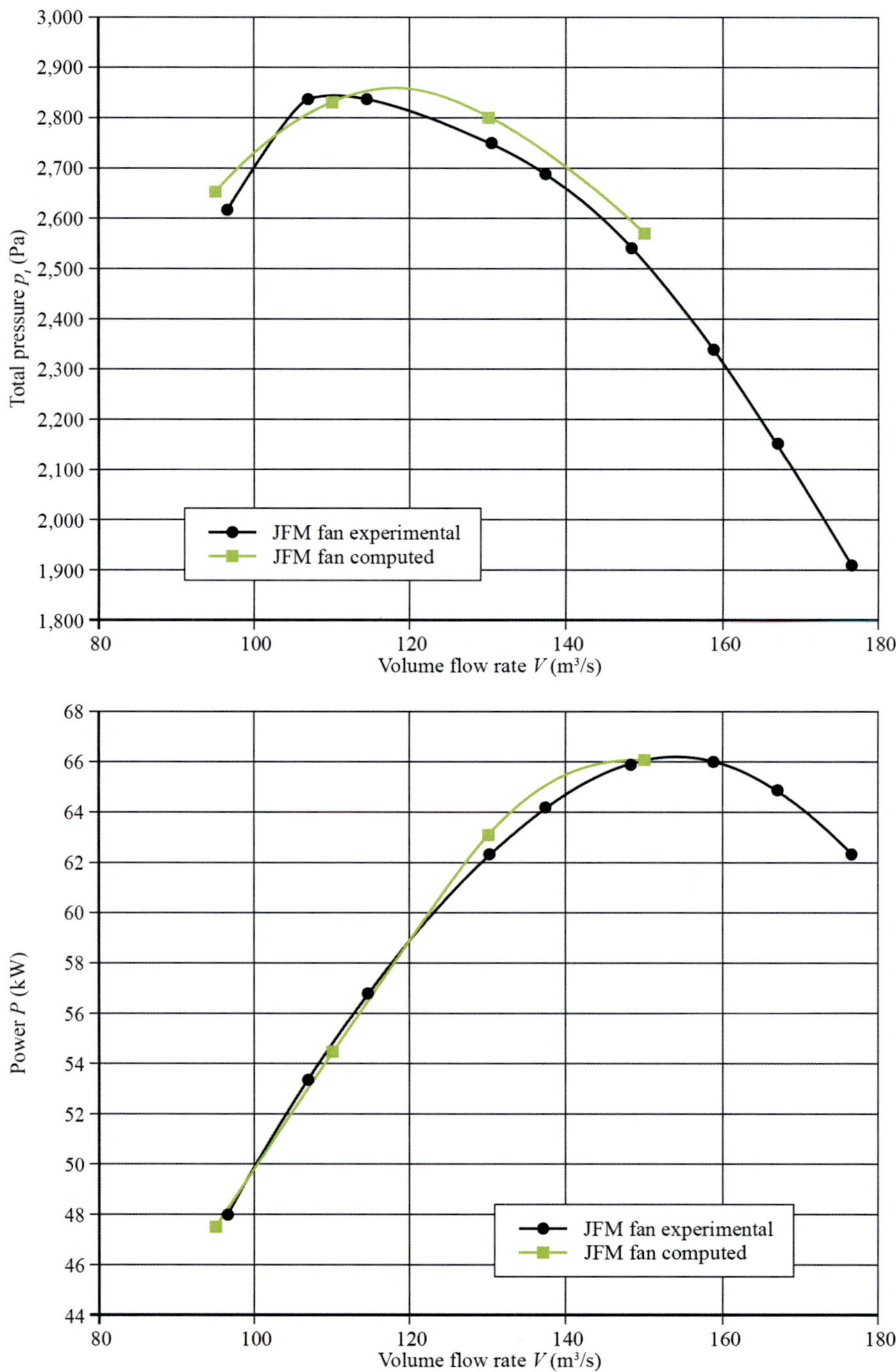

FIGURE 1.5. Numerical prediction and experimentally measured performance characteristics for a tunnel ventilation fan, from Corsini *et al.* (2013b). Computations conducted using a low-Reynolds number cubic k-ε turbulence model, from Lien *et al.* (1996).

results. In practice, the RANS formulations that give the best results are those that design engineers have adapted to the specific application (Speziale *et al.*, 1991; Lien and Durbin, 1996). Therefore, a focus for a researcher working in the field is to identify specific RANS formulation deficiencies when applied into a specific application, with the objective of correctly reproducing the laminar to turbulent transition of boundary layers (Lien and Durbin, 1996; Menter *et al.*, 2004, 2006; Langtry and Menter, 2005). Ability to better model laminar to turbulent boundary layer transition is a key factor in modelling the blade-to-blade flow-field more accurately, and the subject remains an active research area.

The above provides an insight into the difficulty that occurs with using a RANS simulation to predict a fan's blade-to-blade flow-field. The active research that occurs with tailoring RANS simulations to specific applications indicates that we cannot assume 'standard' RANS models to give an accurate prediction of either a fan's blade-to-blade flow-field or its overall performance. In essence, one must select or develop a RANS model to reproduce the flow-field physics relevant to each application. Pope (2000) provides a good overview of the literature characterising RANS solution limitations, and the various ways in which design engineers have addressed those limitations.

UNSTEADY RANS (U-RANS)

A way to improve results with the Reynolds-average approach without adding application specific models is to solve the unsteady RANS equations. An unsteady RANS approach still does not account for the turbulence energy cascade from large scale structures. However, an unsteady RANS is able to reproduce unsteady flow phenomena such as the blade tip-to-casing leakage vortex and the development of secondary flow features in the blade hub region. Piotrowski *et al.* (2009) used an unsteady RANS approach and were able to compute the flow-field through a linear compressor cascade. They concluded that the blade tip-to-casing leakage vortex evolved over time, but with its position remaining fixed. In contrast, the hub secondary flow vortex did not remain in a fixed position. The hub secondary flow vortex development was unstable, shedding at a characteristic frequency. Yang (2012) studied the flow-field through a centrifugal fan using an unsteady RANS approach. He concluded that the predicted fan characteristic was more accurate using the unsteady RANS approach then had been possible previously using a steady RANS approach with the same k-ωSST turbulence model.

The primary drawback that engineers associate with an unsteady RANS approach when compared with a RANS approach is the increase in the required computational effort to compute a solution. The increase in computational effort is a consequence of the need to simulate the flow-field over a period of time to correctly model the time varying fluctuation of velocity, pressure and flow-field features. Courant *et al.* (1967) observed that an additional factor that impacts on the required computational effort for an unsteady RANS solution is the length of time step that one selects for the computation. The Courant–Friedrichs–Lewy (CFL) number limits the time step which is a necessary condition for convergence whilst solving numerically. It arises in the analysis of explicit time-marching schemes. The time step must be less than a certain time otherwise the simulation will produce incorrect results. The CFL number's limiting value is:

$$CFL_{\max} < 1 \tag{8}$$

If the CFL number remains below its maximum allowable value, the flow-field's time-based resolution remains adequate for the mesh's spatial resolution that one uses for the computations. In essence, the finer the mesh, the smaller the maximum allowable time step in an unsteady RANS computation. Within the air movement and control community it is common practice to increase the time step such that $CFL_{\max}<1.5$ without the accuracy of the resultant simulation degrading to the point at which conclusions that one draws from the results become error prone. However, a $CFL_{\max}<1.5$ will only give valid results if the CFL number's average value over the entire computational domain remains below unity.

AXIAL FANS

The challenges that occur with computing both the blade-to-blade flow-field and overall fan characteristic are application specific. We may split the applications into two basic groups, axial and centrifugal fans. First, we will consider axial fans.

Rotor-only computations, pitch-wise periodicity and turbulent structures

When modelling an axial fan's blade-to-blade flow-field the least complex configuration that one can model is the rotor alone. If one neglects static components then one can carry out the computations in the rotating frame of reference, and consequently, there is no requirement to account for the computational mesh's movement. Those researchers who have used a rotor only approach augment the RANS momentum equation with terms that represent Coriolis and centrifugal forces (Pope, 2000; Davidson, 2004; Hirsch, 2007; Kundu *et al.*, 2012). Denton and Dawes (1999) and Denton (1979) observed that one may use periodic boundary conditions at mid-pitch in order to compute the flow-field through a single blade or vane passage (Corsini and Rispoli, 2005; Sheard *et al.*, 2009).

A disadvantage that occurs with applying periodic boundary conditions is that they can fail to model accurately turbulent structures if one associates those structures with long wavelengths or length-scales. The limitations that occur with applying periodic boundary conditions is not an issue with a RANS approach as the approach itself does not model the turbulent structures that periodic boundary conditions fail to model. Consequently, the use of periodic boundary conditions does not degrade the RANS simulation's accuracy.

An unsteady RANS simulation is able to model major unsteady blade-to-blade flow-field features (Horlock and Denton, 2005; Piotrowski *et al.*, 2009). An example of a flow-field feature that an unsteady RANS simulation can model is the vortex shedding that occurs with blade wakes. Vanella *et al.* (2008) observed that when conducting an unsteady RANS simulation it can be advantageous to extend the computational domain to more than a single blade or vane passage. A reason for extending the computational domain when the studied fan has a high solidity is the possibility that turbulent structures that generate in one blade or vane passage may impinge upon the next. Consequently, an unsteady simulation that models only one blade or vane passage will not model the flow-field physics accurately.

It is only necessary to extend an unsteady RANS computational domain to the entire fan when attempting to gain insight into phenomena that affect the entire rotor. Pullan *et al.* (2012) reported an example of phenomena that affects the entire rotor. They studied the onset of stall in an axial compressor using a Spalart-Allmaras unsteady RANS simulation, discretising each rotor's blade-to-blade passage using a total of 74×10^6 cells. They considered this number of cells as the minimum required to adequately model the flow-field physics that occurs with rotor stall. The number of cells that Pullan *et al.* (2012) used is approximately one order of magnitude larger than that designers working within the air movement and control community routinely use, and is consequently beyond that which is practical at the time of writing.

A further consideration when using a rotor only simulation for either a RANS or unsteady RANS simulation is the computational domain's axial extent up and down stream of the blade leading and trailing edge. For a fan with well-conditioned in-flow characteristics, Corsini and Rispoli (2005) concluded that one could achieve a mesh independent solution with a computational domain that extended at least half a blade chord up- and down-stream of the blade leading and trailing edge. When simulating the blade-to-blade flow-field for a fan with a distorted in-flow, Thiart and von Backstrom (1993) concluded that the computational domain must extend more than half a blade chord up-stream of the blade leading edge, with the required distance dependent on the in-flow distortion's extent.

A final consideration when using a rotor only simulation for either a RANS or unsteady RANS simulation is the definition of in-flow conditions. Engineers typically associate low-speed fans intended for air system application with relatively low inlet flow velocities, and therefore it is reasonable to assume that the in-flow is incompressible. It is therefore possible to define an inlet velocity profile and either a fixed turbulence level or a turbulence profile that accounts for the inlet velocity profile. A typical velocity profile would be the profile that engineers associate with fully developed pipe flow. A typical mid-stream turbulence level would be three per cent, rising to ten per cent in the near-wall region as velocity reduces (Sheard *et al.*, 2009).

When considering compressible flows, an alternative approach is to fix static pressure at the outlet and total pressure and mass flow rate at the inlet. Hirsch (2007) found this approach to result in an accurate simulation of the blade-to-blade flow-field, if the computational domain were extended to more than one blade chord length forward of the blade leading edge. The assumption implicit in the assertion that outlet static pressure, inlet total pressure and mass flow rate can be fixed is that the fan is operating in an ideal system. In practice, commercial and industrial fans are more usually operating in a complex system, and consequently, heavily distorted in-flow is the norm rather than the exception. When the in-flow is heavily distorted, an experimentally measured in-flow velocity and turbulence profile are required. If experimentally measured in-flow velocity and turbulence profiles are not available then one must add a portion of the inlet duct system to the simulation if the resulting solution is to be accurate.

Rotor–stator interaction and contra-rotating fans

A practice that has become established in the air movement and control community is to pre-rotate the flow into an axial fan inlet (Fernández Oro *et al.*, 2007). Pre-rotating the air into the fan can facilitate a seven to ten per cent increase in fan pressure developing capability with little or no reduction in fan efficiency. As a fan motor is typically specified with between 10 and 15 per cent reserve, adapting in-service fans to include the in-flow's pre-rotation is a way to increase a fan's pressure developing capability. Increasing a fan's pressure developing capability may be required in the event that one incorrectly specified the fan which needs up to a ten per cent increase in pressure developing capability in its application. One may model the effect of pre-swirl in a RANS or unsteady RANS simulation by distorting the in-flow velocity profile to account for the pre-swirl effect (Corsini *et al.*, 2013b).

When the required pressure developing capability exceeds that of a single fan by a factor of between two and two and a half, a common practice within the industrial fan community is to configure two fans, counter rotating in series (Brailko *et al.*, 2004). The use of two counter-rotating fans is common in tunnel ventilation systems, and engineers occasionally have adopted it in wind tunnel applications. An advantage of the counter rotating configuration is that the second fan removes the exit swirl from the first, resulting in axial flow down-stream of the fan. A disadvantage is the relatively high blade passing velocity of the counter-rotating blades that result in two counter rotating fans producing approximately 15 dB more noise than one fan running alone. Modelling two counter rotating fans in either a RANS or an unsteady RANS simulation requires the model to account for the two blade rows' relative motion. We may model this in three ways, using a:

- mixing plane;
- frozen rotor; or,
- an unsteady coupling with moving mesh.

The mixing plane approach is the approach that researchers first developed when they attempted to model the interaction of counter-rotating blade rows or the interaction between a rotating and static blade row. The approach is based on the assumption that one could undertake separate steady-state simulations for each rotor or the rotor and the stator. One would then calculate circumferentially averaged velocity, pressure and turbulence variables at the rotor's out-flow, and use these calculated values as the in-flow conditions for the second rotor or stator (Denton and Dawes, 1999).

A disadvantage of the mixing plane approach is that wakes from the first rotor mix out at the mixing plane, rather than mixing out gradually as they wash downstream into the second rotor or the stator. Mixing out wakes instantaneously at the mixing plane is responsible for the inability of the mixing plane approach to simulate the impact of wakes' from the first rotor on the second rotor or stator. The effect of wakes from a rotor on a downstream stator is real and one should model them. If one neglects them, the resultant simulation is still capable of predicting with reasonable accuracy the blade-to-blade flow-field and the rotor and stator's overall performance. However, this is not the case with a counter-rotating fan. The effect of wakes from the first rotor on the second is significant enough for a mixing plane approach to reduce the second rotor blade-to-blade flow-field prediction's accuracy so far that it is no longer possible to accurately compute the overall fan performance.

A better approximation of rotor-rotor or rotor-stator interaction than the mixing plane approach is the frozen rotor approach. The primary advantage of the frozen rotor approach is that the rotor to rotor or rotor to stator coupling does not require a moving mesh. The relative position of the two rows is fixed in time, with relative motion managed using different frames of reference and adding Coriolis and centrifugal forces to the momentum equation in the rotating frame of reference (Brailko *et al.*, 2004). In contrast to the mixing plane approach, the frozen rotor approach is able to account for a non-uniform circumferential distribution of velocity and pressure. Non-uniform circumferential distribution of velocity and pressure lead to a more realistic representation of wake mixing, and consequently, a more accurate simulation of the blade-to-blade flow-field in the second rotor or stator.

Designers working in the air movement and control community have favoured the frozen rotor approach, as the resultant RANS or unsteady RANS simulations are able to predict the overall fan performance characteristic with good accuracy. This does not mean that the approach is without drawbacks. Adding Coriolis and centrifugal forces to the momentum equation in the rotating frame of reference constitutes an approximation of the flow-field physics that limits the resulting simulations' accuracy. To accurately simulate the flow-field physics it is necessary to undertake a fully unsteady coupling of the two blade rows. This requires a moving-mesh that can increase the required computational effort by up to an order of magnitude compared to a frozen rotor simulation. The increase in computational effort is a consequence of the required time to account for the interaction between the blade rows. The minimum number of interactions depends on clocking the two rows with Yang *et al.* (2012) proposing that computing one and a half rotor revolutions is sufficient to account for rotor-stator interaction. Despite the advantages of a moving mesh approach, the additional computational effort associated with a moving mesh is the reason designers working in the air movement and control community have accepted the limitations of a frozen rotor approach.

CENTRIFUGAL FANS

The modelling considerations that apply to an axial fan are unchanged for a centrifugal fan, with one exception. When modelling centrifugal fans, one cannot reduce the computational effort by modelling a single blade passage with imposed periodic boundary conditions. Centrifugal fan simulations must be full rotor simulations, and consequently, they inherently require more computational effort than an axial fan single blade passage flow-field simulation.

Issues with in-flow and out-flow boundary conditions are similar for both axial and centrifugal fans. However, centrifugal fans require additional consideration of the computational domain's extent up and down stream of the rotor. At the fan inlet it is customary to add a hemisphere with a diameter two to three times the fan inlet's diameter, or location at which total pressure is specified (Lee *et al.*, 2010; Wolfram and Carolus, 2010). The fan outlet is complicated because centrifugal fans are typically coupled with an exhaust volute that discharges through a diffuser. Consequently, to correctly predict a fan's characteristics it is customary to extend the computational domain three duct diameters down-stream of the diffuser tongue. Fan mass flow rate is then specified at the down-stream limit of the computational domain (Lee *et al.*, 2010; Wolfram and Carolus, 2010).

With a centrifugal fan simulation it is not possible to use the mixing plane approach when modelling the fan rotor and stator. The mixing plane approach relies on circumferentially averaging velocity and pressure. A high circumferential pressure imbalance characterises centrifugal fans, and therefore, any simulation must be capable of modelling this unbalance (Lee *et al.*, 2010). Within the air movement and control community, the established approach for modelling a centrifugal fan impeller is to use the frozen rotor approach. Researchers have used the frozen rotor approach using both RANS and unsteady RANS simulations to predict centrifugal fan characteristics. Son *et al.* (2011) utilised a RANS simulation with frozen rotor and k-ε turbulence model in high-Reynolds formulation to study the effect of varying inlet bell mouth geometry on a centrifugal fan's flow rate characteristics.

In practice, an unsteady RANS simulation is best able to simulate the flow features that occur with fan stall. Yang *et al.* (2012) was able to use an unsteady RANS simulation to predict a centrifugal fan's characteristics up to its stall point. Corsini *et al.* (2013a) were able to demonstrate that a high-Reynolds number formulation, standard k-ε turbulence model and a frozen rotor approach proved to be sufficient to correctly predict a large centrifugal fan's characteristics for process industry applications both at peak efficiency and at peak pressure operating points, Figure 1.6.

SIMULATION OF VENTILATION SYSTEMS WITH SYNTHESISED METHODS

The computational methods that we have discussed so far have been concerned with predicting the blade-to-blade flow-field through the blade passages of either

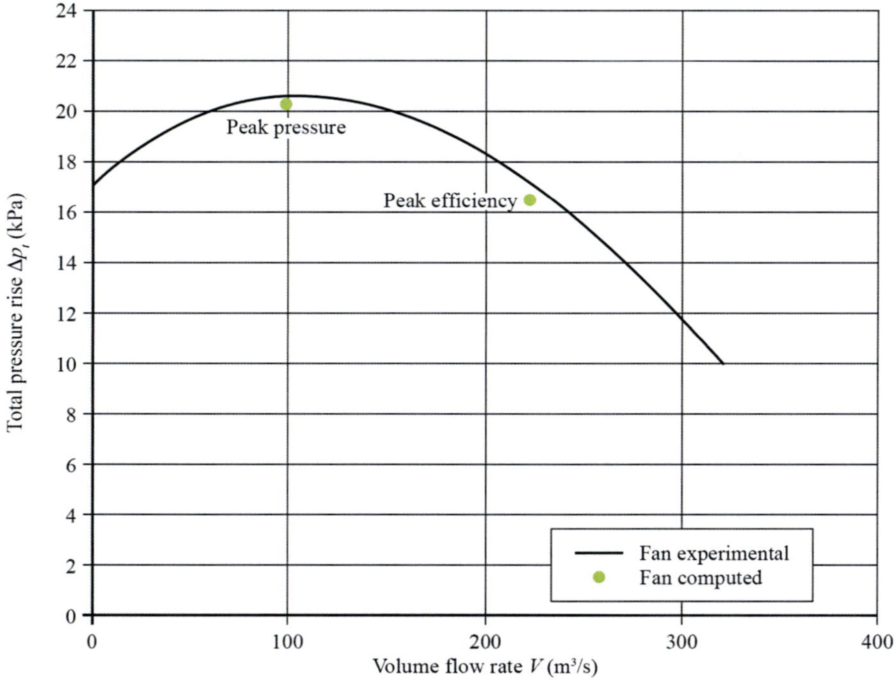

FIGURE 1.6. Numerically predicted and experimentally measured characteristic of a centrifugal fan originally developed for application in the cement industry. Computation conducted using a high-Reynolds number standard k-ε turbulence model and a frozen rotor approach, from Corsini *et al.* (2013a).

axial or centrifugal fans. In a development environment the objective is to facilitate a prediction of the fan's characteristic for either an existing geometry, or a proposed new geometry. Although the computational methods can account for in-flow and out-flow boundary conditions, the simulations are essentially stand-alone fan simulations. One does not account for the system within which the fan is to be embedded.

The majority of low-speed fans intended for air system applications are embedded into a complex system. The complexity occurs as a consequence of multiple fans and dampers interacting and consequently, inducing significant variations in fan in-flow and out-flow conditions. As low-speed fans must operate in a system, there is a need to be able to predict the system's performance and the impact of that system on the fans embedded within it. However, the computational approaches that work well when modelling an axial or centrifugal fan's blade-to-blade flow-field are not suitable for modelling a system. A fan's characteristic time scale is invariably related to the fan's rotation speed, with high-speed centrifugal blowers operating at up to 10,000 rpm. In contrast, the complex systems within which low-speed fans are embedded have characteristic time scales related to the speed of sound, and consequently, in large systems the characteristic time scale is typically between one and ten seconds.

To model the flow-field accurately through a system the computation must extend over a time period longer than the system's characteristic time scale. In practice, this would require one to calculate the fan's blade-to-blade flow-field over hundreds or thousands of revolutions. The required computational effort to calculate a blade-to-blade flow-field over so many revolutions is orders of magnitudes greater than that available within the air movement and control community. Consequently, the computational methods that they apply to an axial or centrifugal blade-to-blade flow-field's predictions are not suitable for predicting system performance. Therefore, one must predict system performance by synthesising the fan's effect on the system within which it is embedded.

Fans as pressure discontinuities

When considering how to model the fan's impact on the system within which it is embedded, the least complex approach is to treat the fan as a discontinuity of static pressure. This requires knowing the fan's pressure and volume characteristic, such that one may relate correctly the flow rate through the fan to the pressure discontinuity across it. The system model then treats the fan as a surface or single layer of cells across which it applies the discontinuity. In this way the effect a fan on a system may by synthesised using a 'volume condition'. Angeli (2012) used the volume condition approach to synthesise the presence of jet fans in the Mont Blanc tunnel. This simulation's aim was to model the entire Mont Blanc tunnel system to enable the impact of different tunnel operating scenarios on the average air velocity induced in the tunnel by the jet fans. In practice, jet fans are required to induce an average air velocity of at least three metres per second if they are to clear smoke and keep escape routes clear in the event of a fire.

Actuator disc and actuator line methods

The least complex approach to modelling a fan installed in a system is the volume condition approach. More complex approaches are the actuator disc and actuator line approaches. Betz (1966) developed these approaches for modelling wind turbines. Engineers have now applied these approaches to fans when installed in a system (Meyer and Kröger, 2001). They involve the introduction of an additional body force in the momentum equation that accounts for the exchange of momentum between blades and fluid. They require that the computational method has incorporated into its code lift and drag coefficients for the fan blades at a series of span-wise blade sections. Engineers use the computational method to compute the incoming fluids angle of attack at each blade span-wise location. They then use the angle of attack in combination with lift and drag coefficients to compute the additional body forces on the fluid, following which the code is able to compute the exchange of momentum between the blades and fluid.

More sophisticated codes are able to compute the exchange of momentum between the blades and fluid at run-time using the boundary element method (BEM) to dynamically recompute the blade's lift and drag as a function of the current velocity profile. With the actuator disc approach, one treats the exchange of momentum as constant at different radial positions and not dependant on the azimuthal coordinates (Betz, 1966). Meyer and Kröger (2001) utilised the actuator disc approach to successfully predict a fan's characteristics. They also reported that the approach resulted in a slight under prediction of fan power consumption that they attributed to the approach's inability to account for tip-clearance phenomena. Moreover, the stream-wise velocity profile downstream of the fan was slightly under predicted due to the modelling assumptions. Despite these reservations, van der Spuy *et al.* (2010) were able to use the actuator disc approach to accurately simulate the performance of an array of 288 axial flow fans, Figure 1.7.

The actuator line approach is similar to the actuator disc approach. However, unlike the actuator disc approach, the actuator line approach considers every fan blade and the fan blades' rotation within the computational domain. Sørensen and Shen (2002) originally developed the actuator line approach for modelling wind turbines. The approach involves mapping each blade with a series of actuator points along actuator lines. One computes body forces at each point using the actuator disc

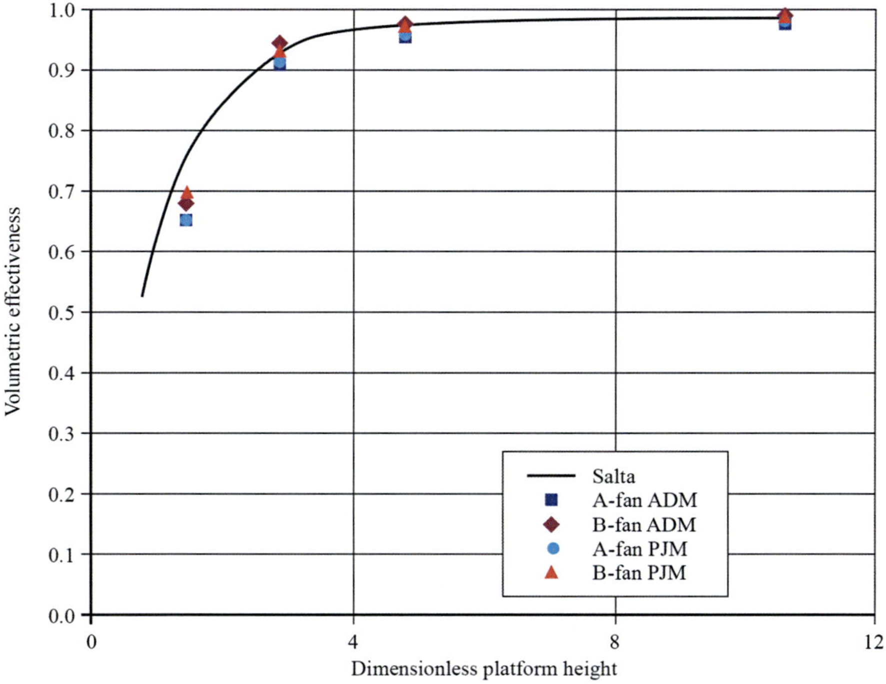

FIGURE 1.7. Numerical predictions of the volumetric effectiveness of an array of fans (symbols) and experimental measured performance of the array (line). Computations based on the synthesised actuator disk model approach, from van der Spuy *et al.* (2010).

approach methodology. One then applies the body force at the actuator point via a three-dimensional Gaussian function to avoid singularities. As the actuating lines rotate inside the computational domain, it is possible to both account for unsteady wake release and simulate the rotating blades' interaction with a stator.

MODELLING OF EROSION

Erosion of both axial and centrifugal fan blades is a major issue for the air movement and control community. Engineers classically associate axial fans in induced-draft power and centrifugal fans in cement application with gas passing through the fan containing erosive particles. These particles both change the aerodynamic profile and reduce the fan blades' chord length. The combined effect is a reduced fan pressure developing capability and efficiency. Predicting erosion patterns is critical during blade design optimisation to minimise susceptibility to erosion and subsequently, to predict the rate of material loss and consequently, the fan maintenance schedule. Therefore, developing computational methods that can predict erosion is critical when developing new fans intended for application in erosive environments.

We may achieve a numerical prediction of erosion in one of two ways: Lagrangian computation of particle dispersion via single particle tracking (Venturini, 2010) or via cloud particle tracking (Baxter, 1989). Single particle tracking involves computing single particle trajectories at run-time using the calculated flow-field parameters at the particle's current location. In contrast, cloud particle tracking involves tracking a statistical representation of an entire particle cloud, only computing the cloud's centre trajectory. Single particle tracking is the more accurate approach. However, Venturini (2010) observed that the computational effort required for single particle tracking is between one and two orders of magnitude larger than that required for cloud particle tracking. For this reason, the air movement and control community favours cloud particle tracking.

Once one has computed particle trajectory, either via single particle tracking or via cloud particle tracking, one may predict impact frequency of particles on the blade pressure and suction surface. Corsini *et al.* (2013c) were able to successfully predict impact frequency distributions, Figure 1.8. They identified the blade pressure surface area (A) and suction surface area (F) that particle impact did not affect. Additionally, they were able to identify the pressure surface areas which particle impact most affected. These included the pressure surface outer 20 per cent of blade span at the leading edge (B), an area affected by the tip leakage vortex (C), an area close to the blade trailing edge (D) and areas close to the blade hub (E). In contrast, the suction surface was primarily affected in the blade tip region (G), along the entire blade span at the leading edge (H) and at the trailing edge in the near-hub area (I).

Tabakoff *et al.* (1979) modelled erosion as a function of particle velocity, angle of impact, impact energy plus particle and blade material properties. Corsini *et al.* (2013c) applied Tabakoff *et al.*'s (1979) model in a RANS code. Ghenaiet (2009) also used Tabakoff *et al.*'s erosion model to study the effect of sand ingestion into a

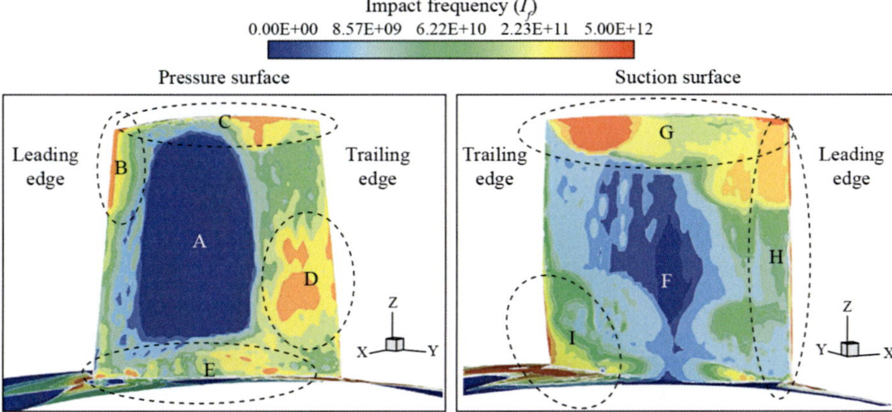

FIGURE 1.8. Impact frequency I_f regions predicted over the blade pressure surface, left and suction surface, right. Corsini *et al.* (2013c) derived the regions by predicting the impact frequency regions for particles from 0.75 μm to 135 μm diameter, combined in proportion to the in-service distribution of particle size.

ventilation system. Consequently, we may conclude that although predicting erosion is challenging, it is possible and the particle cloud tracking approach provides designers with a practical method for doing so.

FLOW INDUCED VIBRATIONS AND COMPUTATIONAL AERO ACOUSTICS

The success of those researchers who have used computational methods to predict erosion illustrates that computational methods have progressed beyond an isolated prediction of the blade-to-blade flow-field through a fan rotor. This progress manifests itself in two primary areas of research, the computation of aero-elastic induced vibration and computation of overall noise level and noise spectrum. Using the three-dimensional computational fluid dynamic code of He and Denton (1994) as a basis for predicting the blade-to-blade flow-field, researchers have calculated the flutter of turbomachinery blades. A full discussion of applying computational methods to both the prediction of aero-elastic induced vibration, overall noise level and noise spectrum is beyond the scope of this chapter. Marshall and Imregun (1996) provide a comprehensive review of aero-elastic computational methods as applied to turbomachinery. Envia *et al.* (2004) provide a comprehensive review of computational aeroacoustics.

FUTURE TRENDS

When modelling the physics of all turbomachinery flow, RANS approach limitations hinder the accuracy with which one can model the blade-to-blade flow-field. Consequently, a RANS approach limits the accuracy with which one may calculate fan performance characteristics. The RANS approach is not able to model the turbulence's spectral content and is not sensitive to the flow-field's multi-scale features. Scholars who have studied the flow-field physics within all turbomachinery classes are increasingly turning to large eddy simulations (LES) to overcome RANS approach limitations.

The principle issue with all large eddy simulations is the formidable computations effort. The computational effort is simply beyond the reach of even the largest manufacturers within the air movement and control community at the time of writing. The required computational hardware to successfully run a large eddy simulation is between two and three orders of magnitude greater than that currently available. Despite this reservation, high-performance computing (HPC) facilities are developing rapidly, exploiting massively parallel clusters of individual processors. Despite the promise of the high-performance computing facilities, at the time of writing the costs to use them is beyond the reach of the air movement and control community.

Additionally, the use of high-performance computing facilities requires that an expert rewrite and optimise the computational methods to run across thousands of individual processors. To further compound matters, the commercial CFD code industry sells commercial CFD codes on a 'per processor' basis that effectively makes the cost of running the codes across more than a hand full of processors prohibitive. Therefore, we may conclude that it likely to be another five years before the air movement and control community can realistically consider a large eddy simulation approach to predict a fan's performance characteristic.

Despite the challenges that large eddy simulation poses, turbomachinery designers are moving towards large eddy simulation-based computational methods. This movement has been difficult, in part, as a consequence of large eddy simulation-based computational methods requiring a higher level of user capability than RANS-based computational methods. However, large eddy simulation-based methodologies offer the possibility to compute unsteady phenomena, such as those that occur with stall inception, unsteady blade and bearing loads, vibration and noise. The air movement and control community widely anticipates predicting the unsteady flow-field parameters responsible for generating overall fan noise and spectrum. A computational method that could give a reliable prediction of not only a fan's aerodynamic characteristics, but also acoustic characteristics would be useful during new fan design, development and optimisation.

Although full large eddy simulation methodologies are still some years away from routine application within the air movement and control community, there is progress with 'hybrid' approaches. These hybrid approaches are based on sensitising an unsteady RANS approach to instabilities. Notable examples are detached eddy simulations (DES) and scalar-adaptive simulations (SAS). Other hybrid methods utilise a combined large eddy simulation and RANS approach, with the actuator disc and actuator line approaches also offering the possibility of combining with a large eddy simulation approach.

The detached eddy simulation approach is a sensitised large eddy simulation method based on the unsteady RANS Spalart-Allmaras model. The detached eddy simulation approach is the most widely used of the available hybrid methodologies as it is able to account for the turbulence spectral content in turbomachinery simulations. At the time of writing, designers working within the air movement and control community have not reported in the literature the use of the detached eddy simulation approach. Therefore, we may conclude that the computational effort that engineers associate with a detached eddy simulation approach remains beyond that available within the community at the present time.

The scalar-adaptive simulation approach is similar to the detached eddy simulation approach as it utilises a methodology based on an unsteady RANS methodology (Menter and Egorov, 2005, 2010a, 2010b). The scalar-adaptive simulation approach is based on a k-ω SST turbulence model sensitised to more than one characteristic length scale by introducing second order derivatives of velocity. Yang *et al.* (2012) utilised the scalar-adaptive simulation approach to compute the flow-field through a centrifugal fan to facilitate the study of unsteady loads on the fan impeller and to predict the pressure fluctuation magnitude. Wolfram and Carolus (2010) also used the scalar-adaptive simulation approach to compute the flow-field through a centrifugal fan. They were able to identify vortical structures that developed in the fan inlet, and went on to study the mechanism by which these vortical structures entered the fan impeller and their role in generating the fan's acoustic signature.

Finally, hybrid large eddy simulation / RANS methodologies solve the unsteady RANS equations in the near-wall region and the large eddy simulation equations in the flow's main core. This hybrid approach reduces the required computational effort for a full large eddy simulation whilst still accounting for large-scale structures in the majority of the flow. This models a part of the turbulence spectrum. Borello *et al.* (2010) observed that this hybrid approach is facilitated as the discretised system of large eddy simulation and RANS equations share the same form and consequently, bridging the two within a single computational code is relatively easy to accomplish.

A challenge of using a hybrid large eddy simulation / RANS methodology is the interface between the two. Borello *et al.* (2009) studied the practicality of interfacing large eddy simulations and RANS computations in a single hybrid simulation. They studied the relative merits of both a moving and a fixed surface to create a buffer zone between the large eddy simulation and the RANS computations, with the buffer zone thickness comprising typically three to five cells. Borello *et al.*'s hybrid approach (2009) is noteworthy in that it is based on an elliptic relaxation RANS model. They demonstrated that this model could correctly reproduce the tip leakage vortex in a linear cascade without an increase in grid resolution when compared to an unsteady RANS computation based on the same RANS closure. Consequently, the required computational effort for the hybrid large eddy simulation / RANS computation does not increase in comparison to an unsteady RANS because of mesh density. However, the hybrid large eddy simulation / RANS computation does require more through-flow time steps than the unsteady RANS computation to achieve convergence and one must reduce the time step amplitude to maintain the Courant– Friedrichs–Lewy (CFL) number below unity. As a result, a hybrid large eddy simulation / RANS simulation required approximately twice the computational effort of a comparable unsteady RANS simulation.

In theory, one can couple the actuator disk and actuator line approaches with a RANS or large eddy simulation computation. This would allow the resulting hybrid simulation to reproduce large-scale structures' unsteady interaction. Sørensen and Shen (2002) noted that at the present time there are no examples of hybrid methodology applications based on either the actuator disk and actuator line approaches applied to fans. However, engineers have successfully applied both approaches to wind turbines and therefore have the potential to apply them into low-speed fan applications.

Finally, Davidson (1997) noted that a theoretically possible hybrid approach would be to use low-resolution grids coupled with advanced sub-grid scale models. Borello *et al.* (2013b) were able to demonstrate the potential of this approach by incorporating a sub-grid scale model sensitive to each grid cell's size. Using this approach, Borello *et al.* (2013b) were able to recover part of the turbulence spectrum that is lost with a RANS approach and were able to analyse the turbulent structures within a tunnel ventilation fan's blade-to-blade passage, Figure 1.9.

As Borello *et al.*'s methodology (2013b) is inherently unsteady, they were able to utilise it to study the effects of train induced pressure pulses on tunnel ventilation fans. Borello *et al.* (2013a) concluded that the quasi-instantaneous changes of the operating conditions did not influence predicted large scale turbulent structures. In contrast, the operating condition's quasi-instantaneous changes did impact small scale turbulent structures, Figure 1.10. However, Borello *et al.*'s (2013a) most significant conclusion was that the operating condition's quasi-instantaneous changes resulted in doubling the unsteady blade loads, despite the fan not driving into stall, Figure 1.11. They concluded that doubling the unsteady blade load was

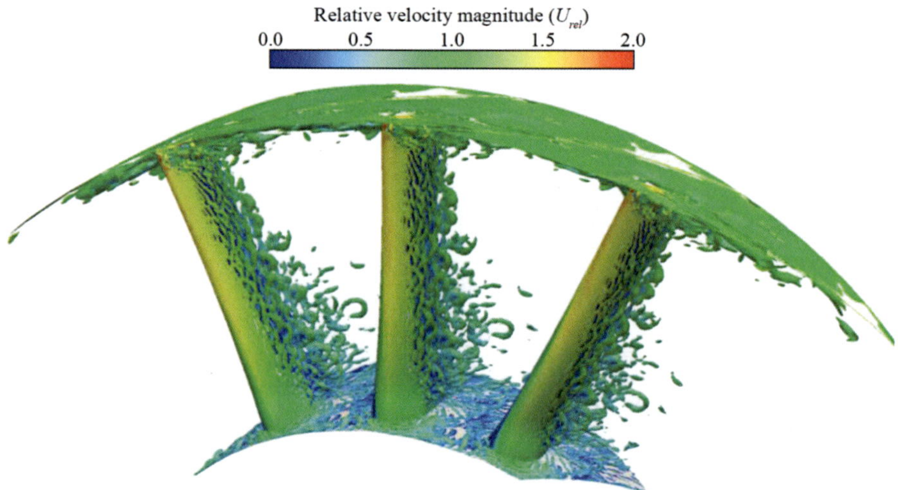

FIGURE 1.9. Numerically predicted vortical structures through the blade passage of a tunnel ventilation fan, computed using a large eddy simulation (LES). These structures are visualised with iso-surfaces of vortical structure deduction parameter (Q) and coloured with relative velocity magnitude (U_{rel}), from Borello *et al.* (2013b).

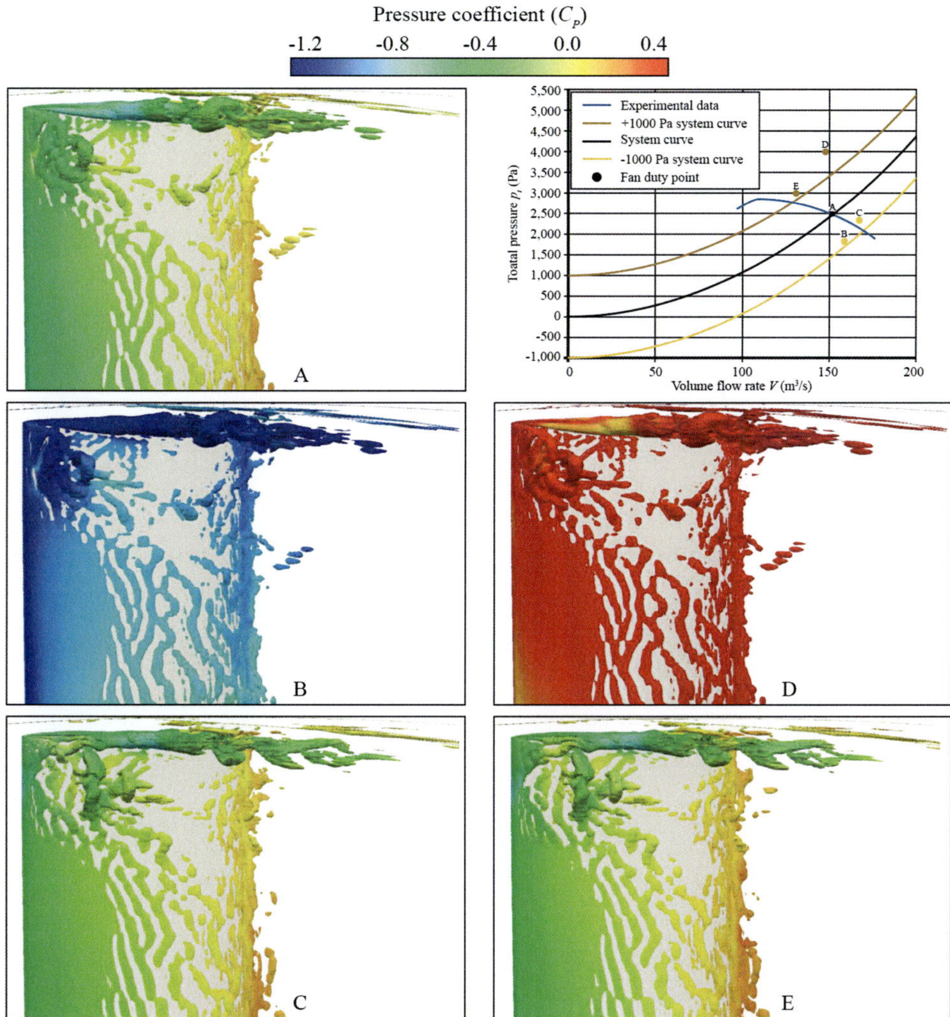

FIGURE 1.10. The numerically predicted vortical structures through a tunnel ventilation fan's blade passage, computed using a large eddy simulation (LES). These structures are visualised with iso-surfaces of vortical structure deduction parameter (Q) and coloured with pressure coefficient (*Cp*), from Borello *et al.* (2013a).

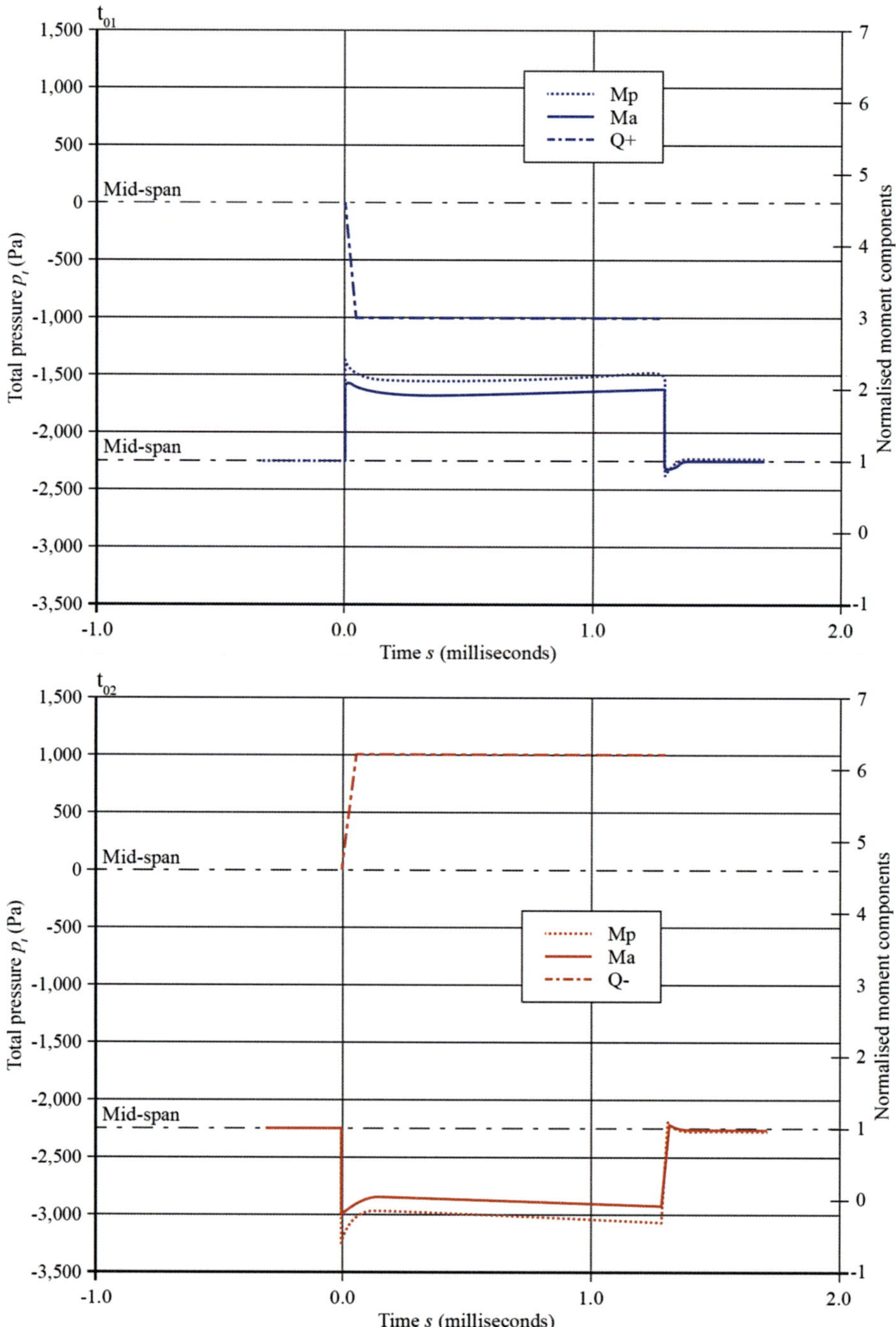

FIGURE 1.11. Time histories of the axial and peripheral moment components M_a and M_p on the blade surface under the effects of increasing (Q+) and decreasing (Q-) volume flow rates. Values are normalised with respect to the corresponding value associated with normal fan operation at its nominal duty point, from Borello *et al.* (2013a).

likely the cause of recent tunnel ventilation fans' in-service mechanical failures in tunnel systems with large pressure pulses.

Authors of commercial CFD codes do not champion hybrid methodology development. They are slow to respond to market demands for more advanced hybrid methodologies, preferring instead to exploit a monopolistic position in the market with well proven RANS-based methodologies that converge reliably. As a direct result, the last decade has seen the widespread adoption of open-source codes. Foremost among the open-source CFD codes that scholars working with the air movement and control community have adopted are OpenFOAM (Jasak, 2010) and CODE_SATURNE (Archambeau *et al.*, 2004). These open-source codes are free and customisable. The air movement and control community is fragmented, with the majority of designers working in small to medium size organisations that do not have the financial resources to pay the multi-processor licence fees associated with commercial codes. Additionally, designers working within the air movement and control community typically develop products for a wide range of applications, and consequently, the ability to customise a code is attractive.

The current generation of open-source codes meets the scalability standards of large high-performance computing facilities and consequently, they can take advantage of rapidly developing massively parallel computation capabilities. An additional advantage of open-source computational fluid dynamic codes is that one can couple them with other codes. For example, the open-source optimising code DAKOTA (Eldred *et al.*, 2006) is able to operate in conjunction with a computational fluid dynamics code, passing back a revised set of parameters to optimise blade geometry through a Pareto optimisation study. When attempting to optimise an aerofoil, Spisso (2012) combined OpenFOAM and Point Wise, a commercial mesh generator with built-in scripting capabilities for automated mesh generation and DAKOTA. The air movement and control community has a long history of producing either design-to-order or heavily adapted-to-order fan designs. Therefore, the emergence of viable optimisation tools represents an exciting development that is likely to become increasingly relevant over the next five years.

CONCLUSIONS

Computational methods have become established within the air movement and control community. Despite the inevitability of separated flow through the low-speed fans intended for air system application, engineers have developed computational methods that simulate both the blade-to-blade flow-field and overall fan performance to an accuracy that is of practical use to designers.

For the first time, the legislative environment within Europe and the planned legislative environment in the USA are setting minimum fan and motor efficiency levels. This change in legislative environment is driving the air movement and control community to adopt computational methods as they are now required to improve fan performance to at least the legal minimum. This legislative imperative is resulting in designers adopting open-source code in an on-going effort to spread the use of computational codes from analysis-based research applications into development and order related engineering application. Within the next five years it is likely that the air movement and control community will have abandoned commercial codes in favour of open source alternatives.

CFD as a design tool

Today, the air movement and control community routinely uses computational fluid dynamic codes as design tools. Commercial codes incorporating a basic RANS methodology with either a k-ε or k-ω wall function approach are able to predict both axial fan and centrifugal fan performance characteristics. Over the stable part of the fan's characteristic these predictions are accurate enough to be of practical use to designers. Further, they achieve the predictions with a computational effort that is realistic for the majority of designers working within the air movement and control community.

As either an axial or centrifugal fan approaches peak pressure, and therefore its stability limit, classical RANS-based methodologies with wall functions reach their capability limit. The hypotheses that engineers used to derive all RANS methodologies are no longer valid. As a fan approaches stall, flow-field structures become more complex and consequently, it is no longer possible to accurately predict the fan's performance characteristics.

Despite the inability of RANS-based methodologies to predict a fan's characteristic when stalled, the numerical prediction is still able to identify that the fan is stalling. Fan designers are primarily focused on ensuring that their designs do not operate in stall. Consequently, it is enough for a computational method to predict that the fan will stall and that is why RANS-based methodologies remain the favoured methodologies within the air movement and control community.

If the computational method is required to predict a fan's characteristic performance when stalled, a methodology that better models the flow-field physics is required. Moving from a high- to low-Reynolds formulation in the RANS methodology improves the flow-field physics' modelling. However, a low-Reynolds formulation requires a substantial increase in grid resolution close to walls. Consequently, a low-Reynolds formulation typically requires twice the computational effort required for a high-Reynolds formulation. Despite the increase in computational effort, today designers working within the air movement a control community favour RANS methodologies with a low-Reynolds formulation.

CFD as an analysis tool

Computational fluid dynamics codes offer fan designers the possibility of simulating in detail the blade-to-blade flow-field. Codes incorporating a basic RANS methodology predict a fan's overall performance characteristics well, but the limitations inherent in their formulation make them less suitable for studying the blade-to-blade flow-field. Real insight requires a code based on an unsteady RANS methodology incorporating advanced low-Reynolds models, a detached eddy simulation (DES) or a scalar-adaptive simulation (SAS). These unsteady RANS-based methodologies have the potential to simulate the blade-to-blade flow-field as a fan approaches stall. They can predict critical flow-field structures within the blade passage, providing fan designers with the necessary insight to develop a design.

Designer working within the air movement and control community would classically develop a design to delay the onset of stall or improve fan efficiency at a desired operating point. However, insight into, for example, the nature of blade tip leakage flow features can facilitate the development of blade tip features that both minimise the tip leakage vortex's intensity whilst maintaining its vorticity above a critical threshold value such that it does not burst. Thus, a designer is able to effectively develop blade tip treatments that minimise fan acoustic emissions. As such, insight into an axial or centrifugal fan's blade-to-blade flow-field facilitates the achievement of a broad range of development goals.

Future trends

Designers working within the air movement and control community have successfully integrated commercial computational fluid dynamic codes into the design process. The cost of the necessary computer hardware to run these codes has become progressively more affordable. Despite the reduced computer hardware costs, commercial computational fluid dynamic codes remain out of reach of many designers working for smaller companies. This is because the cost of a licence for a commercial code remains high. The code author's policy of charging a licence fee for each processor results in a commercial code licence for a multi-processor computer typically larger than the hardware's cost.

Designers within the air movement and control community are increasingly working with the code OpenFOAM and CODE_SATURNE. These codes are licence free and can run on clusters of computers. This offers designers the possibility of networking office computers. Designers may use computers that the engineering department uses during the day with design software at night as part of a cluster to run open-source computational fluid dynamics codes. At the time of writing OpenFOAM and CODE_SATURNE have user interfaces that are less user friendly than commercial codes. Despite this caveat, the open-source movement generally has made dramatic progress over the last five years. Therefore, it is likely that open source computational fluid dynamic codes will become dominant within the air movement and control community, as that community struggles to develop more efficient products in response to current and planned future legislation mandating minimum fan and motor efficiency.

REFERENCES

Angeli, D. (2012), 'OpenFOAM @ Mimesis and Numerical Simulation of the Mont Blanc Tunnel', *HPC Enabling of OpenFOAM for CFD Applications*, CINECA, Casalecchio di Reno, Bologna, Italy, 26 November.

Archambeau, F., Mechitoua, N. and Sakiz, M. (2004), 'Code_Saturne: a Finite Volume Code for the Computation of Turbulent Incompressible Flows – Industrial Applications', *IJFV International Journal on Finite Volumes*, vol. 1, pp. 1–62.

Baxter, L.L. (1989), 'Turbulent Transport of Particles', PhD thesis, Brigham Young University, Provo, UT, USA.

Betz, A. (1966), *Introduction to the Theory of Flow Machines* (D.G. Randall, trans.), Pergamon Press, Oxford, UK.

Bianchi, S., Corsini, A. and Sheard, A.G. (2013), 'Demonstration of a Stall Detection System for Induced Draft Fans', *Proceedings of the Institution of Mechanical Engineers, Part A: Journal of Power and Energy*, vol. 227, pp. 272–284.

Blake, W.K. (1986a), *Mechanics of Flow-induced Sound and Vibration, Volume 1: General Concepts and Elementary Sources*, Academic Press Inc., Orlando, FL, USA.

Blake, W.K. (1986b), *Mechanics of Flow-induced Sound and Vibration, Volume 2: Complex Flow-structure Interactions*, Academic Press Inc., Orlando, FL, USA.

Borello, D., Hanjalić, K. and Rispoli, F. (2005), 'Prediction of Cascade Flows with Innovative Second-moment Closures', *Transactions of the ASME, Journal of Fluid Engineering*, vol. 127(6), pp. 1059–1070.

Borello, D., Hanjalić, K. and Rispoli, F. (2007), 'Computation of Tip-leakage Flow in a Linear Compressor Cascade with a Second-moment Turbulence Closure', *International Journal of Heat and Fluid Flow*, vol. 28(4), pp. 587–601.

Borello, D., Delibra, G., Hanjalić, K. and Rispoli, F. (2009), 'Large-eddy Simulations of Tip Leakage and Secondary Flows in an Axial Compressor Cascade using a Near-wall Turbulence Model', *Proceedings of the Institution of Mechanical Engineers Part A, Journal of Power and Energy*, vol. 223(6), pp. 645–655.

Borello, D., Delibra, G., Hanjalić, K. and Rispoli, F. (2010), 'Hybrid LES/RANS Study of Turbulent Flow in a Linear Compressor Cascade with Moving Casing', *Proceedings of the 55th American Society of Mechanical Engineers Gas Turbine and Aeroengine Congress,* Glasgow, Scotland, 14–18 June, paper no. GT2010-23755.

Borello, D., Corsini, A., Delibra, G. and Sheard, A.G. (2013a), 'Numerical Investigation of Detrimental Aerodynamic Effect of Pressure Pulses on a Metro Tunnel Fan', *Proceedings of the 10th European Turbomachinery Conference*, Lappeenranta, Finland, 15–19 April, pp. 573–582.

Borello, D., Corsini, A., Delibra, G., Fiorito, M. and Sheard, A.G. (2013b), 'Large Eddy Simulation of a Tunnel Ventilation Fan', *Transactions of the ASME, Journal of Fluids Engineering*, vol. 135(7), paper no. 071102, pp. 1–9.

Brailko, I.A., Mileshin, V.I., Nyukhtikov, M.A. and Pankov, S.V. (2004), 'Computational and Experimental Investigation of Unsteady and Acoustic Characteristics of Counter-rotating Fans', *Proceedings of the ASME 2004 Heat Transfer/Fluids Engineering Summer Conference*, Vol. 2, Parts A and B, Charlotte, NC, USA, 11–15 July, paper no. HT-FED2004-56435.

Chen, W.L., Lien, F.S. and Leschziner, M.A. (1998), 'Computational Prediction of Flow Around Highly Loaded Compressor-cascade Blades with Non-linear Eddy-viscosity Models', *International Journal of Heat and Fluid Flow*, vol. 19(4), pp. 307–319.

Commission Regulation (EU) no. 327/2011 (2011), *Official Journal of the European Union,* 1 June, http://www.amca.org/UserFiles/file/COMMISSION%20REGULATION%20%28 EU%29%20No%20327-2011.pdf.

Corsini, A. and Rispoli, F. (2004), 'Using Sweep to Extend Stall-free Operational Range in Axial Fan Rotors', *Proceedings of the Institute of Mechanical Engineering Part A, Journal of Power and Energy*, vol. 218(3), pp. 129–139.

Corsini, A. and Rispoli, F. (2005), 'Flow Analyses in a High-pressure Axial Ventilation Fan with a Non-linear Eddy-viscosity Closure', *International Journal of Heat and Fluid Flow*, vol. 26(3), pp. 349–361.

Corsini, A. and Sheard, A.G. (2007), 'Tip End-plate Concept Based on Leakage Vortex Rotation Number Control', *Journal of Computational and Applied Mechanics*, vol. 8, pp. 21–37.

Corsini, A., Rispoli, F. and Sheard, A.G. (2007), 'Development of Improved Blade-tip End-plate Concepts for Low-noise Operation in Industrial Fans', *Proceedings of the IMechE Part A, Journal of Power and Energy*, vol. 221, pp. 669–681.

Corsini, A., Rispoli, F. and Sheard, A.G. (2009), 'Aerodynamic Performance of Blade-tip End-plates Designed for Low-noise Operation in Axial-flow Fans', *Transactions of the ASME, Journal of Fluids Engineering*, vol. 131, paper no. 081101, pp. 1–13.

Corsini, A., Rispoli, F. and Sheard, A.G. (2010), 'Shaping of Tip End-plate to Control Leakage-vortex Swirl in Axial-flow Fans', *Transactions of the ASME, Journal of Turbomachinery*, vol. 132, paper no. 031005, pp. 1–9.

Corsini, A., Delibra, G., Rispoli, F., Sheard, A.G. and Venturini, P. (2013a), 'Simulation of the Particle-laden Flows in a Large Centrifugal Fan', *Proceedings of the 58th American Society of Mechanical Engineers Gas Turbine and Aeroengine Congress*, San Antonio, TX, USA, 3–7 June, paper no. GT2013-94982.

Corsini, A., Delibra, G. and Sheard, A.G. (2013b), 'Leading Edge Bumps in Reversible Axial Fans', *Proceedings of the 58th American Society of Mechanical Engineers Gas Turbine and Aeroengine Congress*, San Antonio, TX, USA, 3–7 June, paper no. GT2013-94853.

Corsini, A., Rispoli, F., Sheard, A.G. and Venturini, P. (2013c), 'Numerical Simulation of Coal Fly-ash Erosion in an Induced Draft Fan', *Transactions of the ASME, Journal of Fluids Engineering*, vol. 135, paper no. 081303, pp. 1–12.

Courant, R., Friedrichs, K. and Lewy, H. (1967), 'On the Partial Difference Equations of Mathematical Physics', *IBM Journal of Research and Development*, vol. 11(2), pp. 215–234.

Craft, T.J., Launder, B.E. and Suga, K. (1996), 'Development and Application of a Cubic Eddy-viscosity Model of Turbulence', *International Journal of Heat and Fluid Flow*, vol. 17(2), pp. 108–115.

Cumpsty, N.A. (2004), *Compressor Aerodynamics,* Krieger Publishing Company, Malabar, FL, USA.

Davidson, L. (1997), 'Large Eddy Simulation: a Dynamic One-equation Subgrid Model for Three-dimensional Recirculating Flows', *Proceedings of the 11th International Symposium on Turbulent Shear Flow*, Vol. 3, Grenoble, 8–10 September, pp. 26.1–26.6.

Davidson, P.A. (2004), *Turbulence, an Introduction for Scientists and Engineers,* Oxford University Press, Oxford, UK.

Denton, J.D. (1979), *Extension of the Finite Volume Time Marching Method to Three Dimensions,* VKI Lecture Series 1979-7, Rhode-St-Genèse, Belgium.

Denton, J.D. and Dawes, W.N. (1999), 'Computational Fluid Dynamics for Turbomachinery Design', *Proceedings of the Institution of Mechanical Engineers Part C, Journal of Mechanical Engineering Science*, vol. 213(2), pp. 107–124.

Durbin, P.A. (1993), 'A Reynolds Stress Model for Near-wall Turbulence', *Journal of Fluid Mechanics*, vol. 249, pp. 465–498.

Durbin, P.A. (1995), 'Separated Flow Computations with the k-epsilon-v-squared Model', *AIAA Journal*, vol. 33(4), pp. 659–664.

Durbin, P.A. and Pettersson Reif, B.A. (2001), *Statistical Theory and Modelling for Turbulent Flows*, John Wiley & Sons Ltd, Chichester, UK.

Eldred, M.S., Adams, B.M., Gay, D.M., Swiler, L.P., Haskell, K., Bohnhoff, W.J., Eddy, J.P., Hart, W.E., Watson, J.P., Hough, P.D. and Kolda, T.G. (2006), 'DAKOTA, a Multilevel Parallel Object-oriented Framework for Design Optimization, Parameter Estimation, Uncertainty Quantication and Sensitivity Analysis', *Sand Report SAND2006-4056*, Sandia National Laboratories, Livermore, CA, USA.

Envia, E., Wilson, A.G. and Huff, D.L. (2004), 'Fan Noise: a Challenge to CAA', *International Journal of Computational Fluid Dynamics,* vol. 18(6), pp. 471–480.

Fernández Oro, J.M., Díaz, K.M.A., Morros, C.S. and Marigorta, E.B. (2007), 'Unsteady Flow and Wake Transport in a Low-speed Axial Fan with Inlet Guide Vanes', *Transactions of the ASME, Journal of Fluids Engineering*, vol. 129(8), pp. 1015–1029.

Ghenaiet, A. (2009), 'Numerical Study of Sand Ingestion through a Ventilating System', *Proceedings of the World Congress on Engineering 2009 Vol II*, London, UK, 1–3 July.

Gho, S.L. (2013), 'AMCA Grows in Asia', *AMCA inmotion(3)*, p. 27.

Hanjalić, K. and Jakirlić, S. (1993), 'A Model of Stress Dissipation in Second-moment Closures', *Applied Scientific Research*, vol. 51, pp. 513–518.

Hanjalić, K. and Jakirlić, S. (1998), 'Contribution Towards the Second-moment Closure Modelling of Separating Turbulent Flows', *Computers and Fluids*, vol. 27(2), pp. 137–156.

Hanjalić, K. and Launder, B. (2011), *Modelling Turbulence in Engineering and the Environment: Second-moment Routes to Closure,* Cambridge University Press, Cambridge, UK.

Hanjalić, K., Popovac, M. and Hadziabdic, M. (2004), 'A Robust Near-wall Elliptic-relaxation Eddy-viscosity Turbulence Model for CFD', *International Journal of Heat and Fluid Flow*, vol. 25(6), pp. 1047–1051.

Hanjalić, K., Kenjeres, S., Tummers, M.J. and Jonker, H.J.J. (2009), *Analysis and Modelling of Physical Transport Phenomena*, Delft Academic Press, Delft, The Netherlands.

Hauer, A. and Brooks, J. (2012), 'Fan Motor Efficiency Grades in the European Market', *AMCA inmotion(2)*, pp. 14–20.

He, L. and Denton, J.D. (1994), 'Three-dimensional Time-marching Inviscid and Viscous Solutions for Unsteady Flows Around Vibrating Blades', *Transactions of the ASME, Journal of Turbomachinery*, vol. 116(3), pp. 469–476.

Hirsch, C. (2007), *Numerical Computation of Internal and External Flows*, Butterworth-Heinemann, Oxford, UK.

Horlock, J.H. and Denton, J. (2005), 'A Review of Some Early Design Practice using Computational Fluid Dynamics and a Current Perspective', *Transactions of the ASME, Journal of Turbomachinery*, vol. 127, pp. 5–13.

Huang, L. and Wang, J. (2005), 'Acoustic Analysis of a Computer Cooling Fan', *Journal of the Acoustical Society of America*, vol. 118, pp. 2190–2200.

Jasak, H. (2010), 'OpenFOAM: a Year in Review', *Proceedings of the Fifth OpenFOAM Workshop*, Gothenburg, Sweden, 21–24 June.

Jones, W.P. and Launder, B.E. (1972), 'The Prediction of Laminarization with a Two-equation Model of Turbulence', *International Journal of Heat and Mass Transfer*, vol. 15(2), pp. 301–314.

Kobayashi, T. and Yoda, M. (1987), 'Modified k-e Model for Turbulent Swirling Flow in a Straight Pipe', *JSME International Journal*, vol. 30, pp. 66–71.

Kundu, P.K., Cohen, I.M. and Dowling, D.R (2012), *Fluid Mechanics*, 5th edition, Elsevier Academic Press, Oxford, UK.

Lakshminarayana, B. (1996), *Fluid Dynamics and Heat Transfer of Turbomachinery*, John Wiley & Sons, New York, NY, USA.

Langtry, R.B. and Menter, F.R. (2005), 'Transition Modelling for General CFD Applications in Aeronautics', *Proceedings of the 43rd AIAA Aerospace Sciences Meeting and Exhibit*, Reno, NV, USA, 10–13 January, paper no. AIAA 2005-522.

Launder, B.E. and Sandham, N.D. (Eds) (2002), *Closure Strategies for Turbulent and Transitional Flows*, Cambridge University Press, Cambridge, UK.

Lee, S., Heo, S. and Cheong, C. (2010), 'Prediction and Reduction of Internal Blade-passing Frequency Noise of the Centrifugal Fan in a Refrigerator', *International Journal of Refrigeration*, vol. 33(6), pp. 1129–1141.

Lien, F.S. and Durbin, P.A. (1996), 'Non Linear k-ε-v^2 Modelling with Application to High Lift', *Proceedings of the 1996 Centre for Turbulence Research Summer Program*, Stanford University, CA, USA, 24 June–19 July, pp. 2–22.

Lien, F.S., Chen, W.L. and Leschziner, M.A. (1996), 'Low-Reynolds-number Eddy-viscosity Modelling Based on Non-linear Stress-strain/Vorticity Relations', *Proceedings of the 3rd Symposium on Engineering Turbulence Modelling and Measurements*, Crete, Greece, 27–29 May.

Marshall, J.G. and Imregun, M. (1996), 'A Review of Aeroelasticity Methods with Emphasis on Turbomachinery Applications', *Journal of Fluids and Structures*, vol. 10(3), pp. 237–267.

Menter, F.R. (1993), 'Zonal Two-equation k-w Turbulence Model for Aerodynamic Flows', *Proceedings of the 24th Fluid Dynamics Conference*, Orlando, FL, USA, 6–9 July, paper no. AIAA1993-2906.

Menter, F.R. (1994), 'Two-equation Eddy-viscosity Turbulence Models for Engineering Applications', *AIAA Journal*, vol. 32(8), pp. 1598–1605.

Menter, F. and Egorov, Y. (2005), 'A Scale-adaptive Simulation Model using Two-equation Models', *Proceedings of the 43rd AIAA Aerospace Sciences Meeting and Exhibit*, Reno, NV, USA, 10–13 January, paper no. AIAA2005-1095.

Menter, F.R. and Egorov, Y. (2010a), 'The Scale-adaptive Simulation Method for Unsteady Turbulent Flow Predictions. Part 1: Theory and Model Description', *Flow, Turbulence and Combustion*, vol. 85(1), pp. 113–138.

Menter, F.R. and Egorov, Y. (2010b), 'The Scale-adaptive Simulation Method for Unsteady Turbulent Flow Predictions. Part 2: Application to Complex Flows', *Flow, Turbulence and Combustion*, vol. 85(1), pp. 139–165.

Menter, F.R., Langtry, R.B., Likki, S.R., Suzen, Y.B., Huang, P.G. and Völker, S. (2004), 'A Correlation-based Transition Model using Local Variables Part 1: Model Formulation', *Transactions of the ASME, Journal of Turbomachinery*, vol. 128(3), pp. 1–10.

Menter, F.R., Langtry, R. and Völker, S. (2006), 'Transition Modelling for General Purpose CFD Codes', *Flow, Turbulence and Combustion*, vol. 77, pp. 277–303.

Meyer, C.J. and Kröger, D.G. (2001), 'Numerical Simulation of the Flow Field in the Vicinity of an Axial Flow Fan', *International Journal for Numerical Methods in Fluids*, vol. 36, pp. 947–969.

Pinelli, M., Ferrari, C., Suman, A., Morini, M. and Rossini, M. (2012), 'Fluid Dynamic Design and Optimization of a Double Entry Fan Driven by Tractor Power Take Off for Mist Sprayer Applications', *Proceedings of the Fan 2012 Conference*, Senlis, France, 18–20 April.

Piotrowski, W., Elsner, W. and Drobniak, S. (2009), 'Transition Prediction on Turbine Blade Profile with Intermittency Transport Equation', *Transactions of the ASME, Journal of Turbomachinery*, vol. 132(1), pp. 1–10.

Pope, S.B. (2000), *Turbulent Flows*, Cambridge University Press, Cambridge, UK.

Popovac, M. and Hanjalić, K. (2007), 'Compound Wall Treatment for RANS Computation of Complex Turbulent Flows and Heat Transfer', *Flow, Turbulence and Combustion*, vol. 78(2), pp. 177–202.

Pullan, G., Young, A.M., Day, I.J., Greitzer, E.M. and Spakovszky, Z.S. (2012), 'Origins and Structure of Spike-type Rotating Stall', *Proceedings of the 57th American Society of Mechanical Engineers Gas Turbine and Aeroengine Congress*, Copenhagen, Denmark, 11–15 June, paper no. GT2012-68707.

Rábai, G. and Vad, J. (2005), 'Validation of a Computational Fluid Dynamics Method to be Applied to Linear Cascades of Twisted-swept Blades', *Periodica Polytechnica, Mechanical Engineering*, vol. 49(2), pp. 163–180.

Sander, G.F. and Lilley, D.G. (1983), 'The Performance of an Annular Vane Swirler', *Proceedings of the AIAA/SAE/ASME 19th Joint Propulsion Conference*, Seattle, WA, USA, 27–29 June, paper no. AIAA-83-1326.

Šarić, S., Kniesner, B., Mehdizadeh, A., Jakirlić, S., Hanjalić, K. and Tropea, C. (2008), 'Comparative Assessment of Hybrid LES/RANS Models in Turbulent Flows Separating from Smooth Surfaces', in Peng, S.H. and Haase, W. (Eds), *Advances in Hybrid RANS-LES Modelling, Notes on Numerical Fluid Mechanics and Multidisciplinary Design*, vol. 97, pp. 142–151.

Sheard, A.G., Corsini, A., Minotti, S. and Sciulli, F. (2009), 'The Role of Computational Methods in the Development of an Aero-acoustic Design Methodology: Application in a Family of Large Industrial Fans', *Proceedings of the 14th International Conference on Modelling Fluid Flow Technologies*, Budapest, Hungary, 9–12 September, pp. 71–79.

Son, P.N., Kim, J. and Ahn, E.Y. (2011), 'Effects of Bell Mouth Geometries on the Flow Rate of Centrifugal Blowers', *Journal of Mechanical Science and Technology*, vol. 25(9), pp. 2267–2276.

Sørensen, J.N. and Shen, W.Z. (2002), 'Numerical Modelling of Wind Turbine Wakes', *Transactions of the ASME, Journal of Fluids Engineering*, vol. 124(2), pp. 393–399.

Spalart, P.R. and Allmaras, S.R. (1992), 'A One-equation Turbulence Model for Aerodynamic Flows', *Proceedings of the 30th Aerospace Sciences Meeting and Exhibit*, Reno, NV, USA, 6–9 January, paper no. AIAA-92-0439.

Speziale, C.G., Sarkar, S. and Gatski, T.B. (1991), 'Modelling the Pressure-strain Correlation of Turbulence: an Invariant System Dynamic Approach', *Journal of Fluid Mechanics*, vol. 227, pp. 245–272.

Spisso, I. (2012), 'Parametric and Optimization Study: OpenFOAM and Dakota', *HPC Enabling of OpenFOAM for CFD Applications*, CINECA, Casalecchio di Reno, Bologna, Italy, 26 November.

Tabakoff, W., Kotwal, R. and Hamed, A. (1979), 'Erosion Study of Different Materials Affected by Coal Ash Particles', *Wear*, vol. 52(1), pp. 161–173.

Thiart, G.D. and von Backstrom, T.W. (1993), 'Numerical Simulation of the Flow Field Near an Axial Flow Fan Operating under Distorted Inflow Conditions', *Journal of Wind Engineering and Industrial Aerodynamics*, vol. 45(2), pp. 189–214.

Tropea, C., Yarin, A.L. and Foss, J.F. (Eds) (2007), *Springer Handbook of Experimental Fluid Mechanics*, Springer, Paris, France.

US Department of Energy (2013), 'Energy Conservation Standards Rulemaking Framework for Commercial and Industrial Fans and Blowers', US Department of Energy, 1 February.

Vad, J. and Horváth, C. (2013), 'Study on the Effects of Axial Clearance Size on the Operation of an Axial Flow Electric Motor Cooling Fan', *Proceedings of the 10th European Turbomachinery Conference*, Lappeenranta, Finland, 15–19 April, pp. 543–552.

van der Spuy, S.J., von Backström, T.W. and Kröger, D.G. (2010), 'An Evaluation of Simplified Methods to Model the Performance of Axial Flow Fan Arrays', *R & D Journal of the South African Institution of Mechanical Engineering*, vol. 26, pp. 12–20.

Vanella, M., Piomelli, U. and Balaras, E. (2008), 'Effect of Grid Discontinuities on Large-eddy Simulation Statistics and Flow Fields', *Journal of Turbulence*, vol. 9, paper ID 32.

Venturini, P. (2010), 'Modelling of Particle-wall Deposition in Two-phase Gas-solid Flow', PhD thesis, Sapienza Università di Roma, Rome, Italy.

Wilcox, D.C. (1988), 'Reassessment of the Scale-determining Equation for Advanced Turbulence Models', *AIAA Journal*, vol. 26(11), pp. 1299–1310.

Wilcox, D.C. (1993), 'Turbulence Modelling for CFD', DCW Industries Inc., CA, USA.

Wolfram, D. and Carolus, T.H. (2010), 'Experimental and Numerical Investigation of the Unsteady Flow Field and Tone Generation in an Isolated Centrifugal Fan Impeller', *Journal of Sound and Vibration*, vol. 329(21), pp. 4380–4397.

Yang, Y., Lucius, A. and Brenner, G. (2012), '3D Unsteady CFD Simulation of the Pressure Fluctuation in a Centrifugal Fan', *Proceedings of the Fan 2012 Conference*, Senlis, France, 18–20 April.

Chapter 2

Computational Analysis of Noise Reduction Devices in Axial Fans with Stabilised Finite Element Formulations

A. Corsini, F. Rispoli, A.G. Sheard and T.E. Tezduyar

ABSTRACT

The chapter illustrates how air movement fan blade designers can use a Reynolds-averaged Navier–Stokes (RANS) modelling technique to minimise fan far-field noise. The authors use the modelling technique to analyse the blade-to-blade flow-field of an air movement fan when fitted with noise reduction devices. The studied noise reduction devices comprise end-plates fitted to the tip of each blade to control the blade tip-to-casing gap leakage flow. It is the blade tip-to-casing gap leakage flow that drives flow-field physics in the blade-tip region. This flow-field physics includes the generation of near-field aerodynamics flow features. The unsteady pressure associated with these flow features propagates to far-field, contributing to both the fan's overall and tonal far-field noise signature. Control of the blade tip-to-casing gap leakage vortex facilitates control of near-field flow features resulting in a reduction in fan far-field noise. We based our Reynolds-averaged Navier–Stokes modelling technique on the streamline-upwind / Petrov–Galerkin (SUPG) and pressure-stabilising / Petrov–Galerkin (PSPG) methods. The use of any Reynolds-averaged Navier–Stokes modelling technique is challenging when attempting to predict a fan's blade-to-blade flow-field. The challenge is a consequence of the inevitability of separated flow from electric motors, flat-plate motor supports and conduits that make obtaining converged computational solutions difficult. We addressed the challenges presented by separated flow regions by developing a novel stabilisation approach based on a 'diffusion for reaction dominated with jump' (DRDJ) operator. This stabilisation approach facilitated managing the discontinuities in the flow-field associated with regions of separated flow and therefore converged computational solutions.

This chapter is a revised and extended version of Corsini, A., Rispoli, F., Sheard, A.G. and Tezduyar T.E. (2012), 'Computational Analysis of Noise Reduction Devices in Axial Fans with Stabilized Finite Element Formulations'. *Computational Mechanics*, vol. 50, pp. 695–705.

NOMENCLATURE

		Latin letters
$B_{k\tilde{\varepsilon}}$		Dissipation-destruction matrices
$c_{\varepsilon 2}$, c_{μ}, $f_{\varepsilon 2}$, f_{μ}		Closure coefficients for the turbulent model
D_c	[mm]	Casing diameter
D_h	[mm]	Hub diameter
D_{hc}		Hub-to-casing diameter ratio
e		Strain rate
f_r		Rotational frequency
$\mathfrak{I}$		Vector of external forces
$\mathfrak{I}_{k\varepsilon}$		Source vector of turbulence closure equations
h		Element length
J_e		Normalised measure of the variation ('jump') in the solution over the element
k		Turbulent kinetic energy
$k_{DRDJ\text{-}k}$, $k_{DRDJ\text{-}\varepsilon}$		DRDJ diffusivities
$k^{DC}{}_{k\varepsilon}$		Dissipation matrix for the DC scheme
n		Direction normal to the boundary
N_a		Interpolation function associated with node a
n_{el}		Number of elements
nen		Number of element nodes
Δp	[Pa]	Static pressure variation
Δp_{dyn}	[Pa]	Dynamic pressure variation
p^{stab}, $p^{stab}{}_{k\varepsilon}$		Streamline-upwind / Petrov–Galerkin stabilisation operators
P_k		Production of turbulent kinetic energy
r		Solution gradient
Re_t		Turbulence Reynolds number $Re_t=k^2/\nu\tilde{\varepsilon}$
r_k, r_{ε}		Unit vectors in the direction of the solution gradient
s		Unit vector in the velocity direction
s(p,u)		Stress tensor
s_{Re}		Reynolds stress tensor
t		Time abscissa
u		Velocity vector
y^+		Normalised wall distance
$\bar{w}$		Vorticity

Greek letters

α_r		Parameter of integration rule for the element reaction coefficient matrix
β		Exchange coefficient
β_k^2, β_ε^2		Reaction numbers
γ_ϕ		Reaction-advection parameters : γ_k, γ_ε
Γ_g, $\Gamma g_{k\varepsilon}$, Γ_h, $\Gamma_{hk\varepsilon}$		Subset of the boundary Γ
δ	[deg]	Blade tip stagger angle
$\tilde{\varepsilon}$		Homogeneous dissipation
$\kappa_{AR\text{-}\kappa}$		Advection-reaction diffusivity
$\kappa_{DR\text{-}\kappa}$		Diffusion-reaction diffusivity
ν_m		Molecular viscosity
$\nu_{k\varepsilon}$		Diffusivity
ν_t		Turbulent kinematic viscosity
ν_u		Viscosity $\nu_u = \nu + \nu_t$
ρ		Air density
σ_k, σ_ε		Closure coefficients for the turbulent model
τ_t		Turbulence time scale $\tau_t = k/\tilde{\varepsilon}$
τ_{SUPG}, τ_{PSPG}		Stabilisation parameters
ϕ		Vector of turbulence closure variables $\phi = (k, \tilde{\varepsilon})^t$
χ	[% span]	Rotor tip clearance
Ω		Rate of rotation tensor

Acronyms

DC	Discontinuity-capturing
LES	Large eddy simulations
DRDJ	Diffusion for reaction dominated with jump
DRD	Diffusion for reaction dominated
GMRES algorithm	Generalised minimal residual method (GMRES)
PSPG	Pressure-stabilising / Petrov–Galerkin
SUPG	Streamline-upwind / Petrov–Galerkin
TI	Turbulence intensity

Symbols, subscript and superscripts

g, gkε, h, hkε	Given functions representing the essential and natural boundary conditions
^	Invariant
ϕ	Indicates the variables modelled with advection–diffusion-reaction equations
AR-k	Advection-reaction
DR-k	Diffusion-reaction

INTRODUCTION

The design process of fans intended for both air movement and aerospace application is driven by the same requirement for high efficiency and low noise. The flow-field features that originate within the blade tip-to-casing gap comprise both aerodynamic loss mechanisms and near-field noise sources. Consequently, the desire to maximise efficiency and minimise noise results in blade designers attempting to minimise the negative impact of these flow-field features. One may achieve this minimisation through optimising the blade-tip geometry which controls flow-field features and thus, minimises their negative aerodynamic and acoustic impact.

The flow-field features that originate within the blade tip-to-casing gap contribute to fan far-field noise. This is a consequence of the interaction of those features with casing boundary layers and blade-to-blade passage secondary flow features (Quinlan and Bent; 1998; Jang *et al.*, 2003; Fukano and Jang, 2004). Longhouse (1978) studied the acoustic performance of fans and compressors, concluding that the features within the blade tip-to-casing flow-field constitute the most significant far-field noise sources. Marcinowski (1953), the first researcher to study noise associated with tip dynamics, demonstrated that increases in broadband noise levels occur with increasing tip clearance. The largest changes in noise level were apparent at frequencies greater than the blade-passing frequency.

Bianchi *et al.* (2014) reviewed the current state-of-the-art in air movement fan flow and noise control, offering a vision for potential improvements in noise reduction via novel application of flow and noise control technologies. Bianchi *et al.* (2014) examined the interaction between aerodynamic cause and acoustic effect and the application of control technologies that current cause-and-effect theories have inspired. The purpose was two-fold: first, to provide a vision for aerodynamic research; and second, to provide air movement fan designers with an insight into the technology available now that they may apply in an on-going effort to reduce fan noise. Bianchi *et al.* (2014) concluded that industrialists and academics alike must focus their research on elucidating the physics underpinning the link between features in the fan flow-field and fan far-field noise.

Researchers studying fans intended for aerospace application have used casing treatments to improve operating margin (Takata and Tsukuda, 1977; Smith and Cumpsty, 1984; Thompson *et al.*, 1998). Wadia *et al.* (1998) studied tone-noise reduction techniques, identifying the potential of both positive sweep (tip further downstream than hub) and positive lean (tip moved circumferentially in the direction of rotor rotation). Corsini and Rispoli (2004) report that positive sweep and lean results in greater noise reduction. This is because positive sweep and lean maximises the number of wake interactions per vane. However, tone noise is particularly reduced. Corsini and Rispoli (2004) observed a small reduction in broadband noise only at lower fan speed.

The most widely used passive noise control technique within the air movement and control community is the blade-tip end-plate technique which is based on modifications to the blade tip by means of anti-vortex appendages. Quinlan and Bent (1998) have proposed blade-tip end-plates, and others have proposed various solutions in patents for air movement fans (Jensen, 1986; Longet, 2003; Mimura, 2003; Uselton *et al.*, 2008). Ffowcs Williams (1977) first identified the potential of blade-tip end-plates. He observed that because of the role organised structures in turbulent flow play in the noise generation process, controlling these structures may be a key to noise suppression. Experimental and numerical studies have identified that significant improvement can result from adopting tip leakage flow control technologies (Corsini and Sheard, 2007). Sheard *et al.* (2009) applied Corsini and Sheard's (2007) research, patenting a concept for a blade-tip end-plate design. The design methodology was based on exploiting a linkage between end-plate geometry and the blade tip-to-casing leakage vortex swirl (Sheard, 2014).

The use of any Reynolds-averaged Navier–Stokes modelling technique is challenging when attempting to predict an air movement fan's blade-to-blade flow-field. The challenge is a consequence of the inevitability of separated flow that makes obtaining converged computational solutions difficult. One may use the Reynolds-averaged Navier–Stokes modelling techniques if the adopted eddy-viscosity turbulence model is the subject of application specific development. Largely in an effort to avoid the need for application specific development, researchers are increasingly favouring large eddy simulations (LES). Hoffman and Johnson (2004) concluded that when using a large eddy simulation the choice of sub-grid scale closure has little impact on accuracy of the resulting numerical simulation. When conducting a large eddy simulation, the use of anisotropy-resolving turbulence models results in numerical diffusivity schemes that are inherently better able to model the flow-field's non-isotropy.

Although large eddy simulations are attractive, they typically require one to two orders of magnitude more computational effort then a Reynolds-averaged Navier–Stokes modelling technique. The computational effort associated with a large eddy simulation is simply beyond the air movement and control community's reach. Consequently, Reynolds-averaged Navier–Stokes modelling techniques and their associated application specific developments are likely to remain the norm within the air movement and control community for the foreseeable future.

The application specific development of Reynolds-averaged Navier–Stokes modelling techniques is classically focused on the non-equilibrium phenomena's numerical representation. These phenomena originate from the turbulence dissipation mechanisms that govern the flow-field physics. Modelling this flow-field physics requires using scale-determining equations with reaction-like terms which become dominant in stagnation regions, separated boundary layers and recirculating flow regions where the flow velocity approaches zero.

Since the mid 1980s there has been a steady stream of studies reported in the literature focused on stabilised formulations for advection–diffusion-reaction equations. These formulations include equations that model the flow-field physics by incorporating reaction-like terms. Tezduyar and Park (1986) and Tezduyar *et al.* (1987) report a diffusion for reaction dominated (DRD) formulation. Codina (1998) studied the relative merit of different finite element methods for solving the diffusion–convection-reaction equations. Franca and Valentin (2000) presented an improved stabilised finite element method for solving the advective–reactive–

diffusive equations. Corsini *et al.* (2004) reported a stabilising/Petrov–Galerkin (SPG) formulation and Corsini *et al.* (2006a), a diffusion for reaction dominated with jump (DRDJ) formulation. Hughes' (1995) variational multi-scale concept has also been influential, underpinning Hauke (2002), Corsini *et al.* (2005) and Gravemeier and Wall's (2007) stabilisation methods.

This chapter presents the results of a research programme to develop a stabilised formulation that facilitates computing the turbulent flow-field through an air movement fan fitted with blade-tip end-plates. The formulation is based on Brooks and Hughes' (1982) streamline-upwind / Petrov–Galerkin (SUPG) and Tezduyar (1992) and Tezduyar *et al.*'s (1992) pressure-stabilising / Petrov–Galerkin (PSPG) methods. The formulation is stabilised using a developed form of the Corsini *et al.* diffusion for reaction dominated with jump (DRDJ) formulation (Corsini *et al.*, 2006a, 2009, 2010a, 2011, 2012). In the research reported in this chapter, we focused on the stabilisation and discontinuity-capturing parameters that we used with the streamline-upwind / Petrov–Galerkin and pressure-stabilising / Petrov–Galerkin formulations (Tezduyar and Park, 1986; Tezduyar *et al.*, 1987; Tezduyar and Osawa, 2000; Tezduyar, 2003; Catabriga *et al.*, 2005; Rispoli *et al.*, 2007; Takizawa *et al.*, 2011a, 2011b).

REYNOLDS-AVERAGED NAVIER–STOKES FORMULATIONS FOR INCOMPRESSIBLE TURBULENT FLOWS

Reynolds-averaging the Navier–Stokes equations refers to a process of Reynolds-averaging the quantities within the equations. Let $\Omega \subset R^{nsd}$ represent the spatial domain with boundary and (0, *T*) represent the time domain. We may write the unsteady Reynolds-averaged Navier–Stokes equations of incompressible turbulent flows as:

$$\rho\left(\frac{\partial u}{\partial t} + u \cdot \nabla u - \Im\right) - \nabla \cdot \sigma = 0, \tag{1}$$

$$\nabla \cdot u = 0, \tag{2}$$

$$\rho\left(\frac{\partial \phi}{\partial t} + u \cdot \nabla \phi + B_{k\varepsilon}\phi - \Im_{ks}\right) - \nabla \cdot (\rho(\nabla \phi)v_{ks}) = 0, \tag{3}$$

where ρ is the density, u the velocity vector, $\phi = (k, \tilde{\varepsilon})^T$ is the vector of the turbulence closure variable, and k and $\tilde{\varepsilon}$ are respectively the turbulent kinetic energy and homogeneous dissipation.

The symbols $\Im$ and $\Im_{k\varepsilon}$ represent the vector of external forces and the source vector in the turbulence closure equations. Corsini and Rispoli (2005) proposed that $\Im$ accounts for the volume sources related to the second- and third-order terms in

Craft *et al.*'s (1996) non-isotropic stress-strain relation. The external force vector reads as:

$$\begin{aligned}\Im = \nabla.\Big[& -0.1_{v_t \tau}\left(\varepsilon(u)\cdot\varepsilon(u) - \varepsilon(u):\varepsilon(u)\frac{1}{3}I\right) \\ & + -0.1_{v_t \tau}\left(\bar{\omega}(u)\cdot\varepsilon(u) + (\bar{\omega}(u)\cdot\varepsilon(u))^T\right) \\ & + 0.26_{v_t \tau}\left(\bar{\omega}(u)\cdot\bar{\omega}(u) - \bar{\omega}(u):\bar{\omega}(u)\frac{1}{3}I\right) \\ & - 10_{c_\mu^Z v_t \tau^Z}\left(\varepsilon(u)\cdot\varepsilon(u)\cdot\bar{\omega}(u) + (\varepsilon(u)\cdot\varepsilon(u)\cdot\bar{\omega}(u))^T\right) \\ & - 5_{c_\mu^Z v_t \tau^Z}\left(\varepsilon(u):\varepsilon(u)\right)\varepsilon(u) + 5_{c_\mu^Z v_t \tau^Z}\left(\bar{\omega}(u):\bar{\omega}(u)\right)\varepsilon(u)\Big]\end{aligned} \tag{4.1}$$

where $\varepsilon(u) = ((\nabla u) + (\nabla u)^T)$ is twice the strain-rate tensor, $\bar{\omega}(u) = ((\nabla u) - (\nabla u)^T)$ is twice the vorticity tensor and v_t is the turbulent kinematic viscosity. We define the turbulent kinematic viscosity as $v_t = c_\mu f_\mu \tau k$, and $t = \frac{k}{\tilde{\varepsilon}}$ as the turbulence time scale, with c_μ and f_μ and other closure coefficients for Craft *et al.*'s (1996) turbulence model (see Table 2.1). We define the turbulence Reynolds number as:

$$\mathrm{Re}_t = \frac{k^2}{v\tilde{\varepsilon}}$$

where v represents the molecular viscosity, and $\hat{e}$ and $\hat{\bar{\omega}}$ are, respectively, the strain-rate and vorticity invariants which we define as:

$$\hat{e} = \tau\sqrt{0.5\varepsilon(u):\varepsilon(u)} \text{ and } \hat{\bar{\omega}} = \tau\sqrt{0.5\bar{\omega}(u):\bar{\omega}(u)}\,.$$

We define the source vector $\Im_{k\varepsilon}$ as:

$$\Im_{ks} = \begin{bmatrix} P_k - D \\ c_{s1} P_{k\frac{\tilde{\varepsilon}}{k}} + E \end{bmatrix}, \tag{4.2}$$

where $P_k = R : \nabla u$ is the production of turbulent kinetic energy, with R representing the Reynolds stress tensor and:

$$D = 2v\nabla\sqrt{k}\cdot\nabla\sqrt{k_t} \text{ and } E = 0.0022\hat{e}k\tau v_t \,\|\nabla\cdot(\nabla u)\|^2\,.$$

Table 2.1. *Closure coefficients for Craft* et al.*'s (1996) turbulence model.*

c_μ	$\dfrac{0.3\left[1-\exp\left[-\dfrac{0.36}{\exp[-0.75\max(\hat{e},\hat{\bar{\omega}})]}\right]\right]}{1+0.35[\max(\hat{e},\hat{\bar{\omega}})]^{1.5}}$
f_u	$1-\exp\left[-\left(\dfrac{\mathrm{Re}_t}{90}\right)^{0.5}-\left(\dfrac{\mathrm{Re}_t}{400}\right)^{2}\right]$
$c_{\varepsilon 1}$	1.44
$c_{\varepsilon 2}$	1.92
$f_{\varepsilon 2}$	$[1-0.3\exp(-\mathrm{Re}_t^2$
σ_ε	1.3
σ_k	1.0

We define the stress tensor as:

$$\sigma(p,u) = -\left(p+\frac{2}{3}\rho k\right)I + \rho v_u \varepsilon(u) \tag{5}$$

with $v_u = v + v_t$.

The diffusivity matrix defines the diffusion terms in the turbulence closure equations as:

$$v_{k\varepsilon} = \begin{bmatrix} \left(v+\dfrac{v_t}{\sigma_k}\right) & 0 \\ 0 & \left(v+\dfrac{v_t}{\sigma_k}\right) \end{bmatrix} \tag{6}$$

The reaction terms account for the dissipation-destruction matrices which we define as:

$$B_{k\varepsilon}\phi = \begin{bmatrix} B_k & 0 \\ 0 & B_\varepsilon \end{bmatrix}\phi, \tag{7.1}$$

with

$$B_k = \frac{\tilde{\varepsilon}}{k},\ B_\varepsilon = c_{\varepsilon 2} f_{\varepsilon 2}\frac{\tilde{\varepsilon}}{k}. \tag{7.2}$$

We represent the boundary conditions for Equations (1) and (3) as:

$$u = g \text{ on } \Gamma_g, \text{ and } \phi = g_{k\varepsilon} \text{ on } \Gamma_{gk\varepsilon}, \tag{8.1}$$

$$n \cdot \sigma = h \text{ on } \Gamma_h, \text{ and } n \cdot (\rho(\nabla\phi)v_{k\varepsilon}) = 0 \text{ on } \Gamma_{hk\varepsilon} \tag{8.2}$$

where Γ_g, $\Gamma_{gk\varepsilon}$, Γ_h and $\Gamma_{hk\varepsilon}$ are the subsets of the boundary Γ, n is the direction normal to the boundary and g, $g_{k\varepsilon}$ and h are given functions representing the boundary conditions.

STREAMLINE-UPWIND / PETROV–GALERKIN FORMULATION FOR REYNOLDS-AVERAGED NAVIER–STOKES EQUATIONS OF TURBULENT FLOWS

In describing the streamline-upwind / Petrov–Galerkin (SUPG) formulation of Equations (1), (2) and (3), we assume that we have constructed some suitably-defined finite-dimensional trial solution and test function space $\mathbb{S}_u^h$, $\mathbb{S}_p^h$, $\mathbb{S}_{\phi h}$ and $\mathbb{V}_u^h$, $\mathbb{V}_p^h$, $\mathbb{V}_\phi^h$.

The streamline-upwind / Petrov–Galerkin/pressure-stabilising / Petrov–Galerkin formulation reads as follows:

find $u^h \in \mathbb{S}_u^h, p^h \in \mathbb{S}_p^h \phi^h \in \mathbb{S}_\phi^h$ such that $\forall w^h \in \mathbb{V}_u^h, \forall q^h \in \mathbb{V}_p^h$ and $\forall \psi^h \in \mathbb{V}_\phi^h$:

$$\begin{aligned} &\int_\Omega w^h \cdot \rho\left(\frac{\partial u^h}{\partial t} + u^h \cdot \nabla u^h - \Im^h\right) d\Omega + \int_\Omega \varepsilon(w^h) : \sigma(p^h, u^h) d\Omega \\ &- \int_{\Gamma^h} w^h \cdot h^h d\Gamma + \int_\Omega q^h \nabla \cdot u^h d\Omega + \sum_{e-1}^{n_{el}} \int_{\Omega e} p^{stab}(w^h, q^h) \\ &\cdot \left[\mathcal{L}(p^h, u^h) - \rho \Im^h \right] d\Omega = 0 \end{aligned} \tag{9.1}$$

where

$$\mathcal{L}(q^h, w^h) = \rho\left(\frac{\partial w^h}{\partial t} + u^h \cdot \nabla w^h\right) - \nabla \cdot \sigma(q^h, w^h), \tag{9.2}$$

and

$$\begin{aligned} &\int_\Omega \psi^h \cdot \rho\left(\frac{\partial \phi^h}{\partial t} + u^h \cdot \nabla \phi^h + B_{k\varepsilon}\phi - \Im_{k\varepsilon}^h\right) d\Omega + \int_\Omega \nabla \psi^h : (\rho(\nabla\phi^h) v_{k\varepsilon}) d\Omega \\ &- \sum_{e=1}^{n_{el}} \int_{\Omega e} p_{k\varepsilon}^{stab}(\psi^h) \cdot \left[\mathcal{L}_{k\varepsilon}(\phi^h) - \rho \Im_{k\varepsilon}^h \right] d\Omega \\ &+ \sum_{e=1}^{n_{el}} \int_{\Omega^e} K_{k\varepsilon}^{DC} \rho \nabla \psi^h : \nabla \phi^h d\Omega = 0 \end{aligned} \tag{10.1}$$

where

$$\mathcal{L}(\phi^h) = \rho\left(\frac{\partial \phi^h}{\partial t} + u^h \cdot \nabla \phi^h + B_{k\varepsilon}\phi^h\right) - \nabla \cdot (\rho(\nabla \phi^h)\nu_{k\varepsilon} . \tag{10.2}$$

In the source vector's discretisation of the turbulence closure equation $\Im k_{\varepsilon}^h$, we compute the $\nabla \cdot (\nabla u)$ in the E term by first calculating the nodal values of ∇u using a least-squares fit and then taking the divergence of the interpolated value of ∇u.

In Equations (9.1), (9.2), (10.1), and (10.2), p^{stab}, $p_{k\varepsilon}^{stab}$, and $K_{k\varepsilon}^{DC}$ are, respectively, the streamline-upwind / Petrov–Galerkin stabilisation operators and the dissipation matrix for the discontinuity-capturing scheme. The vectors p^{stab}, $p_{k\varepsilon}^{stab}$ take the following forms:

$$p^{stab}(w^h) = \tau_{SUPG}(u^h \cdot \nabla)w^h + \frac{\tau_{PSPG}}{\rho}\nabla q, \tag{11.1}$$

$$pk_{\varepsilon}^{stab}(\psi^h) = \begin{bmatrix} \tau_{SUPG-k} & 0 \\ 0 & \tau_{SUPG-\varepsilon} \end{bmatrix}(u^h \cdot \nabla)\psi^h. \tag{11.2}$$

Here τ_{SUPG} and τ_{PSPG} are the streamline-upwind / Petrov–Galerkin and pressure-stabilising / Petrov–Galerkin stabilisation parameters, and the latter allows us to use equal-order velocity-pressure approximations. We define the discontinuity-capturing dissipations terms for advection–diffusion-reaction equations as:

$$K_{k\varepsilon}^{DC} = \begin{bmatrix} \kappa_{DRDJ-k} & 0 \\ 0 & \kappa_{DRDJ-\varepsilon} \end{bmatrix}. \tag{12}$$

Here $\kappa_{DRDJ\text{-}k}$ and $\kappa_{DRDJ\text{-}\varepsilon}$ are the DRDJ diffusivities.

STABILISATION AND DISCONTINUITY-CAPTURING PARAMETERS

Streamline-upwind / Petrov–Galerkin parameters

We first define the element length (Tezduyar and Park, 1986) at the advection-dominated limit:

$$h_{UGN} = 2\left(\sum_{a}^{nen} | s \cdot \nabla N_a |\right)^{-1} \tag{13.1}$$

where s is the vector in the direction of velocity, *nen* is the number of element nodes and N_a is the interpolation function associated with node a. At the diffusion-dominated limit, we define the element lengths (Tezduyar, 2003; Catabriga *et al.*, 2005) as follows:

$$h_{UGN} = 2\left(\sum_{a}^{nen} | s \cdot \nabla N_a |\right)^{-1} \tag{13.2}$$

$$h_{RGN} = 2\left(\sum_{a}^{nen} | r \cdot \nabla N_a |\right)^{-1} \tag{13.3}$$

$$h_{RGN-\varepsilon} = 2\left(\sum_{a} | r_\varepsilon \cdot \nabla N_a |\right)^{-1} \tag{13.4}$$

where r, r_k and r_ε are the unit vectors in the direction of the solution gradient which we define as:

$$r = \frac{\nabla \| u \|}{\| \nabla \| u \| \|}, r_k = \frac{\nabla | k |}{\| \nabla | k | \|}, r_\varepsilon = \frac{\nabla | \varepsilon |}{\| \nabla | \varepsilon | \|}. \tag{14}$$

Tezduyar (2003) and Rispoli *et al.* (2007) defined the components of τ_{SUPG} corresponding to the advection-, transient- and diffusion-dominated limits as follows:

$$\tau_{SUGN1} = \left(\sum_{a}^{nen} | u \cdot \nabla N_a |\right)^{-1} = \frac{h_{UGN}}{2 \| u \|}, \tag{15.1}$$

$$\tau_{SUGN2} = \frac{\Delta t}{2}, \tau_{SUGN3} = \frac{h_{RGN}^2}{4\nu}, \tau_{SUGN3-k} = \frac{h_{RGN-k}^2}{4\nu_k} \tag{15.2}$$

$$\tau_{SUGN3-\varepsilon} = \frac{h_{RGN-\varepsilon}^2}{4\nu_\varepsilon}. \tag{15.3}$$

From these, we define the stabilisation parameters by using Tezduyar and Osawa's (2000) r-switch:

$$\tau_{SUPG} = \left(\frac{1}{\tau_{SUGN1}^{r_\varepsilon}} + \frac{1}{\tau_{SUGN2}^{r_\varepsilon}} + \frac{1}{\tau_{SUGN3}^{r_\varepsilon}}\right)^{-\frac{1}{r_\varepsilon}} \tag{16.1}$$

$$\tau_{SUPG} + \tau_{SUPG}, \quad , \tag{16.2}$$

$$\tau_{SUPG-\phi} = \left(\frac{1}{\tau_{SUGN1}^{r_\varepsilon}} + \frac{1}{\tau_{SUGN2}^{r_\varepsilon}} + \frac{1}{\tau_{SUGN3-\phi}^{r_\varepsilon}} \right)^{-\frac{1}{r_\varepsilon}} \tag{16.3}$$

Here the subscript ϕ generates the expressions corresponding to k and $\tilde{\varepsilon}$. We define the streamline-upwind / Petrov–Galerkin "diffusivities" as:

$$\nu_{SUPG} = \tau_{SUPG} \|u\|^2, \kappa_{SUPG-\phi} = \tau_{SUPG-\phi} \|u\|^2, \tag{17}$$

with ϕ indicating the variables modelled in the advection–diffusion-reaction equations.

Discontinuity-capturing parameters

Corsini *et al.* (2006a) originally developed the diffusion for reaction dominated with jump (DRDJ) formulation. This is an improved version of Tezduyar and Park's (1986) diffusion for reaction dominated (DRD) method. We used the diffusion for reaction dominated with jump formulation to manage the numerical instabilities at the reaction-dominated limit in Equation (3). The diffusion for reaction dominated with jump multi-dimensional diffusivity tensors read as:

$$\kappa_{DRDJ-\phi} = \kappa_{AR}(\gamma_\phi, J_e)SS + \kappa_{DR}(\beta_\phi, J_e)(tt + vv). \tag{18}$$

Here the advection-reaction operators apply in the stream-wise direction s and read as:

$$\kappa_{AR}(\gamma_\phi, J_e) = \frac{1}{2} u h_{UGN} J_e \left(-\coth \gamma_\phi + \gamma_\phi \left(\frac{1}{\sinh^2 \gamma_\phi} + 4\alpha \right) \right) \tag{19}$$

and the reaction-diffusion operators apply cross-wise, on the t-v plane and read as:

$$\kappa_{DR}(\beta_\phi, J_e) = \beta_\phi \left(\frac{h_{RGN-\phi}}{2} \right)^2 J_e \left(4\alpha + \frac{1}{\sinh^2 \beta_\phi} - \frac{1}{\beta_\phi^2} \right) \tag{20}$$

Again, the subscript ϕ generates the expressions corresponding to k and $\tilde{\varepsilon}$. We define the reaction-advection parameters γ_ϕ as:

$$\gamma_k = \frac{B_k}{\| u \|} \frac{h_{UGN}}{2} \tag{21.1}$$

$$\gamma_\varepsilon = \frac{B_\varepsilon}{\| u \|} \frac{h_{UGN}}{2} \tag{21.2}$$

and define the reaction numbers as:

$$\beta_k^2 = \frac{B_k}{\nu_k}\frac{h_{RGN-k}^2}{4}, \beta_\varepsilon^2 = \frac{B_\varepsilon}{\nu_\varepsilon}\frac{h_{RGN-\varepsilon}^2}{4}. \tag{21.3}$$

In Equations (19) and (20), α is the parameter representing the integration rule for the element reaction coefficient matrix with α equal to 1/6 for two-point Gaussian quadrature and 0 for the 'lumped' case. Also, in Equations (19) and (20), J_e is a normalised measure of the variation ('jump') in the solution over an element.

We define the advection-reaction (AR) and diffusion-reaction (DR) 'diffusivities' as:

$$\kappa_{AR-\phi} = \kappa_{AR}(\gamma_\phi, J_e) \tag{22}$$

$$\kappa_{DR-\phi} = \kappa_{DR}(\beta_\phi, J_e) \tag{23}$$

FAMILY OF FANS

We conducted the reported research on a family of commercially available cooling fans configured for application over a compact cooling unit's tube bank, Figure 2.1. The studied fan configuration, coded AC90/6, incorporates a six-blade unswept rotor, with modified ARA-D profile aerofoil blades. One may set the blade-

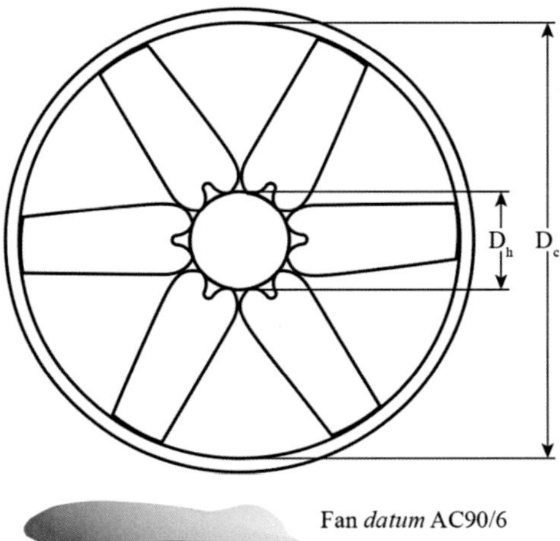

Design specifications	
Blade number	6
Blade tip stagger angle δ (degrees)	62
Hub-to-casing diamter ratio D_{hc}	0.22
Hub diameter D_h (mm)	200
Casing diameter D_c (mm)	907
Rotor tip clearance χ (% span)	1.0
Rotational frequency f_r (rpm)	900-935

FIGURE 2.1. The fan *datum* AC90/6 blade geometry and impeller specification.

pitch angle during final assembly to customise the fan to a desired duty point. We used a direct coupled-induction 400-volt (AC), 3-phase motor to drive the rotor at a constant speed of 950 rpm, resulting in a 44.7 m/s blade tip speed and a 95 Hz blade-passing frequency (BPF). In the research reported in this chapter, we studied fan performance at a nominal design operating flow rate of 7 m^3/s.

In its original embodiment, the studied fan did not include a blade-tip end-plate. We used it as a *datum* against which to assess the performance of fan variants with blade-tip end-plates. Therefore, in the reported research we refer to the fan without blade-tip end-plates as the fan *datum* AC90/6. In addition to the fan *datum* AC90/6, we studied a variant of the fan fitted with a constant thickness blade-tip end-plate, Figure 2.2. When fitted with a constant thickness blade-tip end-plate, we named the fan AC90/6/TF.

The constant thickness blade-tip end-plate design was inspired by the designs developed for tip-vortex control and drag reduction in aircraft wings and catamaran hulls. The constant thickness blade-tip end-plate ran along the blade pressure surface, ending at the blade trailing edge with a square tail. The addition of this constant thickness blade-tip end-plate resulted in the thickness of the fan AC90/6/TF blade tip increasing by a factor of three compared to the fan *datum* AC90/6. We considered Inoue *et al.*'s (1986) size of the blade-tip end-plate for axial compressor blades. They estimated that the optimum blade-tip end-plate size was between 10 and 20 per cent blade span. In practice, we were able to manufacture blades with a blade-tip end-plate size five per cent of blade span. The fan blades were manufactured from injection molded plastic, with the size of the blade-tip end-plate the largest the blade manufacturing technique could produce.

Corsini and Sheard (2007) studied both the fan *datum* AC90/6 and AC90/6/TF using a blade-tip vortex 'breakdown criteria' based on Rossby number (Spall *et al.*, 1987). Following Ito *et al.*'s method (1985), they concluded that there is a threshold value of Rossby number below which vortex rotation cannot reduce if the vortex is to remain stable. We defined this critical Rossby number range using Uchida *et al.* (1985) and Garg and Leibovich's (1979) critical Rossby numbers. Uchida *et al.* (1985) defined a critical Rossby number associated with the breakdown of an axi-symmetric vortex in a swirling flow. Garg and Leibovich (1979) also defined a critical Rossby number associated with an aircraft wing tip vortices' breakdown.

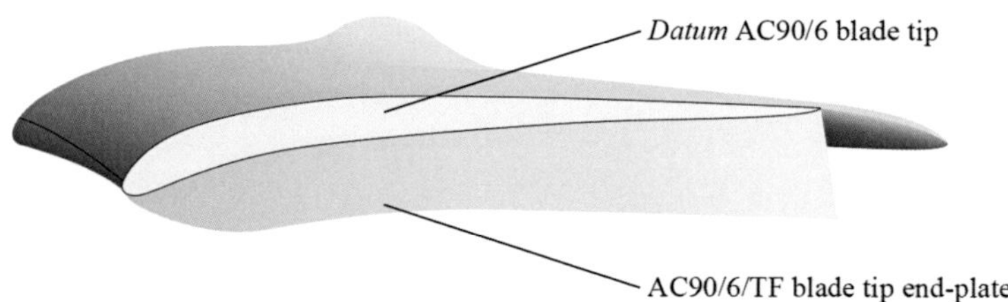

FIGURE 2.2. The studied fan *datum* AC90/6 without a fitted blade-tip end-plate and with a constant thickness blade-tip end-plate, AC90/6/TF.

COMPUTATIONS

In the research reported in this chapter, we solved the Reynolds-averaged Navier–Stokes equations of incompressible turbulent flow in the rotating frame of reference. Consequently, we based the computations on a version of Equation (1) that includes non-inertial terms. When implementing the stabilised formulations, we added these non-inertial terms to the other source terms. We considered adopting the arbitrary Lagrangian-Eulerian (ALE) technique favoured by Bazilevs *et al.* (2011), Hsu and Bazilevs (2012) and Hsu *et al.* (2012). We also considered the space-time technique favoured by Takizawa *et al.* (2011a, 2011b). We chose to solve the Reynolds-averaged Navier–Stokes equations of incompressible turbulent flow in the rotating frame of reference as it resulted in the lowest computational effort.

The approach we adopted when constructing a computational mesh utilised a non-orthogonal body fitted H-type grid system. We split the mesh into two regions, one in the blade-to-blade region and a second in the blade tip-to-casing gap. In the axial direction, 26 per cent of the nodes were distributed upstream and 74 per cent downstream of the blade leading edge. In the span-wise direction, 14 nodes were spaced across the blade tip-to-casing gap. All mesh density regions maintained an adequate computational cell aspect ratio toward solid boundaries, Figure 2.3. The computational grid that we developed for the blade illustrates how mesh density increased in the solid wall's near vicinity, Table 2.2.

We defined the boundary conditions according to Corsini and Rispoli (2004) and Corsini *et al.*'s (2004b) recommendations. They studied the performance of ducted high-solidity fans. The Dirichlet conditions for the relative velocity components are imposed at the inflow section, half a mid-span chord upstream of the leading edge, as obtained from a flow simulation in an annular passage of identical hub-to-casing diameter ratio that includes an upstream spinner cone.

We used a turbulent kinetic energy (k) pitch-wise profile derived from an axi-symmetric turbulence intensity profile previously developed by Corsini and Rispoli (2004). The turbulence intensity profile was a near-uniform six per cent across the annulus, increasing to ten per cent near the casing. We used the turbulent kinetic energy's pitch-wise profile to calculate the radial distribution of dissipation based on a dissipation length scale (l_ε) of one per cent of blade mid-span pitch. We completed

Table 2.2. *Computational mesh data for fan* datum *AC90/6 and fan AC90/6/TF. Both computational meshes are block-structured, constructed from tri-linear hexahedral elements.*

	datum AC90/6	AC90/6/TF
Elements	774,258	773,400
Nodes	797,505	802,589
Element aspect ratio		
Average	1.2	1.9
Max	15.2	11.9

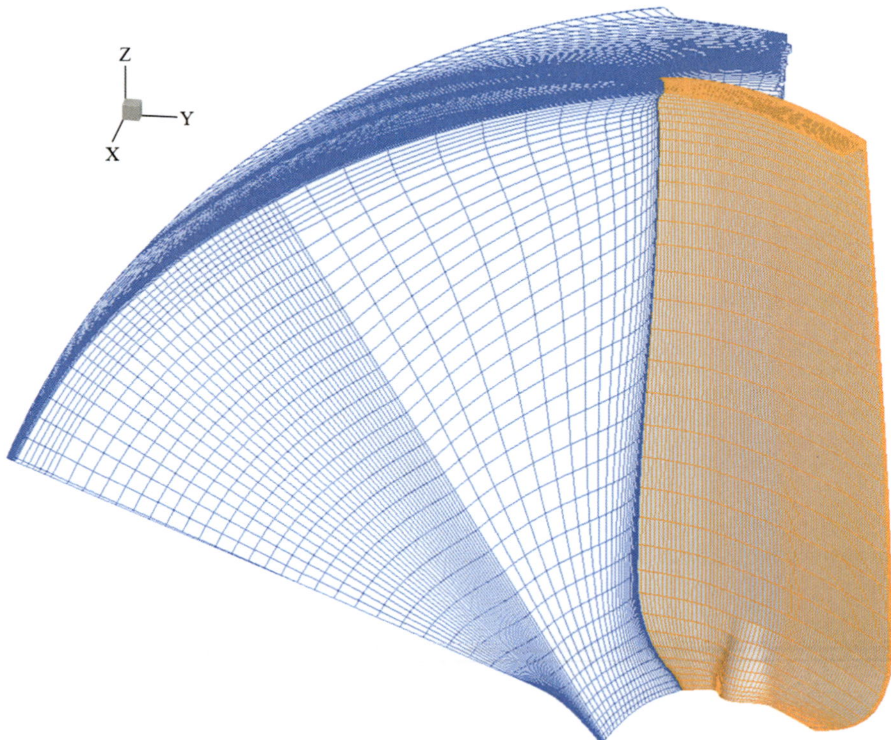

FIGURE 2.3. The computational grid that the authors used in the numerical simulations. They formed the mesh by merging a structured H-type grid through the up-stream, blade-to-blade passage and down-stream region with a second H-type grid in the blade tip-to-casing region.

the boundary conditions by assuming periodicity both up- and down-stream of the fan blades.

We parallelised an iterative Krylov method when smoothing and solving using the adopted multi-grid finite element method scheme with an original additive domain decomposition algorithm. This algorithm managed the message passing operations using the Message Passing Interface (MPI) libraries, a standardised message-passing system. We used the preconditioned GMRes(5) and GMRes(50) algorithms respectively as a smoother and as the core solver. We adopted standard boundary conditions which researchers have previously used when undertaking unswept and forward-swept blade numerical simulation. Corsini and Rispoli (1999) and Corsini *et al.* (2001) have utilised this approach. They were able to predict well both the unswept and forward-swept blade performance.

RESULTS

We validated the quality of the blade-to-blade flow-field computational analysis using Corsini and Sheard's (2007) method. We predicted the overall performance of the studied fan's *datum* AC90/6 and AC90/6/TF and compared the predicted performance with Corsini *et al.*'s (2006b) performance measurement, Table 2.3. We may reasonably assume that if we are able to predict static and dynamic pressure accurately, we accurately predict flow-field features in the blade-to-blade passage. The predicted static and dynamic pressure for the fan's *datum* AC90/6 and AC90/6/TF were within the uncertainty of the measured performance.

Aerodynamic features induced by the blade-tip end-plate

We used our computational analysis to study the flow-field features through the blade-to-blade passage of the fan *datum* AC90/6 and AC90/6/TF. The computational analysis was particularly helpful when visualising the flow-field features that develop through the blade tip-to-casing gap. We may identify structures within the blade tip-to-casing leakage flow by studying the turbulent kinetic energy (k) distributions. We present two-dimensional turbulent kinetic energy contour at 98.5 per cent blade span, within the blade tip-to-casing gap, Figure 2.4.

The chord-wise evolution of turbulent kinetic energy for the studied fans indicates differences between the fan *datum* AC90/6 and AC90/6/TF. The fan AC90/6/TF has a reduced blade tip-to-casing leakage flow rate compared to the fan *datum* AC90/6. This reduction, in turn, reduces the peak turbulent kinetic energy associated with vortex cores. The reduction in peak turbulent kinetic energy was particularly evident for the fan AC90/6 around mid-chord. In contrast, the fan *datum* AC90/6 is associated with higher turbulent kinetic energy in the blade leading edge region and over the first half of the blade chord. It is over the first half of the blade chord that organised vortical structures originate from the roll-up of leakage flow moving from the blade pressure to suction side. This leakage flow is the origin of the primary blade tip-to-casing leakage vortex. This vortex attaches to the blade suction surface before separating at mid-chord and migrating into the blade-to-blade passage.

When we study the fan AC90/6 turbulent kinetic energy distribution, it is apparent that the evolution of blade tip-to-casing leakage flow features is affected by the presence of a blade-tip end-plate. The reduced leakage flow rate results in a reduced momentum transfer over the blade tip that then results in a redistribution of turbulent kinetic energy. A consequence of this redistribution is a peak in suction

Table 2.3. *Experimentally measured (Corsini* et al., *2006b) and numerically predicted fan aerodynamic performance for the fan* datum *AC90/6 and AC90/6/TF at the design operating point.*

	Experimentally measured		Numerically predicted	
	Δp (Pa)	Δp_{dyn} (Pa)	Δp(Pa)	Δp_{dyn} (Pa)
datum AC90/6	134.8	67	133.3	66
AC90/6/TF	126.2	75	126.1	72

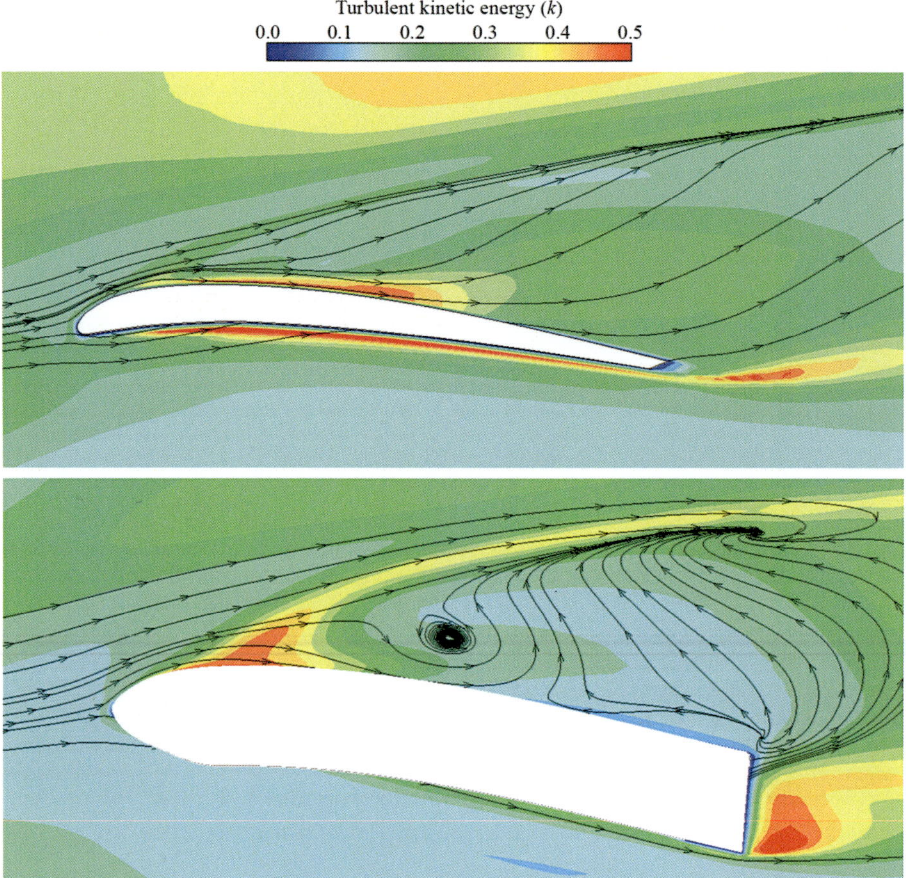

FIGURE 2.4. Two-dimensional turbulent kinetic energy (k) contours numerically predicted at the fan design operating point. The authors present turbulent kinetic energy contours at 98.5 per cent blade span for the fan *datum* AC90/6 and AC90/6/TF. The flow-field streamlines illustrate the effect of fitting the constant thickness blade-tip end-plate.

surface turbulent kinetic energy at approximately 20 per cent chord downstream of the blade leading edge. At approximately mid-chord, the suction surface is characterised by a region of low turbulent kinetic energy distribution. Corsini *et al.* (2010b) observed that this mid-chord region is characterised by the presence of a large separation bubble which forms as a consequence of the breakdown of the blade tip-to-casing leakage vortex.

We may complement our analysis of turbulent kinetic energy with an analysis of turbulent viscosity (v_t) normalised by the molecular viscosity. We present two-dimensional turbulent viscosity contours at 98.5 per cent blade span, within the blade tip-to-casing gap overlaid with blade tip-to-casing leakage core paths, Figure 2.5.

A comparison of the turbulent viscosity contours for the fan *datum* AC90/6 and AC90/6/TF illustrated the impact of the blade-tip end-plate fitted to fan AC90/6/TF.

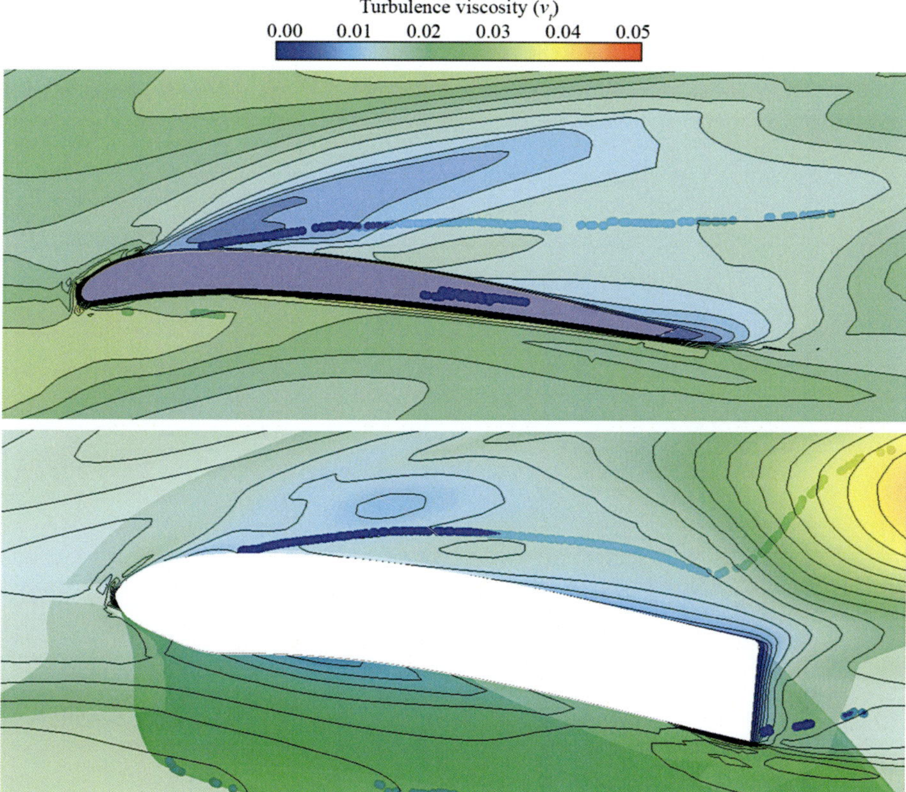

FIGURE 2.5. Contours of two-dimensional turbulence viscosity (ν_t) normalised by the molecular viscosity numerically predicted at the fan design operating point. The authors present contours at 98.5 per cent blade span for the fan *datum* AC90/6 and AC90/6/TF. The blade tip-to-casing leakage vortex core trajectories illustrate the effect of fitting the constant thickness blade-tip end-plate.

The blade-tip end-plate enhances turbulent viscosity. This enhancement manifests itself first as a more uniform distribution of turbulent viscosity and second, as a mixing region immediately downstream of the blade trailing edge where there is a region of higher turbulent viscosity where the blade tip-to-casing leakage vortex interacts with blade-to-blade passage secondary flow features. In contrast to the fan AC90/6/TF, the fan *datum* AC90/6 is characterised by generally lower turbulent viscosity levels along the blade tip-to-casing leakage vortex path.

Working of the numerical diffusivities

Our prediction of turbulent kinetic energy and turbulent viscosity provided an insight into the flow-field physics in the blade tip-to-casing region. The conclusions drawn from our analysis are self-consistent with Corsini *et al.*'s (2010b) conclusions, giving confidence in the accuracy of the numerical simulation.

We now move on to analyse performance of the dissipative model inherent in our choice of streamline-upwind / Petrov–Galerkin (SUPG) and diffusion for reaction dominated with jump (DRDJ) coupling methods. Our analysis is focused on establishing the adaptability of the streamline-upwind / Petrov–Galerkin and diffusion for reaction dominated with jump to the local magnitude of advection–diffusion-reaction. Specifically, we study the issues inherent in a dissipative model adaptively as a numerical counterpart of Hoffman and Johnson's (2004) direct numerical simulations (DNS) and large eddy simulations (LES). We also review the impact of Formaggia *et al.*'s (2004) anisotropic mesh adaptation in advection–diffusion-reaction flows. We illustrate the impact of numerical diffusivity in the blade tip-to-casing region of the fan *datum* AC90/6 and AC90/6/TF.

We present two-dimensional streamline-upwind / Petrov–Galerkin (SUPG) diffusivity ($\kappa_{SUPG\text{-}k}$) contours numerically predicted at 98.5 per cent blade span for the fan *datum* AC90/6 and AC90/6/TF, Figure 2.6. When we study the streamline-upwind / Petrov–Galerkin diffusivity contours for both the fan *datum* AC90/6 and AC90/6/TF, it is apparent that there is a reduction in diffusivity in the blade leading edge region. Diffusivity is also low over the rear part of the blade suction surface, the blade-to-blade region associated with the presence of the blade tip-to-casing leakage vortex. The fan AC90/6/TF blade-to-blade flow-field has particularly low diffusivity, a phenomena that we may associate with breakdown of the blade tip-to-casing vortex. Therefore, we may conclude that diffusivity decreases with decreasing turbulent viscosity, with diffusivity adapting to the local flow-field.

We may complement our analysis of streamline-upwind / Petrov–Galerkin (SUPG) diffusivity ($\kappa_{SUPG\text{-}k}$) with an analysis of advection-reaction diffusivity ($\kappa_{AR\text{-}k}$). We present two-dimensional advection-reaction diffusivity contours at 98.5 per cent blade span, within the blade tip-to-casing gap overlaid with blade tip-to-casing leakage core paths, Figure 2.7. To facilitate insight into the flow-field physics, we also plot blade tip-to-casing leakage vortex paths.

Consider the advection-reaction diffusivity contours for both the fan *datum* AC90/6 and AC90/6/TF. It is apparent that the evolution of the blade tip-to-casing leakage vortex influences the distribution of the reaction dominated with jump (DRDJ) advection-reaction diffusivity which stabilises the turbulent kinetic energy equation, particularly in those flow-field regions where the relative velocity vector aligns with the vorticity vector. This occurs in regions of positive relative helicity.

We may complement our analysis of advection-reaction diffusivity ($\kappa_{AR\text{-}k}$) with an analysis of diffusion-reaction diffusivity ($\kappa_{DR\text{-}k}$). We present two-dimensional diffusion-reaction diffusivity contours at 98.5 per cent blade span, within the blade tip-to-casing gap overlaid with blade tip-to-casing leakage core paths, Figure 2.8. To facilitate insight into the flow-field physics, we also plot blade tip-to-casing leakage vortex paths. Consider the diffusion-reaction diffusivity contours for both the fan *datum* AC90/6 and AC90/6/TF. The diffusivity peaks at the boundary between swirling flow-field structures and the blade-to-blade flow-field. These peaks are a consequence of the high shear rate at the boundary between swirling structures and the flow-field.

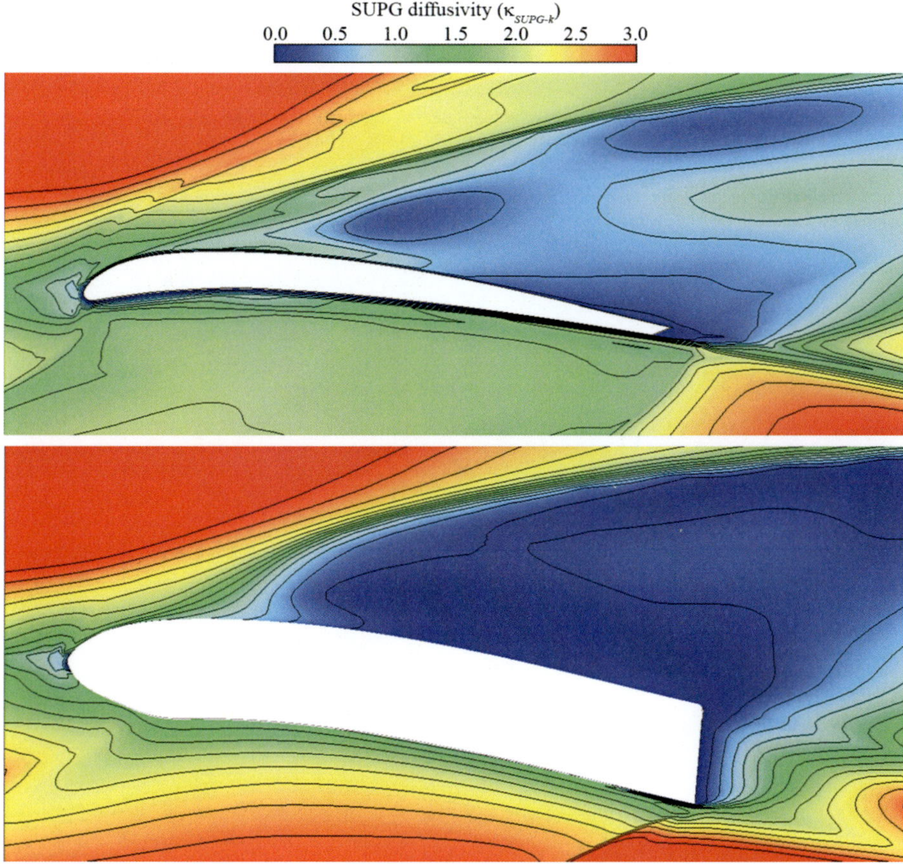

FIGURE 2.6. Two-dimensional streamline-upwind / Petrov–Galerkin (SUPG) diffusivity ($\kappa_{SUPG\text{-}k}$) contours numerically predicted at the fan design operating point. The authors present streamline-upwind / Petrov–Galerkin diffusivity contours at 98.5 per cent blade span for the fan *datum* AC90/6 and AC90/6/TF.

We complete our diffusivity analysis with a comparison of streamline-upwind / Petrov–Galerkin (SUPG) diffusivity ($\kappa_{SUPG\text{-}k}$), advection-reaction diffusivity ($\kappa_{AR\text{-}k}$) and diffusion-reaction diffusivity ($\kappa_{DR\text{-}k}$). We present two-dimensional contours of streamline-upwind / Petrov–Galerkin, advection-reaction and diffusion-reaction diffusivity. The contours are presented at 98.5 per cent blade span, within the blade tip-to-casing gap overlaid with blade tip-to-casing leakage core paths, Figure 2.9. The contours illustrate the difference between the three diffusivities. The streamline-upwind / Petrov–Galerkin diffusivity is characterised by a topology bounded by the blade tip-to-casing leakage vortex. In contrast, the diffusion-reaction diffusivity appears more adaptive to the local flow-field's dynamic behaviour.

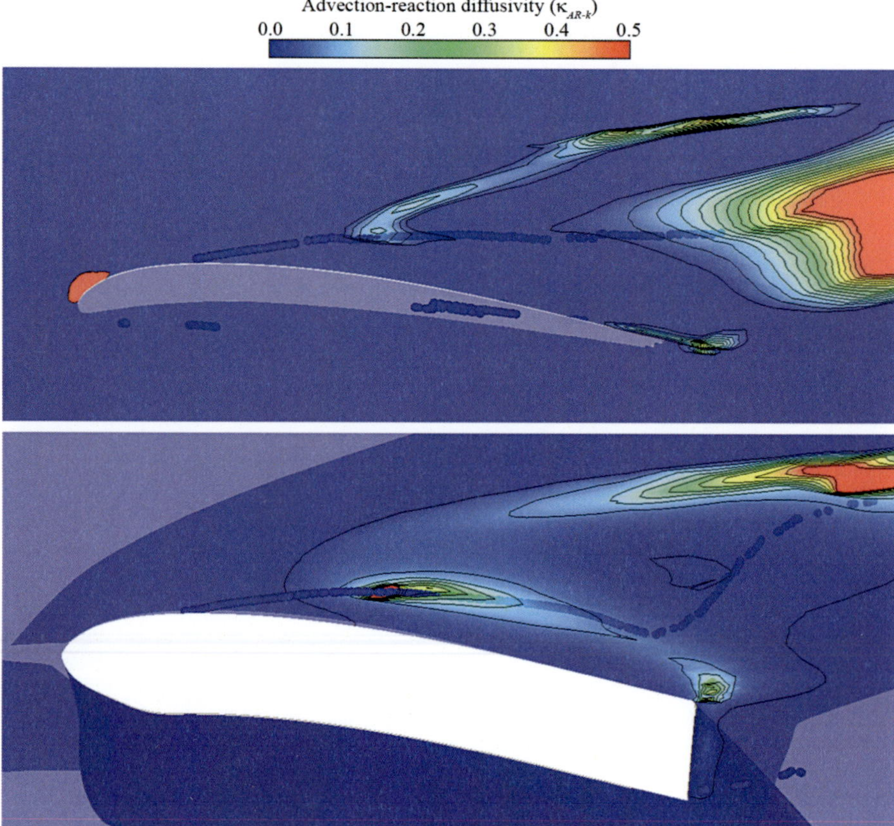

FIGURE 2.7. Two-dimensional advection-reaction diffusivity (κ_{AR-k}) contours numerically predicted at the fan design operating point. The authors present advection-reaction diffusivity contours at 98.5 per cent blade span for the fan *datum* AC90/6 and AC90/6/TF. The blade tip-to-casing leakage vortex core trajectories illustrate the effect of fitting the constant thickness blade-tip end-plate.

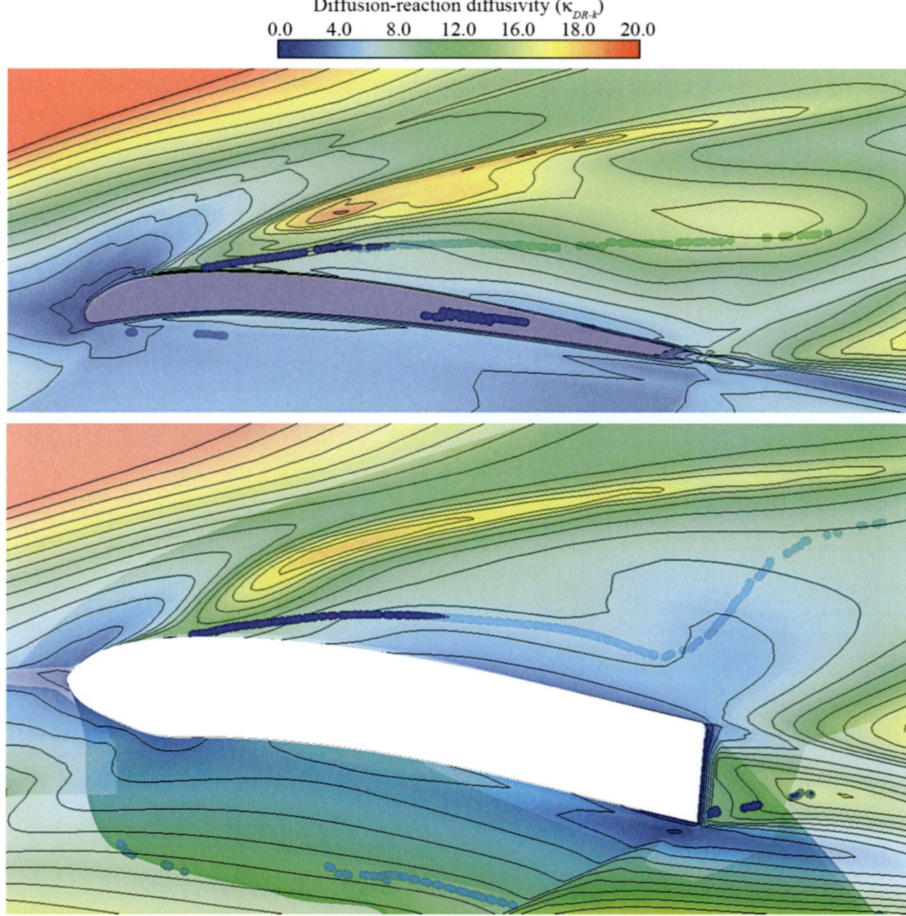

FIGURE 2.8. Two-dimensional diffusion-reaction diffusivity (κ_{DR-k}) contours numerically predicted at the fan design operating point. The authors present diffusion-reaction diffusivity contours at 98.5 per cent blade span for the fan *datum* AC90/6 and AC90/6/TF. The blade tip-to-casing leakage vortex core trajectories illustrate the effect of fitting the constant thickness blade-tip end-plate.

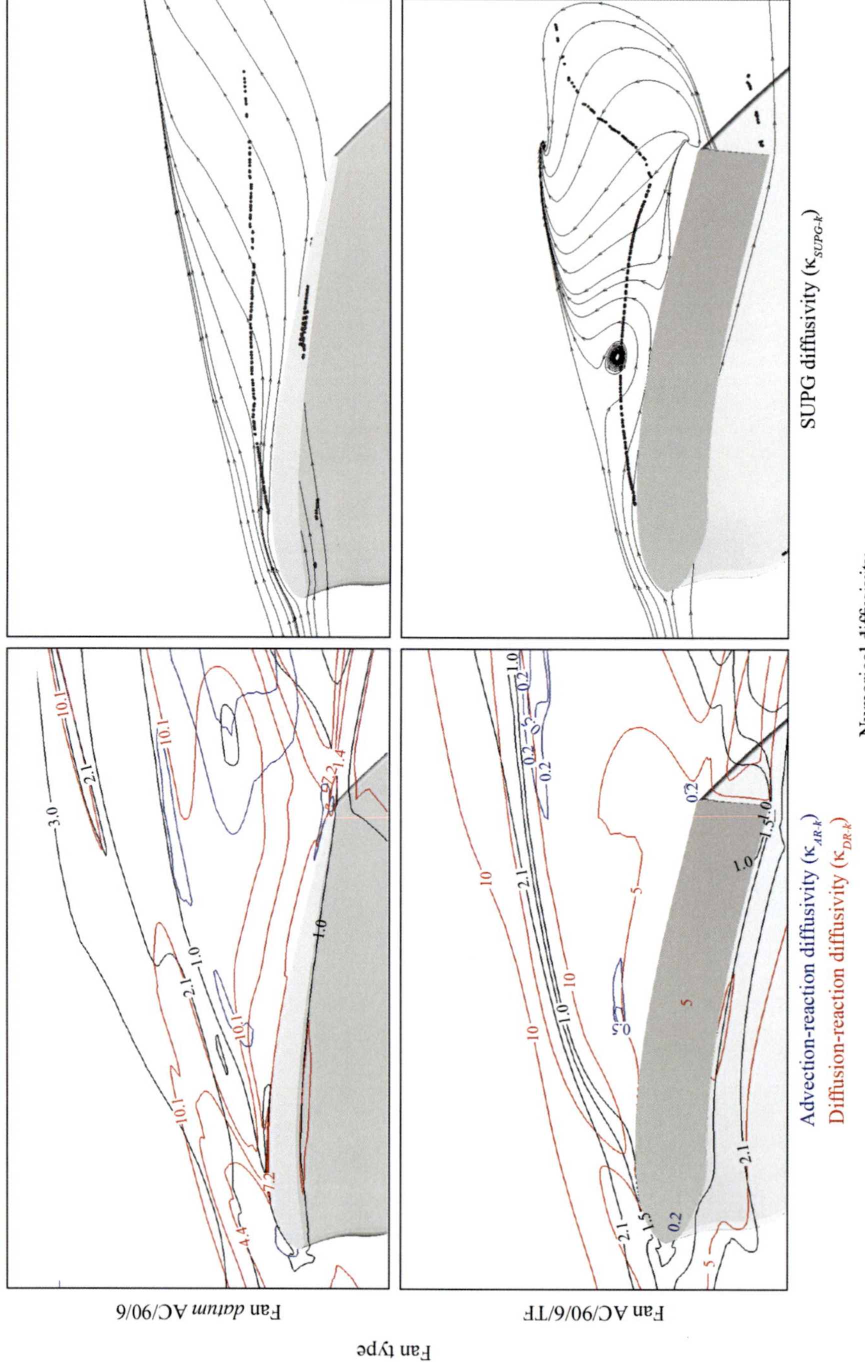

FIGURE 2.9. A comparison of two-dimensional diffusivity contours numerically predicted at the fan design operating point. The authors present diffusivity contours at 98.5 per cent blade span for the fan *datum* AC90/6, top and AC90/6/TF, bottom. They present advection-reaction diffusivity (κ_{AR-k}) and diffusion-reaction diffusivity (κ_{DR-k}) contours to the left in blue and red respectively and streamline-upwind / Petrov–Galerkin diffusivity (κ_{SUPG-k}) contours to the right in black. The blade tip-to-casing leakage vortex core trajectories illustrate the effect of fitting the constant thickness blade-tip end-plate.

CONCLUSIONS

In this chapter we have presented a stabilised formulation for computing the blade-to-blade flow-field of an axial fan both with and without a fitted blade-tip end-plate. The presence of a fitted blade-tip end-plate attenuates the blade tip-to-casing leakage flow rate, and consequently, the magnitude of the flow-field features that constitute aerodynamic noise sources.

We based the stabilisation of our numerical formulation on the streamline-upwind / Petrov–Galerkin (SUPG) and pressure-stabilising / Petrov–Galerkin (PSPG) methods. We stabilised our formulation using reaction dominated with jump (DRDJ) stabilisation. The resultant formulation facilitates numerical simulations with good numerical stability and accuracy, despite the separated flow regions that are an inevitable aspect of air movement fans. The reaction dominated with jump stabilisation is desirable when the reaction-like terms become dominant in the Reynolds-averaged Navier–Stokes model's turbulence closure equations.

We conducted numerical simulations of the flow-field through a fan named *datum* AC90/6 without a fitted blade-tip end-plate and a second fan AC90/6/TF with a fitted blade-tip end-plate. We studied the impact of a dissipative model associated with our choice of streamline-upwind / Petrov–Galerkin method combined with reaction dominated with jump stabilisation. Our study established the capability of the combined model and stabilisation on the local advection–diffusion-reaction magnitude. We conclude that the streamline-upwind / Petrov–Galerkin diffusivity is always distributed following the topology of the blade tip-to-casing leakage vortex boundary. In contrast, the diffusion-reaction diffusivity appears more adaptive to the local flow-field's dynamic behaviour.

REFERENCES

Bazilevs, Y., Hsu, M.-C., Akkerman, I., Wright, S., Takizawa, K., Henicke, B., Spielman, T. and Tezduyar, T.E. (2011), '3D Simulation of Wind Turbine Rotors at Full Scale. Part I: Geometry Modelling and Aerodynamics', *International Journal for Numerical Methods in Fluids*, vol. 65, pp. 207–235.

Bianchi, S., Corsini, A. and Sheard, A.G. (2014), 'A Critical Review of Passive Noise Control Techniques in Industrial Fans', *Transactions of the ASME, Journal of Engineering for Gas Turbines & Power,* vol. 136(4), paper no. 044001, pp. 1–9.

Brooks, A.N. and Hughes, T.J.R. (1982), 'Streamline Upwind / Petrov–Galerkin Formulations for Convection Dominated Flows with Particular Emphasis on the Incompressible Navier–Stokes Equations', *Computer Methods in Applied Mechanics and Engineering*, vol. 32, pp. 199–259.

Catabriga, L., Coutinho, A.L.G.A. and Tezduyar, T.E. (2005), 'Compressible Flow SUPG Parameters Computed from Element Matrices', *Communications in Numerical Methods in Engineering*, vol. 21(9), pp. 465–476.

Codina, R. (1998), 'Comparison of Some Finite Element Methods for Solving the Diffusion–convection-reaction Equation', *Computer Methods in Applied Mechanics and Engineering*, vol. 156, pp. 185–210.

Corsini, A. and Rispoli, F. (1999), 'Numerical Simulation of Three-dimensional Viscous Flow in an Isolated Axial Rotor', *Polish Academy of Sciences, Archive of Mechanical Engineering (Archiwum budowy maszyn)*, vol. XLVI(4), pp. 369–392.

Corsini, A. and Rispoli, F. (2004), 'Using Sweep to Extend the Stall-free Operational Range in Axial Fan Rotors', *Proceedings of the IMechE Part A, Journal of Power and Energy*, vol. 218(3), pp. 129–139.

Corsini, A. and Rispoli, F. (2005), 'Flow Analyses in a High-pressure Axial Ventilation Fan with a Non-linear Eddy-viscosity Closure', *International Journal of Heat and Fluid Flow*, vol. 26, pp. 349–361.

Corsini, A. and Sheard, A.G. (2007), 'Tip End-plate Concept Based on Leakage Vortex Rotation Number Control', *Journal of Computational and Applied Mechanics*, vol. 8, pp. 21–37.

Corsini, A., Rispoli, F., Bencze, F., and Vad, J. (2001), 'Effects of Blade Sweep in a High Performance Axial Flow Rotor', *Proceedings of the 4th European Conference on Turbomachinery*, Florence, Italy, 20–23 March.

Corsini, A., Rispoli, F. and Santoriello, A. (2004), 'A New Stabilized Finite Element Method for Advection–diffusion-reaction Equations using Quadratic Elements', in Vad, J., Lajos, T. and Schilling, R. (Eds), *Modelling Fluid Flow: The State of the Art*, Springer-Verlag, Berlin, Germany.

Corsini, A., Rispoli, F. and Santoriello, A. (2005), 'A Variational Multiscale High-order Finite Element Formulation for Turbomachinery Flow Computations', *Computer Methods in Applied Mechanics and Engineering*, vol. 194, pp. 4797–4823.

Corsini, A., Rispoli, F., Santoriello, A. and Tezduyar, T.E. (2006a), 'Improved Discontinuity-capturing Finite Element Techniques for Reaction Effects in Turbulence Computation', *Computational Mechanics*, vol. 38, pp. 356–364.

Corsini, A., Rispoli, F., Sheard, A.G. and Kinghorn, I.R. (2006b), 'Investigation of Improved Blade-tip Concept for Axial Flow Fan', *Proceedings of the 51st American Society of Mechanical Engineers Gas Turbine and Aeroengine Congress*, Barcelona, Spain, 8–11 May, paper no. GT2006-90592.

Corsini, A., Santoriello, A., Tezduyar, T.E., Menichini, F. and Rispoli, F.(2009), 'A Multiscale Finite Element Formulation with Discontinuity Capturing for Turbulence Models with Dominant Reaction Like Terms', *Transactions of the ASME, Journal of Applied Mechanics*, vol. 76(2), paper no. 021211, pp. 1–8.

Corsini, A., Iossa, C., Rispoli, F. and Tezduyar, T.E. (2010a), 'A DRD Finite Element Formulation for Computing Turbulent Reacting Flows in Gas Turbine Combustors', *Computational Mechanics*, vol. 46(1), pp. 159–167.

Corsini, A., Rispoli, F. and Sheard, A.G. (2010b), 'Shaping of Tip End-plate to Control Leakage Vortex Swirl in Axial Flow Fans', *Transactions of the ASME, Journal of Turbomachinery*, vol. 132(3), paper no. 031005, pp. 1–9.

Corsini, A., Rispoli, F. and Tezduyar, T.E. (2011), 'Stabilized Finite Element Computation of NOx Emission in Aero-engine Combustors', *International Journal for Numerical Methods in Fluids*, vol. 65, pp. 254–270.

Corsini, A., Rispoli, F. and Tezduyar, T.E. (2012), 'Computer Modelling of Wave-energy Air Turbines with the SUPG/PSPG Formulation and Discontinuity-capturing Technique', *Journal of Applied Mechanics*, vol. 79(1), paper no. 010910, pp. 1–8.

Craft, T.J., Launder, B.E. and Suga, K. (1996), 'Development and Application of a Cubic Eddy-viscosity Model of Turbulence', *International Journal of Heat and Fluid Flow*, vol. 17(2), pp. 108–115.

Ffowcs Williams, J.E. (1977), 'Aeroacoustics', *Annual Review of Fluid Mechanics*, vol. 9, pp. 447–468.

Formaggia, L., Micheletti, S. and Perotto, S. (2004), 'Anisotropic Mesh Adaptation in Computational Fluid Dynamics: Application to the Advection–diffusion-reaction and the Stokes Problems', *Applied Numerical Mathematics*, vol. 51(4), pp. 511–533.

Franca, L.P. and Valentin, F. (2000), 'On an Improved Unusual Stabilized Finite Element Method for the Advective–reactive–diffusive Equation', *Computer Methods in Applied Mechanics and Engineering*, vol. 190, pp. 1785–1800.

Fukano, T. and Jang, C.M. (2004), 'Tip Clearance Noise of Axial Flow Fans Operating at Design and Off-design Condition', *Journal of Sound and Vibration*, vol. 275, pp. 1027–1050.

Garg, A.K. and Leibovich, S. (1979), 'Spectral Characteristics of Vortex Breakdown Flowfields', *Physics of Fluids*, vol. 22, pp. 2053–2064.

Gravemeier, V. and Wall, W.A. (2007), 'A 'Divide-and-conquer' Spatial and Temporal Multiscale Method for Transient Convection–diffusion-reaction Equations', *International Journal for Numerical Methods in Fluids*, vol. 54, pp. 779–804.

Hauke, G. (2002), 'A Simple Subgrid Scale Stabilized Method for the Advection–diffusion-reaction Equation', *Computer Methods in Applied Mechanics and Engineering*, vol. 191, pp. 2925–2947.

Hoffman, J. and Johnson, C. (2004), 'Adaptive DNS/LES: a New Agenda in CFD', in Franca, L.P., Tezduyar, T.E. and Masud, A. (Eds), *Finite Element Methods: 1970s and Beyond*, CIMNE, Barcelona, Spain.

Hsu, M.-C. and Bazilevs, Y. (2012), 'Fluid–structure Interaction Modelling of Wind Turbines: Simulating the Full Machine', *Computational Mechanics*, vol. 50(6), pp. 821–833.

Hsu, M.-C., Akkerman, I. and Bazilevs, Y. (2012), 'Wind Turbine Aerodynamics using ALE-VMS: Validation and the Role of Weakly Enforced Boundary Conditions', *Computational Mechanics*, vol. 50(4), pp. 499–511.

Hughes, T.J.R. (1995), 'Multiscale Phenomena: Green's Functions, the Dirichlet-to-Neumann Formulation, Subgrid Scale Models, Bubbles and the Origins of Stabilized Methods', *Computer Methods in Applied Mechanics and Engineering*, vol. 127, pp. 387–401.

Inoue, M., Kuroumaru, M. and Furukawa, M. (1986), 'Behaviour of Tip Leakage Flow Behind an Axial Compressor Rotor', *Transactions of the ASME, Journal of Engineering for Gas Turbines and Power*, vol. 108(1), pp. 7–14.

Ito, T., Suematsu, Y. and Hayase, T. (1985), 'On the Vortex Breakdown Phenomena in a Swirling Pipe-flow', *Memoirs of the Faculty of Engineering, Nagoya University*, vol. 37, pp. 117–172.

Jang, C.M., Fukano, T. and Furukawa, M. (2003), 'Effects of the Tip Clearance on Vortical Flow and its Relation to Noise in an Axial Flow Fan', *JSME International Journal, Series B, Fluids and Thermal Engineering,* vol. 46(3), pp. 356–365.

Jensen, C.E. (1986), 'Axial-flow Fan', patent no. US 4,630,993, 23 December.

Longet, C.M.L. (2003), 'Axial Flow Fan with Noise Reducing Means', patent no. US 2003/0123987 A1, 3 July.

Longhouse, R.E. (1978), 'Control Tip-vortex Noise of Axial Flow Fans by Rotating Shrouds', *Journal of Sound and Vibration*, vol. 58, pp. 201–214.

Marcinowski, H. (1953),'Einfluss des Laufradspalts und der Luftfuehrung bei einem Kuehlgeblaese Axialer Bauart' *Motortechnische Zeitschrift*, vol. 14, pp. 259–262.

Mimura, M. (2003), 'Axial Flow Fan', patent no. US 6,648,598 B2, 18 November.

Quinlan, A.D. and Bent, P.H. (1998), 'High Frequency Noise Generation in Small Axial Flow Fans', *Journal of Sound and Vibration*, vol. 218(2), pp. 177–204.

Rispoli, F., Corsini, A. and Tezduyar, T.E. (2007), 'Finite Element Computation of Turbulent Flows with the Discontinuity-capturing Directional Dissipation (DCDD)', *Computer and Fluids*, vol. 36(1), pp. 121–126.

Sheard, A.G. (Ed.) (2014), *Passive Noise Control in Industrial Fans*, Sigel Press, Cambridge, UK.

Sheard, A.G., Corsini, A. and Rispoli, F. (2009), 'A Meridional Fan', patent no. GB 2,452,104 B, 22 July.

Smith, G.D.J. and Cumpsty, N.A. (1984), 'Flow Phenomena in Compressor Casing Treatment', *Transactions of the ASME, Journal of Engineering for Gas Turbines and Power*, vol. 106(3), pp. 532–541.

Spall, R.E., Gatski, T.B. and Grosch, C.E. (1987), 'A Criterion for Vortex Breakdown', *Physics of Fluids*, vol. 30, pp. 3434–3440.

Takata, H. and Tsukuda, Y. (1977), 'Stall Margin Improvement by Casing Treatment—its Mechanism and Effectiveness', *Transactions of the ASME, Journal of Engineering for Gas Turbines and Power*, vol. 99(1), pp. 121–133.

Takizawa, K., Henicke, B., Montes, D., Tezduyar, T.E., Hsu, M.-C. and Bazilevs, Y. (2011a), 'Numerical-performance Studies for the Stabilized Space-time Computation of Wind-turbine Rotor Aerodynamics', *Computational Mechanics*, vol. 48(6), pp. 647–657.

Takizawa, K., Henicke, B., Tezduyar, T.E., Hsu, M.-C. and Bazilevs, Y. (2011b), 'Stabilized Space-time Computation of Wind-turbine Rotor Aerodynamics', *Computational Mechanics*, vol. 48(3), pp. 333–344.

Tezduyar, T.E. (1992), 'Stabilized Finite Element Formulations for Incompressible Flow Computations', *Advances in Applied Mechanics*, vol. 28, pp. 1–44.

Tezduyar, T.E. (2003), 'Computation of Moving Boundaries and Interfaces and Stabilization Parameters', *International Journal for Numerical Methods in Fluids*, vol. 43(5), pp. 555–575.

Tezduyar, T.E. and Osawa, Y. (2000), 'Finite Element Stabilization Parameters Computed from Element Matrices and Vectors', *Computer Methods in Applied Mechanics and Engineering*, vol. 190, pp. 411–430.

Tezduyar, T.E. and Park, Y.J. (1986), 'Discontinuity-capturing Finite Element Formulations for Nonlinear Convection–diffusion-reaction Equations', *Computer Methods in Applied Mechanics and Engineering*, vol. 59(3), pp. 307–325.

Tezduyar, T.E., Park, Y.J. and Deans, H.A. (1987), 'Finite Element Procedures for Time-dependent Convection–diffusion-reaction Systems', *International Journal for Numerical Methods in Fluids*, vol. 7(10), pp. 1013–1033.

Tezduyar, T.E., Mittal, S., Ray, S.E. and Shih, R. (1992), 'Incompressible Flow Computations with Stabilized Bilinear and Linear Equal-order-interpolation Velocity-pressure

Elements', *Computer Methods in Applied Mechanics and Engineering*, vol. 95(2), pp. 221–242.

Thompson, D.W., King, P.I. and Rabe, D.C. (1998), 'Experimental Investigation of Stepped Tip Gap Effects on the Performance of a Transonic Axial-flow Compressor Rotor', *Transactions of the ASME, Journal of Turbomachinery*, vol. 120(3), pp. 477–486.

Uchida, S., Nakamura, Y. and Ohsawa, M. (1985), 'Experiments on the Axisymmetric Vortex Breakdown in a Swirling Air Flow', *Transactions of the Japan Society for Aeronautical Space Sciences*, vol. 27, pp. 206–216.

Uselton, R.B., Cook, L.J. and Wright, T. (2008), 'Fan with Reduced Noise Generation', patent no. US 7,351,041 B2, 1 April.

Wadia, A.R., Szucs, P.N. and Crall, D.W. (1998), 'Inner Workings of Aerodynamic Sweep', *Transactions of the ASME, Journal of Turbomachinery*, vol. 120(4), pp. 671–682.

Chapter 3

Large-eddy Simulation of a Tunnel Ventilation Fan

D. Borello, A. Corsini, G. Delibra, M. Fiorito and A.G. Sheard

ABSTRACT

In this chapter we present a large eddy simulation (LES) modelling technique using the open-source solver OpenFOAM. Current and forthcoming legislation mandates minimum tunnel ventilation fan efficiency, whilst also requiring the fan to be capable of operating at an elevated temperature. In combination, these legislative requirements imply a level of complexity in the design process that is beyond the historic norm within the tunnel ventilation community.

The developed modelling technique reduces the required computational effort that fan designers classically associate with a large eddy simulation, utilising a course computational grid. The course grid reduces the required computational effort, making it accessible to tunnel ventilation fan designers. The developed modelling technique calculates directly very large scale turbulent structures, and therefore we may characterise it as a very large eddy simulation. We account for the effect of the smaller unresolved turbulence scale using an eddy viscosity turbulence model.

The studied fan aerodynamic design is compromised, resulting in the inevitability of separated flow and intense secondary flow features. The requirement for a mechanical design capable of operating at elevated temperature resulted in the compromised aerodynamic design. We evaluated the very large eddy modelling technique by studying flow-field topology. Despite the limitations of this technique, in combination with the chosen eddy viscosity turbulence model we were able to predict well overall fan performance, flow-field feature location and intensity. We concluded that the modelling technique was capable of modelling separation and secondary flow features in the blade-to-blade flow-field. The accuracy with which we modelled separation and secondary flow features was significantly better than the accuracy achievable with a Reynolds-averaged Navier–Stokes modelling technique. Therefore, the predicted steady and unsteady performance parameters serve as a reference data set for future experimental and numerical analysis. We anticipate that future analysis will be required to refine and develop the modelling technique for application as part of a tunnel ventilation fan design methodology.

This chapter is a revised and extended version of Borello, D., Corsini, A., Delibra, G., Fiorito, M. and Sheard, A.G. (2013), 'Large-eddy Simulation of a Tunnel Ventilation Fan', *Transactions of the ASME, Journal of Fluids Engineering*, vol. 135, paper no. 071102, pp. 1–9.

NOMENCLATURE

Latin letters

b_s	[m]	Blade spacing
c	[mm]	Chord
C_{SGS}		Sub-grid scale (SGS) model coefficient
D_t	[mm]	Blade tip diameter
h	[mm]	Channel height
k		Turbulent kinetic energy
k_{res}		Resolved turbulent kinetic energy
k_{tot}		Total turbulent kinetic energy
L_{ij}, M_{ij}		Resolved stress tensors (for dynamic SGS model)
p_t	[Pa]	Total pressure
Δp_t	[Pa]	Total pressure rise
P	[kW]	Power
$\mathbf{P_i}$		Production term of i-quantity
Q	[1/s]	Parameter for vortex deduction
$Re_{cascade}$		Reynolds number for linear compressor cascade simulations; $Re_{cascade} = c \times U_{bulk}/\nu$
$Re_{channel}$		Reynolds number for channel flow simulations; $Re_{channel} = 2h \times U_{bulk}/\nu$
Re_{fan}		Reynolds number for fan simulations; $Re_{fan} = D_{tip} \times V_{tip}/\nu$
Ro		Rotation number; $Ro = 2h\omega/U_{bulk}$
u^+		Normalised velocity
U	[m/s]	Stream-wise velocity
U_{rel}		Relative velocity magnitude
w^+		Normalised relative velocity magnitude
V	[m^3/s]	Volume flow rate
V_c		Cell volume
x, y, z		Cartesian coordinates
y^+		Normalised wall-distance

Greek letters

α_t	[deg]	Turning angle
γ	[deg]	Pitch angle
δ	[deg]	Stagger angle
ε		Turbulent kinetic energy dissipation rate
ν		Kinematic viscosity
ν_m		Molecular viscosity
ν_{SGS}		Sub-grid scale viscosity
σ		Solidity
$\bar{\phi}$		RANS averaged variable Φ
$\tilde{\phi}$		Filtered variable Φ
χ	[% span]	Tip clearance
ω	[rpm]	Rotational speed
Ω	[1/s]	Rate of rotation tensor

Acronyms

BEMM	Blade element momentum method
CDS	Central difference scheme
CFD	Computation fluid dynamics
CG	Conjugate gradient
DNS	Direct numerical simulations
FTT	Flow through time
ILU	Incomplete lower and upper
LE	Leading edge
PISO	Pressure implicit with splitting of operators
PS	Pressure side
SDS	Scale-adaptive simulations
SGS	Sub-grid scale
SS	Suction side
TLV	Tip-leakage vortex
TVD	Total variation diminishing scheme
U-RANS	Unsteady Reynolds-averaged Navier–Stokes
(V)LES	(Very) large eddy simulations

INTRODUCTION

Tunnel ventilation systems should provide those trapped with smoke-free escape routes in the event of a fire. However, fires in the Mont Blanc and Tauern Tunnels (Lacroix, 1997; Pucher, 1999), the Channel Tunnel (Shipp, 1997) and the Baku underground system (Hedefalk *et al.*, 1998) have demonstrated potentially devastating effects of such underground fires. In the event of a fire, a typical tunnel safety-management system relies on an emergency ventilation system to control and disperse smoke and heat to maintain a smoke-free escape route (Carvel, 1999). Therefore, tunnel ventilation fans must be capable of operating at elevated temperatures to remove hot gas and smoke from the tunnel in the event of a fire. Although no global legislative framework exists for emergency tunnel-ventilation systems, the European Commission (EC), the European Free Trade Association (EFTA) and the International Organisation for Standardisation (ISO) have addressed various aspects of fire safety in such systems.

The EC and EFTA issued mandates to create harmonised standards, and adherence to these standards demonstrates that industries are acting in compliance with European Union (EU) directives. The harmonised standard EN 12101-3 provides specifications against which accredited authorities must evaluate and certify powered emergency smoke and heat ventilators and is legally enforceable within EU member states (Sheard and Jones, 2012).

The International Standards Organisation (ISO), a worldwide federation of national standards bodies, prepares international standards through various technical committees and their working groups. ISO formed a technical committee for the purpose of preparing the international standard ISO 21927-3 for emergency smoke and heat-control equipment. This international standard closely follows the harmonised standard EN 12101-3 (Sheard and Jones, 2008).

The harmonised standard EN 12101-3 and international standard ISO 21927-3 have significant implications for tunnel ventilation fan design praxis. At their most demanding, they require the fan to be capable of operating at 400°C in an emergency whilst clearing smoke and hot gas. A requirement to operate at an elevated temperature has two primary design implications. First, one must increase the blade tip-to-casing clearance during normal operation to ensure that it does not close completely when operating at an elevated temperature. This increase in clearance during normal operation has a negative impact on both fan aerodynamic (Sheard, 2012) and acoustic (Sheard, 2014) performance. Second, the fan mechanical design must account for the reduction in material strength at the emergency operation temperature. This invariably results in blade root profiles that are aerodynamically compromised, an issue exacerbated by unsteady aerodynamic effects. Engineers associate the most significant unsteady aerodynamic effects with the pressure waves induced by trains as they travel along a tunnel. Therefore, tunnel ventilation fans are subjected to a pressure pulse each time a train passes the ventilation shafts within which they are situated (Sheard and Corsini, 2012).

The above requirements imply a level of complexity in the design process that is beyond the historic norm within the tunnel ventilation community. Empirical design methods are no longer appropriate, with tunnel ventilation fan designers increasingly moving from empirical fan to industrial turbomachinery design methodologies. Wallis (1961) and Daly (1985) exemplify the traditional empirical approach to tunnel ventilation fan design and describe a trial-and-error approach that relies on the designer's experience. More recently, tunnel ventilation fan designers have started to utilise computational fluid dynamics analyses at the beginning of the design process (Vad, 2001; Vad *et al.*, 2001). Lee *et al.* (2008) recently applied an inverse approach to the design of cooling fans for electronic appliances. Sheard *et al.* (2009) described a virtual prototyping approach that facilitated early identification and solution of design problems. They based their design methodologies on Horlock and Denton's (2005) design practices which are reliant on computational fluid dynamics to develop appropriate three dimensional blade sections.

An issue when using computational fluid dynamics is that even in the most favourable cases, Reynolds numbers are high resulting in complex flow-fields. A direct numerical simulation (DNS) of the Navier–Stokes equations would require fine computational grids and therefore a computational effort beyond that available to tunnel ventilation fan designers. Consequently, the only viable approach available when attempting to solve the Navier–Stokes equations is a Reynolds-averaged Navier–Stokes simulation. However, as a fan approaches peak pressure, and therefore its stability limit, classical Reynolds-averaged Navier–Stokes modelling techniques reach their capability limit (Corsini *et al.*, 2013). They are not able to model the turbulence's spectral content and are not sensitive to the flow-field's multi-scale features. Therefore, the hypotheses that engineers used to derive all Reynolds-averaged Navier–Stokes methodologies are no longer valid.

Scholars who have studied the flow-field physics within all turbomachinery classes are increasingly turning to large eddy simulations (LESs) to overcome the limitations of Reynolds-averaged Navier–Stokes modelling techniques. The principle issue with all large eddy simulations is the formidable computations effort. Despite the required computational effort, scholars are developing effective large eddy simulation based modelling techniques (Borello *et al.*, 2005, 2007, 2009, 2010; Corsini and Rispoli, 2005; Delibra *et al.*, 2009, 2010a, 2010b; Schneider *et al.*, 2010; Jagannathan *et al.*, 2012; Luo *et al.*, 2012). These modelling techniques are able to predict the flow-field features at play within a tunnel ventilation fans' blade-to-blade passages. However, the computational effort required is simply beyond the reach of even the largest tunnel ventilation fan manufacturer. Despite this reservation, Corsini *et al.* (2013) concluded that high performance computing (HPC) facilities are developing rapidly, exploiting massively parallel clusters of individual processors. Therefore, the necessary computing capability will likely become available at some point in the future.

Despite large eddy simulations being beyond current computing capability of the tunnel ventilation fan community, there is interest in modelling methodologies that overcome at least some of the limitations of Reynolds-averaged Navier–Stokes modelling techniques. A way to improve results is to solve the unsteady Reynolds-averaged Navier–Stokes equations. An unsteady Reynolds-averaged Navier–Stokes methodology still does not account for the turbulence energy cascade from large scale structures. However, it is able to reproduce unsteady flow phenomena such as the blade tip-to-casing leakage vortex and the development of secondary flow features in the blade hub region. Hah and Shin (2012) studied the near-stall flow-field in a transonic fan, conducting unsteady pressure measurements and both an unsteady Reynolds-averaged Navier–Stokes and large eddy simulation of the flow-field. They concluded that the unsteady Reynolds-averaged Navier–Stokes predicted the time-averaged flow-field accurately. However, only the large eddy simulation predicted measured unsteady pressures accurately.

When researchers wish to predict the unsteady features of a flow-field, an unsteady Reynolds-averaged Navier–Stokes helps overcome some of the Reynolds-averaged Navier–Stokes modelling technique limitations. However, Hah and Shin (2012) concluded that a large eddy simulation is necessary if one is to predict accurately unsteady parameters. When researchers do choose a large eddy simulation, they may reduce the computational effort by studying only a component of the flow-field. Luo *et al.* (2012) reported an example of this approach. They used a large eddy simulation to study cavity shedding from a three dimensional twisted hydrofoil. An alternative to studying only one component of the flow-field is to study only boundary layer development through the entire flow-field. Jagannathan *et al.* (2012) adopted this approach, studying boundary layer development during both transition and relaminarisation.

Davidson (1997) studied the alternatives to modelling only one part of the flow-field. He proposed a hybrid approach which used a use low-resolution grid compared to that of a large eddy simulation coupled with an eddy viscosity turbulence model classically used in a Reynolds-averaged Navier–Stokes simulation. Such a hybrid approach has the potential to bridge the gap between the two, affording an intermediate resolution of turbulence scales and therefore having the characteristics of a very large eddy simulation. One directly calculates the very large scale turbulent structures with the eddy viscosity turbulence model accounting for the effects of the unresolved scales. Labois and Lakehal (2011) studied the effectiveness of a very large eddy simulation. They concluded that the very large scale turbulent structures were modelled accurately providing the computational domain was fine enough to resolve the large turbulence scales. The required mesh was relatively fine compared to that required for a mesh independent Reynolds-averaged Navier–Stokes simulation, but relatively course compared to a large eddy simulation.

Realising the potential of a very large eddy simulation requires an adequately fine mesh to resolve the large turbulence scales and an eddy viscosity turbulence model to account for unresolved turbulence. The success of a very large eddy simulation depends on both, and therefore the selection and implementation of an appropriate eddy viscosity turbulence model. The research reported in this chapter studied the potential of two eddy viscosity turbulence models, Germano *et al.*'s dynamic model (1991) and Davidson's (1997) one-equation sub-grid scale model. We characterise the studied eddy viscosity turbulence models, predicting the flow-field features that develop through the tunnel ventilation fan's blade-to-blade passage.

The studied family of fans is intended for application in railway tunnel and metropolitan metro ventilation systems, Table 3.1. The family of fans was designed for emergency operation at 400°C to clear hot gas and smoke in the event of a tunnel fire. The blade tip-to-casing clearance is relatively large, and blade hub sections aerodynamically compromised by mechanical constraints. Therefore, blade tip-to-casing leakage flow in the blade-tip region and separated flow in the blade hub region characterise flow-fields. A traditional Reynolds-averaged Navier–Stokes simulation would find predicting this combination of blade tip and hub flow-field feature challenging. It was the inadequacy of traditional Reynolds-averaged Navier–Stokes simulations when applied to this family of fans that provided the motivation for seeking an alternative.

Table 3.1. *The studied family of fans is intended for application in railway tunnel and metropolitan metro ventilation systems. A requirement of tunnel ventilation fans is the ability to clear hot gas and smoke in the event of a tunnel fire to keep the escape route clear. An ability to operate at elevated temperature invariably requires a robust mechanical design. A consequence of the mechanical design is compromised aerodynamic performance that is challenging to predict accurately.*

Tip diameter D_t	1,400 to 2,800 mm
Performance measurement standard	ISO 5801: 2007
Volume flow rate V	10 to 300 m^3/s
Total pressure rise Δp_t	500 to 3,000 Pa
Rotational speed ω	900 to 1,500 rpm
Blade sections	Unidirectional, truly reversible
High temperature certification	EN 12101-3: 2002 at 200°C, 300°C and 400°C for two hours

NUMERICAL METHOD

Commercial computational fluid dynamic codes do not include hybrid methodologies, instead favouring well proven Reynolds-averaged Navier–Stokes methodologies that converge reliably. The reliability with which the code converges is more important than the accuracy with which the flow-field physics is modelled. As a direct result, the last decade has seen the widespread adoption of open-source codes. Foremost among these is the c++ finite volume based code OpenFOAM, (Jasak, 2010). This open-source code is customisable, and in the programme of work reported in this chapter we have extend the code to account for rotational effects.

We carried out simulations using OpenFOAM1.7.x, solving the filtered Navier–Stokes equations for incompressible fluids using Davidson's (1997) dynamic sub-grid scale model. A dynamic sub-grid scale model requires one to solve an additional transport equation for the local sub-grid scale turbulent kinetic energy k_{SGS} that provides a velocity-scale for the sub-grid turbulence. The transport equation for k_{SGS} reads:

$$\frac{Dk_{SGS}}{Dt} = P_{kSGS} - \varepsilon_{kSGS} + \frac{\partial}{\partial x_j}\left[(\nu + \nu_{SGS})\frac{\partial k_{SGS}}{\partial x_j}\right] \tag{1}$$

where ν is the molecular viscosity, ν_{SGS} is the sub-grid scale viscosity, and P_{kSGS} and ε_{kSGS} are, respectively, the production and the dissipation of the local sub-grid scale turbulent kinetic energy k_{SGS}.

We modelled the sub-grid scale viscosity as:

$$\nu_{SGS} = \langle C_{SGS} \rangle_{xyz} \sqrt{k_{SGS}}\,\Delta \tag{2}$$

where Δ represents a length scale equivalent to the cubic root of cell volume and $\langle C_{SGS} \rangle$ is a coefficient derived by using a procedure similar to Smagorinsky's dynamic model (Germano *et al.*, 1991),

$$\langle C_{SGS} \rangle = -\frac{1}{2}\frac{L_{ij}M_{ij}}{M_{kl}M_{kl}} \text{ where } L_{ij} = \widetilde{\overline{u_i u_j}} - \widetilde{\overline{u_i u_j}} \text{ and } Mij = \Delta \tilde{K}^{\frac{1}{2}} \widetilde{\overline{S_{ij}}} - \Delta \widetilde{k_{SGS}^{\frac{1}{2}} \overline{S_{ij}}} \tag{3}$$

The tilde indicates a filtering operation on the secondary coarser filter and the filtering over-bar filtering on the primary finer filter.

The coefficient $\langle C_{SGS} \rangle$ can be negative, accounting for transfer of turbulent kinetic energy from smaller to larger scale structures, a phenomenon known as backscatter. Backscatter is a temporary phenomenon in any flow-field, classically associated with flow-field structures that are larger than those associated with the inertial sub-range. Delibra *et al.* (2010a) studied the backscattering phenomenon, concluding that it must be modelled if the resulting flow-field simulation is to be accurate. In making this statement, Delibra *et al.* (2010a) assumed that the computational grid was not fine enough to resolve turbulent scales down to the inertial sub-range. In the programme of work reported in this chapter, we account for negative values of the coefficient $\langle C_{SGS} \rangle$. However, we have chosen to specify a lower limit to avoid negative diffusion terms:

$$\nu_{SGS} = \max(-\nu : \nu_{SGS}). \tag{4}$$

We solved the filtered incompressible Navier–Stokes equations in the stationary frame of reference, accounting for Coriolis and centrifugal effects with a modified version of the *piso-Foam* solver. We carried out simulations using a convective central differencing scheme with a second order total variation diminishing (TVD) scheme limit that prevented numerical instabilities (Sweby, 1984). We used a second-order backward scheme for time integration. We solved the equation's system with an incomplete lower and upper (ILU) preconditioned semi-iterative conjugate gradient (CG) solver (Saad, 1996), coupled with a pressure implicit with splitting of operators (PISO) scheme. We set the convergence threshold equal to 10^{-8} for the conjugate gradient solver and to 10^{-5} for the PISO algorithm.

Code validation

As previously stated, realising the potential of a very large eddy simulation requires an adequately fine mesh to resolve the large turbulence scales and an eddy viscosity turbulence model to account for unresolved turbulence. Therefore, we chose to undertake an initial validation of the adopted dynamic sub-grid scale model and rotating solver using two test cases. The first used the code to predict rotating fully developed channel flow. The second used the code to predict the blade tip-to-casing leakage flow over the blades of a linear compressor cascade.

Rotation effects flow development through a channel, resulting in an asymmetric velocity profile across the channel. A stable but turbulent region characterises the channel's pressure side. In contrast, an unstable and stratified region characterises the channel's suction side, Figure 3.1. Grundestam *et al.* (2008)

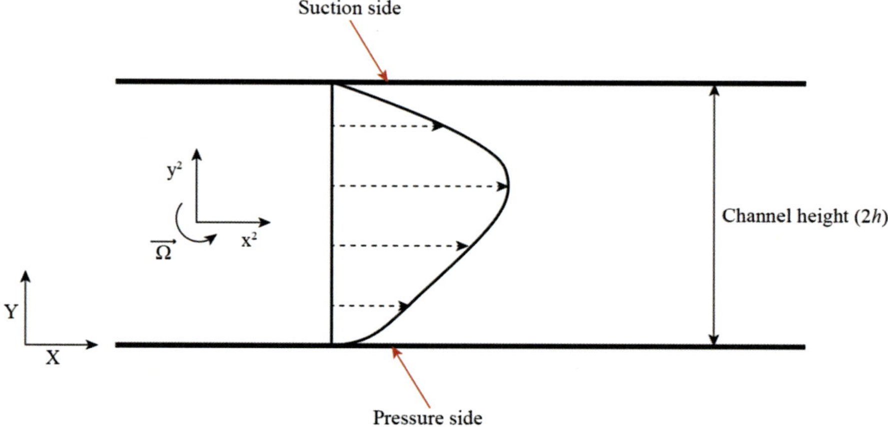

FIGURE 3.1. A test case for the large eddy simulation, characterising the channel flow with rotation. The impact of rotation of the channel flow is to distort its velocity profile within the channel.

concluded that this unstable stratified region results in flow laminarisation. In our test case we selected a rotation number (Ro) and Reynolds number for the channel ($Re_{channel}$) to be respectively 0.98 and 4,026. We were able to undertake a large eddy simulation of the channel flow-field using a relatively simple mesh, Table 3.2. We were able to undertake the large eddy simulation using a pure central difference scheme for the discretisation of convective terms combined with a backward time integration scheme. Therefore, the large eddy simulation is second order accurate in both space and time, and achieves convergence by computing 34 flow through time (FFT) steps.

We used the large eddy simulation to calculate the normalised velocity distribution (u^+) across the channel using five large eddy simulation models:

Smag	Standard Smagorisky large eddy simulation model
DySmag	Dynamic Smagorisky large eddy simulation model
DySmagBound	Dynamic Smagorisky with bounded sub-grid scale viscosity large eddy simulation model
LocDyn	Local dynamic sub-grid scale large eddy simulation model
LocDynBound	Local dynamic sub-grid scale large eddy simulation model with bounded sub-grid scale viscosity

We compared the computed normalised velocity distribution using the five large eddy simulation models with Grundestam *et al.*'s (2008) results using a direct numerical simulation (DNS), Figure 3.2. The four dynamic sub-grid scale models all predicted normalised velocity distributions that matched the direct numerical simulation significantly better than the standard Smagorinsky large eddy simulation model. Therefore, we concluded that the four dynamic sub-grid scale models were all capable of predicting a turbulent flow-field in a rotating flow. The local dynamic sub-grid scale large eddy simulation with bounded sub-grid scale viscosity proved to be the most stable of the four dynamic sub-grid scale models. Thus, we chose to use the local dynamic sub-grid scale large eddy simulation with bounded sub-grid scale viscosity modelling technique for the remainder of the research presented in this chapter.

Table 3.2. *During the first code validation exercise we used a relatively simple computational grid to predict the flow-field through a rotating channel.*

Direction	Domain length	Number of cells
Stream-wise	4π	128
Span-wise	π	64
Wall-normal	2	64

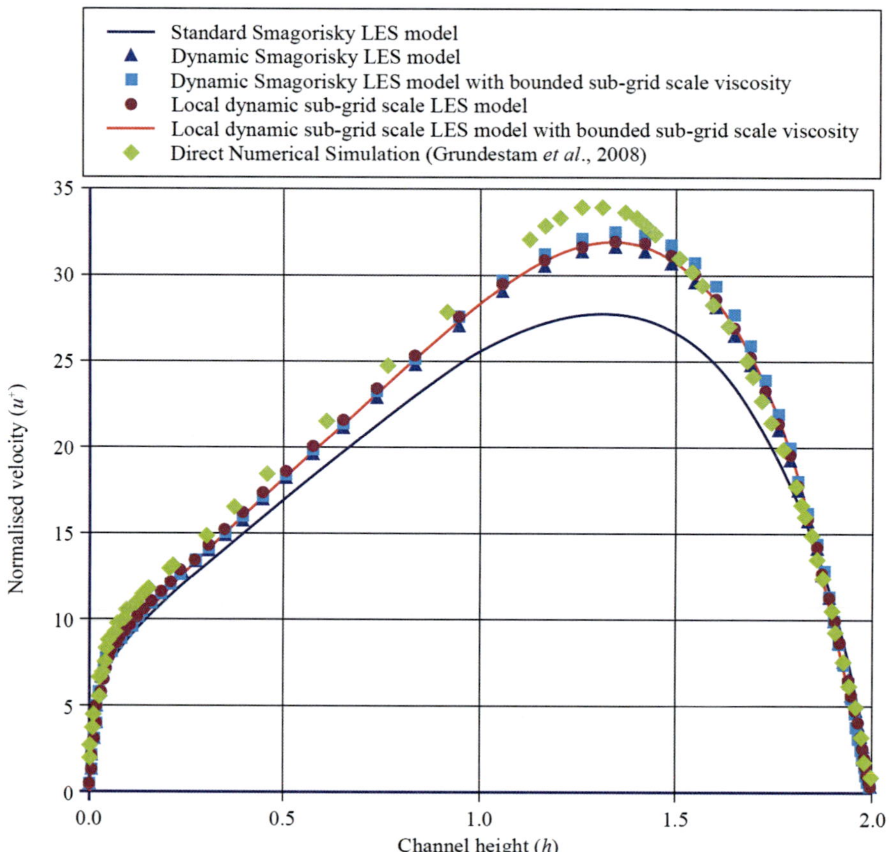

FIGURE 3.2. Rotating channel flow predicted using five large eddy simulation models, compared with Grundestam *et al.*'s (2008) direct numerical simulation.

Having identified the optimum dynamic sub-grid scale modelling technique for predicting turbulent flow-fields in a rotating flow, we analysed the flow-field topology. Dubief and Delcayre's (2000) vortex eduction criterion is based on Q factor with the resulting Q factor iso-lines representing constant vorticity lines. By calculating Q factor iso-lines across the channel with and without rotation, we were able to illustrate the rotation's impact on the channel flow, Figure 3.3. Vortical flow across the entire channel with essentially no mean-line or mid-span quasi two dimensional region of the flow-field characterises the rotating channel flow. These vorical flow-field structures have a first order effect on the flow-field, illustrating the importance associated with their accurate modelling.

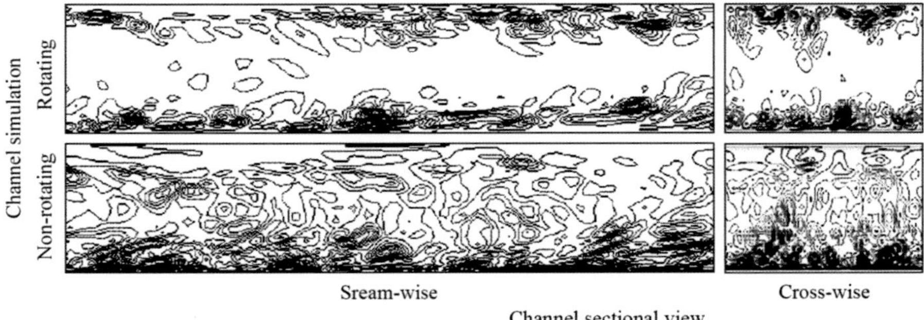

FIGURE 3.3. An analysis of the flow-field topology using Dubief and Delcayre's (2000) vortex eduction criterion based on Q factor with the resulting Q factor iso-lines representing lines of constant vorticity.

Compressor cascade

We continued validating the local dynamic sub-grid scale large eddy simulation with bounded sub-grid scale viscosity modelling technique with a second test case. We used the code to predict the blade tip-to-casing leakage flow over the blades of a linear compressor cascade. Muthanna and Devenport (2004) experimentally studied the linear compressor cascade and also presented its geometry, Figure 3.4. In the programme of work reported in this chapter we predicted the flow-field through the linear compressor cascade. We then compared our predicted results with Muthanna and Devenport's (2004) results, with the difference between the two facilitating insight into prediction accuracy.

Our numerical simulation of the linear compressor cascade utilised a computational domain that extended one blade chord upstream of the blades' leading edge and one blade chord downstream of the blades' trailing edge, Figure 3.5. The computational mesh comprised two blocks, Figure 3.5. We built the first block using hexahedra elements wrapped around the blade. We built the second block using triangular prism elements through the remainder of the computational domain. The total number of cells within the computational domain was 3,325,600. We selected a Reynolds number for the linear compressor cascade ($Re_{cascade}$) of 426,720. Table 3.3 provides our choice of boundary conditions.

We conducted the large eddy simulation using a second order accurate central difference scheme (CDS) coupled with both a total variation diminishing (TVD) scheme and second order implicit time integration scheme. This combination controlled instabilities, and is what we refer to as the local dynamic sub-grid scale large eddy simulation with bounded sub-grid scale viscosity.

The linear compressor cascade included a blade tip-to-cascade wall gap resulting in a blade-tip-to-cascade wall leakage vortex. Therefore, this code validation exercise facilitates studying the accuracy with which one may predict the blade tip-to-cascade wall leakage vortex independent of rotational effects. We assessed the model's ability to predict the linear cascade flow-field through a study of normalised stream-wise velocity (U) across a plane 37 per cent blade chord

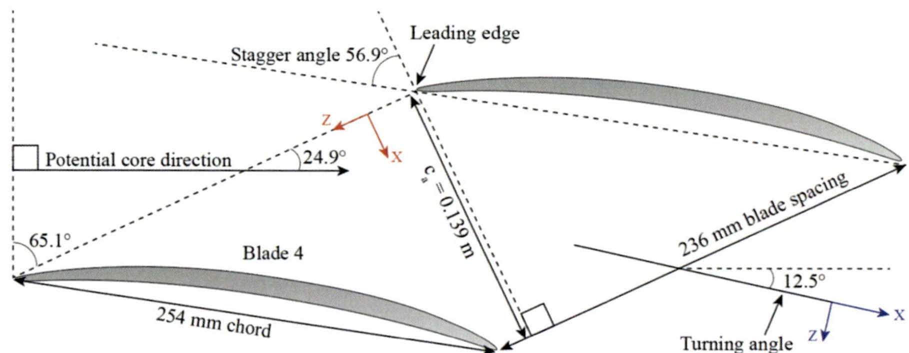

FIGURE 3.4. Muthanna and Devenport's (2004) linear compressor cascade geometry.

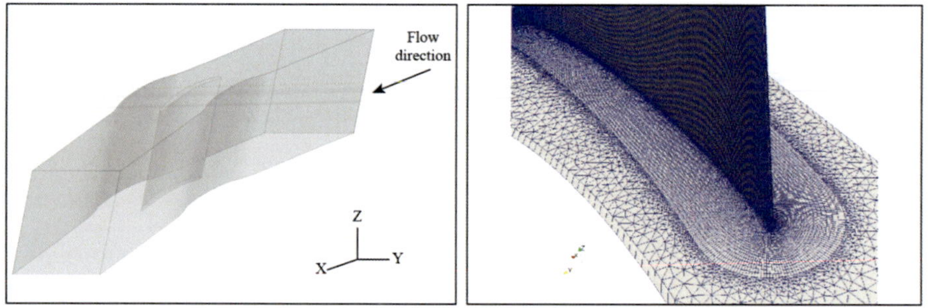

FIGURE 3.5. The computational mesh we developed and used in our numerical simulation of the linear compressor cascade, which Muthanna and Devenport (2004) experimentally studied.

Table 3.3. *Boundary conditions for the second code validation exercise. During the second code validation exercise we predicted the flow-field through a linear cascade of compressor blades.*

Boundary	Condition
Inflow	Mean velocity profile with random fluctuations
Outflow	Zero gradient
Pitch	Periodicity
Walls	No-slip

downstream of the blade trailing edge, Figure 3.6. In addition to Muthanna and Devenport's (2004) experimental results and the results of our large eddy simulation, we also present results from a scale-adaptive simulation (SAS). Egorov *et al.* (2010) and Menter and Egorov (2010) developed this scale-adaptive simulation modelling technique for predicting unsteady turbomachinery turbulent flow regions. To facilitate a direct comparison between our large eddy simulation and the scale-adaptive simulation, we conducted the scale-adaptive simulation using the same computational grid we used for our large eddy simulation.

An advantage of Egorov *et al.* (2010) and Menter and Egorov's (2010) scale-adaptive simulation modelling technique is that one may use it with relatively low density computational grids. Egorov *et al.* (2010) and Menter and Egorov (2010) ran their scale-adaptive simulations on computational grids with mesh densities closer to

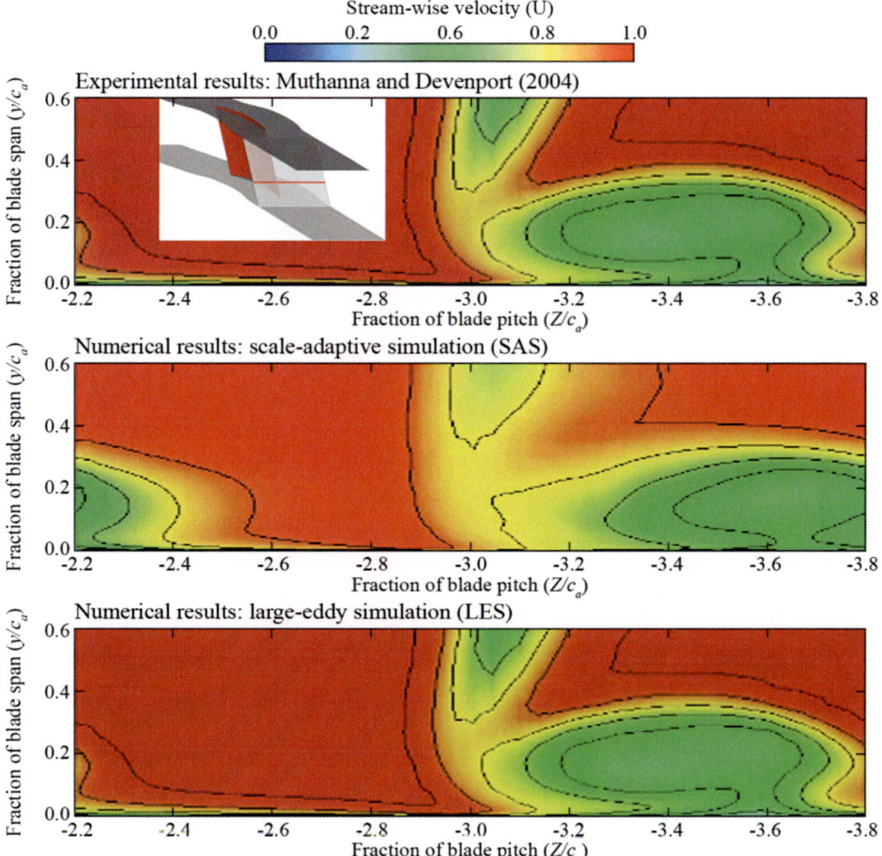

FIGURE 3.6. Normalised stream-wise velocity (U) at a reference plane 37 percent chord downstream of the studied linear cascade blade's trailing edge. We present experimentally measured data from Muthanna and Devenport (2004) and data that we predicted using two numerical simulations. The first is a scale-adaptive simulation and the second a large eddy simulation.

that required for an unsteady Reynolds-averaged Navier–Stokes simulation than a large eddy simulation. Therefore, we may assume the scale-adaptive simulation results presented in Figure 3.6 are less likely to be mesh dependent than our large eddy simulation results.

Consider Muthanna and Devenport's (2004) experimental results, Figure 3.6. There is a clear separation between the blade wake and the blade tip-to-cascade wall leakage vortex. Consider the large eddy simulation results. The large eddy simulation is able to predict accurately the blade tip-to-cascade wall leakage vortex location and intensity. Consider the scale-adaptive simulation results. In contrast to Muthanna and Devenport's experimental results (2004) and the large eddy simulation, the scale-adaptive simulation results are not self-consistent. The blade tip-to-cascade wall leakage vortex has merged with the wake. Borello *et al.* (2010) also studied this linear compressor cascade by using a hybrid large eddy / Reynolds-averaged Navier–Stokes simulation. They concluded that the large eddy simulation is better able to account for an increase in sub-grid scale viscosity that occurs with a relatively course mesh. Therefore, we may conclude that in the research reported in this chapter, the large eddy simulation is able to model the flow-field physics more accurately then the scale-adaptive simulation.

One may test our assertion that the large eddy simulation is better able to account for an increase in sub-grid scale viscosity that occurs with a relatively course mesh by studying normalised sub-grid scale viscosity contours. We define normalised sub-grid scale viscosity as the ratio between sub-grid scale viscosity (ν_{SGS}) and molecular viscosity (ν_m). A study of normalised sub-grid scale viscosity facilitates an evaluation of the relative proportion of sub-grid scale viscosity and molecular viscosity in our large eddy simulation.

Consider the normalised sub-grid scale viscosity contours close to the hub at five per cent, 50 per cent and close to the blade tip at 95 per cent blade span, Figure 3.7. At five per cent blade span the primary turbulent structures develop near each blade's suction surface. These structures extend from approximately mid-chord downstream of each blade trailing edge. This topology indicates that a hub vortex is present that forms as a consequence of boundary layer detachment over the blade chord's aft section. It is here that boundary layer thickness increases due to the influence of the adverse pressure gradient through the blade-to-blade passage. With the exception of the hub vortex, the normalised sub-grid scale viscosity levels are generally low.

At 50 per cent blade span the contours of normalised sub-grid scale viscosity indicate very low levels of normalised sub-grid scale viscosity in the blade-to-blade passage. This indicates that the large eddy simulation modelling technique is modelling well the flow-field physics. It is only in the wake region more than 30 per cent blade chord downstream of the blade trailing edge that we see an increase in normalised sub-grid scale viscosity. The previous analysis of stream-wise velocity utilised results across a plane 37 per cent blade chord downstream of the blade trailing edge, Figure 3.6. The observation that it is only in the wake region more than 30 per cent blade chord downstream of the blade trailing edge that we see an increase in normalised sub-grid scale viscosity may be significant. The increase in normalised

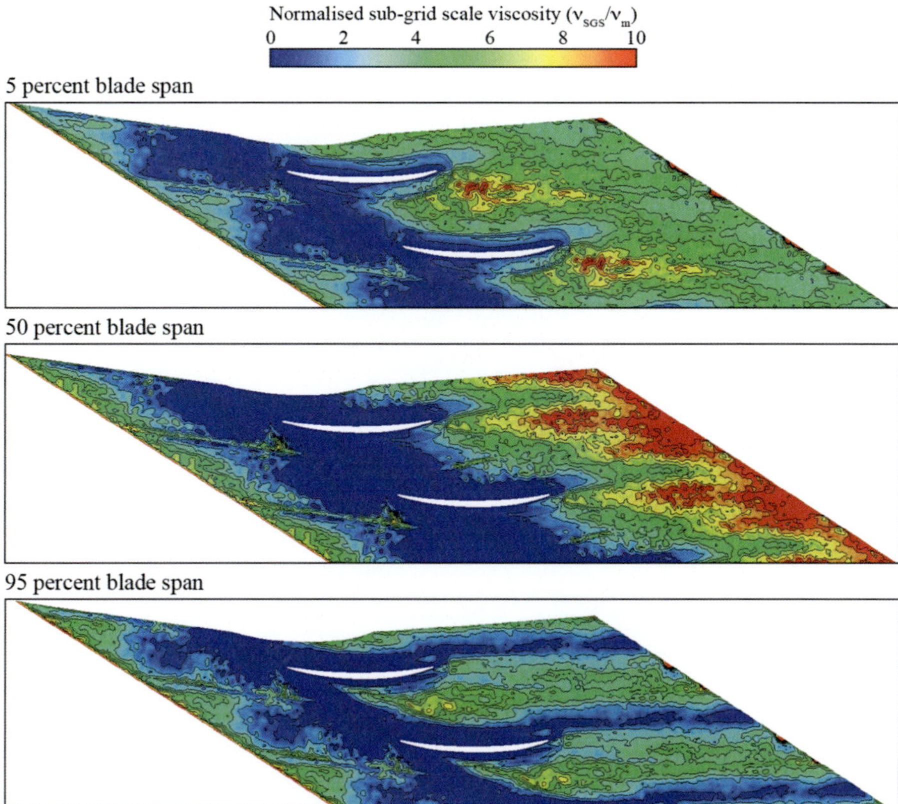

FIGURE 3.7. Normalised sub-grid scale viscosity, the ratio between sub-grid scale viscosity (ν_{SGS}) and molecular viscosity (ν_m). A study of normalised sub-grid scale viscosity facilitates an evaluation of the relative proportion of sub-grid scale viscosity and molecular viscosity in our large eddy simulation.

sub-grid scale viscosity may have degraded the accuracy of the large eddy simulations' predictions of stream-wise velocity.

At 95 per cent blade span the contours of normalised sub-grid scale viscosity indicate that normalised sub-grid scale viscosity starts to increase at approximately 25 per cent blade chord downstream of the blade leading edge. Normalised sub-grid scale viscosity continues to increase up to approximately 70 per cent blade chord downstream of the blade leading edge. This range of blade chord corresponds to the location of the blade tip-to-casing wall leakage vortex. The ability of the large eddy simulation to predict the location of the blade tip-to-casing wall leakage vortex indicates that the large eddy simulation is modelling the flow-field physics. However, we cannot say that it is modelling the flow-field physics well as the magnitude of normalised sub-grid scale viscosity is low. This indicates that the large eddy simulation is not able to model the relatively small scale structures that comprise the blade tip-to-casing wall leakage vortex. This indicates that we may conceptualise more accurately the large eddy simulation as a very large eddy simulation.

DESCRIPTION OF THE TEST CASE

Tunnel ventilation fans are typically between 1.6 and 2.8 metres in diameter, with the most common selection either a 1.8 or a 2.0 metres diameter fan running at 1,500 revolutions per minute. We conducted the reported research on an axial fan, named JFM 224, developed for application in railway tunnel and metro tunnel ventilation systems, Figure 3.8. The fan is 2.24 metres in diameter with 16 blades, a hub-to-tip ratio of 0.5 and a nominal design point Reynolds number (Re_{fan}) of 8.7×10^6, Table 3.4.

This studied fan is unusual as it has a blade tip speed of 175 metres per second in contrast to the generally accepted limit within the tunnel ventilation fan community of 156 metres per second. A 175 metres per second tip speed enables the JFM 224 fan to generate a total pressure rise of 2,500 Pa, Figure 3.9. Engineers generally consider a total pressure rise of 2,000 Pa to be a practical limit for a tunnel ventilation fan with 156 metres per second tip speed. A disadvantage of a 175 metres per second tip speed is that tip generated fan noise increases by approximately 10 dB compared to that of a 156 metres per second tip speed fan. As a consequence, tunnel ventilation design engineers have only utilised the JFM 224 fan in one tunnel ventilation system, the Delhi metro, where pressure developing capability was more important than fan noise. As such, the studied fan constitutes a practical limit in tunnel ventilation fan technology.

HT JFM 224 fan modelling and boundary conditions

We clustered the mesh toward solid boundaries, with the ratio of minimum grid spacing on solid walls to mid-span blade chord set as 7×10^{-4} on the blade tip, casing wall and blade surfaces. The mesh around the blade root illustrates the compromised aerodynamic profile imposed by mechanical constraints, Figure 3.10 right. The compromised aerodynamic profile in the blade root region results in separated flow being an inevitable aspect of the blade-to-blade flow-field. It provides an insight into the difficulty associated with modelling the flow-field physics, and in simply obtaining a converged solution.

The computational domain distribution results in 16 per cent of the elements upstream of the blade leading edge, 50 per cent of the elements within the blade passage and 34 per cent of the elements downstream of the blade trailing edge. In the span-wise direction, 55 nodes were clustered in the blade tip-to-casing gap. In total, the mesh comprises approximately 9 million hexahedral elements distributed with a block-structured topology, Table 3.5.

We clustered the mesh toward solid boundaries, with the ratio of minimum grid spacing on solid walls to mid-span blade chord set as 710^{-4} on the blade tip, casing wall and blade surfaces. The adopted grid refinement towards the solid surfaces sets the normalised wall distance y^+ value to approximately one on the first row of nodes in order to have the first cell within the viscous sub-layer.

FIGURE 3.8. A prototype 2.24 metre diameter unidirectional 1,500 rpm fan named JFM 224 and intended for application in railway tunnel and metropolitan metro ventilation systems. This prototype was designed for duel use, ventilating the tunnel system during routine operation and clearing 400°C gas and smoke for two hours in the event of a tunnel fire (Sheard *et al.*, 2009).

Table 3.4. *The studied fan, named JFM 224 constitutes one fan from a family of fans. We selected this example as a basis for numerical simulation of the flow-field as it is the highest tip speed fan currently applied into any railway tunnel or metropolitan metro ventilation system.*

Blade section	ARA-D	
Blade tip diameter D_t	2,240 mm	
Blade count	16	
Hub-to-tip ratio	0.5	
Design speed	1,500 rpm	
Tip clearance χ	6.5 per cent of the blade span	
	Hub	Tip
Chord c	143	92.5
Solidity σ	0.64	0.21
Pitch angle γ	48	24

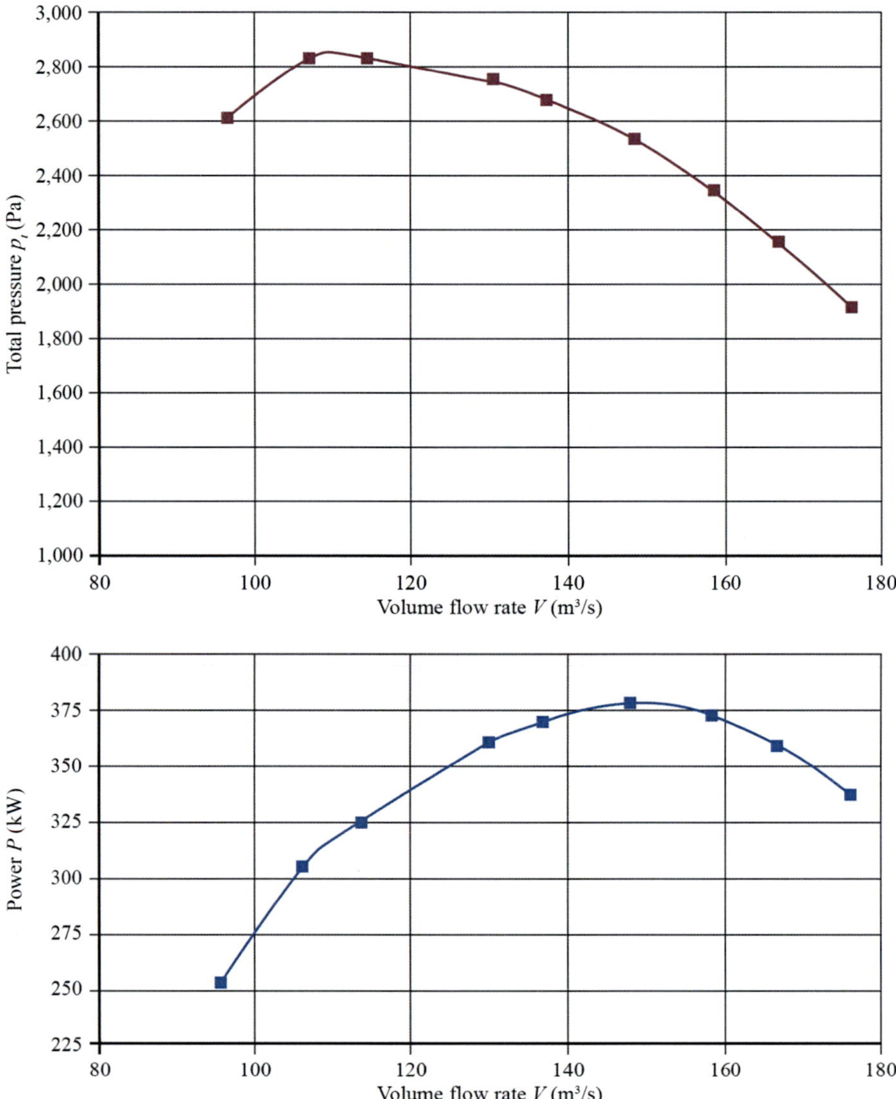

FIGURE 3.9. The fan JFM 224 performance characteristics, measured in accordance with ISO 5801 requirements (2007). The fans nominal design point is at a flow rate of 150 metres per second and a pressure rise of 2,500 pascals. We conducted all numerical simulations at this nominal design point.

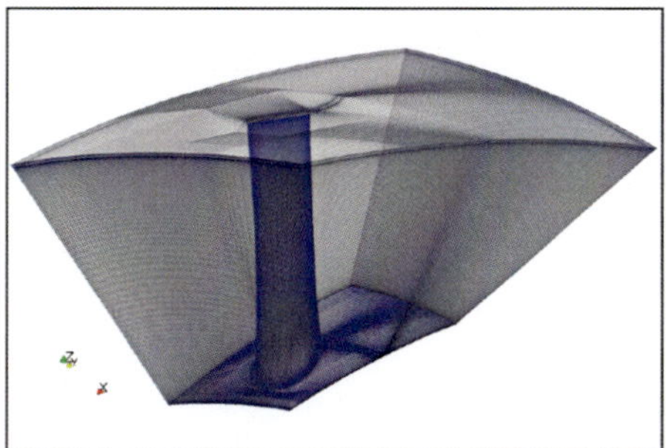
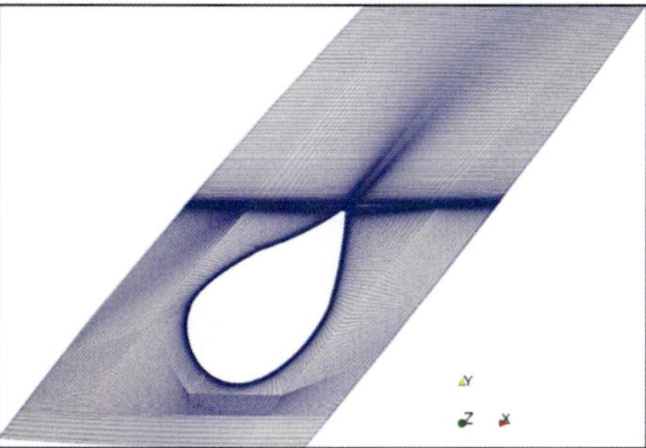

FIGURE 3.10. The fan JFM 224 computational domain extended a half chord upstream of the blade leading edge and one chord downstream from the blade trailing edge, left. The computational domain in the blade root region illustrates the extent to which the aerodynamic design has been compromised. The fan JFM 224 has a blade tip speed of 175 metres per second, over ten per cent higher than any other tunnel ventilation fan. In combination with a requirement to clear 400°C gas and smoke for two hours in the event of a tunnel fire this high tip speed results in the mechanical constraints that compromise the blade root aerodynamic design.

Table 3.5. *For the large eddy simulation of the studied fan we used a relatively fine computational grid. This computational grid was the finest grid we were able to use whilst still obtaining a solution in a reasonable time.*

	Rotor
Nodes	9,021,968
Cells	8,862,550
Tip gap nodes	60,753
Tip gap cells	52,000
Blade surface cells	62,450
Averaged cell aspect ratio	1.34

The inlet flow velocity into the studied fan was low, remaining below a Mach number of 0.2 and therefore we were able to treat the flow as incompressible. We defined boundary conditions following Sheard *et al.* (2009) and Borello *et al.*'s (2011) methods, which advantageously account for the effect of a sharp reduction of flow area into the fan inlet. They also account for the effect wakes generated by five struts upstream of the fan blades, Figure 3.8, that hold the static upstream spinner in place upstream of the fan hub. Flow periodicity, upstream and downstream of the blade, and Neumann outflow conditions complete the boundary conditions.

At the fan inlet we imposed a steady velocity profile, which became unsteady at the fan outlet as a consequence of the vortical flow-field structures that developed through the blade-to-blade passage. These vortical structures were relatively intense compared to those we would expect in a fan designed for aerospace application because of two factors: first, the aerodynamically compromised blade profile in the blade root region; and second, the relatively large blade tip-to-casing gap required to avoid the gap closing in the event of the fan having to operate at an elevated temperature due to a fire. Therefore, separated flow regions, secondary flow features and the blade tip-to-casing leakage vortex dominated the outlet flow-field.

Ideally, we would have imposed an unsteady velocity profile at the fan inlet. However, the fan was not designed for one application, but for a range of tunnel ventilation system applications. As each tunnel ventilation system is different, the resulting inlet conditions into the fan are different. In addition to the variability of inlet conditions in practical application, simply imposing an assumed unsteady velocity profile at the fan inlet can have a negative effect on the flow-field prediction's accuracy through the blade-to-blade passage. Klein *et al.* (2003) have studied this issue, proposing the digital filter based generation of unsteady inlet conditions using both direct numerical and large eddy simulations. Although promising, Klein *et al.*'s (2003) approach requires a prior numerical solution or experimental data set from a similar test case. As no similar test case was available for the studied fan, we accepted a steady velocity profile at the fan inlet.

To minimise computational effort, we undertook a preliminary unsteady Reynolds-averaged Navier–Stokes computation. The computational effort required for an unsteady Reynolds-averaged Navier–Stokes computation is less than that for a large eddy simulation and is a consequence of the need to simulate the flow-field over a period of time to correctly model the time varying fluctuation of velocity, pressure and flow-field features. Courant *et al.* (1967) observed that a factor that impacts on the required computational effort for an unsteady Reynolds-averaged Navier–Stokes solution is the length of time step that one selects for the computation. The Courant–Friedrichs–Lewy (CFL) number limits the time step, which is a necessary condition for convergence whilst solving numerically. If the CFL number remains below 1.0, the flow-field's time based resolution remains adequate for the mesh's spatial resolution that one uses for the computations. In essence, the finer the mesh, the smaller the maximum allowable time step in an unsteady Reynolds-averaged Navier–Stokes computation. For the preliminary unsteady Reynolds-averaged Navier–Stokes computation, we set the time step to 2×10^{-5} resulting in a maximum CFL number below 0.7.

Following convergence of the unsteady Reynolds-averaged Navier–Stokes computation, we used the converged solution as a starting point for our large eddy simulation to compute half a rotor revolution before collecting data over 1.5 flow through time (FTT) steps. The 1.5 flow through time steps corresponded to 3.3 blade-to-blade passages, with the variation in collected data from blade-to-blade passage typically less than one per cent.

RESULTS

Manufacturers typically use the studied fan in tunnel ventilation fan applications. We selected a 2,500 Pa and 150 m^3/s operating point as typical of the design point that the industry associates with the tunnel ventilation systems into which we applied the studied fan. We used the large eddy simulation results to compute pressure rise and absorbed power at the design point flow rate. Agreement between experimentally measured and predicted fan performance was typically no more than one per cent. We estimated the uncertainty in experimentally measured pressure to be three per cent. We estimated the uncertainty in experimentally measured power to be five percent. Therefore, we concluded that the agreement between experimentally measured and predicted parameters was good.

Unsteady flow topology

As we have previously said, the mesh used in the large eddy simulation is too course to result in a fully resolved large eddy simulation and is therefore it is better conceptualised as a very large eddy simulation. We may consider a computational mesh to return a fully resolved large eddy simulation if the resolved part of the turbulent kinetic energy (k_{res}) is a least 80 per cent of the total (k_{tot}). For a very large eddy simulation we can accept that the turbulent kinetic energy's resolved part is lower, but must still be at least 60 per cent of the total.

We may evaluate the ability of the computational grid to return a fully resolved large eddy simulation by considering the ratio of resolved to total kinetic energy, Figure 3.11. The ratio remains above 80 per cent for the majority of the flow-field at both blade mid-height and tip sections, only falling significantly below 60 per cent in

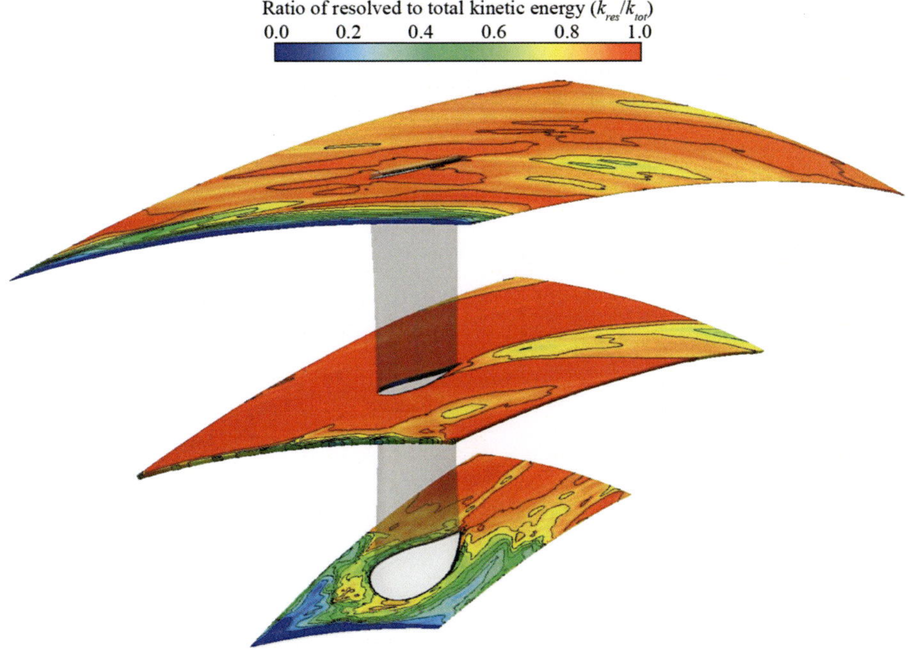

FIGURE 3.11. The ratio of resolved (k_{res}) to total (k_{tot}) turbulent kinetic energy for planes at 5, 50 and 95 per cent blade-span for fan JFM 224. For a fully resolved large eddy simulation, the resolved turbulent kinetic energy must be at least 80 per cent of the total turbulent kinetic energy. The contours of resolved to total turbulent kinetic energy ratio indicate that the resolved turbulent kinetic energy was above 80 per cent of the total over the majority of the computational domain.

the hub region. The steady velocity profile imposed at the fan inlet resulted in an artificially low inlet turbulence level that then remains low due to the interaction of secondary flow features in the hub region. Although at the limit of the available computational capability, we may conclude that the mesh lacks the necessary resolution to resolve the entire anisotropic turbulence spectrum. However, despite this caveat, the mesh is fine enough for a reliable simulation of the flow-field features over the majority of the computational domain. If we account for the simulation's limitations when interpreting the results, it can still provide a useful insight into the flow-field physics.

We may gain an insight into the flow-field topology using Dubief and Delcayre's (2000) vortex eduction criterion based on an instantaneous Q iso-surface of 200 computed at a non-dimensional time of 0.9, Figure 3.12. We have coloured the iso-surface with contours of relative velocity magnitude (U_{rel}) to facilitate interpreting vortical structures through the blade-to-blade passage. The majority of the vortical structures are generated in the blade wake region, by flow through the blade tip-to-casing gap and in the blade hub region. Relatively large scale vortical structures characterise the blade wake region, a feature particularly evident in the blade hub region. We may associate these large scale vortical structures in the blade hub region with the compromised aerodynamic design of the blade root sections. As previously mentioned, these aerodynamic compromises were necessary to produce a mechanically viable design. The large eddy simulation results illustrate the challenges inherent in applying computational methods to air movement fans.

Again as previously mentioned, the mesh that we used for the large eddy simulation was fine enough for a reliable simulation of the flow-field features over

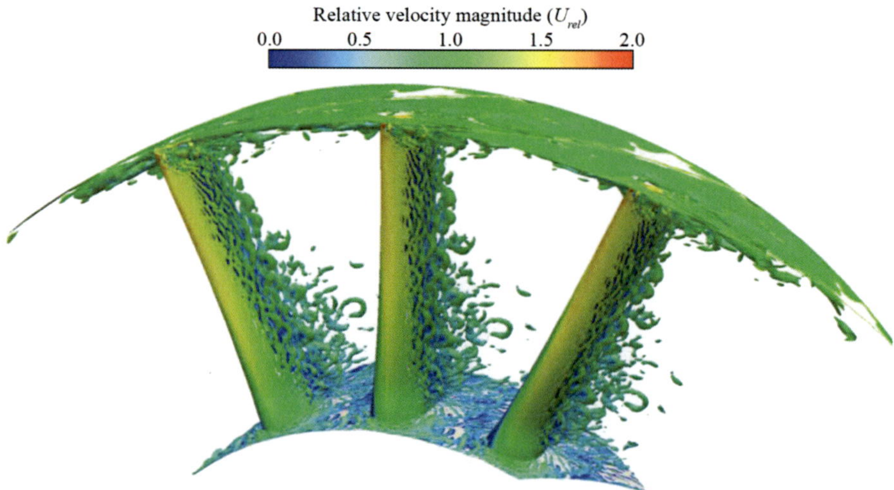

FIGURE 3.12. An analysis of the fan JFM 224 flow-field topology using Dubief and Delcayre's (2000) vortex eduction criterion based on an instantaneous Q iso-surface of 200. We coloured the iso-surface with contours of relative velocity magnitude to facilitate interpreting vortical structures through the blade-to-blade passage.

the majority of the computational domain. The flow-field in the hub region is the region of the computational domain that is least reliable, a factor that we must take into account when interpreting results. Consider the vorical structures that develop in the blade hub region, Figure 3.13. Again, we present the flow-field topology using Dubief and Delcayre's (2000) vortex education criterion based on an instantaneous Q iso-surface of 200. Flow-field streamlines complement the iso-surfaces to facilitate interpreting vortical structures. The horseshoe vortex is clearly visible, highlighted with a red square, as is a system of secondary vortices, highlighted with a red circle in Figure 3.13. We may regard the ability of the large eddy simulation to resolve the basic flow-field features classically associated with the hub region as positive. Although the computational mesh may not be fine enough for a fully resolved large eddy simulation, it is fine enough to resolve the basic flow-field features.

With our second test case we previously validated the large eddy simulations ability to resolve the basic flow-field features that we classically associate with the tip region. We used the code to predict the blade tip-to-casing leakage flow over the blades of a linear compressor cascade, Figure 3.7. In the blade tip region contours of normalised sub-grid scale viscosity indicated that normalised sub-grid scale viscosity starts to increase at approximately 25 per cent blade chord downstream of the blade leading edge. Normalised sub-grid scale viscosity then increased up to approximately 70 per cent blade chord downstream of the blade leading edge. This range of blade chord corresponds to the location of the blade tip-to-casing wall

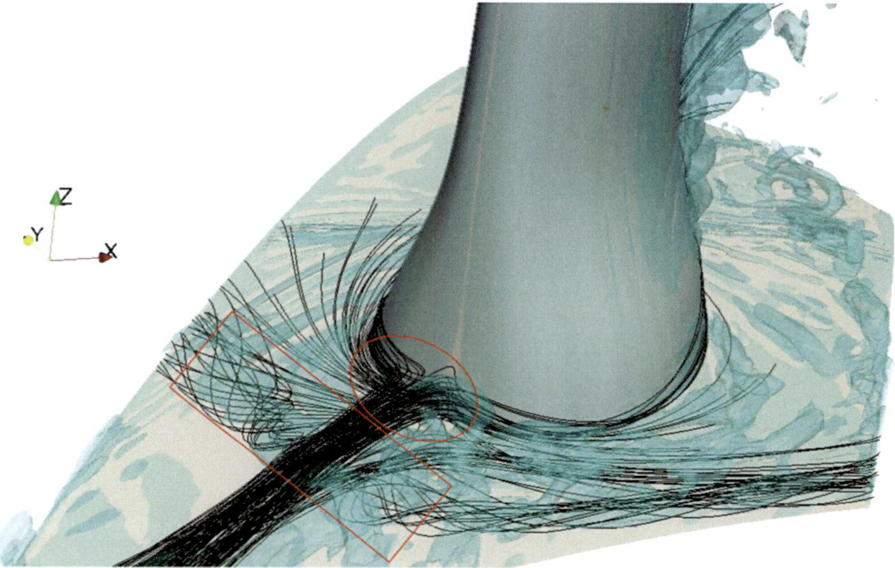

FIGURE 3.13. An analysis of the fan JFM 224 flow-field topology in the hub region using Dubief and Delcayre's (2000) vortex eduction criterion based on an instantaneous Q iso-surface of 200. Flow-field streamlines complement the iso-surfaces to facilitate interpreting vortical structures at the interface between the blade and hub to which it is attached.

leakage vortex. We concluded that the ability of the large eddy simulation to predict the location of the linear compressor cascade blade tip-to-casing wall leakage vortex indicates that it was modelling the flow-field physics. However, we were not modelling the flow-field physics well as the magnitude of normalised sub-grid scale viscosity was low. This indicated that the large eddy simulation was not able to capture the relatively small scale structures that comprise the blade tip-to-casing wall leakage vortex.

We may gain an insight into the effectiveness of the large eddy simulation when modelling the flow-field in the blade-tip region. Consider the studied fan's blade tip-to-casing flow-field topology, Figure 3.14. Once again we present the flow-field topology, this time in the blade tip region using Dubief and Delcayre's (2000) vortex eduction criterion based on an instantaneous Q iso-surface of 200. As with the previous analysis of the blade hub region, flow-field streamlines complement the iso-surfaces to facilitate interpreting vortical structures that develop through the blade tip-to-casing gap.

The blade tip-to-casing leakage vortex is clearly visible, with a clearly defined origin at approximately 25 per cent blade chord, Figure 3.14. This develops as it progresses through the blade-to-blade passage. It separates from the blade suction surface and exits the blade passage, separate and distinct from the blade wake. Bianchi *et al.* (2010, 2011) studied the development of the blade tip-to-casing leakage vortex in an industrial fan. The blade tip-to-casing leakage vortex development is qualitatively similar to that which Bianchi *et al.* (2010, 2011) observed.

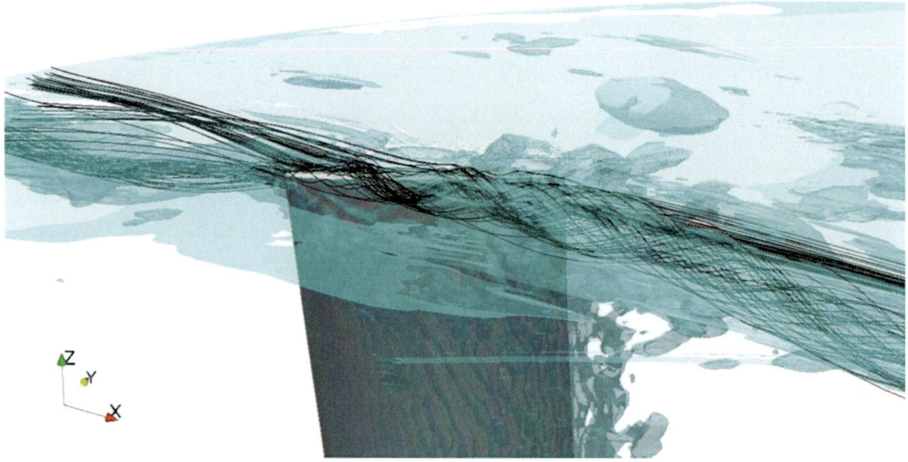

FIGURE 3.14. An analysis of the fan JFM 224 flow-field topology in the blade tip region using Dubief and Delcayre's (2000) vortex eduction criterion based on an instantaneous Q iso-surface of 200. Flow-field streamlines complement the iso-surfaces to facilitate interpreting vortical structures that develop through the blade tip-to-casing gap.

As previously mentioned, the studied fan has a blade tip speed of 175 m/s with the current state of the art in tunnel ventilation fans typically no more than 156 m/s. This results in an increase in tip generated fan noise. The blade tip-to-casing leakage vortex is significantly more intense than that of the fan which Bianchi *et al.* (2010, 2011) studied. Bianchi *et al.*'s (2010, 2011) primary conclusion was that the blade tip-to-casing leakage vortex was the flow-field feature that constituted the dominant far-field noise source. The known elevated noise level of the studied fan is self-consistent with the predicted intensity of the blade tip-to-casing leakage vortex. Therefore, we may conclude that the large eddy simulation is able to resolve the basic flow-field features that the industry classically associates with the blade tip region.

Flow field analysis

We have previously studied contours of normalised sub-grid scale viscosity when evaluating the quality of our large eddy simulation. We defined normalised sub-grid scale viscosity as the ratio between sub-grid scale viscosity (ν_{SGS}) and molecular viscosity (ν_m). A study of normalised sub-grid scale viscosity facilitates an evaluation of the relative proportion of sub-grid scale viscosity and molecular viscosity in our large eddy simulation. Low levels of normalised sub-grid scale viscosity indicate that the large eddy simulation modelling technique is modelling well the flow-field physics. As a qualitative guide we may accept values of normalised sub-grid scale viscosity below ten as indicative of a reasonable quality large eddy simulation. We present normalised sub-grid scale viscosity contours at 5, 50 and 95 per cent blade-span, as time-averaged values and at two time-steps, Figure 3.15.

At five per cent blade-span, normalised sub-grid scale viscosity values above ten characterise the majority of the flow-field. Normalised sub-grid scale viscosity at both time step one and two peak around the blade leading edge in regions we associated with the presence of large velocity gradients induced by the horseshoe vortex. In contrast, the normalised sub-grid scale viscosity is generally low around the blade trailing edge and over the pressure surface. The time averaged contours of normalised sub-grid scale viscosity do not include the very high peaks in normalised sub-grid scale viscosity we observe at both time step one and two. This indicates that it is necessary to study contours of normalised sub-grid scale viscosity at individual time steps.

At 50 per cent blade-span normalised sub-grid scale viscosity values below ten characterise the majority of the flow-field. Normalised sub-grid scale viscosity at both time step one and two peak in the blade wake, a region associated with the presence of large velocity gradients. Once again, the peak values of normalised sub-grid scale viscosity are more apparent at the two studied time steps than in the time averaged values. In contrast to the wake region, very low levels of normalised sub-grid scale viscosity characterise the inlet, blade suction and pressure surfaces. These low values are evident at both the individual time steps and in the time averaged values.

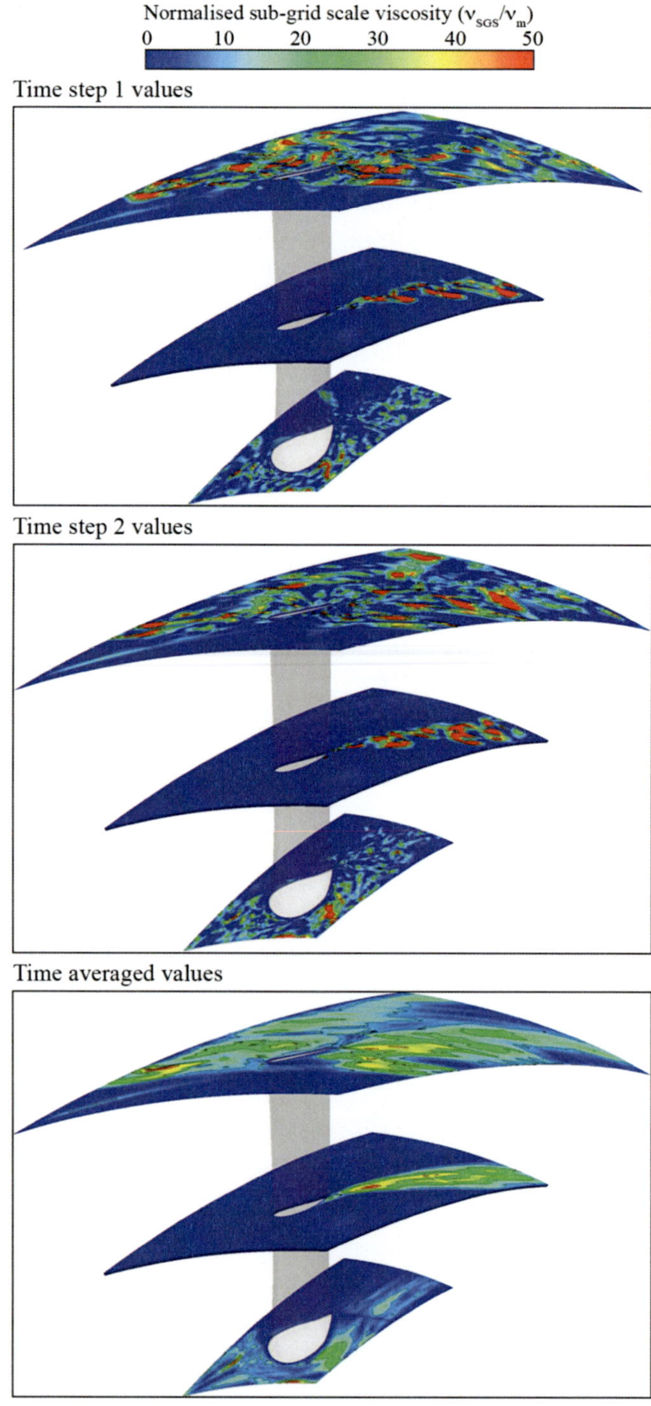

FIGURE 3. 15. Normalised sub-grid scale viscosity, the ratio between sub-grid scale viscosity (ν_{SGS}) and molecular viscosity (ν_m). We present normalised sub-grid scale viscosity contours at 5, 50 and 95 per cent blade-span, as time-averaged values and at two time-steps.

At 95 per cent blade-span normalised sub-grid scale viscosity values above ten again characterises the majority of the flow-field. Normalised sub-grid scale viscosity at time step one and two peak over the blade suction surface, pressure surface and in the wake region. In contrast to the normalised sub-grid scale viscosity at five and 50 per cent blade span, at 95 per cent blade span the time averaged values of normalised sub-grid scale viscosity are better able to resolve the location of the blade tip-to-casing leakage vortex. The normalised sub-grid scale viscosity contours at time step one and two indicate the presence of very high peak regions of normalised sub-grid scale viscosity. We may associate these peak values with swirling structures in both the blade tip-to-casing leakage flow and the wake and therefore may consider them as characteristic of the flow-field.

We may complement our analysis of normalised sub-grid scale viscosity with an analysis of relative velocity magnitude (U_{rel}). Contours of relative velocity magnitude facilitate identifying flow-field structures in the blade hub, tip and trailing edge regions. We present relative velocity magnitude contours at 5, 50 and 95 per cent blade-span, as time-averaged values and at two time-steps, Figure 3.16.

As with our analysis of normalised sub-grid scale viscosity, at five per cent, blade-span contours of relative velocity magnitude illustrate the presence of a separated flow region induced by the presence of the horseshoe vortex. At 50 per cent blade-span the contours of relative velocity magnitude are similar at both time step one, time step two and the time averaged values indicating that the flow-field is close to steady, with the exception of the wake region. At 95 per cent blade span the contours of relative velocity magnitude once again illustrate that the flow-field is highly unsteady.

We may extend our analysis of relative velocity magnitude by studying relative velocity magnitude contours at planes through the blade-to-blade passage, respectively 50 and 95 per cent blade chord downstream of the blade leading edge and 30 per cent downstream of the blade trailing edge, Figure 3.17. Figures 3.15 and 3.16 present the contours of relative velocity magnitude as instantaneous distributions at a time equivalent to time step one.

Consider the contours of relative velocity magnitude at the first plane 50 per cent blade chord downstream of the blade leading edge, Figure 3.17. There is a zone of high relative velocity magnitude in the blade tip region that is characteristic of the blade tip-to-casing leakage vortex that affects the blade tip flow-field across the majority of the blade passage. Consider the contours of relative velocity magnitude at the second plane 95 per cent blade chord downstream of the blade leading edge. The turbulent structures generated by the blade tip-to-casing flow are now primarily located mid-passage, with a low momentum region near the hub, indicating an area of recirculating flow. The blade design results in the 95 per cent blade chord plane located very close to the trailing edge at the hub and less close at the tip. Therefore, we may consider the area of recirculating flow in the hub region to be a wake feature as well as a hub feature. Consider the contours of relative velocity magnitude at the third plane 30 per cent blade chord downstream of the blade trailing edge. The turbulent structures that were evident at the second plane 95 per cent blade chord downstream of the blade leading edge remain evident, but at a generally reduced intensity.

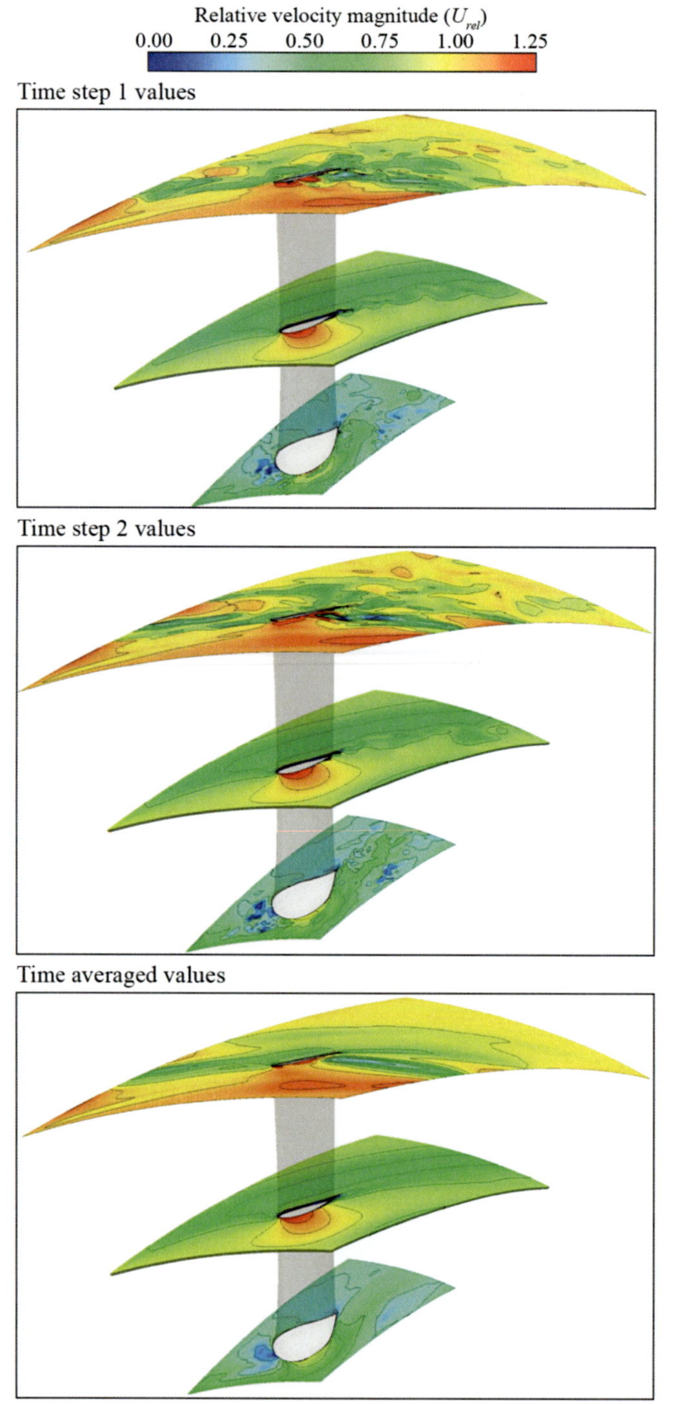

FIGURE 3.16. The relative velocity magnitude (U_{rel}) for planes at 5, 50 and 95 per cent blade-span for fan JFM 224. Contours of relative velocity magnitude facilitate identifying flow-field structures in the blade hub, tip and trailing edge regions.

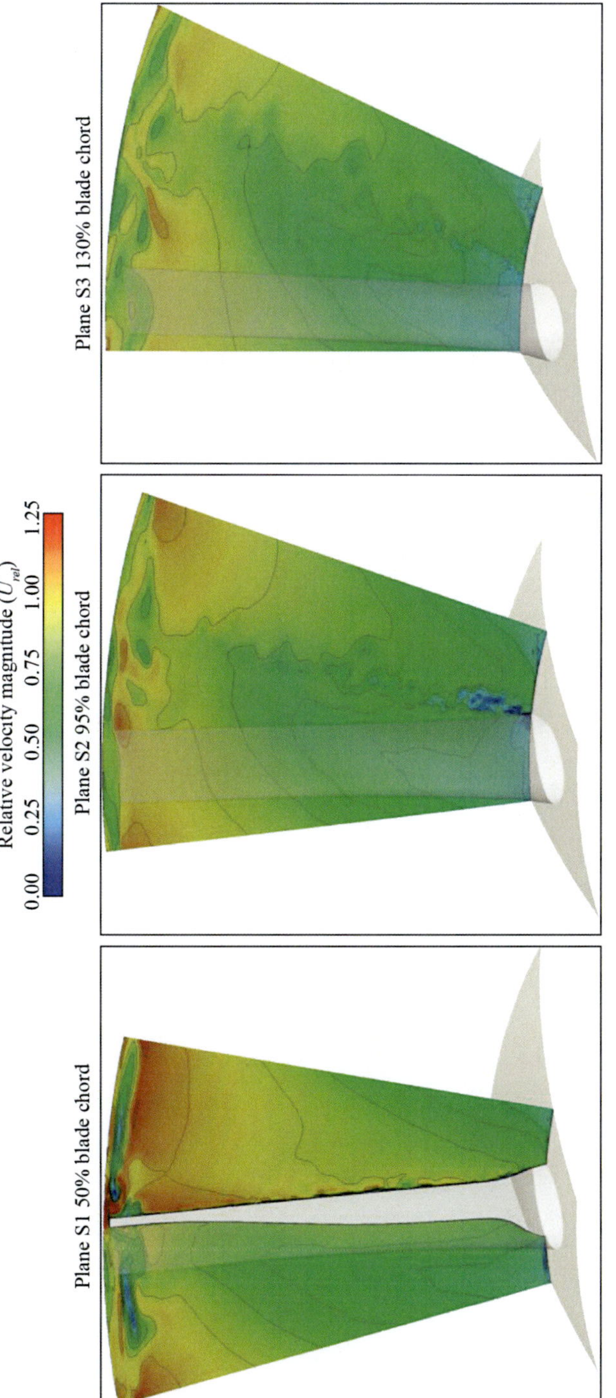

FIGURE 3.17. The relative velocity magnitude (U_{rel}) for the fan JFM 224 at planes S1, S2 and S3 respectively 50, 95 per cent blade-chord downstream of the blade leading edge and 30 per cent blade-chord downstream of the blade trailing edge. Contours of relative velocity magnitude facilitate identifying flow-field structures in the blade hub, tip and trailing edge regions.

We may complete our flow-field analysis by studying normalised total pressure contours, where we define normalised total pressure as total pressure divided by total pressure at the blade tip. We present contours of normalised total pressure at three planes, respectively 50 and 95 per cent blade chord downstream of the blade leading edge and 30 per cent downstream of the blade trailing edge, Figure 3.18. We present the contours of normalised total pressure as instantaneous distributions at a time equivalent to time step one and two in Figures 3.15 and 3.16, plus a third time step facilitates an interpretation of the evolution of flow-field features with time.

Consider the variation of flow-field features with both location within the blade passage and time, Figure 3.18. This form of presentation facilitates insight into the

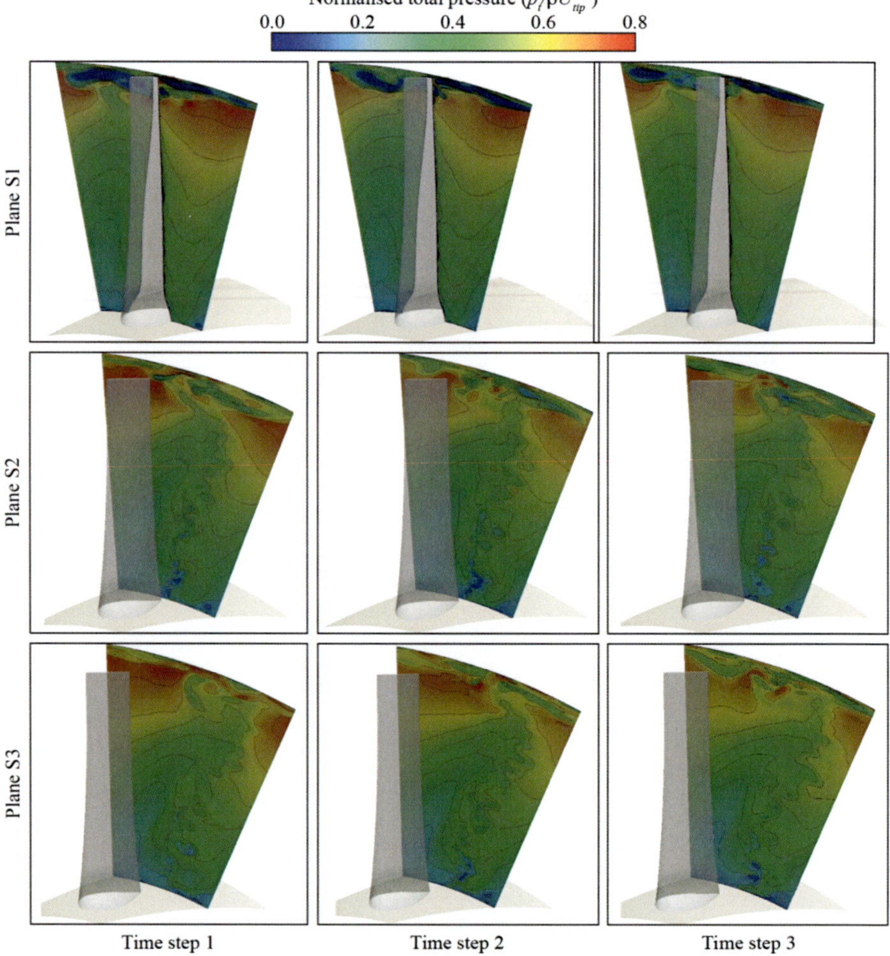

FIGURE 3.18. Normalised total pressure contours for the fan JFM 224. We present contours at planes S1, S2 and S3, respectively 50, 95 per cent blade-chord downstream of the blade leading edge and 30 per cent blade-chord downstream of the blade trailing edge. We also present the contours at three time steps to facilitate interpreting the evolution of flow-field features with time.

evolution of the flow-field through the blade-to-blade passage. The blade tip-to-casing leakage vortex characterises the blade tip-to-casing leakage flow. The flow-field in the blade tip-to-casing region is clearly different at each time step, indicating that the blade tip-to-casing leakage vortex is unsteady. When we study the blade-to-blade passage flow-field from one axial plane to the next it is evident that there is convection and mixing of the turbulent flow-field structures. This is indicative of the radial distribution of total pressure driven by the blades' forced vortex design.

CONCLUSIONS

In the research reported in this chapter we present a large eddy simulation (LES) modelling technique using the open-source solver OpenFOAM. The developed modelling technique reduces the required computational effort that the industry classically associates with a large eddy simulation, utilising a computational grid too course for a classical large eddy simulation. The modelling technique is still able to calculate directly very large scale turbulent structures, which we may characterise as a very large eddy simulation.

We accounted for the effect of the smaller unresolved turbulence scale using an eddy viscosity turbulence model. Therefore, we chose to couple the computational grid with a bounded sub-grid scale (SGS) viscosity model to achieve closure. This modelling approach relied on a one-equation model. We calculated sub-grid scale turbulent kinetic energy by solving a differential transport equation. The use of this closure technique was a remedial strategy that partially recovered the turbulence spectrum when using a low resolution computational grid that alone was unable to resolve the turbulence spectrum.

The resultant very large eddy modelling technique utilised a second order accurate central difference scheme (CDS) coupled with both a total variation diminishing (TVD) scheme and second order implicit time integration scheme. This combination controlled instabilities, and is what we refer to as a local dynamic sub-grid scale large eddy simulation with bounded sub-grid scale viscosity.

We evaluated the very large eddy modelling technique, studying flow-field topology using a vortex eduction criterion based on the Q factor. Despite the limitations of a very large eddy modelling technique, in combination with the chosen eddy viscosity turbulence model we were able to predict well overall fan performance, flow-field feature location and intensity. We concluded that the modelling technique was also capable of modelling separation and secondary flow features in the blade-to-blade flow-field.

The developed modelling technique was better able to model the flow-field physics than a Reynolds-averaged Navier–Stokes approach. Therefore, the predicted steady and unsteady performance parameters serve as a reference data set for future experimental and numerical analysis. We anticipate that future analysis will be required to further refine and develop computational methods for application as part of a tunnel ventilation fan design methodology.

REFERENCES

ISO 5801:2007 (2007), *Industrial Fans: Performance Testing Using Standardised Airways.*

Bianchi, S., Corsini, A. and Sheard, A.G. (2010), 'Sound Generation Mechanism in Low Speed Axial Fans', *Journal of the Acoustical Society of America*, vol. 127, pp. 1837–1856.

Bianchi, S., Corsini, A., Rispoli, F. and Sheard, A.G. (2011), 'Far-field Radiation of Tip Aerodynamic Sound Sources in Axial Fans Fitted with Passive Noise Control Features', *Transactions of the ASME, Journal of Vibration & Acoustics*, vol. 133(5), paper no. 051001, pp. 1–11.

Borello, D., Hanjalić, K. and Rispoli, F. (2005), 'Prediction of Cascade Flows with Innovative Second-moment Closures', *Transactions of the ASME, Journal of Fluids Engineering*, vol. 127(6), pp. 1059–1070.

Borello, D., Hanjalić, K. and Rispoli, F. (2007), 'Computation of Tip-leakage Flow in a Linear Compressor Cascade with a Second-moment Turbulence Closure', *International Journal of Heat and Fluid Flow*, vol. 28(4), pp. 587–601.

Borello, D., Delibra, G., Hanjalić, K. and Rispoli, F. (2009), 'Large-eddy Simulations of Tip Leakage and Secondary Flows in an Axial Compressor Cascade Using a Near-wall Turbulence Model', *Proceedings of the IMechE Part A, Journal of Power and Energy*, vol. 223(6), pp. 645–655.

Borello, D., Delibra, G., Hanjalić, K. and Rispoli, F. (2010), 'Hybrid LES/RANS Study of Turbulent Flow in a Linear Compressor Cascade with Moving Casing', *Proceedings of the 55th American Society of Mechanical Engineers Turbine and Aeroengine Congress*, Glasgow, UK, 14–18 June, paper no. GT2010-23755.

Borello, D., Corsini, A., Minotti, S., Rispoli, F. and Sheard, A.G. (2011), 'U-RANS of a Large Industrial Fan Under Design and Off-Design Operations', *Proceedings of the 9th European Conference on Turbomachinery, Fluid Dynamics & Thermodynamics*, Istanbul, Turkey, 21–25 March, pp. 1665–1676.

Carvel, C. (1999), 'The Effects of Ventilation on Fires in Tunnels', *Proceedings of the International Tunnel Fire and Safety Conference*, Rotterdam, The Netherlands, 2–3 December.

Corsini, A. and Rispoli, F. (2005), 'Flow Analyses in a High-pressure Axial Ventilation Fan with a Non-linear Eddy-viscosity Closure', *International Journal of Heat Fluid Flow*, vol. 26(3), pp. 349–361.

Corsini, A., Delibra, G. and Sheard, A.G. (2013), 'A Critical Review of Computational Methods and Their Application in Industrial Fan Design', *International Scholarly Research Network, Mechanical Engineering*, vol. 2013, article ID 625175, pp. 1–20.

Courant, R., Friedrichs, K. and Lewy, H. (1967), 'On the Partial Difference Equations of Mathematical Physics', *IBM Journal of Research and Development*, vol. 11(2), pp. 215–234.

Daly, B.B. (1985), *Woods Practical Guide to Fan Engineering*, Woods of Colchester Ltd, Colchester, UK.

Davidson, L. (1997), 'Large Eddy Simulation: A Dynamic One-equation Sub-grid Model for Three-dimensional Recirculating Flows', *Proceedings of the 11th International Symposium on Turbulent Shear Flow*, Grenoble, France, 8–10 September, vol. 3, pp. 26.1–26.6.

Delibra, G., Borello, D., Hanjalić, K. and Rispoli, F. (2009), 'U-RANS of Flow and Endwall Heat Transfer in a Pinned Passage Relevant to Gas-turbine Blade Cooling', *International Journal of Heat and Fluid Flow*, vol. 30(3), pp. 549–560.

Delibra, G., Borello, D., Hanjalić, K. and Rispoli, F. (2010a), 'A LES Insight into Convective Mechanism of Heat Transfer in a Wall-bounded Pin Matrix', *Proceedings of the 14th International Heat Transfer Conference (IHTC)*, Washington, DC, USA, 8–13 August, pp. 807–815.

Delibra, G., Hanjalić, K., Borello, D. and Rispoli, F. (2010b), 'Vortex Structures and Heat Transfer in a Wall-bounded Pin Matrix: LES with a RANS Wall-treatment', *International Journal of Heat and Fluid Flow*, vol. 31(5), pp. 740–753.

Dubief, Y. and Delcayre, F. (2000), 'On Coherent-vortex Identification in Turbulence', *Journal of Turbulence*, vol. 1, p. N11.

Egorov, Y., Menter, F.R., Lechner, R. and Cokljat, D. (2010), 'The Scale-adaptive Simulation Method for Unsteady Turbulent Flow Predictions. Part 2: Application to Complex Flows', *Flow, Turbulence and Combustion*, vol. 85(1), pp. 139–165.

Germano, M., Piomelli, U., Moin, P. and Cabot, H. (1991), 'A Dynamic Sub-grid-scale Eddy Viscosity Model', *Physics of Fluids*, vol. 3, pp. 1760–1765.

Grundestam, O., Wallin, S. and Johansson, A.V. (2008), 'Direct Numerical Simulations of Rotating Turbulent Channel Flow', *Journal of Fluid Mechanics*, vol. 598, pp. 177–199.

Hah, C. and Shin, H.-W. (2012), 'Study of Near-stall Flow Behavior in a Modern Transonic Fan with Compound Sweep', *Transactions of the ASME, Journal of Fluids Engineering*, vol. 134(7), paper no. 071101, pp. 1–7.

Hedefalk, J., Wahlstrom, B. and Rohlen, P. (1998), 'Lessons from the Baku Subway Fire', *Proceedings of the 3rd International Conference on Safety in Road and Rail Tunnels*, Nice, France, 9–11 March, pp. 15–28.

Horlock, J.H. and Denton, J.D. (2005), 'A Review of Some Early Design Practice using Computational Fluid Dynamics and a Current Perspective', *Transactions of the ASME, Journal of Turbomachinery*, vol. 127(1), pp. 5–13.

Jagannathan, S., Schwänen, M. and Duggleby, A. (2012), 'Low Pressure Turbine Relaminarization Bubble Characterization using Massively-parallel Large Eddy Simulations', *Transactions of the ASME, Journal of Fluids Engineering*, vol. 134(2), paper no. 021102, pp. 1–13.

Jasak, H. (2010), 'OpenFOAM: A Year in Review', *Proceedings of the 5th OpenFOAM Workshop*, Gothenburg, Sweden, 21–24 June.

Klein, M., Sadiki, A. and Janicka, J. (2003), 'A Digital Filter Based Generation of Inflow Data for Spatially Developing Direct Numerical or Large Eddy Simulations', *Journal of Computational Physics*, vol. 186(2), pp. 652–665.

Labois, M. and Lakehal, D. (2011), 'Very-large Eddy Simulation (V-LES) of the Flow Across a Tube Bundle', *Nuclear Engineering and Design*, vol. 241, pp. 2075–2085.

Lacroix, D. (1997), 'New French Recommendations for Fire Ventilation in Road Tunnels', *Proceedings of the 9th International Conference on Aerodynamics and Ventilation of Vehicle Tunnels: Developments for the 21st Century*, Aosta Valley, Italy, 6–8 October, pp. 103–124.

Lee, K.Y., Choi, Y.S., Kim, Y.L. and Yun, J.H. (2008), 'Design of Axial Fan using Inverse Design Method', *Journal of Mechanical Science and Technology*, vol. 22, pp. 1883–1888.

Luo, X., Ji, B., Peng, X., Xu, H., Nishi, M. and Ji, B. (2012), 'Numerical Simulation of Cavity Shedding from a Three-dimensional Twisted Hydrofoil and Induced Pressure Fluctuation by Large-eddy Simulation', *Transactions of the ASME, Journal of Fluids Engineering*, vol. 134(4), paper no. 041202, pp. 1–10.

Menter, F.R. and Egorov, Y. (2010), 'The Scale-adaptive Simulation Method for Unsteady Turbulent Flow Predictions. Part 1: Theory and Model Description', *Flow, Turbulence and Combustion*, vol. 85(1), pp. 113–138.

Muthanna, C. and Devenport, W.J. (2004), 'Wake of a Compressor Cascade with Tip Gap, Part 1: Mean Flow and Turbulence Structure', *AIAA Journal*, vol. 42(11), pp. 2320–2331.

Pucher, K. (1999), 'Fire in the Tauern Tunnel', *Proceedings of the International Tunnel Fire & Safety Conference*, Rotterdam, The Netherlands, 2–3 December.

Saad, Y. (1996), *Iterative Methods for Sparse Linear Systems*, Society for Industrial and Applied Mathematics, Philadelphia, PA, USA.

Schneider, H., von Terzi, D. and Bauer, H.-J. (2010), 'Large-eddy Simulations of Trailing-edge Cutback Film Cooling at Low Blowing Ratio', *International Journal of Heat and Fluid Flow*, vol. 31, pp. 767–777.

Sheard, A.G. (Ed.) (2012), *Tip Clearance Measurement in Aero and Industrial Turbomachinery*, Sigel Press, Cambridge, UK.

Sheard, A.G. (Ed.) (2014), *Passive Noise Control in Industrial Fans*, Sigel Press, Cambridge, UK.

Sheard, A.G. and Corsini, A. (2012), 'The Mechanical Impact of Aerodynamic Stall on Tunnel Ventilation Fans', *International Journal of Rotating Machinery*, vol. 2012, paper no. 402763, pp. 1–12.

Sheard, A.G. and Jones, N.M. (2008), 'Approval of High-temperature Emergency Tunnel-ventilation Fans: The Impact of ISO 21927-3', *Proceedings of the ITA–AITES World Tunnel Congress and 34th General Assembly*, Agra, India, 19–25 September, pp. 1817–1826.

Sheard, A.G. and Jones, N.M. (2012), 'Powered Smoke and Heat Exhaust Ventilators: The Impact of EN 12101-3 and ISO 21927-3', *Tunnelling and Underground Space Technology*, vol. 28, pp. 174–182.

Sheard, A.G., Corsini. A., Minotti, S. and Sciulli, F. (2009), 'The Role of Computational Methods in the Development of an Aero-acoustic Design Methodology: Application in a Family of Large Industrial Fans', *Proceedings of the 14th International Conference on Modelling Fluid Flow Technologies*, Budapest, Hungary, 9–12 September, pp. 71–99.

Shipp, M. (1997), 'Investigating the Channel Tunnel Fire', *Proceedings of the Seminar on Reducing Fire Risk in Trains*, Heathrow, England, 17 November, pp. 19–29.

Sweby, P.K. (1984), 'High Resolution Schemes using Flux Limiters for Hyperbolic Conservation Laws', *SIAM Journal on Numerical Analysis*, vol. 21(5), pp. 995–1011.

Vad, J. (2001), 'Incorporation of Forward Blade Sweep in the Non-free Vortex Design Method of Axial Flow Turbomachinery Rotors', *Periodica Polytechnica, Mechanical Engineering*, vol. 45(2), pp. 217–237.

Vad, J., Bencze, F., Corsini, A. and Rispoli, F. (2001), 'Non-free Vortex Flow Effects in an Axial Flow Rotor', *Periodica Polytechnica, Mechanical Engineering*, vol. 45(2), pp. 201–216.

Wallis, R.A. (1961), *Axial Flow Fans: Design and Practice*, George Newnes Ltd, London, UK.

Chapter 4

Large-eddy Simulation of the Aerodynamic and Aeroacoustic Performance of a Ventilation Fan

S. Bianchi, D. Borello, A. Corsini, F. Rispoli and A.G. Sheard

ABSTRACT

In this chapter we compute unsteady flow-field parameters and then use them as input to Ffowcs Williams and Hawkings' acoustic model. This enabled us to develop a tool that predicted both fan broadband noise and spectrum. We computed the studied fan's flow-field using the open-source code OpenFOAM. The adopted modelling technique enables OpenFOAM to calculate directly very large scale turbulent structures, which we may characterise as a very large eddy simulation. We account for the effect of the smaller unresolved turbulence scale using an eddy viscosity turbulence model.

Using the predicted unsteady flow-field parameters as input, we were able to use the Ffowcs Williams and Hawkings' acoustic model to resolve the acoustic spectrum up to the fifth harmonic of blade passing frequency. We compared the prediction of broadband noise and spectrum with experimental measurements, and found that they agreed within the uncertainty of the measurements. The agreement between predicted and measured broadband noise and spectrum gives confidence in the accuracy with which the very large eddy simulation modelling technique models the evolution of blade-to-blade flow-field features.

The accuracy of the prediction was restricted by the limitations inherent in the Ffowcs Williams and Hawkings' acoustic model that neglects duct modes. Consequently, the acoustic model did not account for the modal amplification of dipole-like noise sources. Dipole-like noise sources characterise the inter-tonal and broadband noise when one fits a fan in a ducted system. Despite the acoustic model's limitations, the agreement between predicted and measured broadband noise and spectrum was good enough to provide designers with an insight into the acoustic consequences of a change in fan geometry. Therefore, it is a useful tool, as it facilitates the extension of an aerodynamic and mechanical design optimisation process into one that also considers the acoustic consequences of any change in fan geometry.

This chapter is a revised and extended version of Bianchi, S., Borello, D., Corsini, A., Rispoli, F. and Sheard, A.G. (2013), 'Large-eddy Simulation of the Aerodynamic and Aeroacoustic Performance of a Ventilation Fan', *Advances in Acoustics and Vibration*, vol. 2013, article ID 876973, 10 pages.

NOMENCLATURE

Latin letters

c	[mm]	Chord
CFL		Courant-Friedrichs-Lewy Number or Courant number; $CFL = U \times \Delta t / \Delta x$
D_{ht}		Hub to tip diameter ratio
D_t	[mm]	Blade tip diameter
DF		Diffusion factor
f	[kHz]	Frequency
Hn		Normalised helilcity
Lw	[dB]	Sound power level
Δpt	[Pa]	Total pressure rise
p	[Pa]	Static pressure
$\nabla^2 p$		Pressure laplacian
V	[m^3/s]	Volume flow rate
v_c		Convection velocity
w^+		Normalised relative velocity magnitude
z		Vortex vector of leakage flow

Greek letters

α_b	[deg]	Blade angle
γ	[deg]	Pitch angle
ν_{SGS}		Sub-grid scale viscosity
σ	[-]	Solidity
ξ		Normalised enstrophy
ω	[rpm]	Rotational speed

Acronyms, symbols, subscript and superscripts

CFD	Computational fluid dynamics
DF	Diffusion factor
FEA	Finite element analysis
FFT	Fast Fourier transform
FTT	Flow through time
ILU	Incomplete lower and upper
PISO	Pressure implicit with splitting of operators
RANS	Reynolds-averaged Navier–Stokes
SGS	Sub-grid scale
TVD	Total volume diminishing
U-RANS	Unsteady Reynolds-averaged Navier–Stokes
(V)LES	(Very) large eddy simulation

INTRODUCTION

Tunnel ventilation systems have historically incorporated fans that are sold on first cost. An intensely competitive market place has resulted in a tunnel ventilation fan community which is manufacturing led. Specific competence in manufacturing minimises production costs which in turn constitutes a form of competitive advantage. Fan efficiency has typically been a secondary consideration. Borello *et al.* (2013) reported that the recent European and forthcoming US regulation of fan efficiency has resulted in tunnel ventilation fan designers changing their focus. It is now necessary for them to develop tunnel ventilation fans that at least meet current regulatory efficiency requirements and ideally, forthcoming requirements as well.

Although the tunnel ventilation fan community has historically been focused first on cost, the marketplace into which it sells has expectations regarding the acceptability of both fan broadband noise and spectrum. Some market segments are more sensitive than others to fan noise. However, the trend is towards a requirement for lower broadband noise, with some tunnel ventilation fan specifications going further and defining acceptable noise levels in each octave band of the spectrum. Therefore, any design procedure must aim to maintain at least previously achieved acoustic performance, and preferably reduce it.

We classically associate tunnel ventilation fans with relatively high broadband noise levels and tonal peaks relative to other industrial fans with similar operating points. This is a consequence of the need for a larger blade tip-to-casing clearance. This larger blade tip-to-casing clearance is a consequence of current regulatory requirements that require tunnel ventilation systems to provide those trapped with smoke-free escape routes in the event of a fire, (ANSI/ASHRAE Standard 149-2000, 2009; EN 12101-3, 2002). However, fires in the Mont Blanc and Tauern Tunnels (Lacroix, 1997; Pucher, 1999), the Channel Tunnel (Shipp, 1997) and the Baku underground system (Hedefalk *et al.*, 1998) have demonstrated potentially devastating effects of such underground fires.

In the event of a fire, a typical tunnel safety-management system relies on an emergency ventilation system to control and disperse smoke and heat to maintain a smoke-free escape route for people (Carvel, 1999). Therefore, tunnel ventilation fans must be capable of operating at elevated temperatures to remove hot gas and smoke from the tunnel in the event of a fire. It is this requirement to operate at elevated temperature during a tunnel fire that result in the need for larger blade tip-to-casing clearance at ambient temperatures. This ensures clearance does not fall to zero at an elevated temperature (Sheard and Jones, 2005, 2006, 2008, 2012).

The above requirements imply a level of complexity in the design process that is beyond the historic norm within the tunnel ventilation community. Empirical design methods are no longer appropriate, with tunnel ventilation fan designers increasingly moving from empirical fan to industrial turbomachinery design methodologies. Wallis (1961) and Daly (1985) exemplify the traditional empirical approach to tunnel ventilation fan design and describe a trial-and-error approach that relies on the designer's experience. More recently, tunnel ventilation fan designers have started to utilise computational fluid dynamics analyses at the beginning of the design process (Vad, 2001; Vad *et al.*, 2001). Lee *et al.* (2008) recently applied an inverse approach to the design of cooling fans for electronic appliances. Sheard *et al.* (2009) presented a tunnel ventilation fan design methodology utilising computational methods previously developed by the aerospace fan and compressor community. They based their design methodologies on Horlock and Denton's (2005) design practices which are reliant on computational fluid dynamics to develop appropriate three dimensional blade sections.

Sheard *et al.*'s (2009) design methodology combines finite element analysis (FEA) and computational fluid dynamics (CFD) to facilitate creating virtual prototypes. This approach is attractive as one may use it to develop both more efficient and also more highly aerodynamically loaded blade designs. This combination of increased efficiency and increased blade loading offers a duel possibility. New tunnel ventilation fan designs can both meet efficiency targets in current and forthcoming legislation, and be physically smaller. Thus, the new fan designs have a lower manufacturing cost. A disadvantage of this approach is that more highly loaded designs invariably have higher acoustic emissions and therefore fan noise.

Virtual prototyping techniques are not typically employed either by the tunnel ventilation fan community of the wider air movement fan community. However, they are potentially useful as they facilitate the cross-functional evaluation of competing objectives, enabling a designer to evaluate 'downstream design issues' during initial design stages (Jasnoch *et al.*, 1994; Pratt, 1994). Therefore, the virtual prototyping process both reduces the need to build physical prototypes and facilitates the early identification of design problems. Additionally, the ability to evaluate competing objectives is potentially advantageous to tunnel ventilation fan designers. Sheard *et al.* (2009) proposed that a design methodology combining finite element analysis and computational fluid dynamics could be extended. The extension included an acoustic characterisation of a virtual prototype in addition to a structural and aerodynamic characterisation.

Extending a design methodology to include an acoustic characterisation of a virtual prototype required a method for predicting fan broadband noise and spectrum. The majority of the published research on fan noise has focused on high speed fans intended for aerospace application (Cumpsty, 1977). The aerospace community has historically regarded fan noise as a secondary consideration, except when operating at low power (Miles, 2006). Traditionally, the tunnel ventilation fan community has undertaken the characterisation of fan noise using purely empirical techniques. Despite this caveat, extending a design methodology based on virtual prototyping requires a method for predicting the broadband noise and spectrum of a virtual fan. We cannot calculate the unsteady parameters needed to predict broadband noise and spectrum using traditional empirical techniques. The Reynolds-averaged Navier–Stokes based computational methods favoured by the tunnel ventilation fan community are steady-state; they do not predict unsteady parameters. Therefore, we need to use an unsteady computational method.

Borello *et al.* (2013) observed that when researchers wish to predict the unsteady features of a flow-field, an unsteady Reynolds-averaged Navier–Stokes modelling technique does overcome some limitations. Using unsteady Reynolds-averaged Navier–Stokes modelling techniques is well established (Borello *et al.*, 2005, 2007, 2009; Corsini and Rispoli, 2005; Delibra *et al.*, 2009, 2010; Bianchi *et al.*, 2011). However, Hah and Shin (2012) concluded that a large eddy simulation is necessary if one is to predict unsteady parameters. The industry does not widely use large eddy simulation modelling techniques because of the formidable computational effort required to resolve the turbulent spectrum down to the inertial sub-range (Schneider *et al.*, 2010).

When researchers do choose a large eddy simulation, they may reduce the computational effort by studying only a component of the flow-field. Luo *et al.* (2012) reported an example of this approach. They used a large eddy simulation to study cavity shedding from a three dimensional twisted hydrofoil. An alternative to studying only one component of the flow-field is to study only boundary layer development through the entire flow-field. Jagannathan *et al.* (2012) adopted this approach, studying boundary layer development during both transition and relaminarisation.

Davidson (1997) studied the alternatives to modelling only one part of the flow-field. He proposed a hybrid approach which used a low-resolution grid compared to that of a large eddy simulation coupled with a turbulence model that one classically uses in a Reynolds-averaged Navier–Stokes simulation. Such a hybrid approach has the potential to bridge the gap between the two, affording an intermediate resolution of turbulence scales and therefore having the characteristics of a very large eddy simulation. One directly calculates the very large scale turbulent structures, using the eddy viscosity turbulence model to account for the effects of the unresolved scales. Labois and Lakehal (2011) studied the effectiveness of a very large eddy simulation. They concluded that the very large scale turbulent structures were modelled accurately providing the computational domain was fine enough to resolve the large turbulence scales. The required mesh was relatively fine compared to that required for a mesh independent unsteady Reynolds-averaged Navier–Stokes simulation, but relatively course compared to a large eddy simulation.

Realising the potential of a very large eddy simulation requires an adequately fine mesh to resolve the large turbulence scales and a turbulence model to account for unresolved turbulence. Success depends on both, and therefore the selection and implementation of an appropriate turbulence model. When developing a very large eddy modelling technique, Borello *et al.* (2013) used the open-source code OpenFOAM (Jasak, 2010). Borello *et al.* (2013) concluded that Davidson's (1997) one-equation sub-grid scale turbulence model was effective in combination with a very large eddy simulation. It is Borello *et al.*'s (2013) very large eddy simulation, combined with Davidson's (1997) one-equation sub-grid scale turbulence model that we use to predict unsteady flow-field parameters. We then present the development and characterisation of a noise prediction method based on a formulation of Ffowcs Williams and Hawkings' (1969) acoustic model.

By combining Borello *et al.*'s (2013) very large eddy simulation modelling technique with a noise prediction method, we develop an acoustic prediction tool. We may incorporate this acoustic prediction tool into a tunnel ventilation fan design methodology. In the research reported in this chapter, we present our acoustic prediction technique, demonstrating its ability to resolve the spectrum up to the fifth harmonic of the blade passing frequency. The acoustic prediction technique models blade and vane interactions, but does not account for quadrupole noise sources or duct modes. Therefore, a simplifying assumption inherent in the acoustic prediction technique is that the fan is installed in a free-field environment.

DESCRIPTION OF THE FAN

Tunnel ventilation fans are typically between 1.6 and 2.8 metres in diameter, with the most common selection either a 1.8 or a 2.0 metre diameter fan running at 1,500 revolutions per minute. The range of tunnel ventilation fans that is the subject of the research reported in this chapter was designed for application in railway tunnel and metropolitan metro ventilation systems, Table 4.1. As we previously mentioned, tunnel ventilation fans are required to comply with legislative requirements mandating minimum fan efficiency and the ability to operate at an elevated temperature. Additionally, tunnel ventilation fan specifications routinely include an acoustic requirement. Typically, the acoustic requirement would be far-field noise of no more than 50 dBA during normal operation and 80 dBA during emergency operation in the event of a tunnel fire.

We selected a 2.24 metre diameter uni-directional fan from the tunnel ventilation fan range (Sheard *et al.*, 2009). This fan is unusual as it has a blade tip speed of 175 metres per second in contrast to the generally accepted limit within the tunnel ventilation fan community of 156 metres per second, Figure 4.1 and Table 4.2. A 175 metres per second tip speed enables the JFM 224 fan to generate a total pressure rise of 2,500 Pa, Figure 4.2. Engineers generally consider a total pressure rise of 2,000 Pa to be a practical limit for a tunnel ventilation fan with a 156 metres per second tip speed. A disadvantage of a 175 metres per second tip speed is that tip generated fan noise increases by approximately 10 dB compared to that of a 156 metres per second tip speed fan. As a consequence, tunnel ventilation design engineers have only utilised the JFM 224 fan in one tunnel ventilation system, the Delhi metro, where pressure developing capability was more important than fan noise. As such, the studied fan constitutes a practical limit in tunnel ventilation fan technology.

METHODOLOGY

We solved the filtered incompressible Navier–Stokes equations in the stationary frame of reference, accounting for Coriolis and centrifugal effects with a modified version of the *piso-Foam* solver. We carried out simulations using a convective central differencing scheme with a second order total variation diminishing (TVD) scheme limit that prevented numerical instabilities (Sweby, 1984). We used a second-order backward scheme for time integration. We solved the system of

FIGURE 4.1. A prototype 2.24 metre diameter unidirectional 1,500 rpm fan named JFM 224, left and the computer aided design model that we used to define fan geometry, right. We developed the fan JFM 224 for application in railway tunnel and metropolitan metro ventilation systems. We designed this prototype for duel use: ventilating the tunnel system during routine operation and clearing 400°C gas and smoke for two hours in the event of a tunnel fire (Sheard *et al.*, 2009).

Table 4.1. *The studied family of fans is intended for application in railway tunnel and metropolitan metro ventilation systems. A requirement of tunnel ventilation fans is the ability to clear hot gas and smoke in the event of a tunnel fire to keep escape routes clear. An ability to operate at elevated temperature invariably requires a robust mechanical design. A consequence of the mechanical design is compromised aerodynamic performance that is challenging to predict accurately.*

Blade tip diameter D_t	1,400 to 2,800 mm
Performance standard	ISO 5801: 2007
Volume flow rate V	10 to 300 m^3/s
Total pressure rise Δp_t	500–3,000 Pa
Rotational speed ω	900–15,00 rpm
High temperature certification	EN 12101-3: 2002 at 200°C, 300°C and 400°C for two hours

Table 4.2. *The studied fan, named JFM 224 constitutes one fan from a family of fans. We selected this example as a basis for numerical simulation of the flow-field as it is the highest tip speed fan currently applied into any railway tunnel or metropolitan metro ventilation system.*

Blade section	ARA-D	
Blade tip diameter D_t	2,240 mm	
Blade count	16	
Hub-to-tip ratio	0.5	
	Hub	Tip
Chord c (mm)	143	92.5
Solidity σ	0.64	0.21
Pitch angle γ	48	24

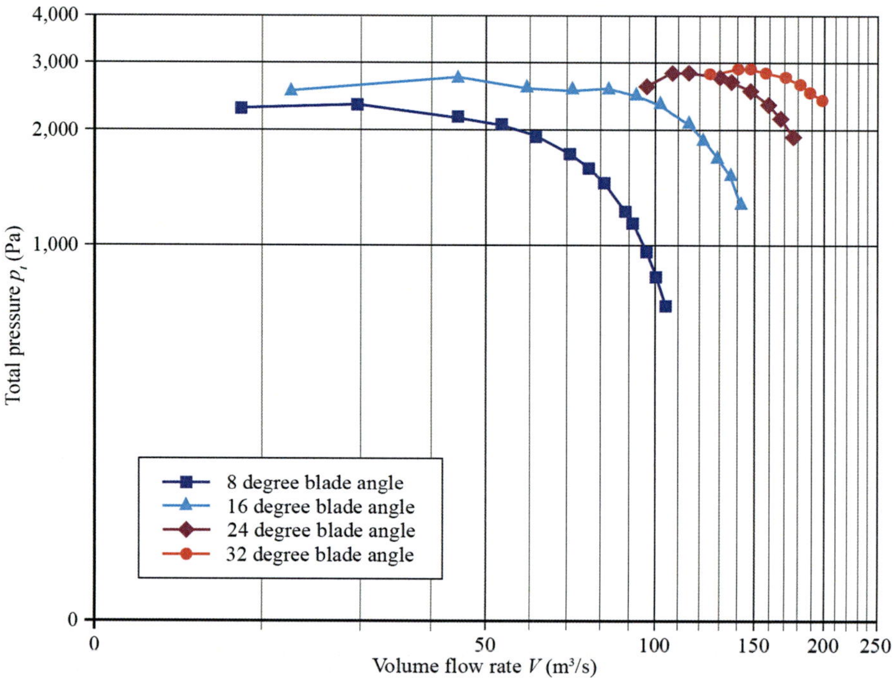

FIGURE 4.2. Performance characteristics of the fan JFM 224 measured in accordance with ISO 5801 requirements (2007). The studied blade angle is 24 degrees.

equation's with an incomplete lower and upper (ILU) preconditioned semi-iterative conjugate gradient (CG) solver (Saad, 1996), coupled with a pressure implicit with splitting of operators (PISO) scheme. We set the convergence threshold equal to 10^{-8} for the conjugate gradient solver and to 10^{-5} for the PISO algorithm.

To minimise computational effort, we undertook a preliminary unsteady Reynold-averaged Navier–Stokes computation. The computational effort required for this is less than that for a large eddy simulation. The computational effort that is required is a consequence of the need to simulate the flow-field over a period of time to correctly model the time varying fluctuation of velocity, pressure and flow-field features. Courant *et al.* (1967) observed that a factor that impacts on the required computational effort for an unsteady Reynolds-averaged Navier–Stokes solution is the length of time step that one selects for the computation. The Courant–Friedrichs–Lewy (CFL) number limits the time step, which is a necessary condition for convergence whilst solving numerically. If the Courant–Friedrichs–Lewy number remains below 1.0, the flow-field's time based resolution remains adequate for the mesh's spatial resolution that one uses for the computations. In essence, the finer the mesh, the smaller the maximum allowable time step in an unsteady Reynolds-averaged Navier–Stokes computation. For the preliminary unsteady Reynold-averaged Navier–Stokes computation, we set the time step to 2×10^{-5} resulting in a maximum Courant–Friedrichs–Lewy number below 0.7.

Following convergence of the unsteady Reynolds-averaged Navier–Stokes computation, we used the converged solution as a starting point for our large eddy simulation. Using the large eddy simulation, we computed half a rotor revolution before collecting data over two flow through time (FTT) steps, which corresponded to 54 blade-to-blade passages, with the variation in collected data from blade-to-blade passage typically less than one per cent.

Aeroacoustic method

We chose to undertake our acoustic prediction as a form of post-processing of large eddy simulation unsteady parameters, programming our prediction method in Matlab. Ffowcs Williams and Hawkings (1969) used experimental data when studying sound generation by turbulence and surfaces in motion. Therefore, in the research reported in this chapter, we chose to treat the unsteady parameters that we predicted using the large eddy simulation as if we were measuring them experimentally. The starting point for our acoustic model was Lighthill's acoustic analogy time resolved formulation (1952). Our implementation of the formulation comprised the Ffowcs Williams and Hawkings (1969) equations combined with Farassat's (1981) solution for rotation surfaces.

We developed our acoustic method, incorporating a procedure to interpolate source terms from the unsteady parameters derived from the large eddy simulation. We identified near-field noise sources using the large eddy simulation to create virtual pressure transducers in close proximity to the blade surface in both chord- and stream-wise directions. We processed the large eddy simulation parameters to derive pressure fluctuations at each virtual pressure transducer location, and then phase-locked averaged them to obtain unsteady pressure data. We then used the time-resolved unsteady pressure data to calculate the fundamental parameters required as input to Farassat's (1981) solution for rotation surfaces. We used the output of each virtual pressure transducer to calculate a local Mach number and the fluctuating lifting force on a blade. We then used both as input to Farassat's (1981) solution for rotation surfaces at each temporal step.

Our acoustic method constitutes a two-step hybrid method that relies on Lighthill's acoustic analogy (1952) that assumes a decoupling of noise generation and noise propagation. The first step comprises identifying noise sources in the near-field using the unsteady parameters derived from the large eddy simulation. The second step involves solving our variation of Lighthill's acoustic analogy in the time domain to calculate the tonal noise radiated to the free-field.

We adapted Farassat's (1981) solution for rotation surfaces for use with the Ffowcs Williams and Hawkings (1969) equations. Our adaptation enabled us to account for subsonic noise sources present in the exhaust flow of the studied fan. Our adaption is valid for an observer located in the studied fan's rotor plane and therefore we chose to locate our observer position at the exit from the fan rotor on the fan casing. When studying tonal noise in a transonic rotor, Bianchi *et al.* (2011)

validated our approach to positioning the observer. They provide a more complete description of the approach. In summary it involves:

- collecting source term time histories;
- building a course acoustic mesh around the blade large enough to include all sources;
- interpolating source terms onto the course acoustic mesh; and
- computing acoustic parameters whilst accounting for spectral volume source terms.

Using the above approach we were able to derive the studied fan's acoustic signature. We derived the acoustic signature by shifting the blade's signature in time by the number of blades and summing the pressures for each observer time within a period based on blade passing frequency. A limitation of the approach was a consequence of us neglecting the duct cut-off effect that occurred with spinning modes.

Numerical grid and boundary conditions

The computational domain extends half a blade chord upstream of the blade leading edge and one chord downstream of the blade trailing edge, Figure 4.3. Following Borello *et al.*'s method (2013), the computational domain distribution resulted in 16 per cent of the elements upstream of the blade leading edge, 50 per cent of the elements within the blade passage and 34 per cent of the elements downstream of the blade trailing edge. In the span-wise direction, 55 nodes were clustered in the blade tip-to-casing gap. In total, the mesh comprised approximately 9 million hexahedral elements distributed with a block-structured topology, Table 4.3. The mesh was clustered toward solid boundaries, with the ratio of minimum grid spacing on solid walls to mid-span blade chord set as 7×10^{-4} on the blade tip, casing wall and blade surfaces. The adopted grid refinement towards the solid surfaces sets the normalised wall distance y^+ value to approximately one on the first row of nodes in order to have the first cell within the viscous sub-layer.

The inlet flow velocity into the studied fan was low, remaining below a Mach number of 0.2 and therefore we were able to treat the flow as incompressible. We

Table 4.3. *Computational mesh data for the flow-field numerical simulations.*

	Rotor
Nodes	9,021,968
Cells	8,862,550
Tip gap nodes	60,753
Tip gap cells	52,000
Blade surface cells	62,450
Averaged cell aspect ratio	1.34

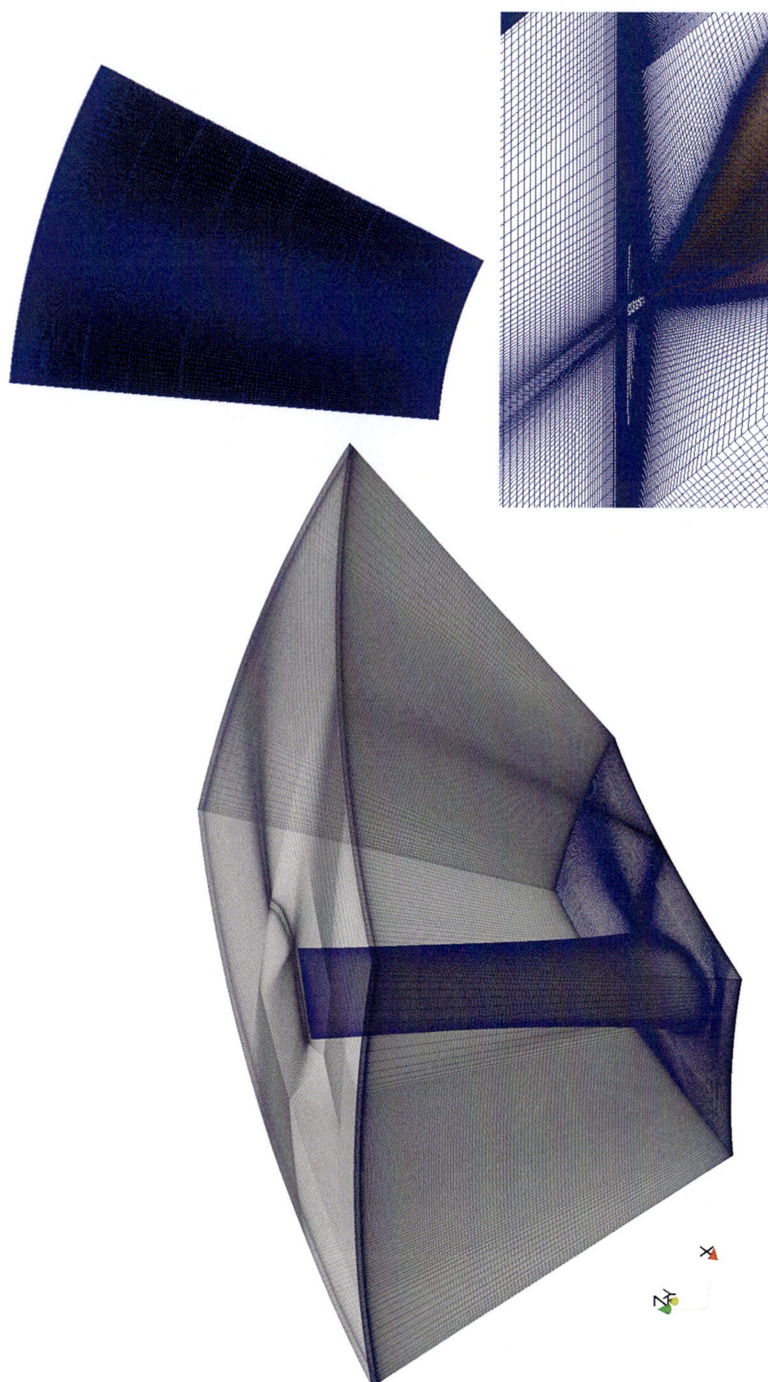

FIGURE 4.3. The fan JFM 224 computational domain extended a half chord upstream of the blade leading edge and one chord downstream from the blade trailing edge.

defined boundary conditions following Sheard *et al.* (2009) and Borello *et al.*'s (2011) methods. An advantage of these methods was that they account for the effect of a sharp reduction of flow area into the fan inlet. They also account for the wake effect generated by five struts upstream of the fan blades, Figure 4.1, that hold the static upstream spinner in place upstream of the fan hub. Flow periodicity, upstream and downstream of the blade, and Neumann outflow conditions complete the boundary conditions.

RESULTS

We conducted our analysis of the studied fan's flow-field with both a blade pitch angle of 24 degrees and at the fan's peak pressure operating point. Stall is incipient at the peak pressure operating point. The 24 degree peak pressure operating point corresponds to a 120m^3/s volume flow rate and a total pressure rise of 2,800 Pa, Figure 4.2. We first study the flow-field topology through the blade-to-blade passage and then consider the evolution of unsteady parameters.

Flow topology

We may gain an insight into the flow-field topology using Dubief and Delcayre's (2000) vortex eduction criterion, which is based on Q factor with the resulting Q factor iso-lines representing lines of constant vorticity. Therefore, the vortex eduction criterion is a useful parameter when attempting to identify coherent vortices in separated flow regions. In our analysis, we present an instantaneous Q iso-surface of 1,000, Figure 4.4. We present the results of the large eddy simulation at three time steps. We have coloured the iso-surface with contours of relative velocity magnitude to facilitate interpreting the vortical structures.

The majority of the vortical structures are generated in the blade wake, blade tip and blade hub regions. Relatively large scale vortical structures characterise the blade wake region, a feature particularly evident at the blade hub. We may associate these large scale vortical structures in the blade hub region with the compromised aerodynamic design of the blade root sections. These aerodynamic compromises were necessary to produce a mechanically viable design. Therefore, the large eddy simulation results illustrate the challenges inherent in applying computational methods to air movement fans.

Despite the compromised blade aerodynamic profile in the blade root region, the Q iso-surface indicates the presence of a horseshoe vortex, with both suction and pressure surface legs evident. A closer inspection of the Q iso-surface indicates the presence of a smaller vortex close to the blade leading edge and a larger vortex upstream of the horseshoe vortex. This vortex system appears to influence positively the stability of the hub end-wall flow-field.

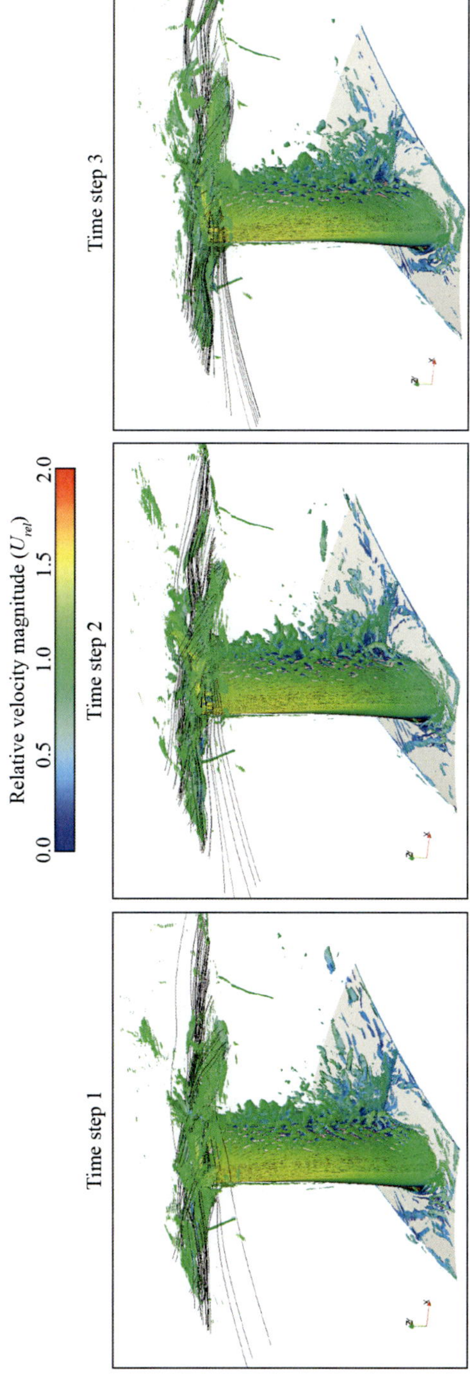

FIGURE 4.4. An analysis of the fan JFM 224 flow-field topology using Dubief and Delcayre's (2000) vortex eduction criterion based on an instantaneous Q iso-surface of 1,000. We present the results of the large eddy simulation at three time steps. We have coloured the iso-surface with contours of relative velocity magnitude to facilitate interpreting vortical structures.

If we now consider the Q iso-surface at the blade mid-span, is it evident that wake shedding in the blade hub region still influences vortex shedding in the wake region. As we move from the blade hub to mid-span, the intensity of the blade wake vortex shedding decreases. This decreasing trend continues as we move from the blade mid-span to the blade tip, with the blade tip region dominated not by wake vortex shedding, but by the blade tip-to-casing clearance flow. The blade tip-to-casing leakage vortex forms near the blade leading edge and migrates across the blade-to-blade passage. The location of the blade tip-to-casing leakage vortex origin close to the blade leading edge indicates that stall is incipient.

We may gain additional insight into the evolution of the blade tip-to-casing leakage vortex by studying two-dimensional normalised helicity (H_n) contours. When studying tip-clearance flow-field features, Furukawa *et al.* (1999) first studied normalised helicity (H_n), which is a useful parameter as it facilitates visualising blade tip-to-casing leakage vortex core trajectories. We present two-dimensional helicity contours at eight chord-wise locations through the fan blade-to-blade passage at three time steps separated by a time interval equivalent to 0.3 normalised time, Figure 4.5.

The development of the blade tip-to-casing leakage vortex core is characterised by higher normalised helicity. Studying the normalised two-dimensional helicity contours reveals that the blade tip-to-casing flow is not characterised by a single blade tip-to-casing leakage vortex. The blade tip-to-casing flow is characterised by the presence of multiple vortices that are able to develop as a consequence of the large blade tip-to-casing gap associated with the studied fan. A study of vortex helicity from one time step to the next indicates that helicity does not vary with time at the blade leading edge, but is periodic in the blade trailing edge region.

We may continue our analysis of the flow-field topology with a study of two-dimensional normalised enstrophy (ξ) contours. Enstrophy is a parameter related to kinetic energy that we define as the integral of the square of vorticity. We present enstrophy contours at three chord-wise locations in the blade hub region through the fan blade-to-blade passage at three time steps separated by a time interval equivalent to 0.3 normalised time, Figure 4.6. The contours of enstrophy have been augmented with flow-field streamlines to facilitate interpreting vortical structures generally, and specifically the hub horseshoe vortex. The enstrophy contours illustrate the evolution of the horseshoe vortex's pressure surface leg. When we study the enstrophy contours at the different time steps, it is apparent that the horseshoe vortex is stable.

Unsteady load analysis

We continued our analysis of the flow-field topology with a study of diffusion factor based on Lieblein's (1956) definition, but unlike Lieblein, we used local quantities. By defining diffusion factor based on local quantities, we were able to derive information about the mixing process downstream of the studied fan. We define local diffusion factor (*DF*) as:

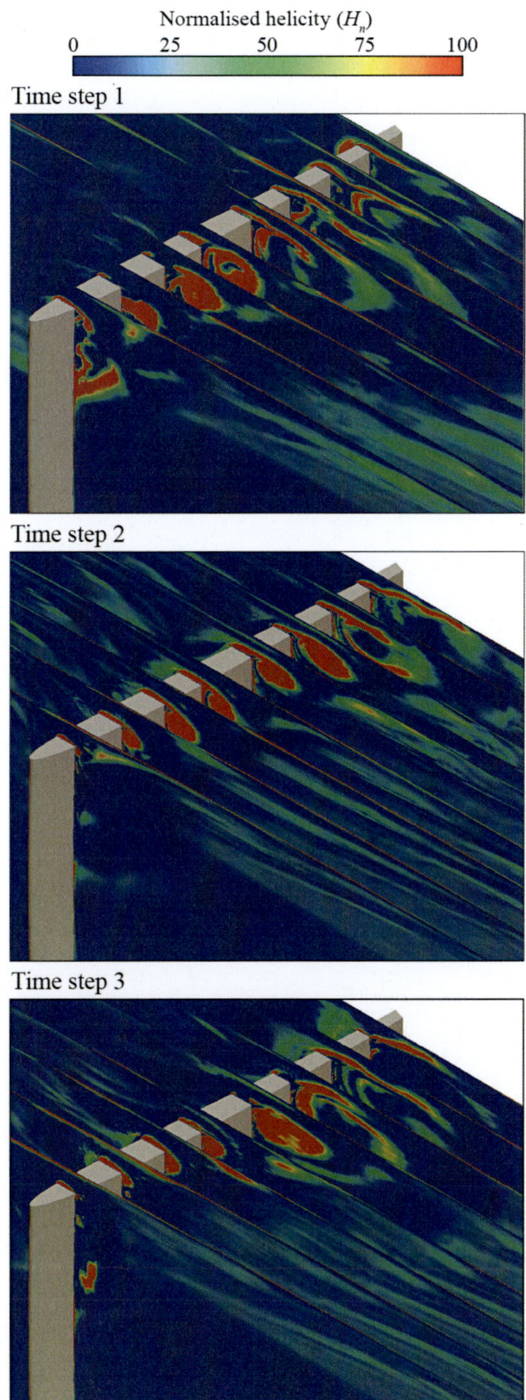

FIGURE 4.5. Three-dimensional normalised helicity (H_n) contours numerically predicted at the fan peak pressure operating point. We present helicity contours at eight chord-wise locations through the fan blade-to-blade passage. Higher normalised helicity characterises development of the blade tip-to-casing leakage vortex core. We present the results at three time steps.

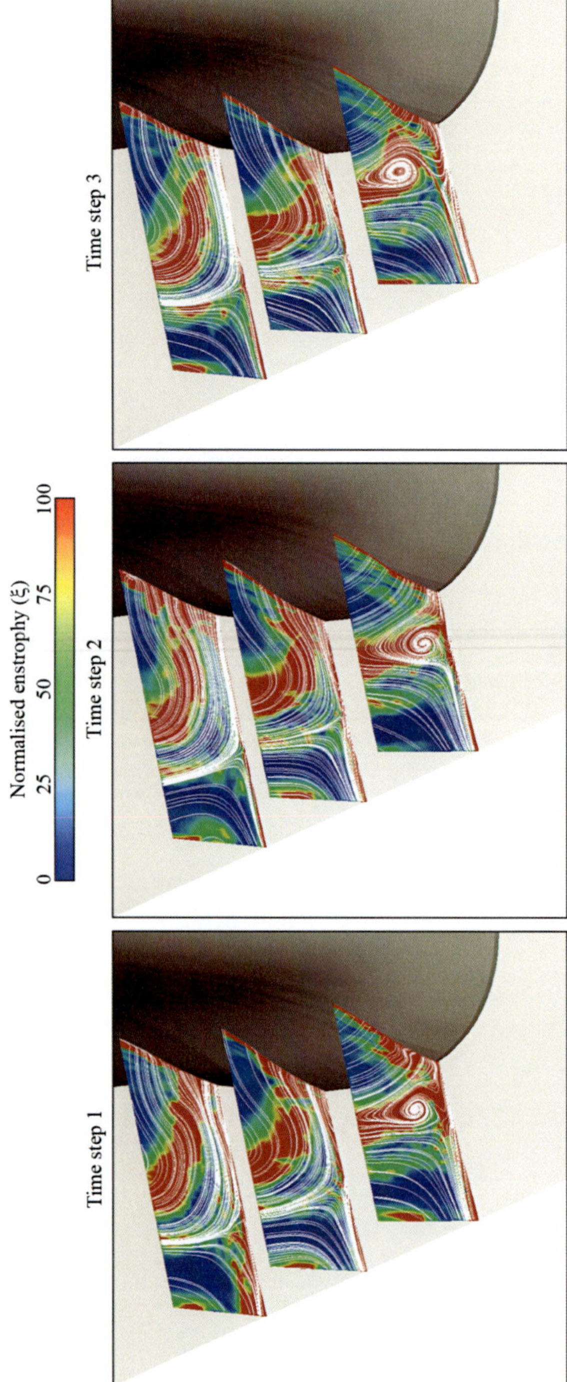

FIGURE 4.6. Two-dimensional normalised enstrophy (ξ) contours numerically predicted at the fan peak pressure operating point. Enstrophy is a parameter related to kinetic energy that we define as the integral of the square of vorticity. We present enstrophy contours at three chord-wise locations in the blade hub region through the fan blade-to-blade passage. We have augmented the enstrophy contours with flow-field streamlines to facilitate interpreting vortical structures generally, and specifically the hub horseshoe vortex.

$$DF = 1 - \frac{V}{V_{in}} + \frac{V^{tg} - V_{in}^{tg}}{2\Sigma V_{in}} \tag{1}$$

where Σ is the fan's solidity, V is the local velocity, V^{tg} is the tangential velocity and the '*in*' subscript refers to average values at the inflow. We present contours of diffusion factor at the peak pressure operating point 20 per cent blade chord downstream of the blade trailing edge. We present contours of diffusion factor at three time steps separated by a time interval equivalent to 0.3 normalised time, Figure 4.7.

Consider the contours of diffusion factor at each time step, Figure 4.7. A high diffusion core at approximately 90 per cent blade span characterises the blade tip-to-casing leakage vortex. Diffusion factor falls close to zero over the outer 20 per cent of the blade span, indicating that stall is incipient. The secondary flow structures through the blade-to-blade passage affects the span-wise distribution of diffusion factor. The secondary flow structures in the blade hub region occur with high levels of diffusion. The distribution of diffusion within the blade wake region is both distorted and unsteady. The blade's prescribed radial work distribution results in the unloading of the blade tip. The vortices generated through the blade tip-to-casing gap occur with diffusion factors that change from one time step to the next. Once again, we may interpret this as indicative of incipient stall.

Aeroacoustic analysis

Our analysis of the flow-field topology indicates that the large eddy simulation is predicting the anticipated flow-field features, and providing an insight into their stability. We now move on to use the unsteady parameters generated by the large eddy simulation to predict fan noise. As we have previously mentioned, we chose to undertake our acoustic prediction as a form of post-processing of large eddy simulation unsteady parameters. We present and compare the resulting prediction of noise spectrum with Borello *et al.*'s (2011) in-duct acoustic measurements, Figure 4.8. Borello *et al.*'s (2011) made measurements with a blade angle of 24 degrees and at the fan design operating point, 2,500 Pa and 150 m^3/s. We therefore present the predicted and measured data at the studied fan's design operating point.

A comparison of the predicted and measured spectrum indicates that the acoustic prediction is a good match at the blade passing frequency and then up to the third harmonic of blade passing frequency. The predicted spectrum under predicts the experimental measurements between harmonics of blade passing frequency. Defining broadband noise as noise above 1,000 Hz, we see that predicted broadband noise oscillates around the experimentally measured values. After approximately 7,000 Hz, the prediction is generally lower than the experimentally measured values. This under prediction may be a consequence of the simplifying assumptions inherent in the acoustic model. Specifically, neglecting the modal amplification of dipole-like noise sources may account for the under prediction of inter-tonal and high frequency noise.

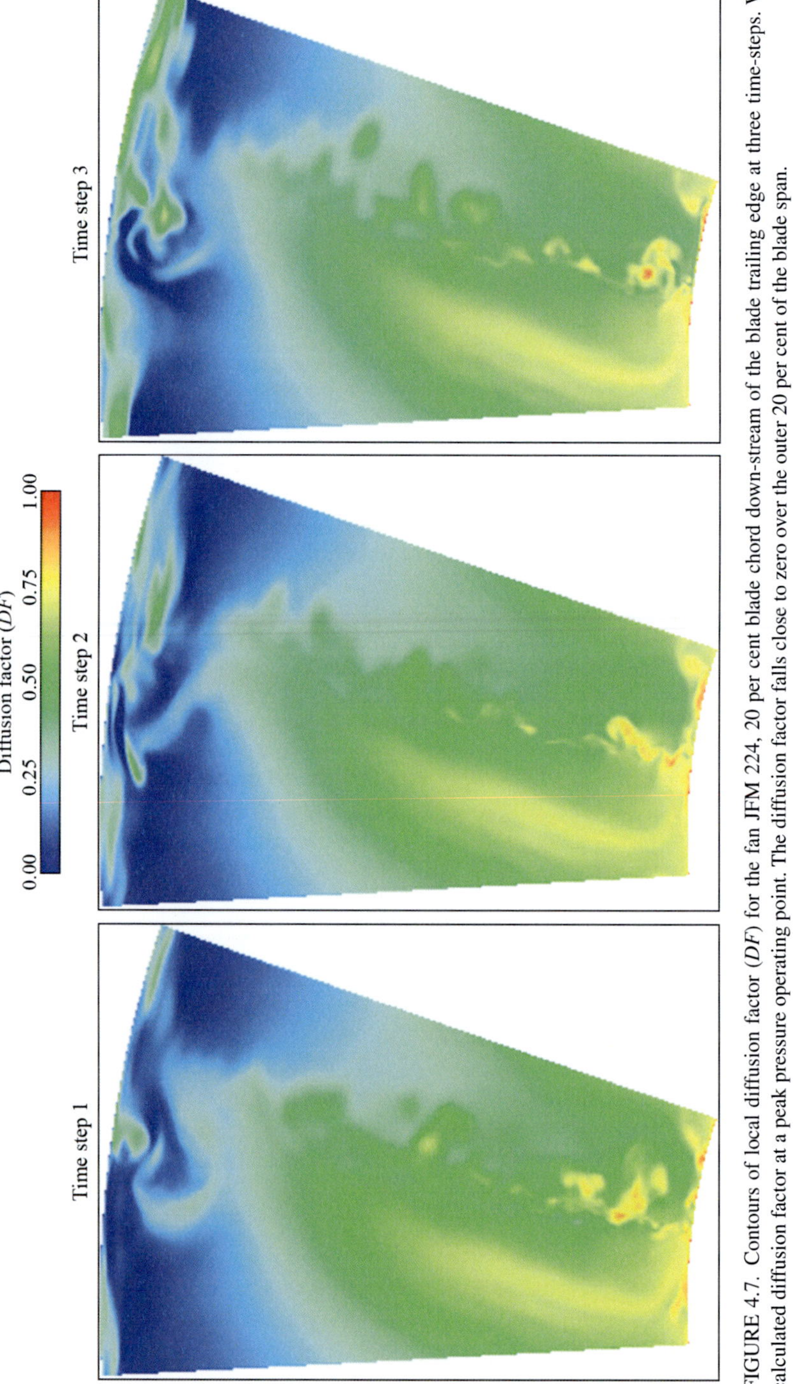

FIGURE 4.7. Contours of local diffusion factor (*DF*) for the fan JFM 224, 20 per cent blade chord down-stream of the blade trailing edge at three time-steps. We calculated diffusion factor at a peak pressure operating point. The diffusion factor falls close to zero over the outer 20 per cent of the blade span.

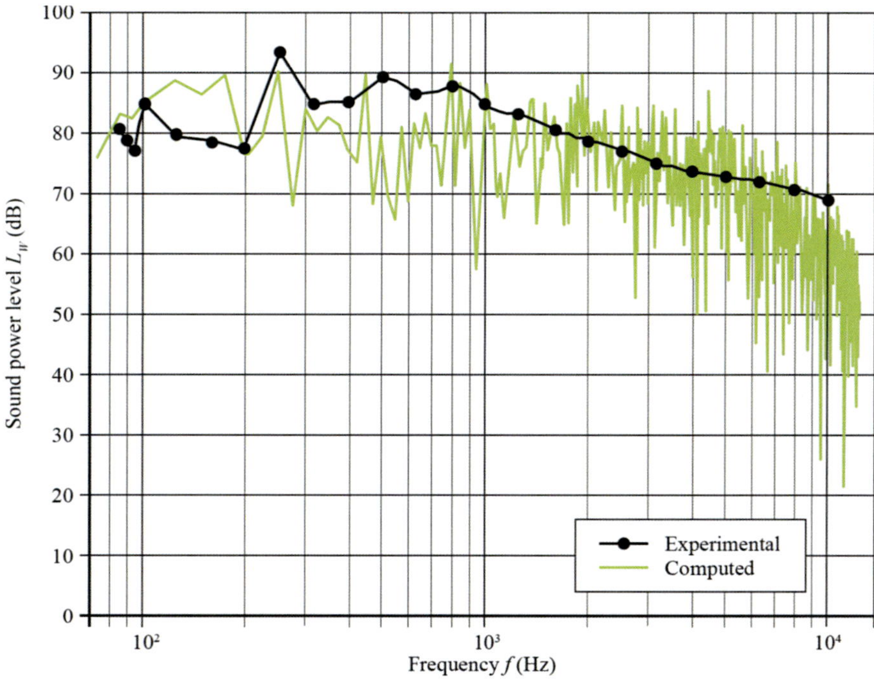

FIGURE 4.8. A comparison of the sound power level (L_W) of the fan JFM 224, which Borello *et al.* (2011) experimentally measured with the fan operating at its design operating point. We also predicted the sound power level by using the large eddy simulation to generate the necessary parameters as input into Ffowcs Williams and Hawkings's (1969) acoustical model.

The analysis presented so far has focused on parameters that characterise aerodynamic aspects of the blade-to-blade and blade tip-to-casing flow-field. Presenting these parameters in the qualitative form of three-dimensional images facilitates an insight into the flow-field physics. However, although we may infer acoustic significance of flow-field features, we are not able to quantify their relative significance. In an effort to quantify the relative acoustic significance of flow-field features, Powell (1963, 1964) characterised the turbulent flow-field motion that is responsible for acoustic emissions. Powell (1963, 1964) discovered that vortex formation and motion is the fundamental noise generation mechanism. The fluctuating pressure that occurs with vortex formation and motion exerted on a rigid body can generate sound as a distribution of dipole sources. This dipole source interpretation is self-consistent with Curle's (1955) conclusions. Curle (1955) also identified that fluctuating pressure exerted on a rigid body results in dipole type noise sources.

For a relatively high Reynolds number flow with no heat release, we can neglect the entropy and viscous-stress terms in Lighthill's (1952, 1954) acoustic tensor. Consequently, the Reynolds-stress factor becomes the dominant contributor to sound generation. This provides the following sound source term:

$$\nabla \cdot (\omega \times v_c), \quad (2)$$

where ω is the leakage vortex flow's vector and v_c is its convection velocity. Powell (1963) used the same source term to characterise the acoustic productivity of flow-field features.

Powell (1964) developed the concept of a survey to identify the most relevant vortex-generated sound sources in a low-speed flow-field. Arakawa *et al.* (2005) used this sound-source survey technique to generate sound-source intensity maps. Following Arakawa *et al.*'s (2005) method, we numerically predicted Powell's (1964) three-dimensional sound-source distributions. We predicted the evolution of Powell's (1964) three-dimensional sound-source distributions and present them at two time steps and as time averaged values, Figure 4.9. We calculated Powell's (1964) three-dimensional sound-source distributions at the studied fan's design operating point to facilitate comparison with the predicted spectrum presented in Figure 4.8. We normalised the Powell (1964) sound sources by a global flow time scale based on the casing diameter and the blade tip velocity. This normalisation facilitated a direct comparison of Powell's (1964) sound-source distributions.

Consider the Powell (1964) sound sources in the blade wake region, Figure 4.9. The most productive acoustic sources are located immediately behind the blade trailing edge. A vortex street characterises the wake region, with vortices coming alternatively from each side of the aerofoil. This vortex motion stretches the sheer layer that separates from the blade surface at the trailing edge. The roll-up of this shear layer into vortices' is acoustically productive. As a consequence of the unstable structure of the wake region, the magnitude of individual sound sources changes with time.

Consider the Powell (1964) sound sources in the blade tip region. The most intense sound sources correlate with the blade tip-to-casing leakage vortex location. This studied fan's blade tip-to-casing leakage vortex is particularly intense as a consequence of the large blade tip-to-casing gap. However, the Powell (1964) sound source intensity indicates that it is the sound sources in the blade wake, and not the blade tip region that are dominant. This counter-intuitive conclusion is a consequence of the controlled and steady roll-up of the blade tip-to-casing leakage vortex.

CONCLUSION

In the research reported in this chapter, we present an acoustic method that one may use to predict fan broadband noise and spectrum. We use a large eddy simulation (LES) to predict unsteady flow-field parameters. Using the predicted unsteady flow-field parameters as input, we were able to use Ffowcs Williams and Hawkings' acoustic model to resolve the acoustic spectrum up to the fifth harmonic of blade passing frequency. We compared the prediction of broadband noise and spectrum with experimental measurements, and found that they agreed within the uncertainty of the measurements. The agreement between predicted and measured

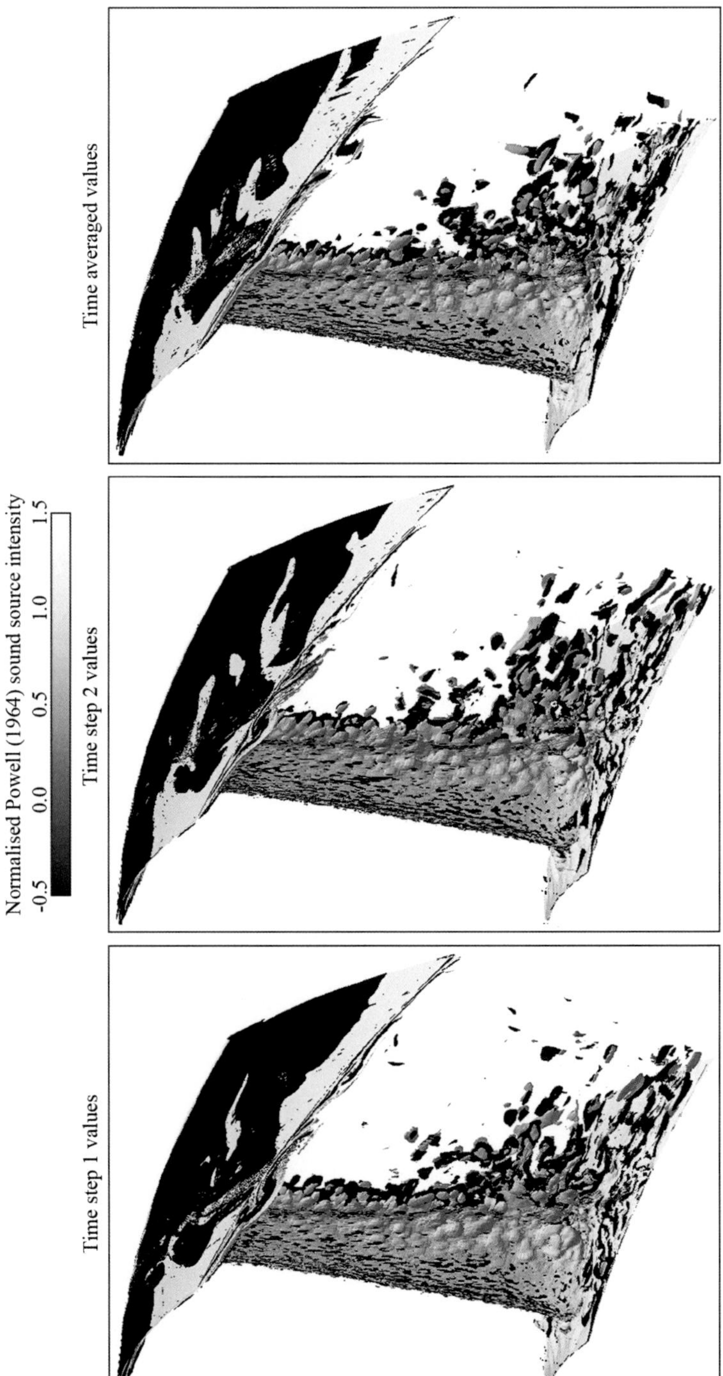

FIGURE 4.9. Three-dimensional Powell (1964) sound source distributions numerically predicted at the fan design operating point. We present Powell's (1964) sound source distributions at two time steps and also as time averaged values.

broadband noise and spectrum gives confidence in the accuracy with which the large eddy simulation modelling technique models the evolution of blade-to-blade flow-field features.

A comparison of the predicted and measured spectrum indicated that the acoustic prediction was a good match at the blade passing frequency and then up to the third harmonic of blade passing frequency. The predicted spectrum under predicted the experimental measurements between harmonics of blade passing frequency. Above 1,000 Hz the prediction oscillated around the experimentally measured values. Above 7,000 Hz the prediction was generally lower than the experimentally measured values. The accuracy of the prediction is probably limited by the limitations inherent in the Ffowcs Williams and Hawkings' acoustic model, which neglects duct modes. As a consequence, our acoustic method does not model the modal amplification of dipole-like noise sources. Dipole-like noise sources characterise the inter-tonal and broadband noise when one fits a fan in a ducted system.

Despite the limitations of our acoustic method, we concluded the agreement between predicted and measured broadband noise and spectrum to be good. It can provide tunnel ventilation fan designers with an insight into the acoustic consequences of a change in fan geometry. Therefore, it is a useful tool that one may apply into a fan design optimisation process. Thus, one would extend the optimisation process to include a characterisation of the acoustic consequences of any change in fan geometry.

REFERENCES

ANSI/ASHRAE Standard 149-2000 (2009), *Laboratory Methods of Testing Fans Used to Exhaust Smoke in Smoke Management Systems.*

EN12101-3 (2002), *Smoke and Heat Control Systems. Specification for Powered Smoke and Heat Exhaust Ventilators.*

ISO 5801:2007 (2007), *Industrial Fans: Performance Testing Using Standardised Airways.*

Arakawa, C., Fleig, O., Makoto, I. and Masakazu, S. (2005), 'Numerical Approach for Noise Reduction of Wind Turbine Blade Tip with Earth Simulator', *Journal of the Earth Simulator*, vol. 2, pp. 11–33.

Bianchi, S., Corsini, A. and Paniagua, G. (2011), 'Amplification of the Force and Tonal Noise in Transonic HPT', *Proceedings of the 9th European Conference on Turbomachinery, Fluid Dynamics & Thermodynamics*, Istanbul, Turkey, 21–25 March, pp. 1213–1223.

Borello, D., Hanjalić, K. and Rispoli, F. (2005), 'Prediction of Cascade Flows with Innovative Second-moment Closures', *Transactions of the ASME, Journal of Fluids Engineering*, vol. 127(6), pp. 1059–1070.

Borello, D., Hanjalić, K. and Rispoli, F. (2007), 'Computation of Tip-leakage Flow in a Linear Compressor Cascade with a Second-moment Turbulence Closure', *International Journal of Heat and Fluid Flow*, vol. 28(4), pp. 587–601.

Borello, D., Delibra, G., Hanjalić, K. and Rispoli, F. (2009), 'Large-eddy Simulations of Tip Leakage and Secondary Flows in an Axial Compressor Cascade Using a Near-wall Turbulence Model', *Proceedings of the IMechE Part A, Journal of Power and Energy*, vol. 223(6), pp. 645–655.

Borello, D., Corsini, A., Minotti, S., Rispoli, F. and Sheard, A.G. (2011), 'U-RANS of a Large Industrial Fan Under Design and Off-Design Operations', *Proceedings of the 9th European Conference on Turbomachinery, Fluid Dynamics & Thermodynamics*, Istanbul, Turkey, 21–25 March, pp. 1665–1676.

Borello, D., Corsini, A., Delibra, G., Fiorito, M. and Sheard, A.G. (2013), 'Large-eddy Simulation of a Tunnel Ventilation Fan', *Transactions of the ASME, Journal of Fluids Engineering*, vol. 135, paper no. 071102, pp. 1–9.

Carvel, C. (1999), 'The Effects of Ventilation on Fires in Tunnels', *Proceedings of the International Tunnel Fire and Safety Conference*, Rotterdam, The Netherlands, 2–3 December.

Corsini, A. and Rispoli, F. (2005), 'Flow Analyses in a High-pressure Axial Ventilation Fan with a Non-linear Eddy-viscosity Closure', *International Journal of Heat Fluid Flow*, vol. 26(3), pp. 349–361.

Courant, R., Friedrichs, K. and Lewy, H. (1967), 'On the Partial Difference Equations of Mathematical Physics', *IBM Journal of Research and Development*, vol. 11(2), pp. 215–234.

Cumpsty, N.A. (1977), 'Review — a Critical Review of Turbomachinery Noise', *Transactions of the ASME, Journal of Fluids Engineering*, vol. 99(2), pp. 278–293.

Curle, N. (1955), 'The Influence of Solid Boundaries upon Aerodynamic Sound', *Proceedings of the Royal Society of London A*, vol. 231, pp. 505–514.

Daly, B.B. (1985), *Woods Practical Guide to Fan Engineering*, Woods of Colchester Ltd. Colchester, UK.

Davidson, L. (1997), 'Large Eddy Simulation: A Dynamic One-equation Sub-grid Model for Three-dimensional Recirculating Flows', *Proceedings of the 11th International Symposium on Turbulent Shear Flow*, Grenoble, France, 8–10 September, vol. 3, pp. 26.1–26.6.

Delibra, G., Borello, D., Hanjalić, K. and Rispoli, F. (2009), 'U-RANS of Flow and Endwall Heat Transfer in a Pinned Passage Relevant to Gas-turbine Blade Cooling', *International Journal of Heat and Fluid Flow*, vol. 30(3), pp. 549–560.

Delibra, G., Borello, D., Hanjalić, K. and Rispoli, F. (2010), 'An LES Insight into Convective Mechanism of Heat Transfer in a Wall-bounded Pin Matrix', *Proceedings of the 14th International Heat Transfer Conference*, Washington, DC, USA, 8–13 August, paper no. IHTC14-23205, pp. 807–815.

Dubief, Y. and Delcayre, F. (2000), 'On Coherent-vortex Identification in Turbulence', *Journal of Turbulence*, vol. 1, p. N11.

Farassat, F. (1981), 'Linear Acoustic Formulas for Calculation of Rotating Blade Noise', *AIAA Journal*, vol. 19(9), pp. 1122–1130.

Ffowcs Williams, J.E. and Hawkings, D.L. (1969), 'Sound Generation by Turbulence and Surfaces in Arbitrary Motion', *Philosophical Transactions of the Royal Society of London A*, vol. 264, pp. 321–342.

Furukawa, M., Inoue, M., Saiki, K. and Yamada, K. (1999), 'The Role of the Tip Leakage Vortex Breakdown in Compressor Rotor Aerodynamics', *Transactions of the ASME, Journal of Turbomachinery*, vol. 121, pp. 469–480.

Hah, C. and Shin, H.-W. (2012), 'Study of Near-stall Flow Behavior in a Modern Transonic Fan with Compound Sweep', *Transactions of the ASME, Journal of Fluids Engineering*, vol. 134(7), paper no. 071101, pp. 1–7.

Hedefalk, J., Wahlstrom, B. and Rohlen, P. (1998), 'Lessons from the Baku Subway Fire', *Proceedings of the 3rd International Conference on Safety in Road and Rail Tunnels*, Nice, France, 9–11 March, pp. 15–28.

Horlock, J.H. and Denton, J.D. (2005), 'A Review of Some Early Design Practice using Computational Fluid Dynamics and a Current Perspective', *Transactions of the ASME, Journal of Turbomachinery*, vol. 127(1), pp. 5–13.

Jagannathan, S., Schwänen, M. and Duggleby, A. (2012), 'Low Pressure Turbine Relaminarization Bubble Characterization using Massively-parallel Large Eddy Simulations', *Transactions of the ASME, Journal of Fluids Engineering*, vol. 134(2), paper no. 021102, pp. 1–13.

Jasak, H. (2010), 'OpenFOAM: A Year in Review', *Proceedings of the 5th OpenFOAM Workshop*, Gothenburg, Sweden, 21–24 June.

Jasnoch, U., Kress, H. and Rix, J. (1994), 'Towards a Virtual Prototyping Environment', in Rix, J. (Ed.), *Virtual Prototyping: Virtual Environments and the Product Design Process*, Chapman and Hall, London, UK, pp. 173–183.

Labois, M. and Lakehal, D. (2011), 'Very-large Eddy Simulation (V-LES) of the Flow Across a Tube Bundle', *Nuclear Engineering and Design*, vol. 241, pp. 2075–2085.

Lacroix, D. (1997), 'New French Recommendations for Fire Ventilation in Road Tunnels', *Proceedings of the 9th International Conference on Aerodynamics and Ventilation of Vehicle Tunnels: Developments for the 21st Century*, Aosta Valley, Italy, 6–8 October, pp. 103–124.

Lee, K.Y., Choi, Y.S., Kim, Y.L. and Yun, J.H. (2008), 'Design of Axial Fan using Inverse Design Method', *Journal of Mechanical Science and Technology*, vol. 22(10), pp. 1883–1888.

Lieblein, S. (1956), 'Experimental Flow in Two-dimensional Cascades', in NACA RME 56B03, reprinted in NASA SP-36, *Aerodynamic Design of the Axial-Flow Compressor*, pp. 183–226.

Lighthill, M.J. (1952), 'On Sound Generated Aerodynamically. I. General Theory', *Proceedings of the Royal Society of London A*, vol. 211, pp. 564–587.

Lighthill, M.J. (1954), 'On Sound Generated Aerodynamically. II: Turbulence as a Source of Sound', *Proceedings of the Royal Society A*, vol. 222, pp. 1–32.

Luo, X., Ji, B., Peng, X., Xu, H., Nishi, M. and Ji, B. (2012), 'Numerical Simulation of Cavity Shedding from a Three-dimensional Twisted Hydrofoil and Induced Pressure Fluctuation by Large-eddy Simulation', *Transactions of the ASME, Journal of Fluids Engineering*, vol. 134(4), paper no. 041202, pp. 1–10.

Miles, J.H. (2006), 'Procedure for Separating Noise Sources in Measurements of Turbofan Engine Core Noise', *Proceedings of the 12th AIAA/CEAS Aeroacoustics Conference (27th AIAA Aeroacoustic Conference)*, Cambridge MA, USA, paper no. AIAA 2006-2580.

Powell, A. (1963), 'Mechanisms of Aerodynamic Sound Production', AGARD Report No. 466.

Powell, A. (1964), 'The Theory of Vortex Sound', *Journal of the Acoustical Society of America*, vol. 33, pp. 177–195.

Pratt, M.J. (1994), 'Virtual Prototypes and Product Models in Mechanical Engineering', in Rix, J. (Ed.), *Virtual Prototyping: Virtual Environments and the Product Design Process*, Chapman and Hall, London, UK, pp. 113–128.

Pucher, K. (1999), 'Fire in the Tauern Tunnel', *Proceedings of the International Tunnel Fire & Safety Conference*, Rotterdam, The Netherlands, 2–3 December.

Saad, Y. (1996), *Iterative Methods for Sparse Linear Systems*, Society for Industrial and Applied Mathematics, Philadelphia, PA, USA.

Schneider, H., von Terzi, D. and Bauer, H.J. (2010), 'Large-eddy Simulations of Trailing-edge Cutback Film Cooling at Low Blowing Ratio', *International Journal of Heat and Fluid Flow*, vol. 31(5), pp. 767–775.

Sheard, A.G. and Jones, N.M. (2005), 'Emergency Ventilation for Vehicular, Rail and Metro Tunnels', *Proceedings of the International Congress: Safety Innovation Criteria Inside Tunnels*, Gijon, Spain, 29 June–1 July, pp. 303–312.

Sheard, A.G. and Jones, N.M. (2006), 'High-temperature Certification of Large Fans for Emergency Ventilation', *Proceedings of the 12th International Symposium on Aerodynamics and Ventilation of Vehicle Tunnels*, Portoroz, Slovenia, 11–13 July, pp. 123–140.

Sheard, A.G. and Jones, N.M. (2008), 'Approval of High-temperature Emergency Tunnel-ventilation Fans: The Impact of ISO 21927-3', *Proceedings of the ITA–AITES World Tunnel Congress and 34th General Assembly*, Agra, India, 19–25 September, pp. 1817–1826.

Sheard, A.G. and Jones, N.M. (2012), 'Powered Smoke and Heat Exhaust Ventilators: The Impact of EN 12101-3 and ISO 21927-3', *Tunnelling and Underground Space Technology*, vol. 28, pp. 174–182.

Sheard, A.G., Corsini. A., Minotti, S. and Sciulli, F. (2009), 'The Role of Computational Methods in the Development of an Aero-acoustic Design Methodology: Application in a Family of Large Industrial Fans', *Proceedings of the 14th International Conference on Modelling Fluid Flow Technologies*, Budapest, Hungary, 9–12 September, pp. 71–99.

Shipp, M. (1997), 'Investigating the Channel Tunnel Fire', *Proceedings of the Seminar on Reducing Fire Risk in Trains*, Heathrow, England, 17 November, pp. 19–29.

Sweby, P.K. (1984), 'High Resolution Schemes using Flux Limiters for Hyperbolic Conservation Laws', *SIAM Journal on Numerical Analysis*, vol. 21(5), pp. 995–1011.

Vad, J. (2001), 'Incorporation of Forward Blade Sweep in the Non-free Vortex Design Method of Axial Flow Turbomachinery Rotors', *Periodica Polytechnica, Mechanical Engineering*, vol. 45(2), pp. 217–237.

Vad, J., Bencze, F., Corsini, A. and Rispoli, F. (2001), 'Non-free Vortex Flow Effects in an Axial Flow Rotor', *Periodica Polytechnica, Mechanical Engineering*, vol. 45(2), pp. 201–216.

Wallis, R.A. (1961), *Axial Flow Fans: Design and Practice*, George Newnes Ltd, London, UK.

Chapter 5

A Numerical Investigation into the Aerodynamic Effect of Pressure Pulses on a Tunnel Ventilation Fan

L. Cardillo, A. Corsini, G. Delibra, F. Rispoli and A.G. Sheard

ABSTRACT

Tunnel ventilation fans are subjected to pressure pulses as a consequence of trains passing the ventilation shafts within which they are installed. These pressure pulses alter the volume flow rate through a ventilation fan, and consequently the static pressure field around fan blades. Today's trains typically travel through railway tunnels faster than has been the historic norm. Additionally, platform screen doors are now a standard feature of metro systems. Both result in the trains involved inducing larger pressure pulses than fan designers have traditionally assumed. Although tunnel ventilation fan in-service failures are rare, engineers increasingly associate those failures with new or renovated tunnel systems with larger pressure pulses. Consequently, we require insight into the aerodynamic and mechanical consequences that occur with subjecting a tunnel ventilation fan to a pressure pulse.

In the research reported in this chapter, we model the pressure pulse induced by a train in a tunnel system as a rapidly changing fan volume flow rate. We computed the fan operating point by means of a large eddy simulation with a one-equation sub-grid scale turbulence model. We undertook the computation using the open-source computational fluid dynamic code OpenFOAM. An analysis of the computational results provides an insight into the effects of turbulent structures that develop within the fan blade passage. The analysis indicates that a pressure pulse too brief to drive a tunnel ventilation fan into stall still results in doubling the unsteady aerodynamic and mechanical forces to which the blade is subjected. This is a consequence of a pressure pulse increase significant enough to contribute to the tunnel ventilation fan's in-service mechanical failure.

This chapter is a revised and extended version of Cardillo, L., Corsini, A., Delibra, G., Rispoli, F. and Sheard A.G. (2014), 'A Numerical Investigation into the Aerodynamic Effect of Pressure Pulses on a Tunnel Ventilation Fan', *Proceedings of the IMechE Part A, Journal of Power and Energy*, vol. 228(3), pp. 285–299.

NOMENCLATURE

Latin letters

a		Undisturbed flow's speed of sound
c	[mm]	Blade chord
$<c_{SGS}>$	[-]	Smagorinsky coefficient
C_p	[-]	Pressure coefficient: $C_p = {}^1/_2(p-p_0)/U_{tip}$
D_t	[mm]	Blade tip diameter
k_{SGS}	$[m^2/s^2]$	Sub-grid turbulent kinetic energy
l_{LES}	[m]	Large eddy simulation length-scale $l_{LES}= V_c^{1/3}$
M_a		Axial moment component
M_p		Peripheral moment component
p	[Pa]	Static pressure
p_t	[Pa]	Total pressure
Δp	[Pa]	Static pressure rise
Δp_{pulse}	[Pa]	Pressure pulse magnitude
P_{kSGS}	$[m^2/s^3]$	Production of ksgs
Q	$[1/s^2]$	Vortex deduction parameter: $Q = 0.5(\Omega_{ij}\,\Omega_{ij} - s_{ij}s_{ij})$
R	[m]	Radius
S_{ij}	[1/s]	Rate of strain tensor
Δt	[s]	Time variation
t	[s]	Time
ΔU_{bulk}	[m/s]	Change in bulk velocity
U_{bulk}	[m/s]	Bulk velocity
U_{tip}	[m/s]	Tip velocity
V_c	$[m^3]$	Cell volume
V	$[m^3/s]$	Volume flow rate
w	[m/s]	Relative velocity
w_t		Tip leakage vortex tangential velocity
x, y, z		Cartesian coordinates
y^+	[-]	Normalised wall distance

Greek letters

γ	[deg]	Pitch angle
ε_{kSGS}	$[m^2/s^3]$	Sub-grid scale dissipation of k
ν	[m2/s]	Kinematic viscosity
ν_{SGS}	$[m^2/s]$	Sub-grid scale viscosity
σ		Solidity
ρ	$[kg/m^3]$	Air density
ω		Rotational speed

Acronyms

CG	Conjugate gradient
ILU	Incomplete lower and upper
LES	Large eddy simulation
PISO	Pressure implicit with splitting of operators
SGS	Sub-grid scale
TLV	Tip leakage vortex
TVD	Total volume diminishing

Symbols and subscript

< >	ensemble average
i, j, k	indices

INTRODUCTION

Large axial flow fans are an essential part of both railway and metro tunnel ventilation systems. Typically, tunnel ventilation systems have a series of operating scenarios, with each requiring a unique ventilation system configuration in response to a specific tunnel condition. The complexity of tunnel operating scenarios results in the majority of tunnel ventilation fans operating in both forward and reverse directions. In this regard tunnel ventilation fans are unique as other applications do not require reversibility. A need to operate in either a forward or reverse direction results in symmetrical fan blades, with guide vanes reduced to flow-straighteners. A consequence of a requirement for reversibility is that separated flow is an inevitable aspect of even the most carefully developed aerodynamic design. It is therefore challenging to accurately predict tunnel ventilation fan performance.

The aerodynamic challenges posed by a requirement for reversibility are compounded by legislative requirements. Tunnel ventilation fans are required to comply with the European Union Regulation 327 mandating a minimum fan and motor efficiency during routine tunnel operation and simultaneously with the Euro Norm EN 12101-3 requirements (Sheard and Jones, 2012) and its associated international standard ISO 21927-3 (Sheard and Jones, 2008) during emergency operation in the event of a tunnel fire. The requirements of both Regulation 327 and EN 12101-3 are legally binding within the European Community and consequently, tunnel ventilation fan designers must build their design process around compliance with both.

A requirement for tunnel ventilation fans to comply with both Regulation 327 and EN 12101-3 is compounded by a trend towards railway tunnels and metro systems that are longer and deeper than has been the historic norm (Sheard *et al.*, 2013). Longer and deeper tunnels have a larger associated ventilation system pressure drop, and consequently require ventilation fans capable of developing a higher pressure for a given flow rate. Reversible fans are able to generate typically no more than 2,000 Pa in a single stage (Sheard and Daneshkhah, 2012), with the most demanding tunnels now requiring ventilation fans to generate 3,400 Pa (Sheard and

Daneshkhah, 2012). A result of the increase in required pressure developing capability is an aerodynamic design process that is becoming more demanding.

The situation faced by tunnel ventilation fan designers is further complicated by a shift towards higher speed trains in railway tunnels and the use of platform screen doors in metro systems. Both result in a significant increase in the magnitude of pressure pulses imposed upon ventilation fans. A train that passes along a tunnel as it first approaches and then moves away from the ventilation shaft within which ventilation fans are situated generates pressure pulses (Borello *et al.*, 2013b). Historically, tunnel ventilation fans would be subjected to pressure pulses of no more than ± 300 Pa. Today a pressure pulse of ± 700 Pa is typical, with some recent specifications requiring the tunnel ventilation fans to accommodate a ± 1,000 Pa pressure pulse.

The legislative requirements, combined with shifting market requirements, have resulted in the constraints and demands placed upon tunnel ventilation fan designers becoming significantly more onerous over the past decade. In response, they have developed design methodologies based on computational fluid dynamic analysis and virtual prototyping techniques derived from those developed by the aerospace community (Sheard *et al.*, 2009). This is a natural evolution for tunnel ventilation fan designers. Both fans designed for aerospace application and tunnel ventilation application are limited by the onset of aerodynamic instability. Rippl observed (1995) that the alternating stresses induced in blades by aerodynamic forces during rotating stall increases by a factor of five. A more aggressive aerodynamic design is characterised by higher aerodynamic forces generally, and consequently, higher induced mechanical stresses during rotating stall that are more likely to result in a mechanical failure.

Both tunnel ventilation and aerospace fan designers historically produced a mechanical design that could withstand the alternating loads imposed on fan blades during rotating stall, and therefore mechanical failure during a stall event is usually not instantaneous (Sheard and Corsini, 2012). However, if the stall event lasts long enough, the alternating loads may result in initiating a fatigue crack in one or more blades that then goes on to propagate after the fan has returned to stable operation. This is the classical in-service failure mechanism for tunnel ventilation fans that occurs during stable operation, but as a consequence of operating in a stalled condition (Sheard and Corsini, 2012). Some within the air movement and control community believe that tunnel ventilation scenarios that unintentionally result in a tunnel ventilation fan's stalled operation account for 80 per cent of all tunnel ventilation fans' in-service failures.

When one studies historic in-service failures of tunnel ventilation fans, researchers associate the majority with a blade fatigue failure. They classically associate fatigue failure with the large unsteady aerodynamic loads imposed upon fan blades as a consequence of engineers operating them in a stalled condition. This is a consequence of a closed damper sticking or some other mechanical failure in the tunnel ventilation system. However, researchers less easily associate more recent in-service failures with a specific mechanical failure. An apparently fully functional tunnel ventilation system, with no evidence of the fan driving into a stalled condition during any ventilation system operating scenario, can still suffer a fan blade mechanical failure. When one inspects the blade, it typically shows the classic signs that occur with a fatigue failure.

A fatigue failure indicates that alternating stress levels induced in the blades by unsteady aerodynamic forces have been significantly higher than those that occur with stable operation. Historically, researchers have associated significantly higher aerodynamic forces with operation in a stalled condition; however, if recent in-service failures did not occur with the fan operating in a stalled condition, than some other mechanism must be responsible.

When considering why aerodynamic forces may be higher than during stable operation, it is noteworthy that a combination of recent legislative and market demands have resulted in progressively higher pressure tunnel ventilation fan aerodynamic designs. These aerodynamic designs are then subjected to larger pressure pulses in-service than has been the historic norm. This combination of higher pressure and larger pressure pulse will result in larger unsteady aerodynamic loads on fan blades if the fan stalls, and a higher probability that it will stall as a train passes the tunnel ventilation shaft within which it is situated.

Scholarly research on the effect of trains on tunnel systems has focused primarily on the pressure waves induced by a train's motion through a tunnel. The objective is to predict the drag variation on the train when entering, passing through and departing from a station. There is little published literature on the effect of pressure waves on ventilation system performance generally, or the performance of tunnel ventilation fans specifically. One notable exception is Vardy's work (1980a). He conducted an experimental investigation, measuring the magnitude of pressure waves in an underground station on London Transport's Victoria Line. Vardy's work provides an insight into the variation in pressure along both main tunnels and secondary airways that is induced by a single train passing. Vardy made these experimental measurements whilst developing and validating an analytic model that he used to predict the variation in drag on a train as a function of blockage ratio as it enters and leaves a station (Vardy, 1980b).

Other researchers who have studied the aerodynamic performance of tunnel systems include Bellenoue *et al.* (2011) who studied the compression wave that originates when a high-speed train enters an underground station. Bellenoue *et al.* (2011) made both measurements in a tunnel system and modelled the compression wave numerically. William-Louis and Tournier (2005) proposed and validated a method for predicting pressure wave evolution in underground tunnels with one or more trains operating. Recently, Huang and Gao (2010) developed, tested and validated a dynamic computational fluid dynamics method to predict unsteady flows inside subway tunnels, focusing on underground systems with natural and forced ventilation.

Whilst the research literature has been of interest to tunnel ventilation fan designers, it has not been directly applicable to tunnel ventilation fan design. Designers typically still rely on one of three classical approaches during tunnel ventilation fan selection and specification (Sheard and Corsini, 2012). The first approach is the most conservative: to select a fan with a non-stalling blade angle such that the fan cannot stall irrespective of the pressure pulse's magnitude. The second strategy is to select a fan with a high enough pressure developing capability to operate without stalling when subjected to a pressure pulse (Sheard and Corsini, 2012). The third approach is to fit the fan with a stabilisation ring, allowing the selected fan to manage the mechanical consequence of stall when subjected to a pressure pulse (Sheard and Corsini, 2012).

The first strategy is increasingly less favoured as it results in larger fan selections that require larger plant rooms that increase plant room cost. The third strategy is increasingly less favoured as the stabilisation ring typically reduces fan efficiency by between three and five per cent and consequently, is in conflict with the European Union Regulation 327 requirements. Engineers are increasingly favouring the second strategy, and therefore they increasingly associate it with in-service mechanical failures. These in-service mechanical failures occur despite a reasonable degree of certainty that tunnel ventilation system scenarios do not result in inadvertently operating a fan in a stalled condition. There is also a reasonable degree of certainty that the pressure pulses to which a fan is subjected do not result in a fan inadvertently driving transiently into a stalled condition. There is therefore an increasing certainty that operators are not inadvertently operating tunnel ventilation fans in a stalled condition and simultaneously, in-service failures exhibit symptoms that classically occur with operation in a stalled condition.

There is a need to better understand the mechanical consequences that occur with operating high-pressure tunnel ventilation fans in the presence of pressure pulses that are larger than the historic norm. The reported research models the compression and expansion wave effects as they pass through a tunnel ventilation fan. We modelled and simulated the pressure waves using a large eddy simulation (LES). Full large eddy simulations are challenging as a consequence of the formidable computational requirements to compute the turbulence spectrum up to the inertial sub-range (Schneider *et al.*, 2010). The research reported in this chapter therefore utilises a more practical approach, reducing the level of discretisation by means of an improved sub-grid scale model.

Researchers predicting turbomachinery flow have previously established that in internal flow computations large eddy simulations are able, in the majority of the computational domain, to provide sufficient resolution to solve the larger turbulence scales (Labois and Lakehal, 2011). Resolving the larger turbulence scales requires a dynamic sub-grid scale model (Germano *et al.*, 1991; Davidson, 1997). The sub-grid scale model that we chose for this work (Davidson, 1997) is able to account for the transfer rate of energy from small to large scale vortical structures (backscatter). Thus, the sub-grid scale model is able to partly compensate for the computational grid's effect that is too coarse for a fully resolved large eddy simulation. The result is a computational grid small enough to use in practice that is able to produce a reasonably accurate prediction of an unsteady flow-field. We implemented the adopted numerical methodology using an open-source computational fluid dynamics code (Jasak, 2010). We used the open-source code OpenFOAM, a code previously validated by researchers studying the flow-field in both aerospace and industrial turbomachinery applications (Borello *et al.*, 2005, 2007, 2009, 2010; Corsini and Rispoli, 2005; Delibra *et al.*, 2009, 2010; Bianchi *et al.*, 2013).

In this chapter, we first describe the tunnel ventilation fan that is the subject of the reported research and consider the tunnel ventilation system applications into which we had previously installed it. We then define the pressure pulses to which the fan was subjected and clarify the computational approach that occurs with their modelling. We then present the developed numerical model and discuss the results. We analyse the static pressure field's transient evolution over the fan blades, identify the change in the three-dimensional flow-field through the blade passage and summarise the mechanical consequences that occur with pressure pulses.

TEST CASE DESCRIPTION

Railway tunnel and metro tunnel ventilation fans are typically between 1.6 and 2.8 metres in diameter, with the most common selection either a 1.8 or a 2.0 metres diameter fan running at 1,500 revolutions per minute. We conducted the reported research on an axial fan, named JFM 224, developed for application in railway tunnel and metro tunnel ventilation systems, Table 5.1. The fan is 2.24 metres in diameter with 16 blades and a hub-to-tip ratio of 0.5.

This studied fan is unusual as it has a blade tip speed of 175 metres per second in contrast to the generally accepted limit within the tunnel ventilation fan community of 156 metres per second. A 175 metres per second tip speed enables the JFM 224 fan to generate a total pressure rise of 2,500 Pa. Engineers generally consider a total pressure rise of 2,000 Pa to be a practical limit for a tunnel ventilation fan with 156 metres per second tip speed. A disadvantage of a 175 metres per second tip speed is that tip generated fan noise increases by approximately 10 dB compared to that of a 156 metres per second tip speed fan. As a consequence, tunnel ventilation design engineers have only utilised the JFM 224 fan in one tunnel ventilation system, the Delhi metro, where pressure developing capability was more important than fan noise. As such, the studied fan constitutes a practical limit in tunnel ventilation fan technology, with the highest pressure developing capability and consequently, the largest unsteady aerodynamic blade forces when stalled.

Table 5.1. *The studied fan, JFM 224, Borello* et al. *(2013a).*

Blade section		ARA-D	
Blade tip diameter D_t		2,240 mm	
Blade count		16	
Hub-to-tip ratio		0.5	
Duty point pressure		2,500 rpm	
Duty point flow		150 m^3/s	
	Hub		*Tip*
Chord c	143		92.5
Solidity σ	0.64		0.21
Pitch angle γ	48		24

AIR PRESSURE FLUCTUATIONS IN TUNNELS

Trains within a tunnel system generate a positive pressure pulse in front and a negative pressure pulse behind the train. The effect of the positive and negative pressure pulses on a tunnel ventilation fan depends on if it is operating in supply mode, supplying air into the tunnel system, or if it is operating in extract mode, extracting air from the tunnel system. When operating in extract mode the direction of flow is from the tunnel system, and consequently, static pressure is higher at the fan outlet than in the tunnel system. Therefore, as a train approaches the fan there is an increased flow rate through and reduced pressure across the fan. Conversely, as the train departs there is a reduced flow rate through and an increase in pressure across the fan.

Vardy (1980a) conducted a systematic experimental study on the effect of the unsteady flows that occur with train movement in metro tunnel systems in the London Victoria line. On the tunnel wall Vardy (1980a) was able to measure the variation in static pressure that occurred with a train moving through the tunnel, Figure 5.1. He recorded measurements with a train passing through a station at 70 kilometres per hour, not with the train stationary in the station and then accelerating into the tunnel. The reason was that a train passing through a station and entering the tunnel produced the most dramatic change in static pressure, which Vardy (1980a) concluded to be the worst case.

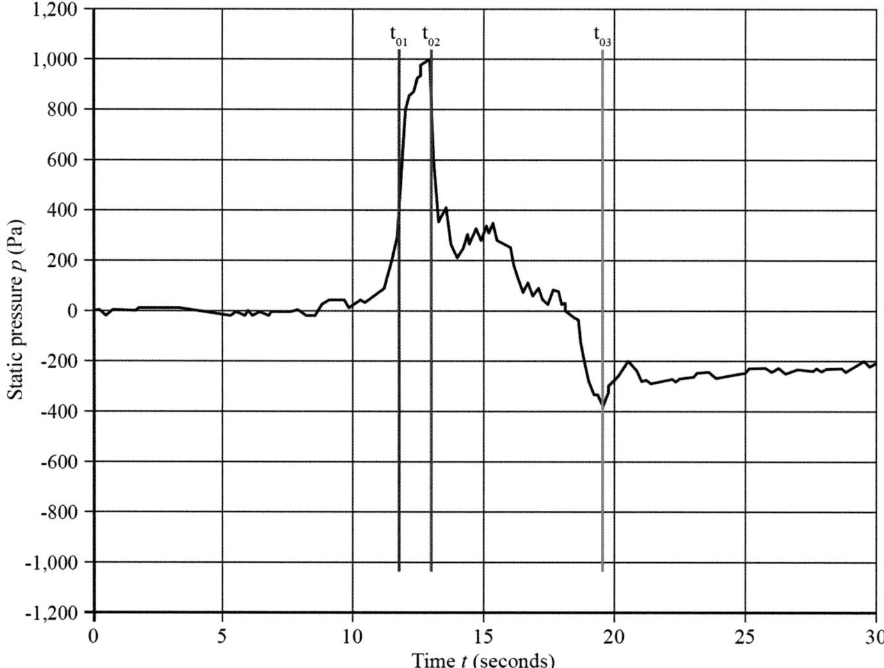

FIGURE 5.1. Change in static pressure in a London Transport Victoria Line metro tunnel (Vardy, 1980a).

Analysing measured static pressure data enabled Vardy (1980a) to conclude that as a train passed through the station and entered the tunnel, a shock wave travelled down the tunnel at the speed of sound. The shock wave arrived at the pressure transducer at time t_{01}, and resulted in a sharp increase in measured static pressure, Figure 5.1. At time t_{02} the train's nose draws level with the pressure transducer, resulting in a sharp decrease in measured static pressure. Between time t_{02} and t_{03} the train passes the pressure transducer, with static pressure falling due to friction effects. At time t_{03} the train's tail draws level with the pressure transducer, with the measured static pressure falling to a minimum due to a low-pressure zone immediately behind the train. After time t_{03} the static pressure gradually increases, as the flow-field through the tunnel stabilises following the train's passage.

Within the air movement and control community it is custom and practice to model a pressure pulse as a shift in system resistance. The fan duty point is the point at which the fan's characteristic intersects the system resistance curve, Figure 5.2. Engineers then assume that shifting the system resistance curve up and down 1,000 Pa models the effect of a positive and negative 1,000 Pa pressure pulse. They then assume the point at which the fan characteristic intersects the 'plus 1,000 Pa system curve' to be the fan duty point when subjected to a +1,000 Pa pressure pulse. Engineers then assume the point at which the fan characteristic intersects the

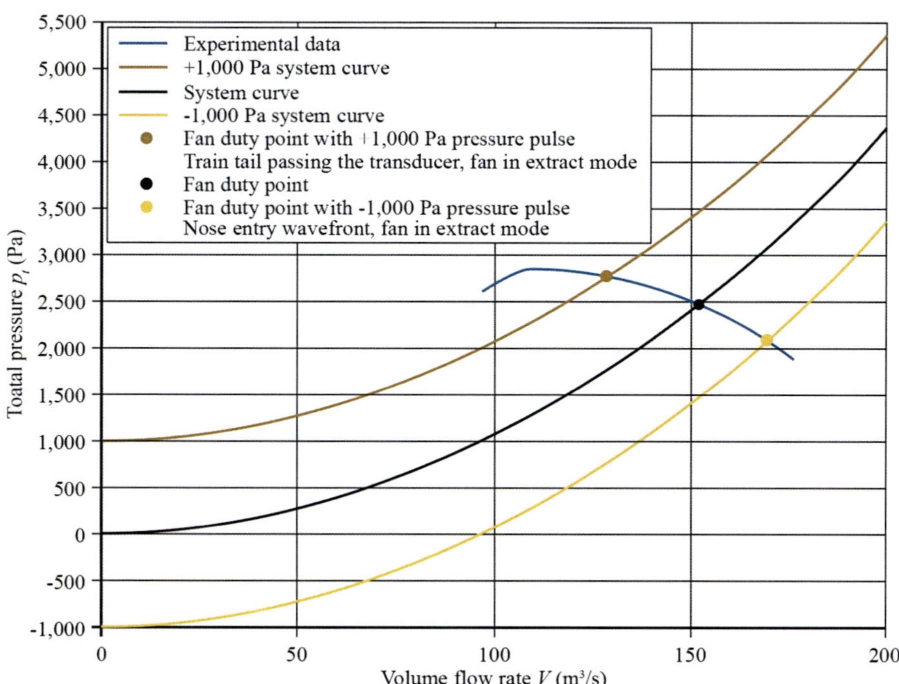

FIGURE 5.2. Fan characteristic curve (blue) with a system curve (black) and the assumed +1,000 Pa system curve associated with a +1,000 Pa pressure pulse (brown) and the assumed −1,000 Pa system curve associated with a −1,000 Pa pressure pulse (yellow).

'negative 1,000 Pa system curve' to be the fan duty point when subjected to a −1,000 Pa pressure pulse, Figure 5.2. Further, they assume that the time scales involved with applying pressure pulses is relatively slow compared to the fan's response time. Consequently, they assume that when operating in extract mode the fan duty point moves smoothly down and then up its characteristic moving from its nominal duty point, down to the −1,000 Pa duty point, back to its nominal duty point, up to the +1,000 Pa duty point and then finally back to its nominal duty point.

Vardy's data (1980a), Figure 5.1, indicates that there is a rapid increase followed by a rapid decrease in static pressure as the train's nose and then tail draw level with the pressure transducer used to measure static pressure. This indicates that the assumption within the air movement and control community when applying pressure pulses may not be realistic. However, Vardy's data (1980a) is not directly applicable to studying ventilation fan performance. The effect of switching tunnel ventilation fans on is to induce a velocity in the air within the tunnel system. Vardy made his pressure measurements (1980a) with the tunnel ventilation fans switched off. As such, Vardy's measurements (1980a) only apply in a tunnel system with tunnel ventilation fans running if we assume that the fan's effect and the train's effect are independent and therefore, we may linearly combine them. Further, Vardy (1980a) measured changes in static pressure. Within the air movement and control community it is custom and practice to characterise tunnel ventilation fan performance in terms of total pressure. If we are to use Vardy's data we must assume that we may neglect total pressure losses in the tunnel system, and consequently, we may characterise the change in static pressure which Vardy measured (1980a) as a total pressure change.

In the research reported in this chapter the above assumptions have enabled us to use Vardy's data (1980a) to characterise the pressure pulse applied to a tunnel ventilation fan as an instantaneous pressure discontinuity. We modelled this pressure discontinuity as a discontinuity of the bulk velocity within a tunnel ventilation shaft. We calculated the change in bulk velocity from the pressure pulse's magnitude:

$$\Delta p_{pulse} = \rho a \Delta U_{bulk} \tag{1}$$

where, ρ and a are the reference air density and the undisturbed flow's speed of sound.

The change in pressure which Vardy measured (1980a) was up to 1,000 Pa and therefore similar to the largest pressure pulses ventilation system designers are specifying. In this chapter's programme of work, we modelled the effect of a ± 1,000 Pa pressure pulse. Following Vardy (1980a), we modelled the pressure wave fronts as impulses applied over 4 milliseconds, Table 5.2. If we assume that the tunnel ventilation fan is operating in extract mode, the approaching train will reduce pressure across the fan, and the departing train will increase pressure across the fan. Therefore, the effect of a train approaching will be to reduce the duty point from its nominal 2,500 Pa over 4 milliseconds to 1,500 Pa at an assumed starting time of t_{01}, Figure 5.1. At a starting time of t_{02}, we then increased the 2,500 Pa duty point over 4 milliseconds to 3,500 Pa, Figure 5.1.

Table 5.2. *Pressure pulse characteristics*

Static pressure rise Δp	±1,000 Pa
Time for static pressure rise Δt_{ramp}	1.3×10^{-5} s
Duration of pressure pulse Δt_{pulse}	4 ms

We modelled each pressure pulse as a separate event. We first applied the pressure pulse, and then computed for two seconds after applying the pressure pulse. We considered modelling the evolution of pressure between pressure pulses beyond the scope of the reported research. The reported research focuses on establishing the impact associated with applying a pressure pulse.

COMPUTATIONAL METHODOLOGY

We conducted computational analysis using the open-source finite volume c^{++} computational fluid dynamic code OpenFOAM 1.7.x. Using this code, we computed the large eddy simulation by solving the filtered Navier-Stokes incompressible equations using a dynamic sub-grid scale model which Davidson (1997) proposed. This sub-grid scale model requires the code to solve an additional transport differential equation for the sub-grid turbulent kinetic energy k_{SGS} which then enables the evaluation of the velocity-scale at the sub-grid level. According to Davidson (1997), we can compute the sub-grid scale viscosity as:

$$\nu_{SGS} = < c_{SGS} >_{xyz} \sqrt{k_{SGS} l_{LES}} \tag{2}$$

where we obtain the coefficient $<c_{SGS}>$ using the dynamic Smagorinsky model (Germano *et al.*, 1991). The fluid kinematic viscosity ν limited the sub-grid scale viscosity. The adopted sub-grid scale model accounts for backscatter, and partly compensates for the effect of a computational grid that is too coarse for a fully resolved large eddy simulation.

We carried out computations using a central difference scheme that incorporated a total value diminishing (TVD) scheme to prevent numerical instabilities and a second-order implicit approach for the time marching solution. Borello *et al.* (2013a) found that using a total value diminishing scheme in combination with the dynamic large eddy simulation controlled the high-frequency turbulent energy's dissipation rate. OpenFOAM's linear solver makes use of an incomplete lower and upper (ILU) preconditioned semi-iterative conjugate gradient (CG) algorithm, in combination with a pressure implicit with splitting of operators (PISO) segregation scheme. We set the convergence threshold to 10^{-8} for the algebraic CG solver and to 10^{-5} for the PISO algorithm. We solved the rotating flow-field in the relative frame of reference, accounting for the effects of Coriolis and centrifugal forces using Borello *et al.*'s (2013a) developed version of the *pisoFoam* solver. Borello *et al.* (2013a) provide a more complete assessment of the methodology, reporting the results of a large eddy simulation of the fan that is the subject of the research reported in this chapter. Borello *et al.* (2013a) conducted a mesh independence study, demonstrating the computational results were independent of the mesh. In the research reported in this chapter, we utilised the same mesh as Borello *et al.* (2013a).

NUMERICAL GRID AND BOUNDARY CONDITIONS

The computational domain extends half a blade chord upstream of the blade leading edge and one chord downstream of the blade trailing edge, Figure 5.3. This distribution results in 16 per cent of the elements upstream of the blade leading edge, 50 per cent of the elements within the blade passage and 34 per cent of the elements downstream of the blade trailing edge. In the span-wise direction, 55 nodes were clustered in the blade tip-to-casing gap. In total, the mesh comprises approximately 9 million hexahedral elements distributed with a block-structured topology, Table 5.3.

The mesh was clustered toward solid boundaries, with the ratio of minimum grid spacing on solid walls to mid-span blade chord set as 7×10^{-4} on the blade tip, casing wall and blade surfaces. The adopted grid refinement towards the solid surfaces sets the normalised wall distance y^+ value to approximately one on the first row of nodes in order to have the first cell within the viscous sub-layer. We derived the inflow pitch-averaged profiles of both axial and radial velocity by numerically simulating the flow-field in the fan's inlet spinner cone region. This method of generating boundary conditions enabled the numerical simulation to account for the effect of the flow area's sharp reductions at the fan inlet. Table 5.4 summarises boundary conditions on each boundary surface.

RESULTS

Modelling pressure pulses as a discontinuity of the bulk velocity within a tunnel ventilation shaft facilitated a prediction of the aerodynamic consequence for the tunnel ventilation fan. Initially, the fan is modelled at its duty point, Point A, Figure 5.4. As the negative pressure pulse that occurs with the approaching train impacts the fan there is a sharp pressure drop across the fan that we may associate with choking the annulus, Point B, Figure 5.4. As the fan adapts to the increased volume flow, the pressure rises and we may consider the fan operating point to have stabilised on its characteristic, Point C, Figure 5.4. In contrast to the negative pressure pulse, we may characterise the positive pressure pulse that occurred with the departing train by a sharp increase in pressure across the fan, Point D, Figure 5.4, followed by a reduction in bulk velocity through the fan, Point E, Figure 5.4.

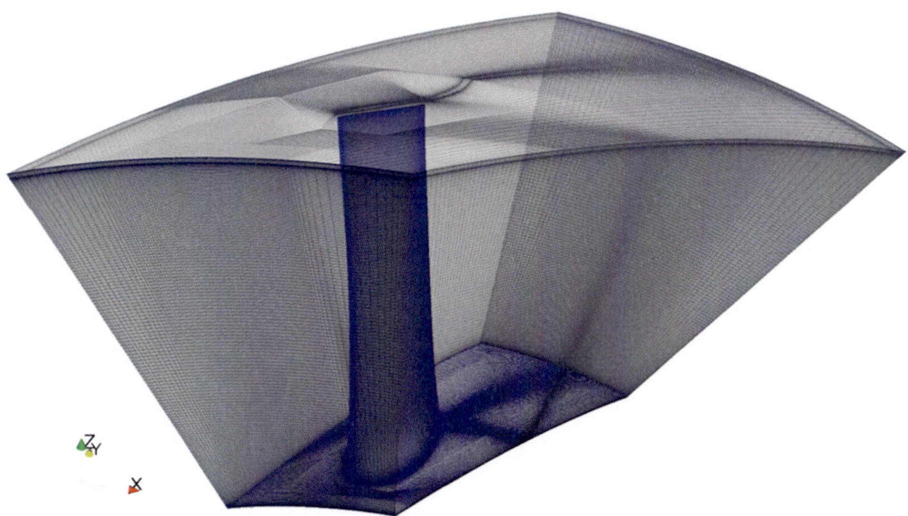

FIGURE 5.3. The studied fans mesh and computational domain, Borello *et al.* (2013a).

Table 5.3. *Computational mesh data, Borello* et al. *(2013a).*

Nodes	9,021,968
Cells	8,862,550
Tip gap nodes	60,753
Tip gap cells	52,000
Blade surface cells	62,450
Average cell aspect ratio	1.34

Table 5.4. *Boundary conditions, Borello* et al. *(2013a).*

Inflow	Velocity profile with ±4.8% variation of U_{bulk} to simulate the effect of pressure pulse following Equation (1)
Outflow	Zero gradient on all quantities
Rotor	Non-slip
Stator	Relative velocity w = ωR
	Periodic boundaries at mid-pitch

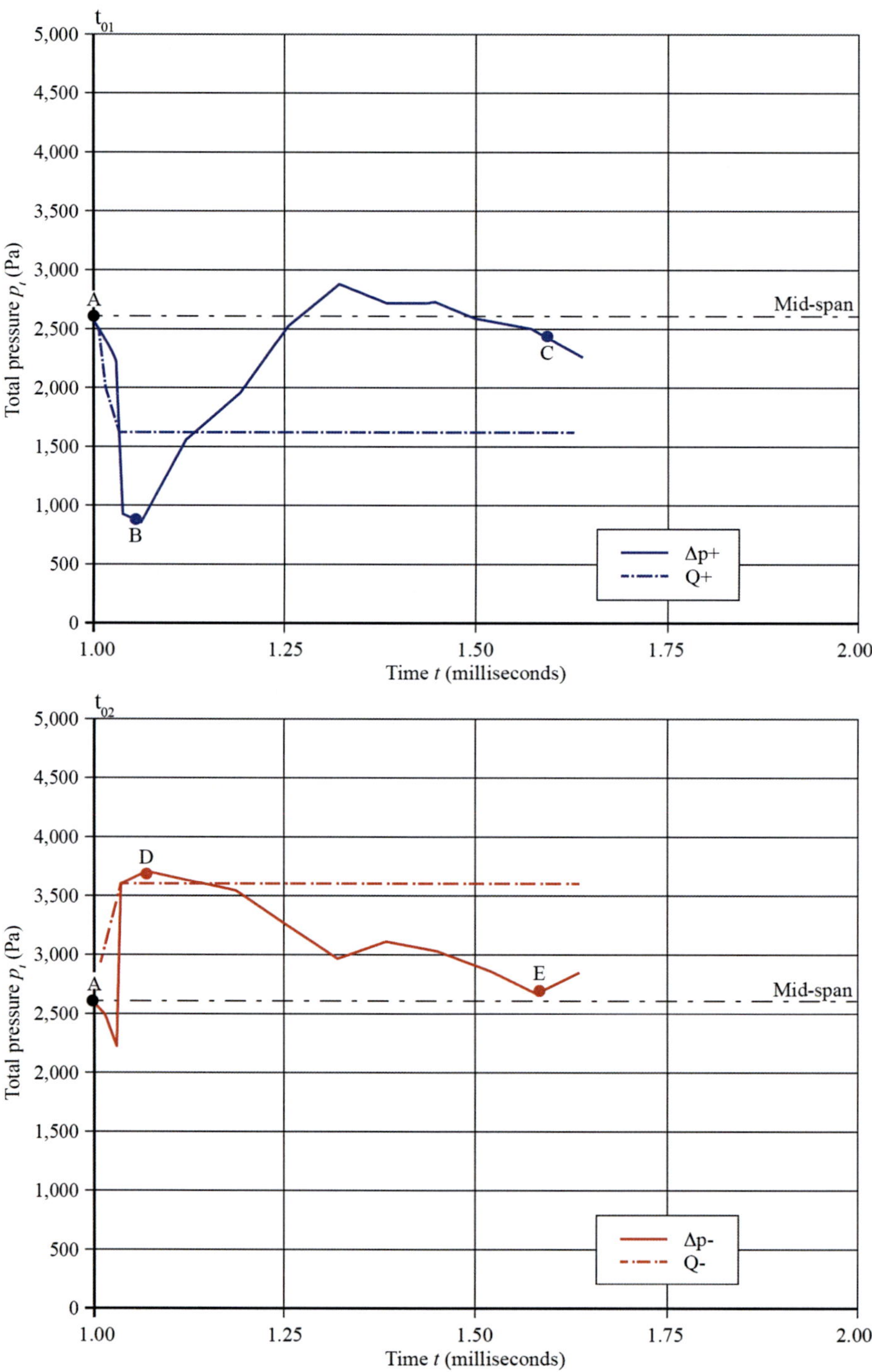

FIGURE 5.4. Total pressure rise evolution after a negative pressure pulse (blue) and positive pressure pulse (red). Solid lines: time varying computational results, chain lines: modelled pressure wave fronts.

Both negative and positive pressure pulses result in the predicted pressure rise over-shooting the step change in pressure that occurred with the bulk velocity's discontinuity. The negative pressure pulse results in a predicted minimum pressure across the fan of 1,800 Pa, Point B, Figure 5.5. The point at which the fan characteristic intersects the 'negative 1,000 Pa system curve' is 2,100 Pa, Figure 5.5, and therefore the predicted minimum pressure across the fan of 1,800 Pa is 300 Pa lower than expected.

The positive pulse results in a predicted maximum pressure rise across the fan of 4,000 Pa, Point D, Figure 5.5. The point at which the fan characteristic intersects the 'positive 1,000 Pa system curve' is 2,750 Pa, Figure 5.5, and therefore the predicted minimum pressure across the fan of 4,000 Pa is 1,250 Pa higher than expected.

Some within the air movement and control community generally assume that when subjected to a pressure pulse, a fan's operating point will move smoothly along its characteristic from its nominal duty point to the point where the system line that occurs with the negative pressure pulse and positive pressure pulse cross the fan's characteristic. When we consider the shift in a fan's operating point from Point A to Point C and from Point A to Point E, Figure 5.5, they are reasonably close to the point at which the negative and positive pressure pulse system lines cross the fan's characteristic. This indicates that the general assumptions within the air movement and control community do apply over time scales of more than half a second.

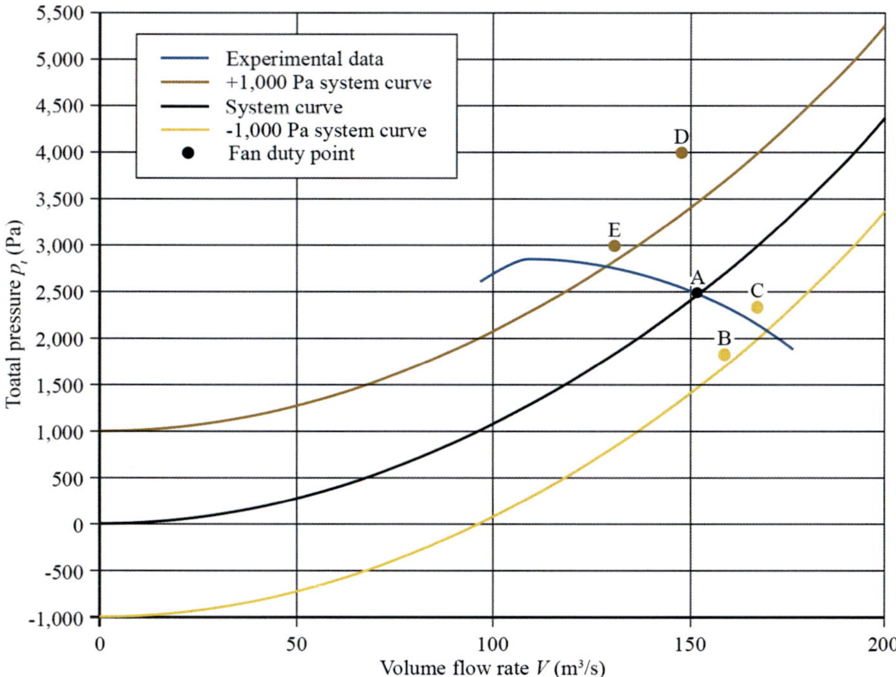

FIGURE 5.5. Evolution of the fan duty point during the negative pressure pulse (Points A, B and C) occurring with a fan approaching a tunnel ventilation shaft and positive pressure pulse (Points A, D and E) occurring with negative pressure pulse. The tunnel ventilation fan is operating in extract mode, extracting air from the tunnel system.

However, they do not apply during the transitional experience that occurs with the rapid pressure changes which Vardy measured (1980a). The data point of primary concern is the positive pressure pulse occurring when a train departs that increases the pressure across the fan, Point D, Figure 5.5. The instantaneous pressure across, and flow rate through the fan takes Point D beyond the fan's peak pressure developing capability. Under steady state operating conditions, attempting to operate the studied fan at Point D would result in the fan stalling.

TRANSIENT BLADE LOADING

Considering the blade loading's transient evolution provides insight into the mechanical impact of the transitional experience of adapting to first negative and second positive pressure pulses. The evolution of bending moment components in both the peripheral (Mp) and in the axial (Ma) direction indicates that applying both negative and positive pressure pulses is associated with a significant increase in the bending moment on fan blades, Figure 5.6.

After applying the negative pressure pulse (t_{01}, Figure 5.6), the pressure across the fan reduces, flow through the fan increases and the resultant fan choking results in a doubling of both peripheral and axial bending moment components. After applying the positive pressure pulse (t_{02}, Figure 5.6), the axial bending moment component drops to almost zero. A consequence of change in bending moment is that the unsteady load to which the fan blades are subjected doubles. Sheard and Corsini (2012) studied the mechanical impact of aerodynamic stall. They conducted an experimental programme to measure the increase in unsteady load to which fan blades were subjected as the studied fan stalled. They concluded that the unsteady forces increased by a factor of seven during stall. Therefore, the doubling of unsteady load, which we calculated in the reported research is low compared to what one sees when a fan stalls, thus we may conclude that the pressure pulse does not drive the studied fan into stall.

Sheard and Corsini's research (2012) considered the fan blades mechanical safety factor, with safety factor defined as the distance from the Gerber line for the material used to manufacture fan blades. A safety factor of one constitutes a combination of steady and unsteady forces that will result in an infinite fatigue life for the blade. A safety factor of less than one will result in a finite fatigue life, and therefore the potential for the blade to fail in fatigue. The fan that Sheard and Corsini (2012) studied had a safety factor of 2.3 under normal operating conditions, and 0.3 when operated in stall. From this we may conclude that the fan had an infinite fatigue life when operated under normal operating conditions, but would suffer a fatigue failure if operated in stall.

One may use Sheard and Corsini's method (2012) to calculate the impact on the safety factor associated with doubling unsteady loads. Assuming a safety factor of 2.3 under normal operation, doubling unsteady loads will result in a safety factor dropping to approximately 1.1. As a safety factor of 1.1 is still greater than one, the fan should still have an infinite fatigue life. However, Sheard and Corsini (2012) note that a safety

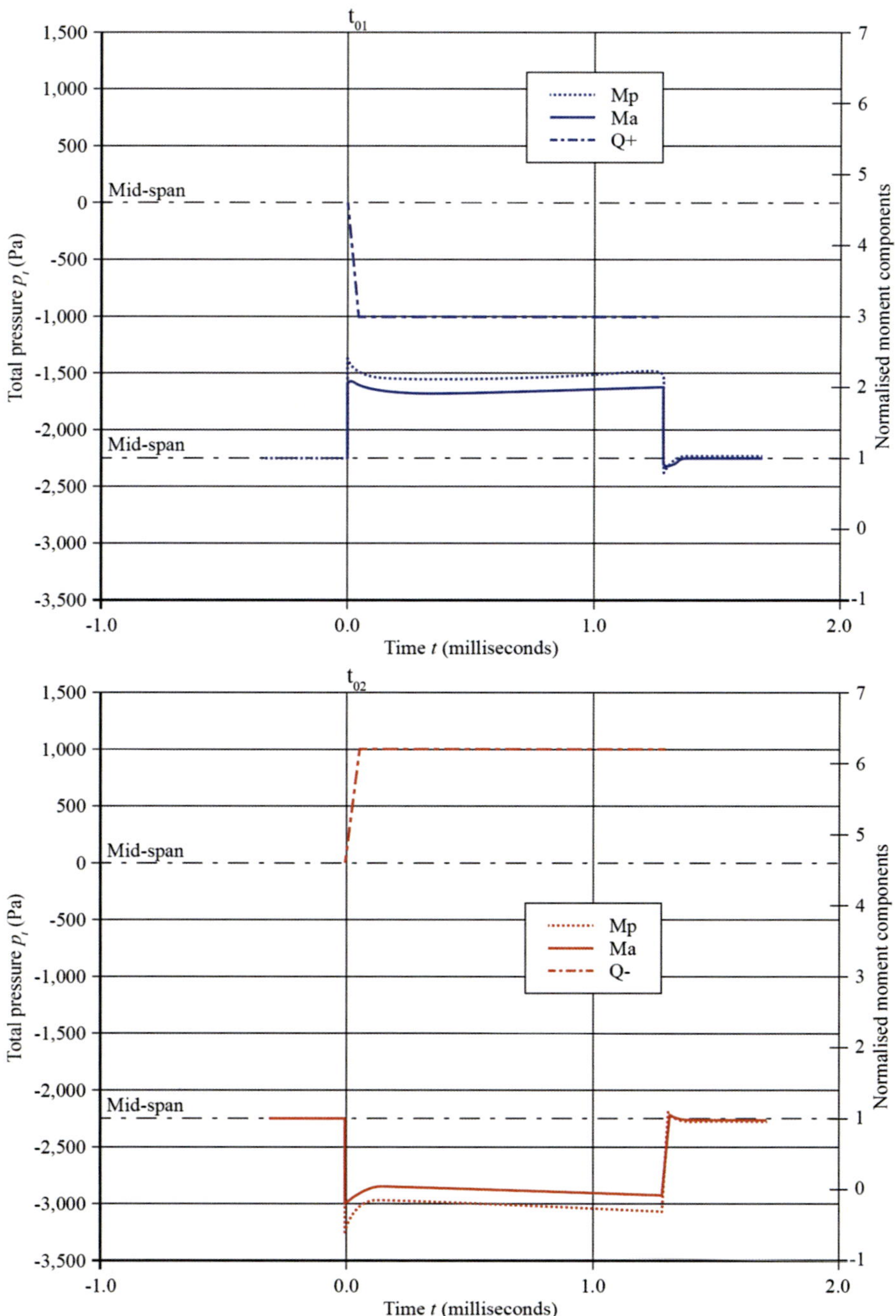

FIGURE 5.6. Time histories of the axial and peripheral moment components M_a and M_p on the blade surface under Q+ and Q-. Values are normalised with respect to the corresponding value in normal operations (labelled as A in Figure 5.4 and 5.5).

factor of two is the norm within the industrial fan community. If the safety factor under normal operation was slightly below two, or unsteady loads were to slightly more than double, then the safety factor could fall below one, resulting in a finite blade fatigue life and therefore, the potential for a blade's in-service fatigue failure.

TRANSIENT BLADE PRESSURE COEFFICIENTS

The mechanical impact of both negative and positive pressure pulses occurs as a consequence of the blade-to-blade flow-field's transient evolution. It is the change in flow-field that results in the change in bending moment on the fan blades. We may characterise the blade-to-blade flow-field by considering the blade static pressure coefficient distributions' transient evolution when subjected to first the negative, Figure 5.7, and second the positive, Figure 5.8, pressure pulse. For both the negative and positive pressure pulse, we have presented the calculated blade tip, mid-span and

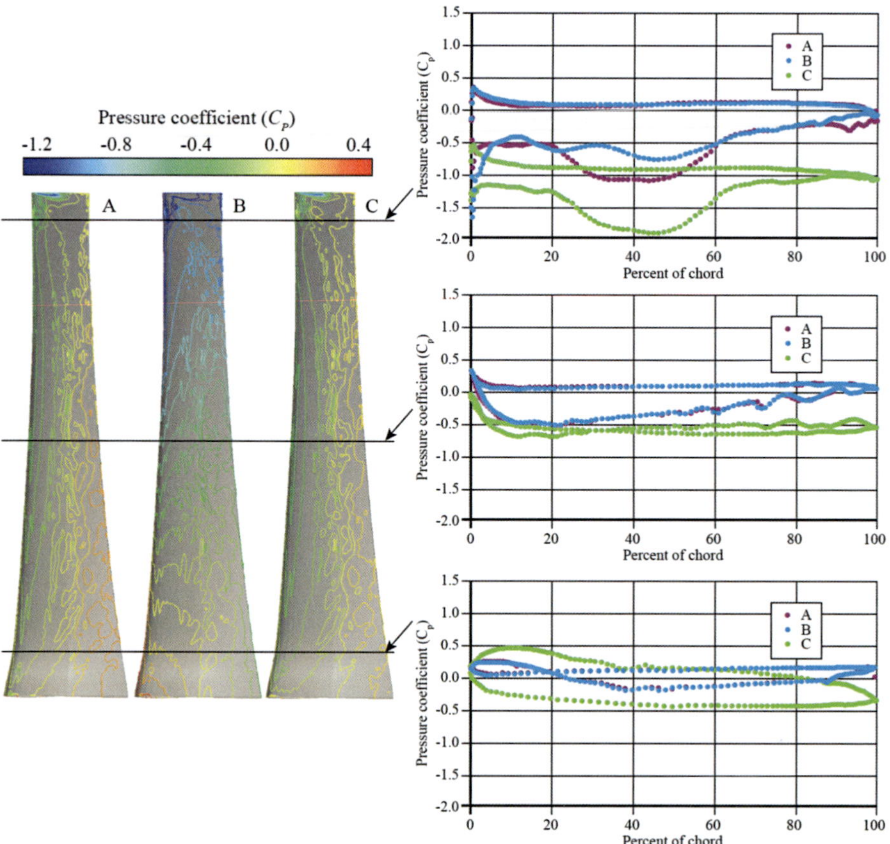

FIGURE 5.7. Left: Pressure coefficient on the rotor's suction surface, for instance A, B, C. Right: Pressure coefficient along the blade at tip (top), mid-span (middle) and hub (bottom).

hub pressure coefficient distributions. For the negative pressure pulse, Figure 5.7, we provide results for operating Point A, Point B and Point C, Figure 5.5. For the positive pressure pulse, Figure 5.8, we provide results for the operating Point A, Point D and Point E, Figure 5.5.

Consider first the change in the static pressure coefficient's iso-contours as the fan moves from its steady state duty point, Point A, Figure 5.7, to the Point B, Figure 5.7, under the influence of the negative pressure pulse. The static pressure coefficient's iso-contours at Point A are almost radial along the full blade span. In contrast, at Point B the static pressure coefficient's iso-contours are almost axial at mid-span. This change from radial to axial static pressure coefficient iso-contours indicates a flow separation that appears to extend from approximately one- to two-thirds of the blade span. This shift at mid-span indicates that transiently the fan has stalled at this point. The blade span's inner and outer third is still characterised by radial static pressure coefficient iso-contours, from which we may conclude that the blade span's inner and outer third are still capable of contributing to the fan's pressure developing capability.

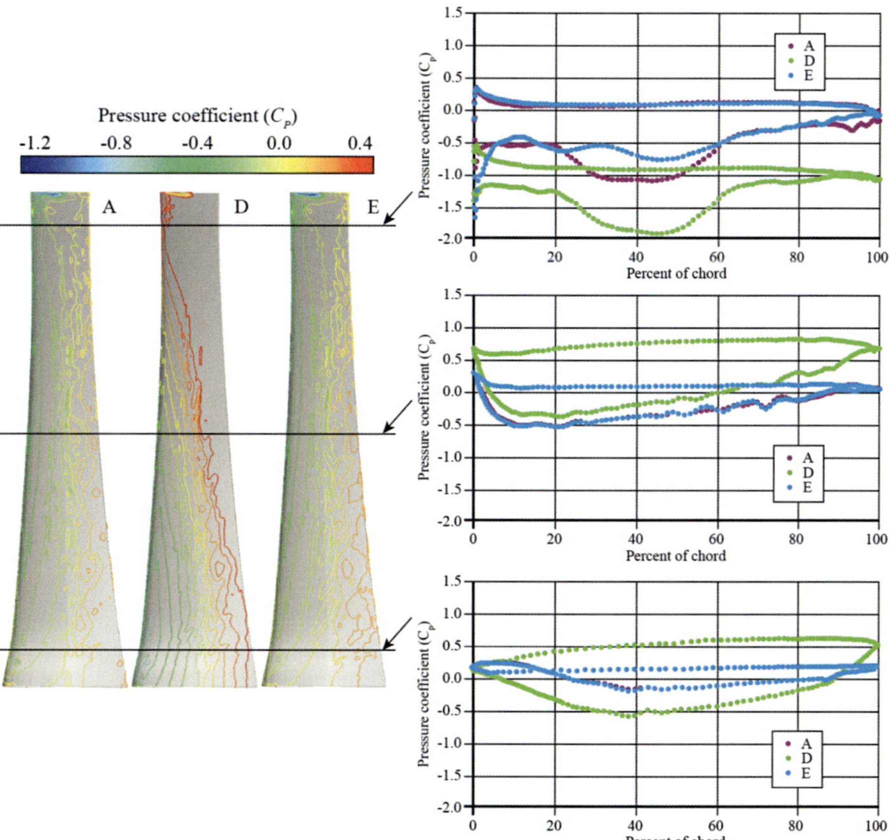

FIGURE 5.8. Left: Pressure coefficient on the rotor's suction surface, for instance A, D, E. Right: Pressure coefficient along the blade at tip (top), mid-span (middle) and hub (bottom).

Despite this transient change in the static pressure coefficient's iso-contours that occur with the shift from Point A to Point B, the fan is able to adapt to the new flow rate, Point C, Figure 5.7. As the fan stabilises at the new flow rate, the static pressure coefficient's iso-contours return to a near radial distribution over the full blade span. A review of the calculated blade tip, mid-span and hub distributions of static pressure coefficient distributions, Figure 5.7, illustrates that the blade tip, mid-span and hub distributions change significantly between Point A and Point B. By Point C, the mid-span and hub distributions have returned to a form essentially similar to that at Point A. The blade tip distribution returns to a form similar to that at Point A, but with some variation that we may attribute to a change in the blade tip-to-casing leakage flow rate that occurs with the increase in flow rate through the fan.

Now consider the change in the static pressure coefficient's iso-contours as the fan moves from its steady state duty point, Point A, Figure 5.8, to the Point D, Figure 5.8, under the influence of the positive pressure pulse. The static pressure coefficient's iso-contours at Point D remain radial over the full blade span. However, a review of the calculated blade tip, mid-span and hub distributions of static pressure coefficient distributions, Figure 5.8, indicates that the drop in mass flow rate through, and associated pressure rise across the fan has resulted in an increase in lift at the blade mid-span and hub. This increase in lift indicates that the fan is doing an increased amount of work at the mid-span and hub sections. In contrast, the blade tip distribution of static pressure coefficient indicates that the blade tip is overloaded and has stalled. As was the case with the negative pressure pulse, despite this transient change in the static pressure coefficient's iso-contours that occurred with the shift from Point A to Point D, the fan is able to adapt to the new flow rate, Point E, Figure 5.8. As the fan stabilises at the new flow rate, the static pressure coefficient's iso-contours return to a near radial distribution over the full blade span.

TRANSIENT VORTICAL STRUCTURES

An analysis of the vortical structures that develop over the fan blades in response to the negative and positive pressure pulses, Figure 5.9, may complement the transient blade loading and static pressure coefficient analysis. We can illustrate the vortical structure development as a series vortex deduction parameter (Q) iso-surfaces during the negative pressure pulse transient (Point A to Point C via Point B) and positive pressure pulse transient (Point A to Point E via Point D). The previous transient blade loading and static pressure coefficient analysis demonstrated that the blade loading and static pressure coefficients were different at the transient points Point B and Point D. In contrast, the large-scale vortical structures, Figure 5.9, are not affected by either the negative or positive pressure pulse, Point B and Point D respectively. There is some evidence of a minor modification of the vortical structure that occurs with the blade tip leakage vortex and a leading edge separation, but overall, the blade vortical structure appears to have been unaffected by either negative and positive pressure pulses.

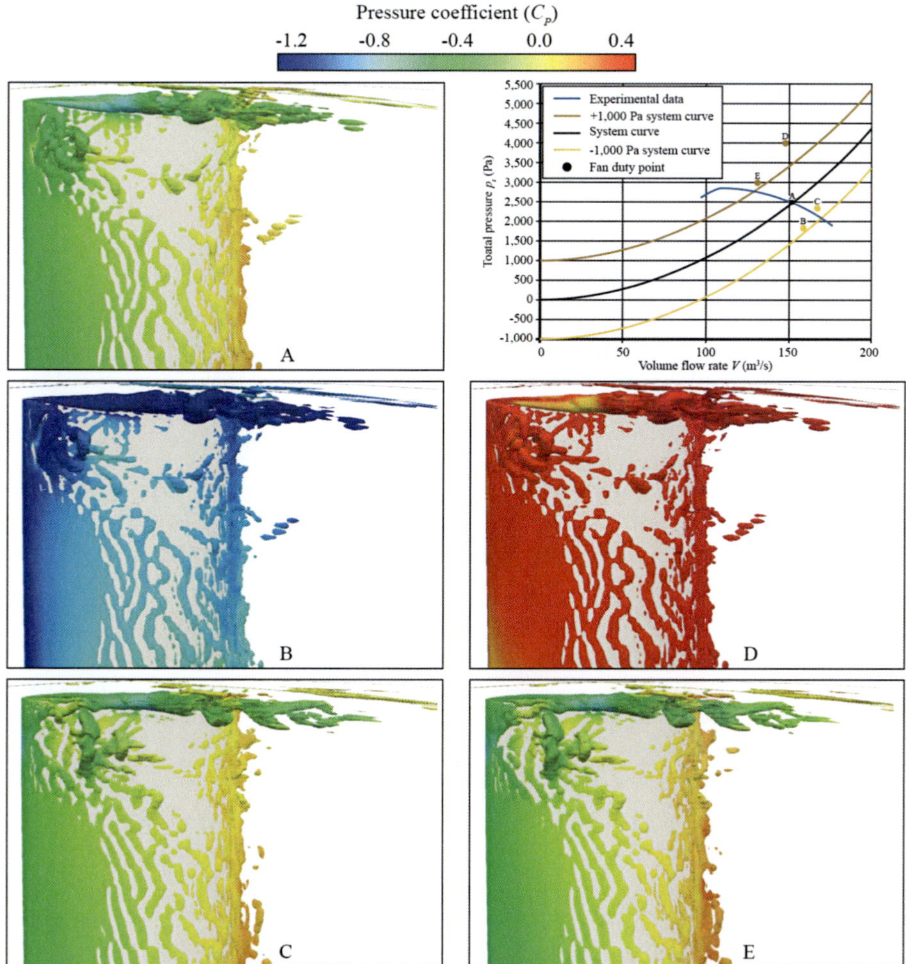

FIGURE 5.9. Turbulent structures visualised with iso-surfaces of Q=10,000 coloured with pressure coefficient. The characteristic curve on the top-right shows the five operating conditions.

We may associate vortical structures that do not change during either the negative or positive pressure pulses with a pressure pulse timescale that is short in comparison to the vortical structures' reaction time. From this we may speculate that the large scale vortical structures' intrinsic coherence requires more time than is available to adapt to the conditions imposed transiently during the negative pressure pulse (Point B) or positive pressure pulse (Point D). When we study the vortex deduction parameter iso-surfaces further, it is apparent that the large scale vortices that occur with the blade tip leakage vortex are largely unaffected by either negative or positive pressure pulses. In contrast, the smaller scale vortical structures that occur with the blade trailing edge flow show some identifiable differences between those at Point A and those at Point C and Point E. This change in small scale vortical structures indicates that they react over a time scale comparable with that of the pressure pulse time scale.

TRANSIENT TIP LEAKAGE VORTEX DEVELOPMENT

The final aspect of our transient analysis is the evolution of the blade tip leakage vortex under the influence of both the negative and positive pressure pulses, Figure 5.10.

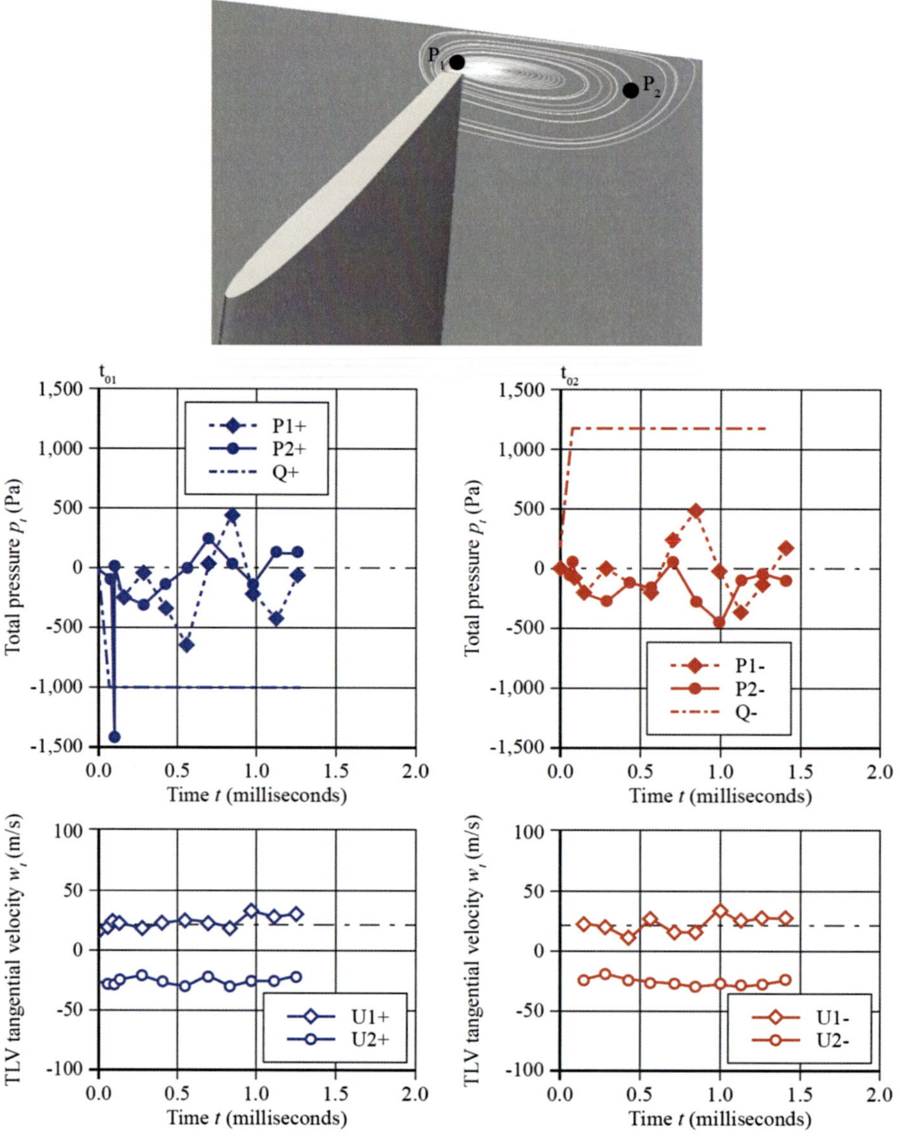

FIGURE 5.10. Time histories of the tip leakage vortex (TLV) tangential velocity component and pressure on points P1 and P2 (see insert) during the negative and positive pressure pulses.

We may illustrate the tip leakage vortex's evolution by considering the tip leakage vortex tangential velocity (w_t) plotted against the local static pressure distributions during both a negative and positive pressure pulse. It is noteworthy that the tip leakage vortex tangential velocity and local static pressure distribution are relatively unaffected by the passage of either the negative or positive pressure pulse. This consistency is self-consistent with previous transient vortical structure analysis conclusions, indicating that the tip leakage vortex structure reacts over a time scale longer than the time scale that occurs with either the negative or positive pressure pulse. There is some evidence of a change in static pressure rise at the start of both negative and positive pressure pulses. However, this change in static pressure rise decays in three rotor revolutions at a frequency of approximately 15 per cent of the fan's rotational frequency. Therefore, we may conclude that both the negative and positive pressure pulse have little impact on the blade tip leakage vortex structure.

CONCLUSIONS

A combination of recent legislative and market demand have resulted in a need for higher pressure tunnel ventilation fans. The magnitude of pressure pulses to which tunnel ventilation fans are subjected in-service is also larger than has been the historic norm. In-service failures of tunnel ventilation fans are relatively uncommon, and have classically occurred with a ventilation system operating scenario that inadvertently drives the fan into a stalled condition that then results in a fatigue failure of one or more fan blades. Recent developments in tunnel ventilation control systems and tunnel fan condition monitoring have made inadvertent stalled operation less likely. However, recent in-service failures still occur with a fatigue failure of one or more fan blades. If we accept that the elevated stress levels in fan blades needed to induce a fatigue failure do not occur with operation in a stalled condition, then some other mechanism must be responsible.

The research reported in this chapter studies the effect of subjecting the highest pressure tunnel ventilation fan currently in-service to a negative 1,000 Pa pressure pulse followed by a positive 1,000 Pa pressure pulse to establish the mechanical consequences. We presented the results of a large eddy simulation in which we simulated the effect of the pressure pulses by assuming a discontinuous change of volume flow rate. We presented the pressure rise's transient evolution, and analysed the fan blades' pressure field variation during the transient passage of both negative and positive pressure pulses through the blade passage. We studied the evolution of small-scale vortical structures, and calculated the resultant peripheral and axial aerodynamic forces induced in the fan blades.

The computational results indicate that the studied fan is able to aerodynamically adapt quickly to the new operating conditions and that the large-scale vortical structures released from the blade are not affected by either the negative or positive pressure pulse. However, both negative and positive pressure pulses do affect small-scale structures in the flow that are characterised by a time-scale comparable with those of the incoming pressure pulse. The negative pressure pulse leads to chocking the flow, with evidence that the fan mid-span section stalls transiently. The positive pressure pulse occurs with an increase in blade loading, but no span-wise redistribution of the blade static pressure field. A consequence of the blade loading increase is that the fan tip section stalls transiently.

We chose the studied fan as it is the highest pressure single-stage reversible fan currently in-service in a tunnel ventilation system. The design has a high enough pressure developing capability to accommodate the imposed 1,000 Pa pressure pulse without stalling, if one applies the pressure pulse as a gradual change in duty point pressure. However, when one models the negative and positive pressure waves as a change in volume flow rate, the numerical analysis indicates that the negative pressure pulse induces a transient blade mid-span stall and the positive pressure pulse induces a transient blade tip stall. The consequence of blade mid and tip section stall is a doubling of the aerodynamic forces applied to the fan blade, and consequently a doubling of the mechanical stress induced in the blade. This doubling of induced mechanical stress each time a train passes the ventilation shaft within which a tunnel ventilation fan is situated may, over time, induce a fatigue crack in one or more fan blades. Once a fatigue crack is induced in at least one blade, it will propagate under the influence of the unsteady mechanical stress induced in a blade by stable operation, ultimately resulting in an in-service fatigue failure.

It is possible that higher pressure tunnel ventilation fan aerodynamic designs, in combination with large pressure pulses, are responsible for some recent in-service tunnel ventilation fan failures. Tunnel ventilation fan designers must respond by modelling pressure pulses' aerodynamic effect, calculating the mechanical consequences and adapting the mechanical design of blades to withstand the actual imposed loads.

REFERENCES

Bellenoue, M., Moriniere, V. and Kageyama, T. (2011), 'Experimental 3-D Simulation of the Compression Wave, Due to Train-tunnel Entry', *Journal of Fluids and Structures*, vol. 16, pp. 581–595.

Bianchi, S., Borello, D., Corsini, A., Rispoli, F. and Sheard, A.G. (2013), 'Large-eddy Simulation of the Aerodynamic and Aeroacoustic Performance of a Ventilation Fan', *Advances in Acoustics and Vibration,* vol. 2013, article ID 876973, pp. 1–10.

Borello, D., Hanjalić, K. and Rispoli, F. (2005), 'Prediction of Turbulence and Transition in Turbomachinery Flows Using an Innovative Second Moment Closures', *Transactions of the ASME, Journal of Fluids Engineering*, vol. 127(6), pp. 1059–1070.

Borello, D., Hanjalić, K. and Rispoli, F. (2007), 'Computation of Tip-leakage Flow in a Linear Compressor Cascade with a Second-moment Turbulence Closure', *International Journal of Heat and Fluid Flow*, vol. 28(4), pp. 587–601.

Borello, D., Delibra, G., Hanjalić, K. and Rispoli, F. (2009), 'Large-eddy Simulations of Tip Leakage and Secondary Flows in an Axial Compressor Cascade Using a Near-wall Turbulence Model', *Proceedings of the Institute of Mechanical Engineers, Part A: Journal of Power and Energy*, vol. 223(6), pp. 645–655.

Borello, D., Delibra, G., Hanjalić, K. and Rispoli, F. (2010), 'Hybrid LES/RANS Study of Turbulent Flow in a Linear Compressor Cascade with Moving Casing', *Proceedings of the 55th American Society of Mechanical Engineers Turbine and Aeroengine Congress,* Glasgow, UK, 14–18 June, paper no. GT2010-23755.

Borello, D., Corsini, A., Delibra, G., Fiorito, M. and Sheard, A.G. (2013a), 'Large-eddy Simulation of a Tunnel Ventilation Fan', *Transactions of the ASME, Journal of Fluids Engineering*, vol. 135(7), paper no. 071102, pp. 1–9.

Borello, D., Corsini, A., Delibra, G. and Sheard, A.G. (2013b), 'Numerical Investigation of Detrimental Aerodynamic Effect of Pressure Pulses on a Metro Tunnel Fan', *Proceedings of the 10th European Turbomachinery Conference*, Lappeenranta, Finland, 15–19 April, pp. 573–582.

Corsini, A. and Rispoli, F. (2005), 'Flow Analyses in a High-pressure Axial Ventilation Fan with a Non-linear Eddy Viscosity Closure', *International Journal of Heat and Fluid Flow*, vol. 26(3), pp. 349–361.

Davidson, L. (1997), 'Large Eddy Simulation: a Dynamic One-equation Sub-grid Model for Three-dimensional Recirculating Flows', *Proceedings of the 11th International Symposium on Turbulent Shear Flow*, Grenoble, France, 8–10 September, pp. 26.1–26.6.

Delibra, G., Borello, D., Hanjalić, K. and Rispoli, F. (2009), 'URANS of Flow and Endwall Heat Transfer in a Pinned Passage Relevant to Gas-turbine Blade Cooling', *International Journal of Heat and Fluid Flow*, vol. 30(3), pp. 549–560.

Delibra, G., Borello, D., Hanjalić, K. and Rispoli, F. (2010), 'An LES Insight into Convective Mechanisms of Heat Transfer in a Wall-bounded Pin Matrix', *Proceedings of the 14th International Heat Transfer Conference*, Washington DC, USA, 8–13 August, paper no. IHTC14-23205, pp. 807–815.

Germano, M., Piomelli, U., Moin, P. and Cabot, W.H. (1991), 'A Dynamic Subgrid-scale Eddy Viscosity Model', *Physics of Fluids*, vol. 3(7), pp. 1760–1765.

Huang, Y.D. and Gao, W. (2010), 'A Numerical Study of the Train-induced Unsteady Airflow in a Subway Tunnel with Natural Ventilation Ducts Using the Dynamic Layering Method', *Journal of Hydrodynamics*, vol. 22(2), pp. 164–172.

Jasak, H. (2010), 'OpenFOAM: a Year in Review', *Proceedings of the 5th OpenFOAM Workshop*, Gothenburg, Sweden, 21–24 June.

Labois, M. and Lakehal, D. (2011), 'Very-large Eddy Simulation (V-LES) of the Flow Across a Tube Bundle', *Nuclear Engineering and Design*, vol. 241(6), pp. 2075–2085.

Rippl, A. (1995), 'Experimentelle Untersuchungen zuminstationaren Betriebsverhahen an der Stabilitarsgrenze eines mehrstufigen transsonischen Verdichters', PhD thesis, Ruhr-Universität Bochum, Germany.

Schneider, H., von Terzi, D. and Bauer, H.J. (2010), 'Large-eddy Simulations of Trailing-edge Cutback Film Cooling at Low Blowing Ratio', *International Journal of Heat and Fluid Flow*, (5), pp. 767–775.

Sheard, A.G. and Corsini, A. (2012), 'The Mechanical Impact of Aerodynamic Stall on Tunnel Ventilation Fans', *International Journal of Rotating Machinery*, vol. 2012, paper no. 402763, pp. 1–12.

Sheard, A.G. and Daneshkhah, K. (2012), 'The Conceptual Design of High Pressure Reversible Axial Tunnel Ventilation Fans', *Advances in Acoustics and Vibration*, vol. 2012, article ID 562309, pp. 1–11.

Sheard, A.G. and Jones, N.M. (2008), 'Approval of High-temperature Emergency Tunnel-ventilation Fans: the Impact of ISO 21927-3', *Proceedings of the ITA–AITES World Tunnel Congress and 34th General Assembly*, Agra, India, 19–25 September, pp. 1817–1826.

Sheard, A.G. and Jones, N.M. (2012), 'Powered Smoke and Heat Exhaust Ventilators: the Impact of EN 12101-3 and ISO 21927-3', *Tunnelling and Underground Space Technology*, vol. 28, pp. 174–182.

Sheard, A.G., Corsini, A., Minotti, S. and Sciulli, F. (2009), 'The Role of Computational Methods in the Development of an Aero-acoustic Design Methodology: Application in a Family of Large Industrial Fans', *Proceedings of the 14th International Conference on Modelling Fluid Flow Technologies*, Budapest, Hungary, 9–12 September, pp. 71–79.

Sheard, A.G., Daneshkhah, K. and Corsini, A. (2013), 'Fan Conceptual Design as Applied to the Marmaray Tunnel Ventilation System', *Proceedings of the 58th American Society of Mechanical Engineers Gas Turbine and Aeroengine Congress,* San Antonio, TX, USA, 3–7 June, paper no. GT2013-94548.

Vardy, A.E. (1980a), 'Unsteady Airflows in Rapid Transit Systems Part 1: Measurements on the London Transport Victoria Line', *Proceedings of the Institution of Mechanical Engineers*, vol. 194(1), pp. 341–348.

Vardy, A.E. (1980b), 'Unsteady Airflows in Rapid Transit Systems Part 2: Theoretical Background and Design Parameters', *Proceedings of the Institution of Mechanical Engineers*, vol. 194(1), pp. 349–356.

William-Louis, M. and Tournier, C. (2005), 'A Wave Signature Based Method for the Prediction of Pressure Transients in Railway Tunnels', *Journal of Wind Engineering and Industrial Aerodynamics*, vol. 93(6), pp. 521–531.

Chapter 6

A Parametric Study of Reversible Jet-fan Blades' Aerodynamic Performance

K. Daneshkhah and A.G. Sheard

ABSTRACT

This chapter presents a fully reversible blade parametric design methodology. The blades are for application in jet-fans that ventilate and provide emergency smoke control in road tunnels. The blade design variables are tip solidity, twist and camber distribution. We base the design methodology on a sensitivity analysis which we derived from a response surface approximation. We construct the latter using a computational analysis of four experimental cases which we generated using an experimental design approach. The sensitivity analysis calculates a rank and a weight for each design variable that affects the jet-fan performance parameters thrust and efficiency. Thus the sensitivity analysis facilitates insight into each variable's relative importance. Finally, we present a design of an existing 'baseline' jet-fan blade by following the design guidelines that we obtained from the sensitivity analysis. We study the redesigned blade's aerodynamic characteristics and compare them to that of the baseline design. We then built and tested a redesigned blade prototype. We present thrust, power and acoustic measurements for both a jet-fan fitted with the baseline blade, and the same jet-fan when fitted with the redesigned blade.

This chapter is a revised and extended version of Daneshkhah, K. and Sheard, A.G. (2013), 'A Parametric Study of Reversible Jet-fan Blades Aerodynamic Performance', *Transactions of the ASME, Journal of Engineering for Gas Turbines and Power*, vol. 135, paper no. 024503, pp. 1–4.

NOMENCLATURE

Latin letters

A		Fan cross section at the outlet
C_t		Camber distribution at the tip
D_h	[m]	Hub diameter
D_i	[m]	Impeller diameter
f	[Hz]	Frequency
L_w	[dB]	Sound power level
N		Blade number
P	[kW]	Shaft power
p_t	[Pa]	Total pressure
r_t		Tip radius
T		Thrust downstream of the impeller
T*		Calculated thrust
$\bar{v}$	m/s	Mass averaged stream-wise velocity
V	[m^3/s]	Volume flow rate
V_{ref}	[m^3/s]	Reference volume flow rate
X_i		Blade design variables
Y		Performance parameter

Greek letters

β_o, β_i		Coefficients of the response surface model
γ		Blade pitch angle
γ_h		Blade pitch angle at hub
γ_t		Blade pitch angle at tip
η		Efficiency
η^*		Calculated efficiency
θ		Blade camber angle
λ		Blade twist
ρ		Air density
σ		Blade solidity
χ		Tip clearance

Acronyms

DoE		Design of experiments

INTRODUCTION

When designing road tunnel ventilation systems, engineers use longitudinal ventilation when possible, as it provides the lowest installation cost. Typically, road tunnel ventilation systems have a series of operating scenarios, with each requiring a unique ventilation system configuration in response to a specific tunnel condition. The complexity of tunnel operating scenarios results in the majority of jet-fans operating in both forward and reverse directions. Cardillo *et al.* (2014) observed that in this regard tunnel ventilation fans are unique as other air movement and control applications do not require reversibility. A need to operate in either a forward or reverse direction results in symmetrical fan blades, with guide vanes reduced to flow-straighteners. A consequence of a requirement for reversibility is that separated flow is an inevitable aspect of even the most carefully developed aerodynamic design. Therefore, the efficiency of the resultant aerodynamic designs is compromised.

The aerodynamic challenges posed by a requirement for reversibility are compounded by legislative requirements. Jet-fans are required to comply with the European Union Regulation 327 mandating a minimum fan and motor efficiency during routine tunnel operation and simultaneously with the Euro Norm EN 12101-3 requirements (EN12101-3, 2002; Sheard and Jones, 2012) and its associated international standard ISO 21927-3 (ISO 21927-3, 2006; Sheard and Jones, 2008) during emergency operation in the event of a tunnel fire. The requirements of both Regulation 327 and EN 12101-3 are legally binding within the European Community and consequently, tunnel ventilation fan designers must build their design process around compliance with both.

Scholars have studied the application of jet-fans for longitudinal ventilation. Karki and Patankar (2000) describe a tunnel ventilation model and its validation for longitudinal ventilation systems using jet-fans. Mutama and Hall (1996) performed an experimental investigation of jet-fan aerodynamic performance in a wind tunnel and showed that the jet-fan's ability to entrain air depends on its proximity to the tunnel wall. Giesen *et al.* (2011) investigated free and near-wall positioning of jet-fans in the tunnel using a computational fluid dynamics analysis and assessed the efficiency of their positioning in large enclosures through a case study. Although a useful contribution to the community, the reported research focuses on predicting the performance of ventilation systems incorporating jet-fans, not on improving jet-fan performance.

There is a need to improve jet fan performance. European Union Regulation 327 mandates minimum fan and motor efficiencies that many reversible jet-fan designs do not achieve. The Euro Norm EN 12101-3 requires a larger blade tip-to-casing gap during routine operation to prevent the gap falling to zero when clearing hot gas and smoke in the event of a tunnel fire. Increasing the blade tip-to-casing gap reduces fan efficiency. A way to improve jet-fan efficiency would be to design larger, lower speed jet-fans. However, tunnel ventilation designers invariably minimise tunnel cross-sectional area to minimise tunnel cost. Therefore, the market demands a required thrust from the minimum diameter jet-fan.

Despite legislative requirements and market demands, there has been little research within the air movement and control community on jet-fan design. Traditionally, designers have used blade design methods based on scaling laws and designers' experience to define blade geometry (Daly, 1985). Corsini *et al.* (2013) commented that since the turn of the twenty-first century computational methods developed within the aerospace community have become established within the air movement and control community. Although useful analysis tools, tunnel ventilation fan designers cannot use these computational methods as a design tool. They facilitate a prediction of the blade-to-blade flow-field, providing the designer with an insight into the effect of a change in geometry. However, the designer must still rely on trial and error and previous experience to define an optimum design configuration.

A way to reduce the need for a trial and error approach is inverse design (Daneshkhah and Ghaly, 2007) and optimisation methods (Mengistu *et al.*, 2007; Mengistu and Ghaly, 2008). These methods can help engineers during the design process. Numerical optimisation methods iteratively modify a set of parameters to achieve a specific objective function. One can compute the latter using a computational fluid dynamic code to predict the fan flow-field. These methods are effective in obtaining an optimum blade shape and they are flexible in terms of formulating a multi-objective function subject to various design constraints. However, they require a computational effort that has historically been beyond that available within the air movement and control community.

A way to reduce the computational effort needed to optimise a design is to use a response surface model. Lian and Liou (2005) successfully applied this approach for multi-objective optimisation of a NASA rotor 67 transonic compressor blade. They maximised the stage pressure ratio and minimised compressor weight. Lian and Liou's (2005) approach was noteworthy as they used an evolutionary algorithm to facilitate the multi-objective optimisation.

In the programme of work reported in this chapter, we present a parametric design methodology for reversible jet-fan blades. We characterise the blade design using three variables with a linear response surface model. We constructed the latter using four experimental test cases which we distributed in the 'design space' using an experimental design approach. We carried out a sensitivity analysis using the response surface model, which gives a rank and weight for each blade design variable affecting jet-fan performance parameters' thrust and efficiency at two operating conditions. We calculated the thrust and efficiency for the four experimental test cases using a commercially available computational fluid dynamics code. We discuss the primary outcomes of the sensitivity analysis results and develop guidelines for the blade redesign process. Finally, we present a redesigned blade with increased thrust compared to that of the original baseline blade.

BLADE DESIGN VARIABLES

A parametric model that fully describes a three-dimensional blade requires a large number of parameters (Mansour and Ghaly, 2003). For the purpose of this study, we did not fully parameterise the blade geometry. Instead, we selected a reduced number of blade design parameters which we anticipated to have the highest influence on blade aerodynamic performance. We then studied the influence of the selected parameters on the jet-fan's overall performance.

The blade tip solidity is the first blade design parameter that we considered in this study:

$$\sigma = \frac{N\,C_t}{2\pi r_t} \tag{1}$$

An axial fan's pressure development capability is directly related to blade tip solidity. For axial fans used in tunnel ventilation, solidity values typically vary from 0.3 to 0.7.

The second design parameter that we used is the blade camber distribution. We constrained a reversible axial fan blade to a symmetrical profile to ensure the same flow condition in both forward and reverse operation. Traditionally, one uses a straight or very low camber profile. In the present study we have described the camber distribution using a Bezier curve with five control points. This definition allows us to define a symmetric camber with minimum parameters. We selected the blade camber at the tip as the design variable and reduced the subsequent sections' camber linearly toward a zero camber at the hub section.

The blade twist is the third parameter that we studied. The blade twist is given by:

$$\lambda = \gamma_t - \gamma_h \tag{2}$$

where γ is the blade pitch angle. We used a linear twist distribution. We could have created a more complex profile by increasing the number of design variables, but we chose a linear twist distribution as our objective was to demonstrate the potential of a linear response surface model as the basis for a blade design methodology.

DESIGN METHODOLOGY

We performed a parametric study of the three selected blade design variable using a design of experiments (DoE) approach. The experimental design allows for an optimum distribution of test cases in the design space enabling us to obtain the maximum information from the minimum number of experiments. We used Beachkofski and Grandhi's (2002) method to generate an optimum distribution of design variables, Table 6.1. The range over which blade solidity and twist are distributed are typical of those that engineers use in jet-fan blade design. As reversible blades classically have no camber, we selected the range following a review of in service camber distributions that engineers have used in historic blade designs.

We selected the minimum number of test cases necessary to construct a linear response surface model. Although a linear model cannot reproduce non-linear aerodynamic characteristics, it can provide insight into the significance of each design variable. A linear response surface model has the following form:

$$Y = \beta_0 + \sum_{i=1}^{N} \beta_i X_i \tag{3}$$

where Y is the performance parameter and X_i are the blade design variables. We generated blade geometry for four experimental cases, Table 6.1, and computed the resultant jet-fan thrust and efficiency using a commercially available computational fluid dynamics code. We then solved the above linear system of equations to obtain the coefficients β_i of the response surface model. Next, we performed a sensitivity analysis using the linear response surface model to study the effect of each design variable on each performance parameter and thus defined the design guideline.

AERODYNAMIC ANALYSIS OF THE BASELINE BLADE DESIGN

We selected a 'baseline' reversible blade that first entered service around the turn of the twenty-first century. This baseline blade was originally designed by engineering using Daly's (1985) scaling laws and the engineers' experience. We built this into an impeller comprising 12 blades mounted on a disk and clamp type hub. We used aerodynamic pods to guide the flow upstream and downstream of the impeller. We installed cylindrical silencers on either side on the impeller duct. The resultant specification of the jet fan incorporating the baseline blade is typical of the specification favoured by tunnel ventilation system designers, Table 6.2.

Table 6.1. *Matrix of design parameters used to define blade geometry for each of the 'design of experiment' test cases.*

Test case	Blade tip solidity σ	Blade camber θ	Blade twist λ
1	0.53	0.0	25
2	0.73	6.7	20
3	0.43	10.0	15
4	0.63	3.3	10

Table 6.2. *Specification of the jet-fan first tested with a fitted baseline and then redesigned blade.*

Impeller material	Aluminum
Design temperature	300°C for 2 hours
Number of blades	12
Hub diameter D_h	500 mm
Impeller diameter D_i	1,250 mm
Tip clearance χ	3.75 mm
Max. rotational speed	1,460 rpm

We used the commercial computational fluid dynamics code ANSYS CFX (ANSYS, 2009) to calculate the thrust and efficiency generated for each of the four experimental cases, Table 6.1. The code uses an element-based finite volume method with second-order discretisation schemes in space and time to solve the pressure and velocity systems of incompressible Reynolds-averaged Navier–Stokes equations. We selected a k–ε turbulence model with scalable wall-functions for the turbulence closure. We based convergence on reducing the maximum of normalised residuals of the momentum and continuity equations to less than 10^{-5}.

We discretised the computational domain using H-C and J type grid blocks. We selected grid dimensions after performing a grid dependency analysis. The selected mesh in the bladed region has 101 nodes in the stream-wise directions, 81 in the circumferential direction, and 101 in the span-wise direction, Figure 6.1. We positioned eight cells across the blade tip-to-casing gap to account for the clearance flow through the blade tip-to-casing gap. The computational domain extended about one fan diameter upstream and one fan diameter downstream of the impeller.

FIGURE 6.1. The computational mesh used to predict baseline jet-fan blade performance.

In order to facilitate convergence of the computational fluid dynamic code, we assumed that the fan was operating in a fully ducted system. This assumption enabled us to applying non-reflecting boundary conditions at inflow and outflow boundaries. At the inflow boundary we specified the flow angles and atmospheric total pressure. At the outflow boundary we specified the static pressure. We also assumed steady state and axi-symmetric flow and therefore were able to specify periodic boundary conditions for a single blade passage computational domain.

We obtained a fan performance curve at each blade pitch angle setting by varying the static pressure at the outflow. We determined the jet-fan performance parameters thrust and efficiency at the operating point that we obtained from the fan curve's intersection and the jet-fan system curve. We calibrated the latter using experimental results at each blade angle setting.

We used the computational fluid dynamics code to calculate thrust downstream of the impeller from:

$$T = \rho A \bar{v} \tag{4}$$

where A is the fan cross section area at the outlet and $\bar{v}$ is the mass averaged stream-wise velocity. We used the computational fluid dynamics code to calculate efficiency from:

$$\eta = 0.515 \frac{T^{\frac{3}{2}}}{2 r_t P} \tag{5}$$

where P is the shaft power calculated from pressure and shear stress forces acting on the blade and hub surfaces.

We used the computational fluid dynamics code to calculate jet-fan performance with the baseline blade at four blade pitch angles, Figure 6.2. Also, note the system resistance curves at each blade angle. We computed thrust and efficiency from the computational fluid dynamics results at the jet-fan operating point associated with each blade pitch angle. We compared these computed results with experimental measurements, Figure 6.3. We made the experimental measurements on a thrust rig in accordance with ISO 13350: 1999 standard requirements (ISO 13350, 1999).

There was a good agreement between computed and experimental values of thrust and efficiency up to a pitch angle of 40 degrees, Figure 6.3. At the higher 44 degree pitch angle the jet-fan operated in the unstable part of its characteristics, Figure 6.2. As the flow became unstable, it separated and recirculated, resulting in the computed thrust and efficiency matching the experimental values less well.

Some scholars working with the air movement and control community have successfully predicted the flow-field through tunnel ventilation fans in the unstable region of their operating range. Notably, Bianchi *et al.* (2013) used a large eddy simulation to predict a tunnel ventilation fan's aerodynamic and aero-acoustic performance. However, the large eddy simulation required two orders of magnitude

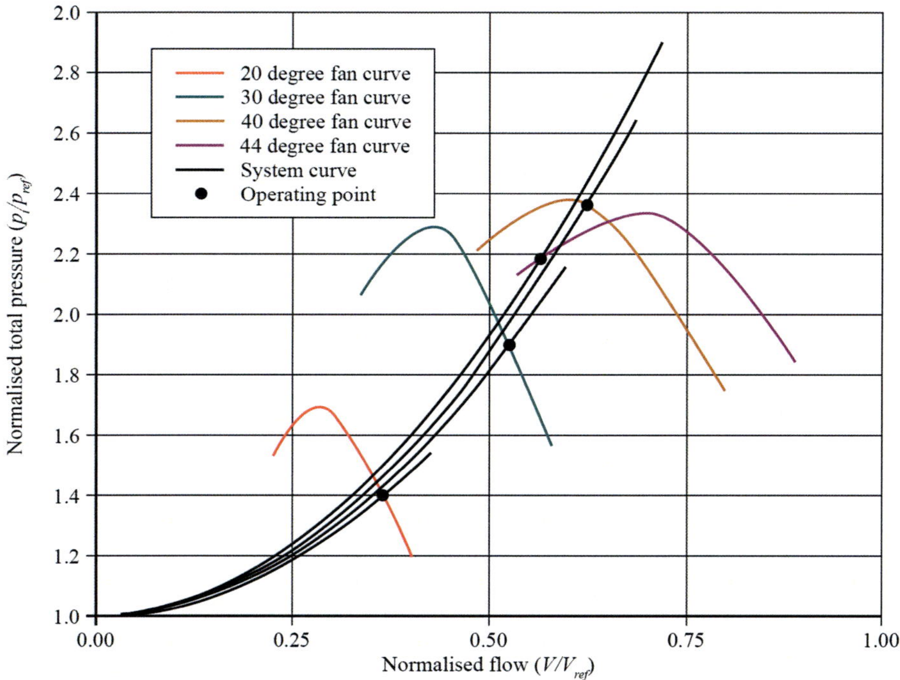

FIGURE 6.2. The baseline blade's fan performance characteristics in a fully ducted system predicted using the commercial computational fluid dynamics code ANSYS CFX (ANSYS, 2009).

more computational effort than that associated with the Reynolds-averaged Navier–Stokes approach that we used in the reported research. Therefore, we accepted that the predicted values were only reliable over the stable region of the jet-fan's operating range.

ANALYSIS OF THE EXPERMENTAL CASES

We conducted eight computational fluid dynamic analysis; two for each of the four test cases, Table 6.1. For each test case we computed thrust and efficiency at a blade pitch angle of 30 degrees, corresponding to a peak efficiency operating condition and 40 degrees, corresponding to a peak thrust operating condition. Using the computed values of thrust and efficiency, we calculated a linear response surface model for each performance parameter by solving Equation 3. The result is a matrix of 'effect plots' that we were able to deduce from the response model approximation, Figure 6.4.

A study of the effect plots indicates that at a 30 degree blade pitch angle, blade-tip solidity has a dominant effect on thrust, Figure 6.4. Increasing the tip solidity improves the fan pressure development and results in an increased flow rate and

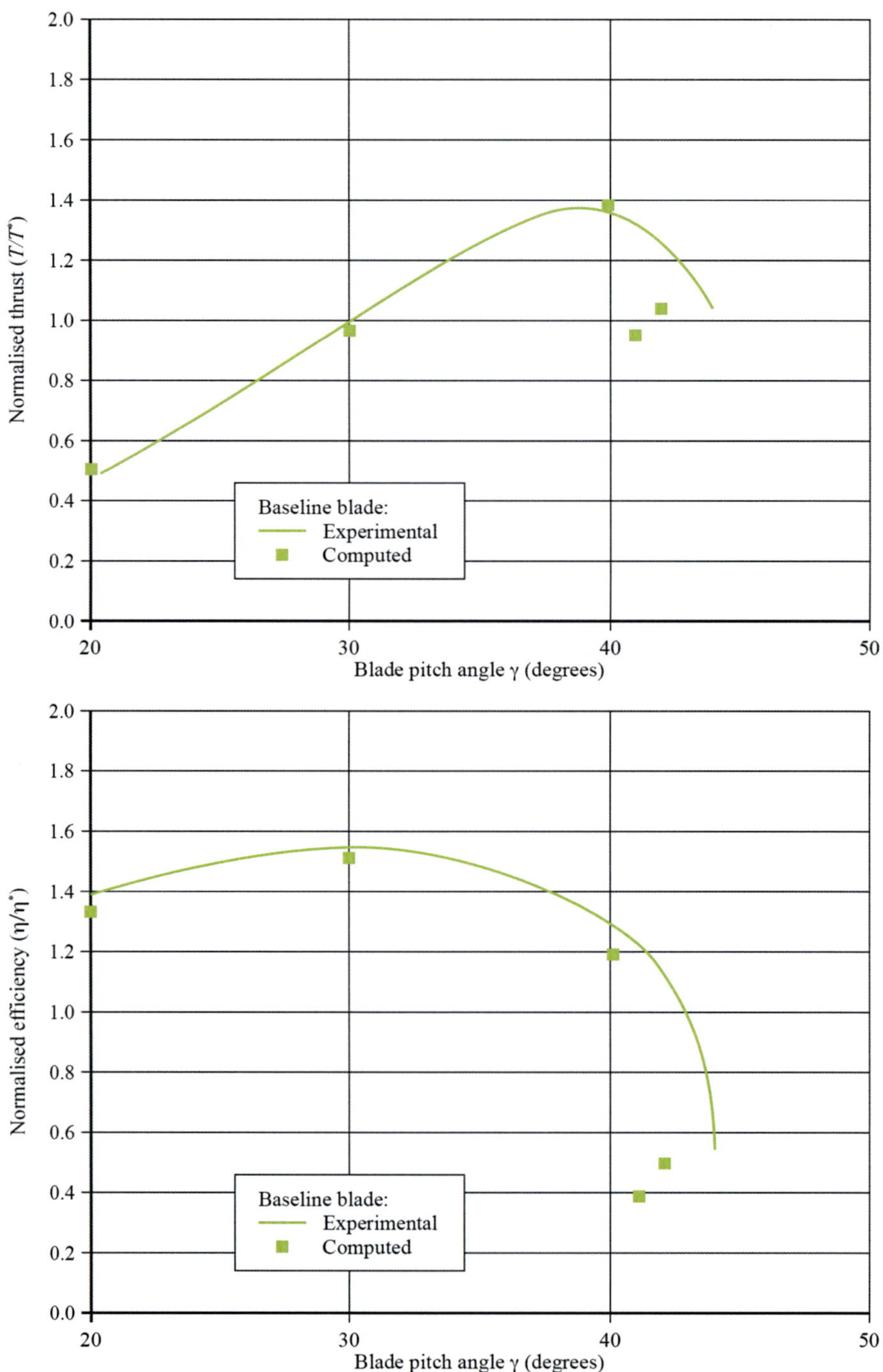

FIGURE 6.3. A comparison of predicted and measured thrust and efficiency for the baseline blade.

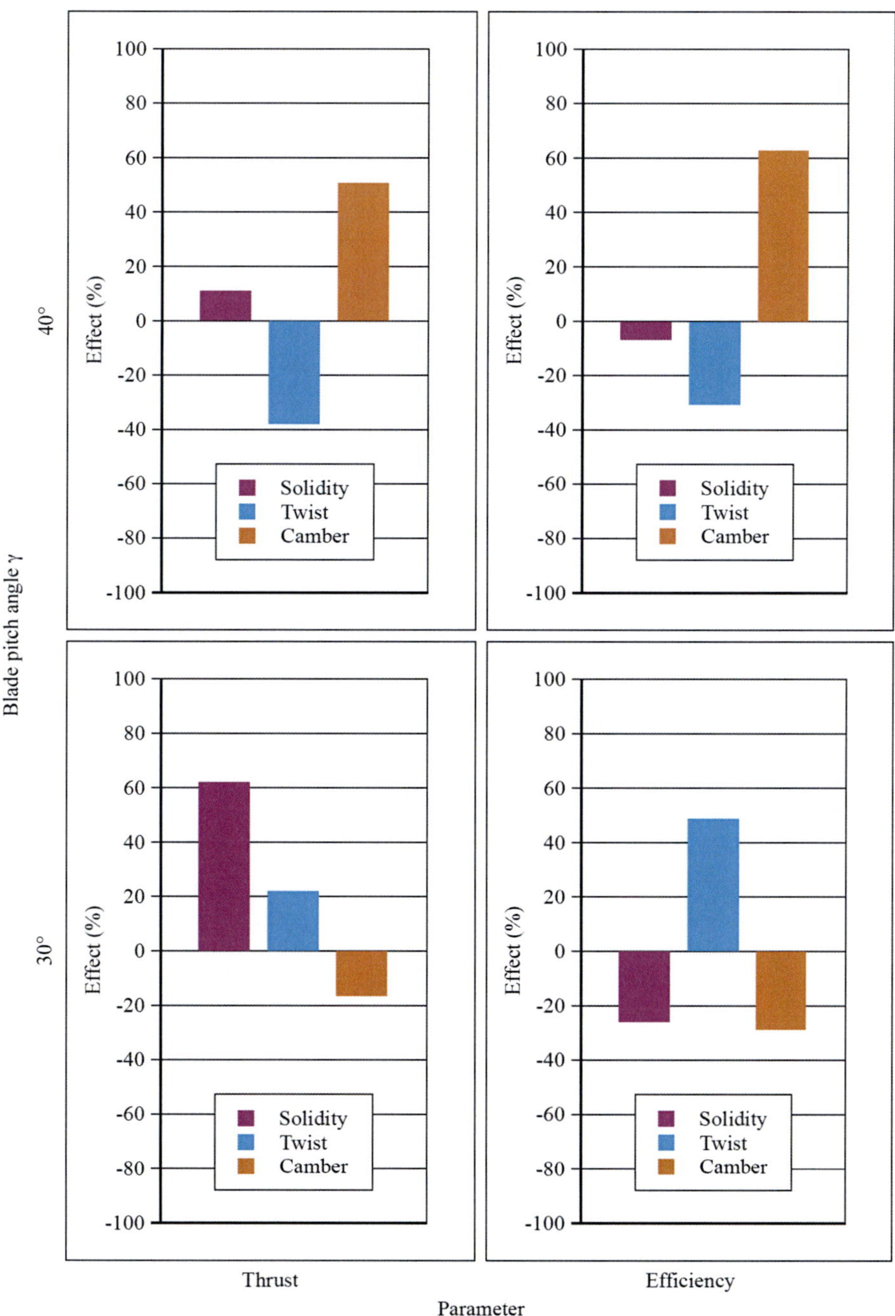

FIGURE 6.4. Thrust and efficiency 'effect plots' clarifying the impact of the blade design parameters' blade tip solidity, blade twist and blade camber distribution.

thrust when the jet-fan operates against a system resistance. However, the aerodynamic losses increase with increasing solidity as a result of an increased blade surface area. Consequently, increasing solidity has a negative effect on efficiency.

Increasing blade twist also has a positive effect on jet-fan thrust at a 30 degree blade pitch angle, Figure 6.4. Increasing blade twist results in an increased flow incidence angle in the blade hub region, which results in a more uniform span-wise work distribution and consequently, an increase in both thrust and efficiency.

Increasing the blade camber distribution has a negative effect on the thrust and efficiency at 30 degree blade pitch angle, Figure 6.4. Increasing camber reduces the incidence angle and unloads the blade which, in turn, results in a lower flow and therefore lower thrust. The negative impact of increasing camber on efficiency is due to the associated increase in boundary layer thickness on the suction surface and the resultant increase in wake thickness downstream of the blade. This increase in boundary layer thickness and wake thickness constitute an increase in aerodynamic losses and therefore a reduced efficiency for a more highly cambered blade.

We may illustrate the effect of camber by considering the blade-to-blade velocity contours at 90 per cent blade span for Test Case 1 and 3, Table 6.1. With a 30 degree blade pitch angle and no camber (Test Case 1), the blade wake is well defined and narrow, Figure 6.5. When we increase camber to 10 degrees (Test Case 3), the increase in boundary layer thickness results in a significantly thicker wake that we may correlate with an increase in aerodynamic loss and therefore reduction of efficiency. The thicker wake for the high camber case contributes to a reduction in jet fan thrust, Figure 6.4.

A study of the effect plots indicates that at a 40 degree blade pitch angle, blade-tip solidity has less impact on thrust then it did at 30 degrees, Figure 6.4. At a 40 degree blade pitch angle, the incidence angle of flow onto the blade leading edge is relatively high as this blade pitch angle is associated with a peak pressure operating point close to the blade's maximum pressure developing capability. As a consequence, an increase in solidity has little effect on blade pressure development and therefore thrust.

Increasing blade twist has a negative effect on jet-fan thrust at a 40 degree blade pitch angle, Figure 6.4. Increasing blade twist results in the incidence angle of flow onto the blade leading edge exceeding that at which the blade stalls and consequently, increasing blade twist has a negative effect on both thrust and efficiency.

Increasing camber has a positive effect on the thrust and efficiency at 40 degree blade pitch angle, Figure 6.4. Increasing camber reduces the incidence angle of flow onto the blade leading edge, and consequently, increases the maximum blade pitch angle before the blade stalls. Increasing the maximum blade pitch angle will result in increasing the maximum thrust. Operating in the stable region of the blade's characteristic as opposed to in the stalled region will result in increased efficiency.

Once again, we may illustrate the effect of camber through considering the blade-to-blade flow-field at 90 per cent blade spans, Figure 6.5. With a 40 degree blade pitch angle and no camber, the flow separates from the blade leading edge due to the high incidence angle, indicating that the blade has stalled. When we increase

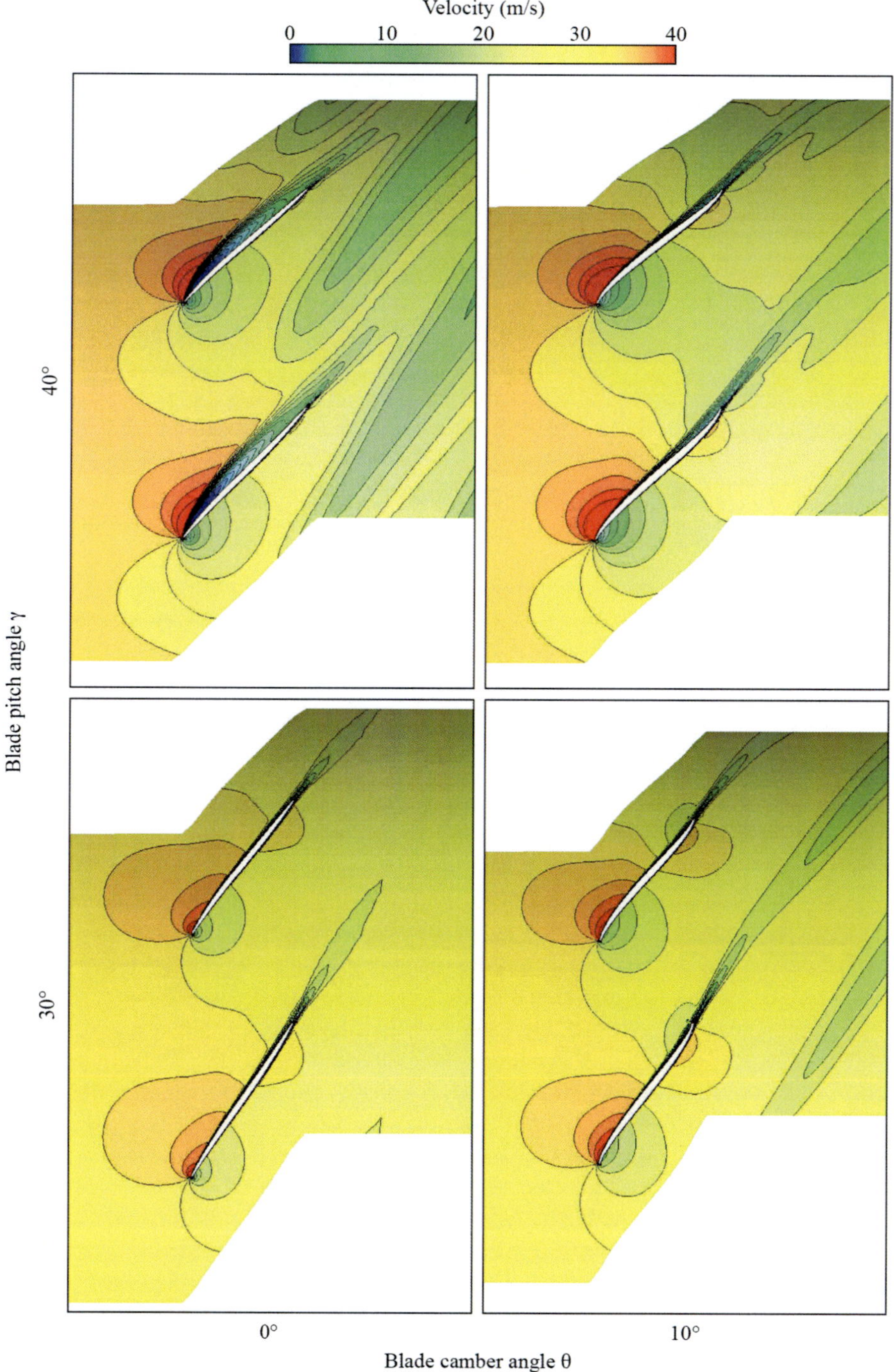

FIGURE 6.5. Predicted velocity contours at 90 per cent blade span illustrating the effect of blade pitch and camber on the blade-to-blade flow-field.

camber to 10 degrees, the effective incidence angle onto the blade reduces and therefore eliminates the leading edge separation. Although boundary layers are thick, the blade wake is well defined, indicating that although highly loaded, the blade is not stalled, still operating in the stable part of its characteristic.

DESIGN OPTIMISATION

The sensitivity analysis provided an insight into the relative importance of the design variables blade tip solidity, blade twist and blade chamber distribution. This insight enabled us to derive optimum values of each variable and define the redesigned blade's geometry. The redesigned blade has 15 per cent higher tip solidity than the baseline design. Blade twist increases to the limit for stable operation without flow separation near the hub. We used a symmetrical cambered distribution to define the section profiles where the maximum camber was 7.5 per cent chord at the blade tip. The camber reduced linearly towards the hub section.

The increase in blade tip solidity resulted in an increase in blade area. The baseline blade solidity was the maximum possible within the mechanical limit of the hub into which we fitted it. Therefore, it was only possible to increase the blade tip solidity if we could achieve the increase in blade area without increasing blade volume, and thus weight. Engineers designed the baseline blade around the turn of the twenty-first century without the aid of finite element analysis. Consequently, the baseline design was conservative. We conducted a finite element analysis of the baseline blade design and identified the peak stress in the blade root. We then conducted a finite element analysis of the redesigned blade, reducing blade thickness by 20 per cent at the hub and 40 per cent at the tip. The resultant redesigned blade has the same weight as the baseline blade, and a slightly lower peak stress in the blade hub region despite the increase in blade chord.

We manufactured the redesigned blade and fitted it to the jet-fan that we previously used to test the baseline blade. Once again, we made experimental measurements on a thrust rig in accordance with ISO 13350: 1999 requirements (ISO 13350, 1999). The experimental and computed thrust and efficiency for both the baseline and redesigned blade indicated that the redesign has resulted in both improved thrust and efficiency, Figure 6.6. The predicted thrust and efficiency at blade pitch angle above 40 degrees did not match the experimental values. The difficulty in predicting thrust and efficiency at higher blade pitch angles is a consequence of flow instability. Despite this caveat, the resultant design methodology was successful as the redesigned blade did have both higher thrust and efficiency than the baseline blade.

The improvement in thrust and efficiency is most pronounced at 40 blade pitch angle, Figure 6.6. This is primarily due to the redesigned blade's increased camber. The increase enables the blade to continue to operate in the stable region of its characteristic. The experimental measurements indicate that the redesigned blade may operate at up to a 42 degree blade pitch angle, with a higher associated thrust and no reduction in efficiency when compared to the baseline blade operating at a 40

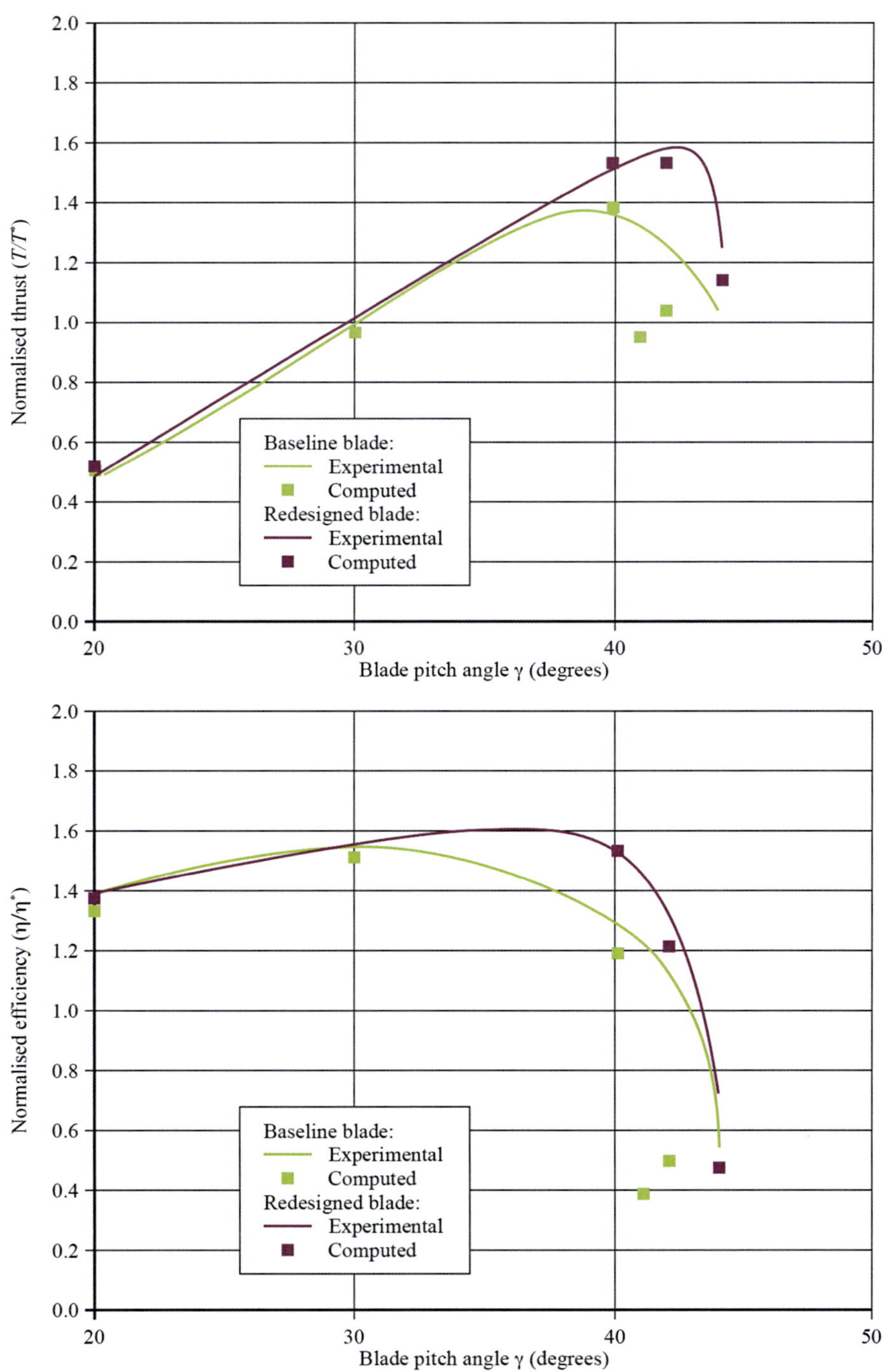

FIGURE 6.6. A comparison of predicted and measured thrust and efficiency for both the baseline and redesigned blade.

degree blade pitch angle. Although a two degree increase in blade pitch angle may seem modest, it does represent a 25 per cent increase in thrust, Figure 6.6. In practical application that will enable four jet fans to do what previously required five, and therefore represents a useful reduction in ventilation system cost.

We achieved the increase in thrust by increasing blade loading. When one increases blade loading we would expect an increase in fan noise (Daly, 1985). Whilst making experimental measurements of thrust and efficiency, we also made acoustic measurement in accordance with ISO 10302: 1996 requirements (ISO 10302, 1996). The redesigned blade broadband noise was 2 dB higher at both 30 and 40 degree blade pitch angles, Figure 6.7. The increase in broadband noise was probably a consequence of the increase in boundary layer thickness and associated thickness of blade wakes. Although a negative consequence of the blade redesign, we can reduce the increase in broadband noise with the addition of longer silencers. Therefore, we conclude that the redesigned blade still represents a significant improvement over the baseline blade design.

CONCLUSION

We presented a blade design methodology for a reversible jet-fan blade. The blade design parameters were blade tip solidity, blade camber and blade twist. We studied the effect of the blade design parameters on jet-fan performance by using a sensitivity analysis derived from a parametric model. We used a 'design of experiment' method to generate an optimum distribution of design parameters in the design space, defining four combinations of design parameters and thus, the geometry of four blades.

We used a commercial computational fluid dynamics code to compute the thrust and efficiency of four blade geometries at both peak efficiency and peak pressure operating points. We used the results to construct a response surface model that facilitated a sensitivity analysis that provided an insight into the underlying flow physics. We then used the response surface model to define the optimum values of the three design parameters that we used to define the redesigned blade's geometry.

We manufactured the redesigned blade, and first tested a jet fan with the baseline blade and then with the redesigned blade. The redesigned blade proved capable of achieving 25 per cent more thrust than the baseline blade at a comparable efficiency. Acoustic measurements indicated that this increase in thrust was associated with a 2 dB increase in broadband noise. Therefore, we conclude that the parametric design study has resulted in a design methodology that successfully captures the design rules. We used the design methodology to generate a redesign that increased thrust within the baseline blade's structural limitations. Therefore, the redesigned blade constitutes a viable alternative to the baseline blade in those applications where increased broadband noise is not a primary consideration.

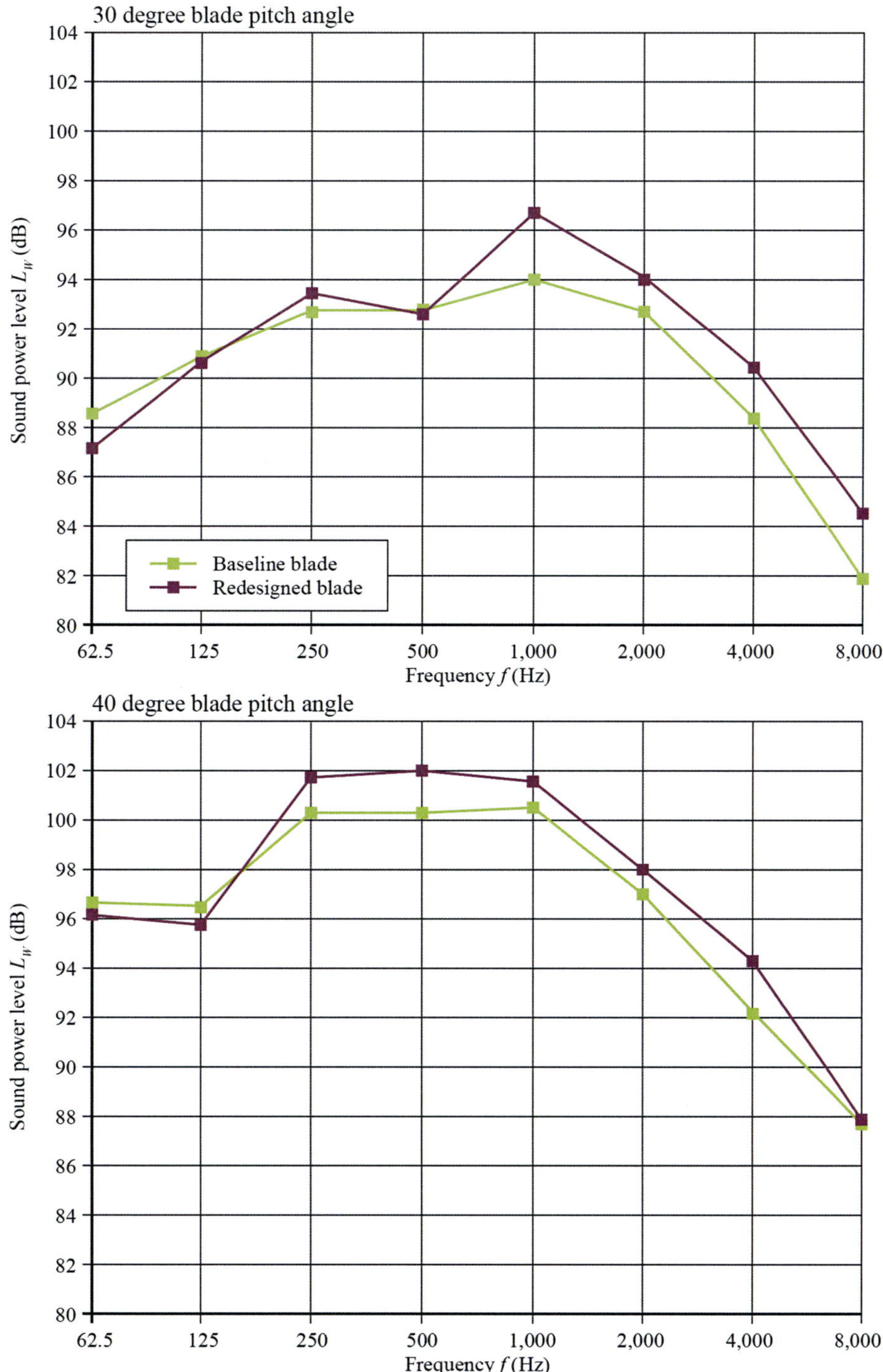

FIGURE 6.7. A comparison of measured fan acoustic spectrum for both the baseline and redesigned blade.

REFERENCES

EN12101-3 (2002), *Smoke and Heat Control Systems. Specification for Powered Smoke and Heat Exhaust Ventilators.*

ISO 10302 (1996), *Fans for General Purposes. Methods of Noise Testing.*

ISO 13350 (1999), *Industrial Fans - Performance Testing of Jet Fans.*

ISO 21927-3 (2006), *Smoke and Heat Control Systems – Part 3: Specification for Powered Smoke and Heat Exhaust Ventilators.*

ANSYS (2009), *CFX-Solver Theory Guide*, ANSYS Inc., Canonsburg, PA, USA.

Beachkofski, B.K. and Grandhi, R.V. (2002), 'Improved Distributed Hypercube Sampling', *Proceedings of the 43rd AIAA/ASME/ASCE/AHS/ASC Structures, Structural Dynamics, and Materials Conference*, Denver, CO, 22–25 April, paper no. AIAA 2002-1274.

Bianchi, S., Borello, D., Corsini, A., Rispoli, F. and Sheard, A.G. (2013), 'Large-eddy Simulation of the Aerodynamic and Aeroacoustic Performance of a Ventilation Fan', *Advances in Vibration and Acoustics*, vol. 2013, article ID 876973, pp. 1–10.

Cardillo, L., Corsini, A., Delibra, G., Rispoli, F. and Sheard, A.G. (2014), 'A Numerical Investigation into the Aerodynamic Effect of Pressure Pulses on a Tunnel Ventilation Fan', *Proceedings of the IMechE Part A, Journal of Power and Energy,* vol. 228(3), pp. 285–299.

Corsini, A., Delibra, G. and Sheard, A.G. (2013), 'A Critical Review of Computational Methods and Their Application in Industrial Fan Design', *International Scholarly Research Network, Mechanical Engineering*, vol. 2013, article ID 625175, pp. 1–20.

Daly, B.B. (1985), *Woods Practical Guide to Fan Engineering*, Woods of Colchester Ltd, Colchester, UK.

Daneshkhah, K. and Ghaly, W.S. (2007), 'Aerodynamic Inverse Design for Viscous Flow in Turbomachinery Blading', *AIAA Journal of Propulsion and Power*, vol. 23(4), pp. 814–820.

Giesen, B.J.M., Penders, S.H.A., Loomans, M.G.L.C., Rutten, P.G.S. and Hensen, J.L.M. (2011), 'Modelling and Simulation of a Jet Fan for Controlled Air Flow in Large Enclosures', *Environmental Modelling and Software*, vol. 26(2), pp. 191–200.

Karki, K.C. and Patankar, S.V. (2000), 'CFD Model for Jet Fan Ventilation Systems', *Proceedings of the 10th International Symposium on Aerodynamics and Ventilation of Vehicle Tunnels Principles, Analysis and Design*, Boston, MA, USA, 1–3 November.

Lian, Y. and Liou, M.S. (2005), 'Multi Objective Optimization using Coupled Response Surface Model and Evolutionary Algorithm', *AIAA Journal*, vol. 43(6), pp. 1316–1325.

Mansour, T. and Ghaly, W.S. (2003), 'An Implicit Geometric Representation of Turbine Blades using NURBS', *Proceedings of the 11th CFD Conference of the Canadian Society of CFD*, Vancouver, Canada, 28–30 May, vol. 1, pp. 238–243.

Mengistu, T. and Ghaly, W.S. (2008), 'Aerodynamic Optimization of Turbomachinery Blades using Evolutionary Methods and ANN-based Surrogate Models', *Journal of Optimization and Engineering*, vol. 9(3), pp. 239–255.

Mengistu, T., Ghaly, W.S. and Manour, T. (2007), 'Aerodynamic Shape Optimization of Turbine Blades using a Design-parameter-based Shape Representation', *Proceedings of the 52nd American Society of Mechanical Engineers Gas Turbine and Aeroengine Congress*, Montreal, Canada, 14–17 May, paper no. GT2007-28041.

Mutama, K.R. and Hall, A.E. (1996), 'The Experimental Investigation of Jet Fan Aerodynamics using Wind Tunnel Modeling', *Transactions of the ASME, Journal of Fluids Engineering*, vol. 118(2), pp. 322–328.

Sheard, A.G. and Jones, N.M. (2008), 'Approval of High-temperature Emergency Tunnel-Ventilation Fans: the Impact of ISO 21927-3', *Proceedings of the ITA–AITES World Tunnel Congress and 34th General Assembly*, Agra, India, 19–25 September, pp. 1817–1826.

Sheard, A.G. and Jones, N.M. (2012), 'Powered Smoke and Heat Exhaust Ventilators: the Impact of EN 12101-3 and ISO 21927-3', *Tunnelling and Underground Space Technology*, vol. 28, pp. 174–182.

Chapter 7

The Effects of Inlet Box Aerodynamics on the Mechanical Performance of a Variable Pitch in Motion Fan

A.G. Sheard

ABSTRACT

This chapter describes research involving an in-service failure of a 'variable pitch in motion' fan's blade bearing. Variable pitch in motion fans rotate at a constant speed, with the changing blade angle varying the load. A pitch-change mechanism facilitates the change in blade angle. A blade bearing supports each blade enabling it to rotate. I observed that as the fan aerodynamic stage-loading progressively increased, so did the rate of blade bearing wear.

The reported research addressed two separate, but linked needs. First, the on-going need to increase fan pressure development capability required an increase in fan loading. This increase was within the context of an erosive operating regime which systematically reduced fan pressure development capability. The second need was to identify the root cause of blade bearing failures. I addressed the linked needs using a computational analysis, improving the rotor inflow aerodynamic characteristics through an analysis of the inlet box and design of inlet guide vanes to control flow non-uniformities at the fan inlet. The results of the improvement facilitated both an increase in fan pressure developing capability and identification of the root cause of the blade bearing failures.

This chapter is a revised and extended version of Sheard, A.G. (2012), 'The Effect of Inlet Box Aerodynamics on the Mechanical Performance of a Variable Pitch in Motion Fan', *Advances in Acoustics and Vibration*, vol. 2012, article ID 278082, pp. 1–10.

NOMENCLATURE

Latin letters

c	[mm]	Blade chord
D_{hc}		Hub to casing diameter ratio
D_t	[mm]	Blade tip diameter
p	[Pa]	Static pressure
p_t	[Pa]	Total pressure
Δp	[Pa]	Static pressure rise
Δp_t	[Pa]	Total pressure rise
Δt	[s]	Time variation
U	[m/s]	Velocity
V	[m^3/s]	Volume flow rate
r^+		Non-dimensional annular radius
u	[m/s]	Tangential velocity

Greek letters

α_b	[deg]	Blade angle
γ	[deg]	Blade stagger angle
η		Efficiency
θ	[deg]	Blade camber angle
σ		Solidity
χ	[% span]	Blade tip clearance

Acronyms

LU	Low upper
MLPS	Modified linear profile scheme
PFS	Pan Fläkt Single

INTRODUCTION

In induced-draft applications, the fan extracts combustion products from a coal-fired power plant's boiler. Current legislation requires power plant operators to fit boilers with emission control systems. These systems reduce the sulphur-dioxide, nitrox-oxide and, in some cases, mercury emissions from the power plant. They also have an associated pressure loss, which the induced-draft fans must overcome.

Increasingly stringent legislation has resulted in emission control systems becoming progressively more complex over the last decade. This results in a market requirement to upgrade in-service induced-draft fans to cope with the associated increased system resistance. Although not reported in the open literature, I was aware that the blade bearings situated under each blade of the variable pitch in motion induced-draft fans were wearing more rapidly with each increase in fan pressure developing capability. Following the most recent upgrade, it had been necessary to replace worn out blade bearings before the scheduled five year major overhaul. I speculated that a further increase in fan pressure developing capability was likely to exacerbate an already unacceptable situation.

The cost to upgrade an existing power plant's emission control system to meet current and foreseen future regulatory requirements can be as high as \$500 million. Existing power plants have a finite life, and in order for the upgrade to make commercial sense, the value of generated electricity must be greater than the cost of upgrading and operating the power plant. The largest single cost when operating a coal-fired power plant is the cost of coal. Perhaps counter-intuitively, upgrading a power plant's emission control system facilitates a move to lower quality coal. Without an emission control system or with a low specification emission control system, a power plant must burn higher quality low-sulphur coal to remain within existing emission limits. Once a power plant fits or upgrades an emission control system, it can burn lower quality, high-sulphur coal as the emission control system will clean the exhaust gas.

The business case for fitting or upgrading an emission control system is linked to a move from higher cost high-quality coal to lower cost low-quality coal. Low-quality coal includes not only more sulphur, but also more silica. Silica is hard, resulting in increased wear in the ball-mills that grind the coal before it enters the boiler. Wear in the boiler ball-mills results in larger carbon particles that are more likely to pass through the boiler and into the induced-draft fan. Consequently, the shift to low-quality coal results in a shift towards an increase in the silica and un-burnt carbon passing through induced-draft fans. As such, an unwanted consequence of switching from high- to low-quality coal is to increase the rate at which fan blades erode, decreasing the fan's pressure developing capability.

The addition or upgrading of emission control equipment increases the system resistance against which an induced-draft fan must operate. An increase in the fan blades' erosion rate results in the need for a further increase in fan pressure developing capability to ensure that as blades erode over the fan's service life, stall margin does not reduce to a point at which the fan stalls in-service. A way to increase an axial flow fan's pressure developing capability is to increase fan loading by fitting inlet guide vanes. Increased fan loading is, in practice, at the expense of fan stall margin and is therefore an inherently risky approach to increasing fan pressure developing capability in an erosive application.

Concern for maintaining a safe operating stall margin is more pressing when one places the fan downstream of an intricate duct system, as is typically the case with induced-draft fans. The inlet flow distortion associated with an intricate duct system results in an 'installation effect' that reduces fan stall margin. When examining the consequence of installation effect, the bulk of published works has focused on noise generation (Guédel, 2006; Bianchi *et al.*, 2012). Installation effects invariably degrade cooling fans' acoustic performance; however, acoustic considerations are not of primary concern for fans in induced-draft application. We can lag induced-draft fans to reduce acoustic emissions. The primary consideration for an induced-draft fan is the aerodynamic impact of a non-asymmetric inflow, with a particular concern for reduced fan efficiency and stall margin. Some researchers, notably Roslyng (1984) and Staiger and Stetter (1993), have studied the aerodynamic impact of fan installation effects. Despite the undoubted quality of the reported research, neither was able to generalise their results in a way that facilitated application beyond their work.

The impact of erosion on induced-draft fan performance has also received relatively little attention in the literature. However, Corsini *et al.* (2011) have studied the impact of erosion on induced-draft fan leading edge geometry. They concluded that three factors dominate the erosion process: (i) the inflow's incidence angle, (ii) the flow deviation correlated to the blade profile and (iii) the blade tip speed. Corsini *et al.* (2012a) continued their work, developing a computational method to predict erosion rate and the time it takes to erode the induced-draft fan blades to the point where they were no longer able to generate the necessary pressure rise to avoid the fan stalling at its design point. Corsini *et al.* (2012b) reported the eroded blade profile at different blade angles, and were able to computationally establish the characteristics of both the *datum* fan, with blades as per the design intent and eroded blades. The erosion occurred over approximately 12,000 hours' in-service operation, corresponding to less than two years continuous operation. Corsini's research team removed the eroded blades when the fan efficiency fell so far that the motor driving the fan was no longer able to supply sufficient power to maintain the required fan duty point.

In the present study, I modelled the inlet duct, inlet guide vanes, fan rotor and stator. I initiated this in response to an in-service issue. A facility operator had incorrectly specified a 1,767 mm diameter single-stage induced-draft fan with too low a pressure developing capability. The blade bearings also suffered premature failure. The fan's installed inlet configuration was complex, leading to the inlet flow's significant distortion which reduced fan performance. In combination with the fan's under-specification, the performance reduction that occurred as a consequence of the poorly conditioned inlet flow resulted in the fan limiting the power station power output. I initiated the computational programme of work reported in this chapter in an effort to both increase the fan pressure developing capability and identify the root cause of the premature blade bearing failure.

Induced-draft fans are classically configured with a rectangular inlet box that accepts flow coming vertically down. The inlet box turns the flow 90 degrees and into a round and horizontal fan inlet. The study used computational analysis as a 'numerical laboratory'. This provided insight into the fan's actual inlet flow-field. The objective was to facilitate inlet guide vane design improvement. The desired outcomes were first, an increase in fan pressure development capability and, second, an understanding of the aerodynamic mechanism that was driving the fan blade bearings' premature failure.

COMPUTATIONAL ANALYSIS

I analysed the geometry of a 1,767 mm single-stage induced-draft fan as installed in a power plant, but with a standardised air inlet system. Greenzweig *et al.* (2011) experimentally studied the effect of actual in-service air inlet system installations on induced-draft fan performance and concluded that the effect of inlet flow distortion on fan performance was significant, with the effect varying with the duty point's position along the fan curve. They found distortion effects greater at higher flow conditions and lower at lower flow conditions. This observation led Greenzweig *et al.* (2011) to conclude that the actual impact of inlet distortion is dependent on the relationship between the fan design parameters, and further that engineers currently do not understand the relationship between these parameters.

The relationship between fan design parameters and the impact of inlet flow distortion on fan performance is a subject in its own right. In the current study my focus was not an experimental research programme aimed at the elucidation of physical mechanisms underpinning the effect of flow distortion on fan performance. My focus was developmental, aimed at applying computational methods to improve an existing in-service fan's performance. In addition, I was able to use the computational results to provide insight into the circumferential variations in the flow-field that then assisted me in identifying the root cause of the observed in-service premature blade bearing failure. As such, I chose to study the in-service fan with a standardised inlet. I used the standardised inlet when model testing to establish the fan's characteristics. This more readily enabled me to utilise my chosen computational code.

The fan that is the subject of this research is designated Pan Fläkt Single (PFS), where 'pan fläkt' in Swedish means 'boiler fan' and 'single' designates a single-stage fan, Table 7.1. I derived the studied fan's characteristics, Figure 7.1, from data that I obtained using a low-speed 1,250 mm aerodynamic fan model. The aerodynamic

Table 7.1. *In-service duty point and specification of the PFS fan studied in the reported research.*

Fan in-service duty point		
Static pressure p (Pa)		2,245
Volume flow rate V (m^3/s)		54.3
Blade angle α_b (degrees)		56.0
Fan rotor data		
Blade number		18
Hub to casing diameter ratio D_{hc}		0.7
Blade tip diameter D_t (mm)		1,767
Blade tip clearance χ (% span)		6
Fan design speed (rpm)		890
Fan blade geometry	**Hub**	**Tip**
Blade chord c (mm)	487	446
Solidity σ	1.48	0.96
Blade stagger angle γ (degrees)	38	60
Blade chamber angle θ (degrees)	45	36

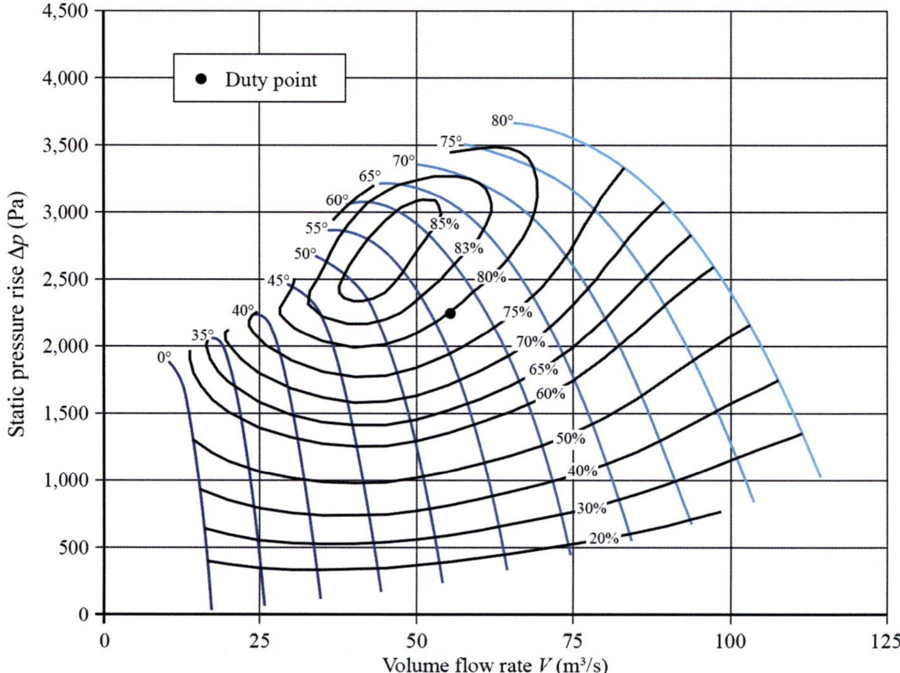

FIGURE 7.1. The performance characteristics of the studied 1,767 mm diameter PFS fan. This fan chart has been scaled from experimental data which the author obtained using a 1,250 mm low-speed aerodynamic model, scaled for speed and diameter using the fan laws.

model had the correct tip solidity and hub ratio, with model test data scaled to the studied fan speed and diameter using the fan laws. I selected the studied fan for induced-draft application, with a duty point that is typical for the application, Figure 7.1 and Table 7.1. The duty point falls below the fan's operating characteristic peak efficiency region as induced-draft applications are classically an erosive application. Fan blade erosion reduces the fan's pressure developing capability, and therefore one must select the fan conservatively to accommodate a reduction in stall margin over the fan's service life.

I modelled the studied fan with a standardised inlet box, rotor, stator and diffuser, plus a section of an expansion chamber, Figure 7.2. The inlet box comprised a square section vertical intake angled through 90 degrees and converted into a circular inlet. A shaft fairing traversed the inlet duct's lower part and expanded in a conical form onto the fan annulus. Following Teixeira *et al.*'s method (2003), I placed a fan with outlet guide vanes downstream of the inlet box discharging into a conical diffuser.

In the original configuration, I fitted five long chord sheet metal stays connecting the fan's hub surface to the outer casing, Figure 7.3. This original configuration was symmetrical with respect to a vertical plane through the fan centre line, with the stays' primary purpose as the provision of structural strength. The stays themselves were intended to present as near a neutral aerodynamic effect as possible.

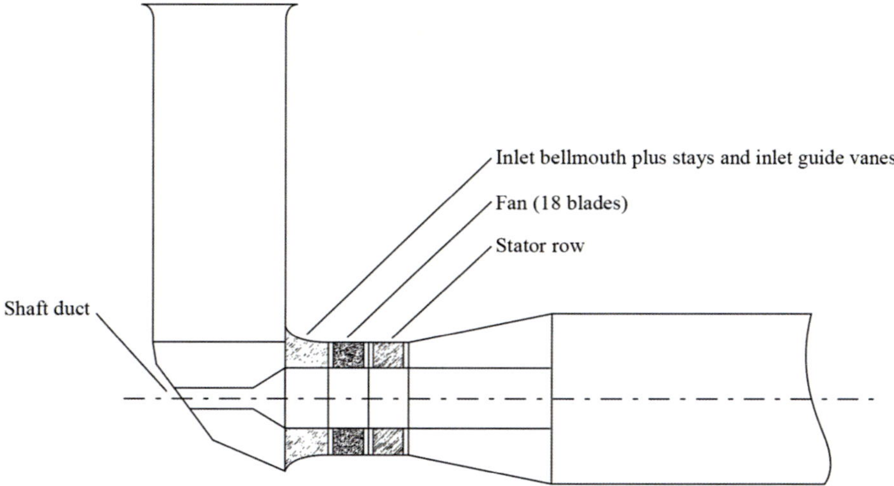

FIGURE 7.2. Side elevation of the fan installation arrangement.

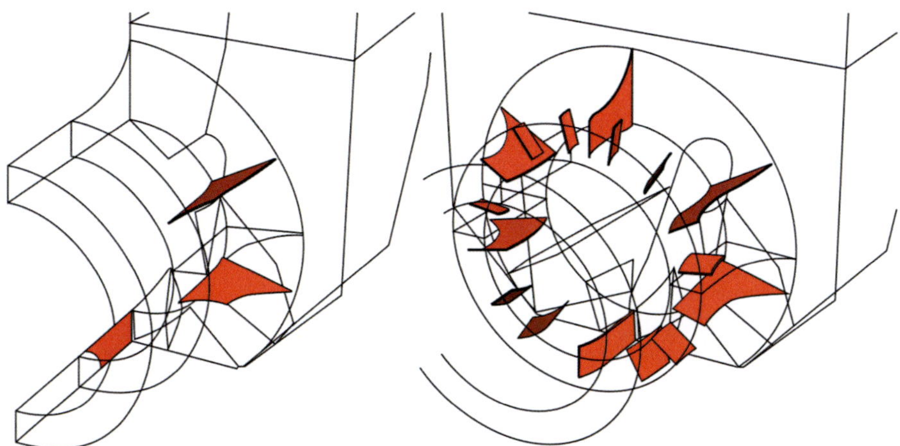

FIGURE 7.3. The original five stay configuration, left, and improved configuration with eight stays plus eight guide vanes, right.

I studied multiple variations of stay and inlet guide vane arrangements. I obtained the greatest increase in fan pressure developing capability and reduction in circumferential variation in pressure when a further three stays and eight short chord inlet guide vanes supplemented the original five stays, Figure 7.3. This improved eight stays plus eight inlet guide vane configurations were circumferentially symmetrical, with the additional stays and guide vanes requiring no original stay modification. In this way I could retrofit the improved configuration to the original configuration without compromising the fan's structural integrity.

Aerodynamic details

The inlet box introduced the initial aerodynamic complexity, where vertical inlet flow turned through 90 degrees from a square-section duct into an annular intake, whilst accommodating a sizeable shaft fairing. The annular intake incorporates the original five stays. The improved configuration supplements the original five stays with an additional three cambered stays, plus cambered chord-wise extensions to the original five stays. The eight additional inlet guide vanes associated with the improved configuration were also cambered. The resultant effect was that the improved configuration pre-swirled the inlet flow into the fan and increased fan aerodynamic loading. Additionally, by varying the camber on each stay and inlet guide vane, I aimed to minimise circumferential variations in the fan rotor inlet flow which the inlet box configuration introduced.

For the fan, an effect of additional loading is a shift of the characteristic upwards and to the right. Hence, for a given flow rate, the fan produces a higher pressure by operating further along its characteristic. This last point implies an unavoidable reduction in stall margin. The additional loading will also result in some decrease in fan efficiency. However, for a well-designed blade that was, in practice, operating at a duty point below peak efficiency, the increased loading may bring the actual fan operating point closer to the peak efficiency operating point. Operating closer to the fan's peak efficiency will go some way towards offsetting the efficiency decrease associated with increased fan loading.

Objective of the simulations

The study's primary objective was to safely maximise the fan loading through the addition of stays and inlet guide vanes. I sought evidence of separation throughout the blade span, as well as information about the rotor inlet flow conditions' possible pitch-wise variations. Additionally, I used the simulations to gain a deeper understanding of the flow-field in the inlet box and ultimately to provide information to eliminate or, at least, minimise sources of inefficiency. I obtained solutions for four flow conditions: duty point, 10 per cent and 20 per cent above and 5 per cent below the duty point condition.

I computed two forms of the inlet box, the original five stay configuration and the improved eight stay, plus eight inlet guide vane configurations. The original configuration was symmetrical around a vertical plane through the fan centre line, and therefore I was able to model only half the inlet box. This simplifying assumption is not applicable to the improved configuration. I varied the camber that was associated with each stay and inlet guide vane in an attempt to minimise the circumferential variation in outlet flow from the stays and inlet guide vanes. Consequently, each stay and inlet guide vane had a different camber, and therefore I was required to model the full annulus.

COMPUTATIONAL TOOL

The computational fluid dynamics code that I employed in the present study was CFX-TASCflow, Version 2.11.1. I used a strongly conservative finite element-based finite volume discretisation method to partition the physical domain into a set of hexahedral control volumes. Out of the possible discretisation schemes available, the current simulations employed the modified linear profile scheme (MLPS). The MLPS gives second order error reduction in most instances (TASCflow, 2003). For steady-state computations, the code employs a transient formulation as a convenient means of introducing relaxation into the iterative non-linear solution (Galpin and Raithby, 1986). If only the steady-state solution is of interest, one can use the time step Δt as a free parameter through which it optimises the convergence rate. This corresponds to the use of pseudo-time which one employs solely to advance the solution in an iterative manner.

The relaxation scheme which one employs in the code belongs to the lower-upper (LU) factorisation family. The code makes use of a multi-grid solution acceleration technique designed to reduce the global linear residuals in between the solution of the non-linear terms. The rationale underpinning multi-grid acceleration is rooted in the observation that iterative solvers tend to reduce errors which correlate with short grid spacing much faster than they do to those which correlate with longer lengths.

When running the inlet box simulations, it was appropriate to displace the outlet boundary conditions location away from the bell-mouth discharge and for this purpose I added an annular ducting grid section to the model. The grid that I employed in the second set of simulations includes rotor and stator blocks, a conical diffuser and a long expansion chamber, Figure 7.4. Although the mesh's extent is large in this last component, it is useful to have the outflow boundary located far downstream of a backward step, thereby preventing recirculation flow occurrence in the boundary's proximity since this can act as a source of overall convergence problems. The grid sizes are about 600,500 nodes for the original configuration of five stays and 1,200,000 nodes for the improved eight stay plus eight inlet guide vane configuration. The nodes that I needed for the improved configuration were

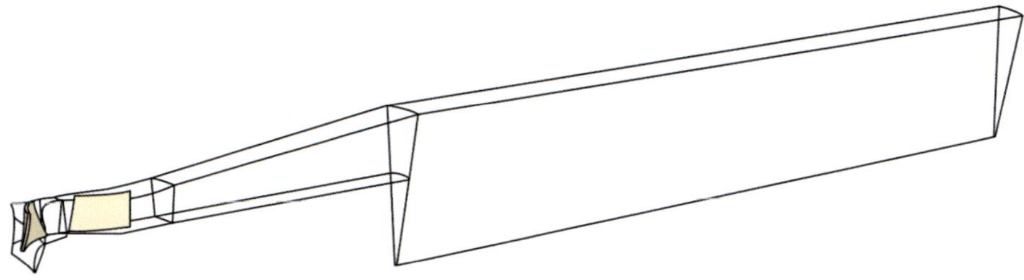

FIGURE 7.4. Rotor, stator, diffuser and expansion chamber computational domains.

approximately double that of the original as a consequence of the requirement to extend the original configuration's half-annulus model to a full-annulus model. I augmented both simulations with an inlet box grid that comprised 325,000 nodes.

SIMULATION PARAMETERS

I specified the boundary conditions applicable to the two simulations as total pressure and flow direction at inlet and mass flow at outlet. I ran both the original and improved inlet box configuration simulations with prescribed inlet profiles. The prescribed inlet profiles' function was strictly that of aligning the flow with the local geometric features given that earlier runs had shown that a prescribed uniform direction, taking no account of these features, induces the appearance of lengthy low momentum flow structures along the inlet duct walls and corners. With a prescribed inlet profile the flow direction's local distribution in the form of a vector plot is indicative of a well-conditioned inlet, Figure 7.5.

I performed the computations in this study on six 1.5 GHz twin-processor PCs networked as a Linux cluster. In this platform, and employing two CPUs in parallel, each computation of the original five stays configuration took approximately 45 CPU hours to perform and 1,700 iterations to reach a maximum root mean square residual error of 3E-06.

INLET BOX SIMULATION RESULTS

An analysis of simulation results of the original five stay configuration illustrated that the flow was generally well behaved without any significant areas of separation and that downstream of the stay outflow plane the flow is predominantly

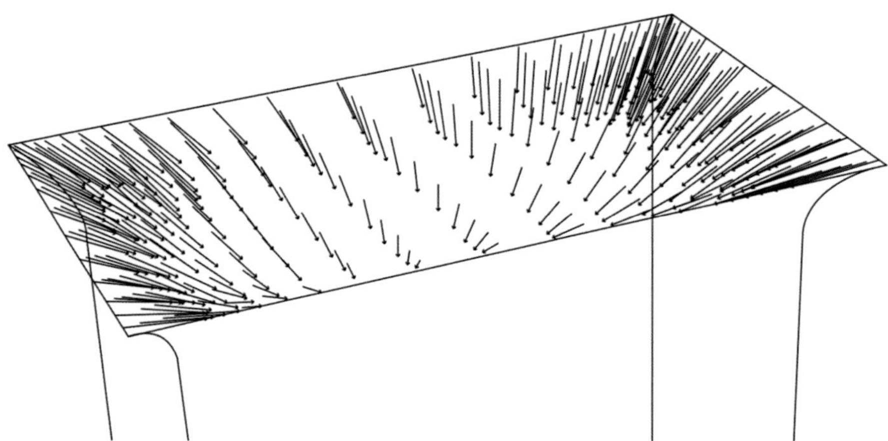

FIGURE 7.5. Prescribed flow direction vectors at the fan duct inlet.

axial, notwithstanding the presence in places of a small amount of tangential secondary flow. This translates into a two to four degree departure from a zero rotor inflow angle at given pitch-wise locations. For the purposes of supplying the corresponding fan assembly model with an inlet profile, I divided the inlet box outflow into four 45-degree sectors, Figure 7.6. Figure 7.7 presents a comparative plot of the sector-averaged tangential velocity. The flow's positive direction is in a clockwise sense when looking upstream of the discharge plane.

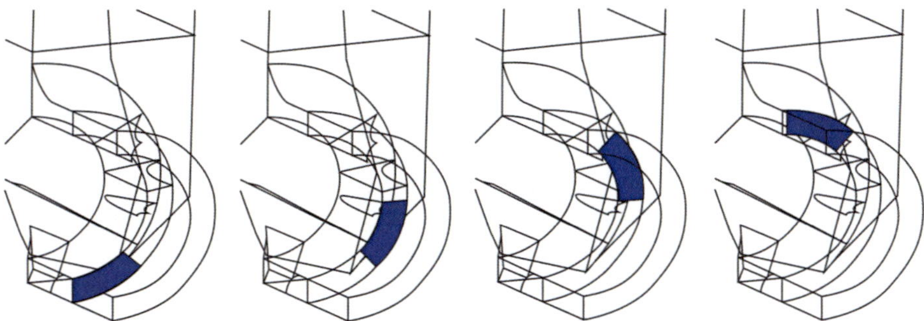

FIGURE 7.6. Flow averaging angular sectors for the original five stay configuration.

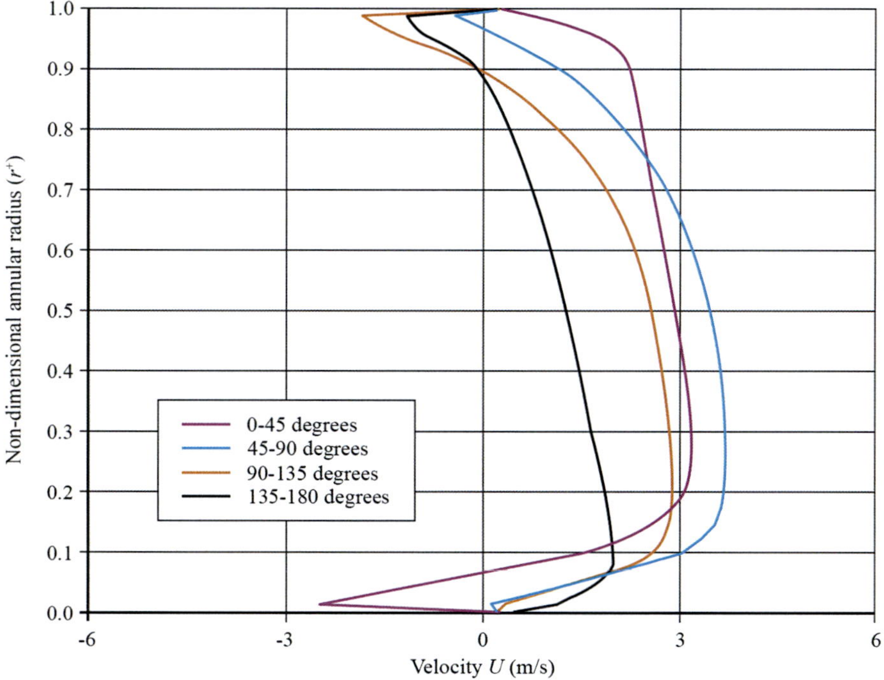

FIGURE 7.7. Tangential velocity radial distribution at the inlet box discharge for the original five stay configuration at the duty point condition.

I applied a similar post-processing method to the improved eight stays plus eight inlet guide vane configuration and therefore the equivalent tangential velocity plot shows results for eight pitch-wise sectors of 45 degrees, Figure 7.8. The origin of angular displacement measurement is as for the five stays case.

The effect of fitting the cambered stays and inlet guide vanes to the improved configuration is apparent, but the comparisons with the original configuration are limited to equivalent sectors alone. Unlike the complete inlet box arrangement, the tangential flow distribution in the five stays geometry is valid only for the side of the box which I computed. Under the flow symmetry assumption, the opposite side will present a similar distribution, but flowing in the opposite direction. One must then superimpose the inlet box's 'handed' effect in its original configuration onto the overall spinning motion which the added turning effects induce, but it is not immediately obvious what this should be locally.

A closer analysis of the results, Figure 7.8, can help clarify the issue associated with the handed effect since in general the velocity distribution in related sectors for the two plots is broadly similar and corresponds mostly to the highest values. Conversely, the lowest values are those which would match the geometric opposite's sectors of the computed five stays domain confirming, therefore, the notion that the flow's right hand side in the inlet box contributes towards the swirling motion, whilst that of the left hand side tends to oppose it.

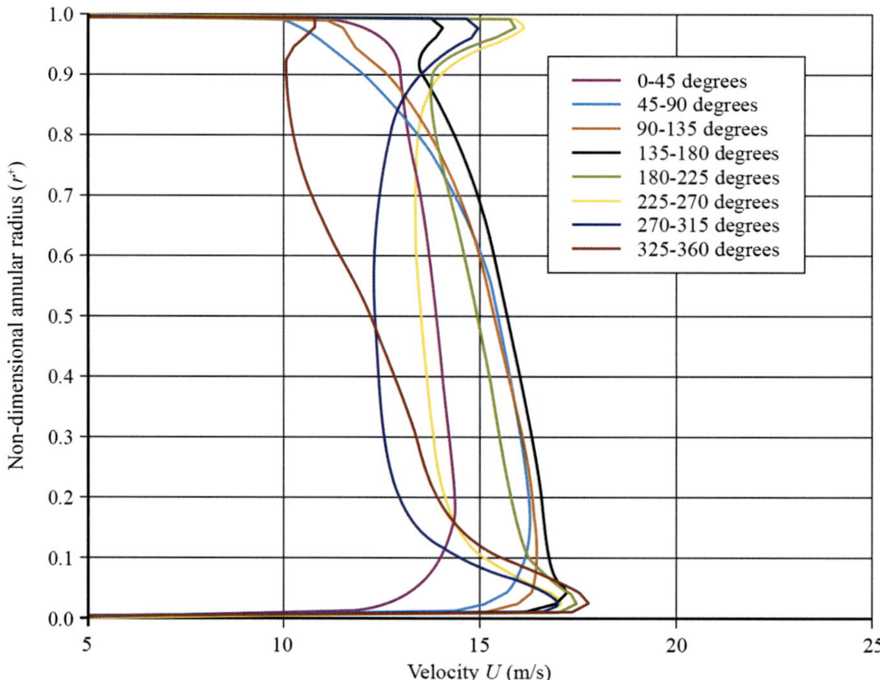

FIGURE 7.8. Tangential velocity radial distribution at the inlet box discharge for the improved eight stay plus eight inlet guide vane configuration at the duty point condition.

FAN SIMULATION RESULTS

The second simulation set addressed the combined geometries of the rotor, stator, diffuser and downstream channel with the inflow boundary conditions supplied to the simulations which I obtained from the inlet box computations at the appropriate flow rates. The original five stay configuration computation involved the study of four flow conditions corresponding to the pitch-wise sectors described above, whilst I carried out that of the eight stay plus eight inlet guide vanes for the corresponding points at the eight circumferential positions.

I presented the fan simulation results for the original five stay configuration on the fan characteristic derived from model test data, Figure 7.9. The duty point pressure is slightly over predicted, and the slope of the simulation results is lower than the model test data. The over prediction of the duty point pressure is small enough for us to consider within the combined uncertainty of the model test data and simulation results. The lower slope of the simulation results implies that the simulation is under predicting the fan's peak pressure developing capability, and is therefore conservative and consequently acceptable.

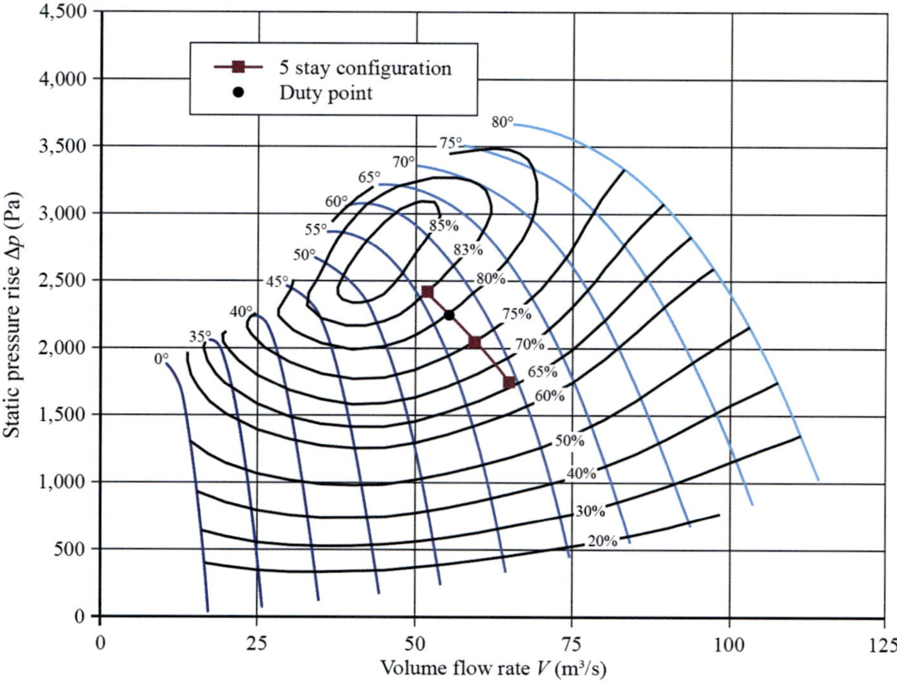

FIGURE 7.9. Experimentally derived characteristics for the studied 1,767 mm diameter PFS fan, with simulation results for the original five stay configuration at four flow conditions, duty point, 10 per cent and 20 per cent above and 5 per cent below the duty point condition.

I neither model tested nor performance tested in-service the improved eight stay plus eight guide vane configuration. Consequently, I presented the improved configuration fan simulation results in a comparative manner between the simulation results for the original and improved configurations. I compared the rotor relative inflow angles for the two configurations for equivalent angular sectors at the duty point, Figure 7.10. I predicted that the mid-height rotor relative inflow angle would increase by approximately five degrees with the change from original to improved configuration. I estimated that in-service, the fan's motor had a margin on peak power sufficient to allow inflow angle to increase by up to seven degrees, with five representing a conservative maximum.

The effect of increasing the inflow angle on fan pressure rise is to increase predicted duty point pressure from 2,320 Pa to 3,300 Pa, Figure 7.11. It is apparent that besides the expected increase in pressure with the change from original to improved configuration, there is a marked change in the characteristic line's gradient which is consistent with a movement towards stall. Despite moving towards stall, the simulation results do not predict that the improved configuration will stall at the duty point. As the original configuration did predict the fan's stall margin conservatively, the improved configuration's stall margin was acceptable.

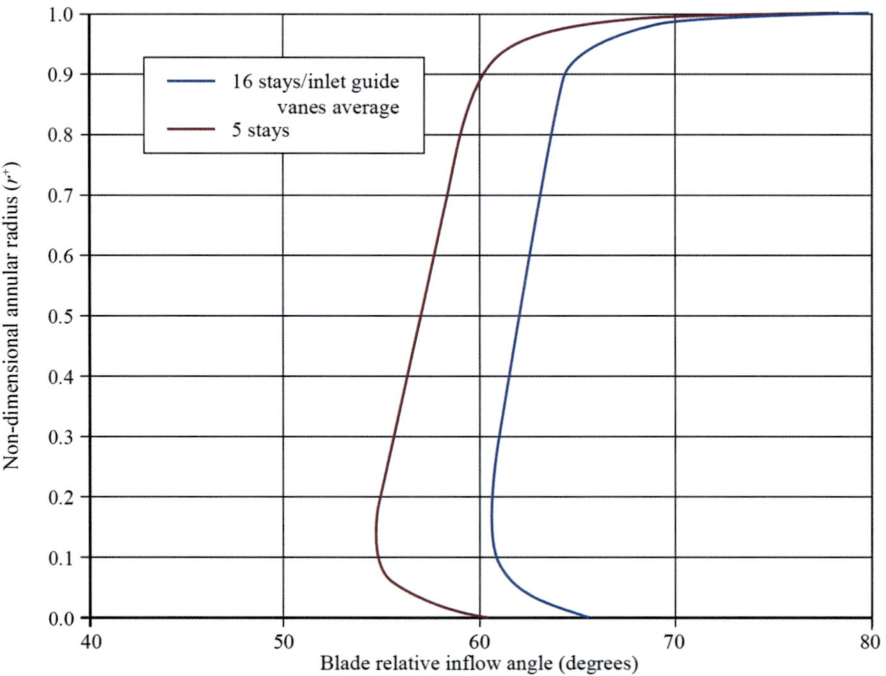

FIGURE 7.10. Radial distribution of the blade relative inflow angle for the original five stay and improved eight stay, plus eight inlet guide vane configurations at the duty point condition.

An analysis of the simulations' adiabatic efficiency results for both the original and improved configurations indicates that the improved configuration has a duty point efficiency approximately 1 per cent less than the original configuration, Figure 7.12. This is consistent with the principle that for a well-designed blade increasing the rotor relative inflow angle should equate to some local increase in losses. Although the simulation results for the duty point confirm increase in losses, the simulation results indicate that at higher flow, the improved configuration is able to maintain efficiency more effectively than the original configuration. The improved configuration's ability to maintain efficiency at higher flow suggests that when considering the rotor in isolation, there is a small reward for operation with a higher relative inflow angle.

The improved eight stay plus eight inlet guide vane configuration increases the fan aerodynamic loading. Increasing fan aerodynamic loading also increased fan acoustic emissions (Pascu *et al.*, 2009). When we analyse the improved eight stay plus eight inlet guide vane configuration using Pascu *et al.*'s method (2009), we can predict an overall increase in fan acoustic emissions of between 3 and 5 dB. In practice, I installed the original configurations and then later retrofitted the improved configuration into the same fan. The fan was located next to a coal-fired power plant's boiler, and in close proximity to other heavy duty rotating equipment. It

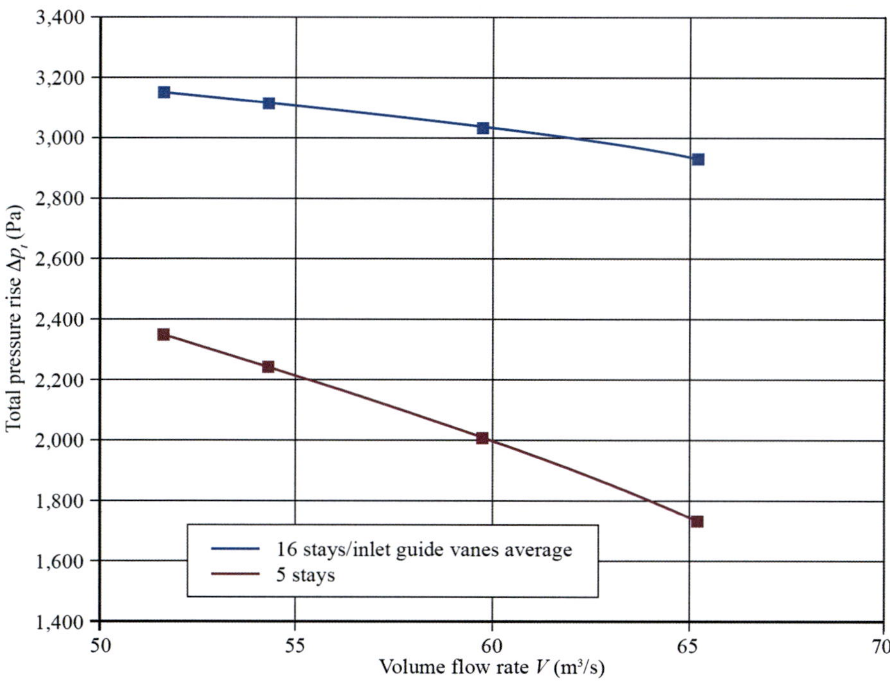

FIGURE 7.11. Total pressure rise for the original five stay and improved eight stay plus eight inlet guide vane configurations for four flow conditions: duty point, 10 per cent and 20 per cent above and 5 per cent below.

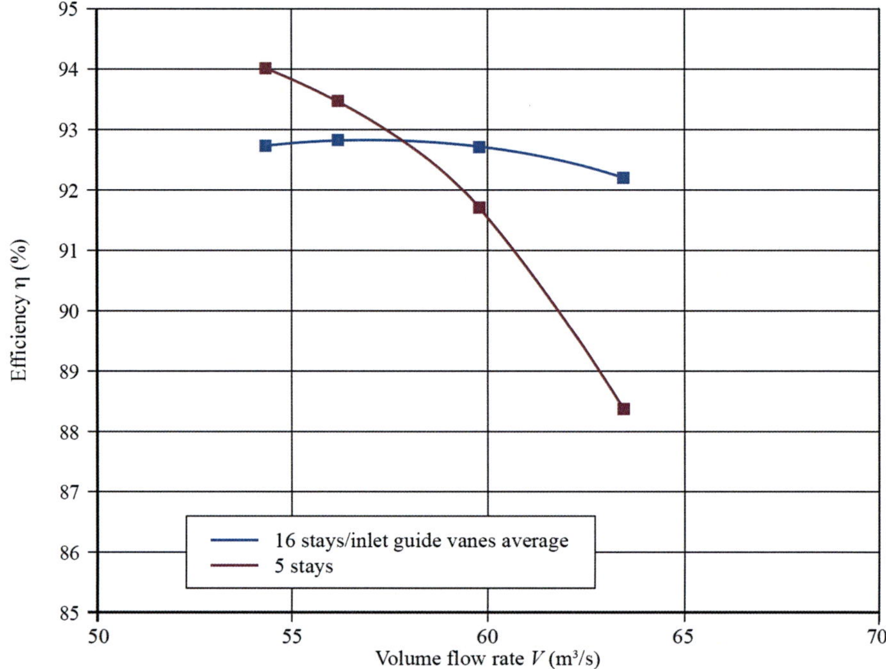

FIGURE 7.12. Normalised efficiency variation relative to the duty point for the original five stay and improved eight stay plus eight inlet guide vane configurations for four flow conditions: duty point, 10 per cent and 20 per cent above and 5 per cent below.

proved impossible in practice to accurately measure on-site fan acoustic emissions with either the original or improved configuration. As such, I was not able to ascertain that the change in stay and guide vane geometry resulted in an increase in fan acoustic emissions. However, it is likely that if the background noise in the fan's vicinity had been lower, an increase would have been apparent.

MECHANICAL IMPACT OF FAN SIMULATION RESULTS

I undertook the computational analysis in response to an in-service issue with an induced-draft fan. The induced-draft fan was not generating the required pressure. The inability to produce the required pressure was a consequence of the fan's under-specification for the application in the first place. It was subject to a distorted inlet flow and also suffered significant erosion as a consequence of a poorly installed and maintained air filtration system upstream of the fan. The improved stay and inlet guide vane configuration increased the fan pressure developing capability, and additionally reduced the circumferential variation in velocity entering the fan blades. I considered this reduction in circumferential velocity as important as a consequence of a second in-service issue with the induced-draft fans generally: premature blade bearing failure.

Collected data downstream of the rotor for the improved eight stay plus eight inlet guide vane configuration, Figure 7.13, shows the predicted cyclic variation in total pressure due to the uneven pressure distribution emanating from the inlet box. The graph's scale exaggerates the true magnitude of the dissimilarity between different sector values. The maximum deviation from the average value is 2 per cent. This deviation, due to the inlet box conditions, is small enough for us to regard as aerodynamically negligible.

Despite the observation that the uneven pressure deviation's aerodynamic effect due to the inlet box conditions may be neglected, we must consider it within the context of the studied fan: a constant speed variable pitch in motion fan. A blade bearing that facilitates change in blade angle supports each fan blade to enable fan-load adjustment. The deviation in aerodynamic load results in a once-per-revolution cyclic load and moment on each blade which then changes the load distribution on the blade bearing, Figure 7.14. This once-per-revolution load results in the bearing balls' cyclic contact angles on the bearing raceway. The cyclic twisting moment causes an infinitesimal rotation in pitch change direction, which then results in bearing ball orbital motion which is different for each ball.

A consequence of the ball orbital motion is that bearing balls tend to migrate towards one location on the bearing raceway (Perkinson *et al.*, 2012). The balls push until they remove all clearance, with the resultant ball to ball separator forces becoming large. This ball to ball separator force results in increased bearing friction due to the friction between bearing balls and ball separator. This results in

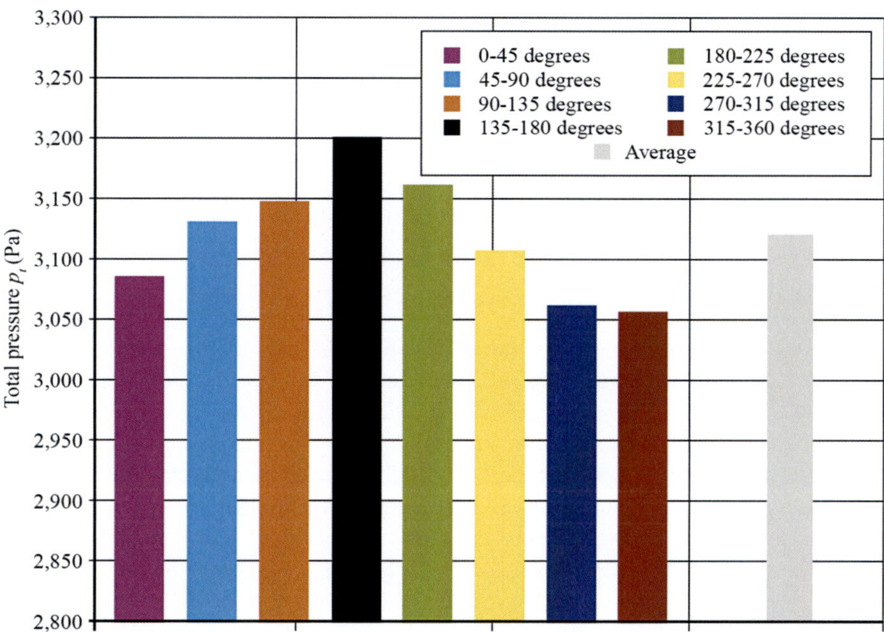

FIGURE 7.13. Total pressure variation with pitch-wise sector for the improved eight stay plus eight inlet guide vane configuration at the duty point condition.

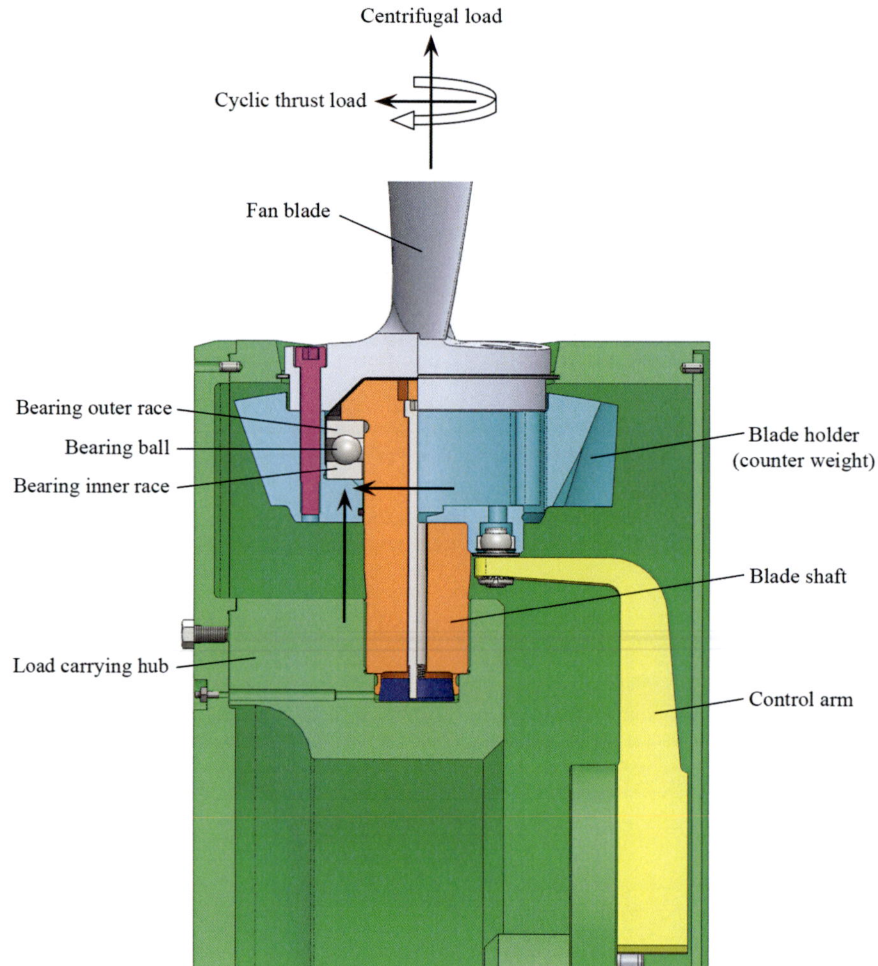

FIGURE 7.14. Mechanical loads induced in the blade bearing by the pitch-wise total pressure variation.

accelerated ball separator wear, Figure 7.15, and consequently, bearing ball wear. The result is the bearing's rapid mechanical deterioration, and consequently, it rapidly increases the force required to change the blade angle.

Reducing the cyclic variation in total pressure due to the uneven pressure distribution emanating from the inlet box below 2 per cent would be difficult to achieve in practice in an industrial fan application. Rather than attempting to eliminate the uneven pressure distribution, Sheard and Rhoden (2012) proposed a novel bearing separator comprising individual bearing separator elements, Figure 7.16, manufactured from a low-friction carbon impregnated Teflon. The resulting bearing configuration, Figure 7.17, has the same imposed mechanical loads that previously resulted in bearing separator wear (Figure 7.15). However, the use of separate bearing separator elements results in the imposed loads inducing rotation in

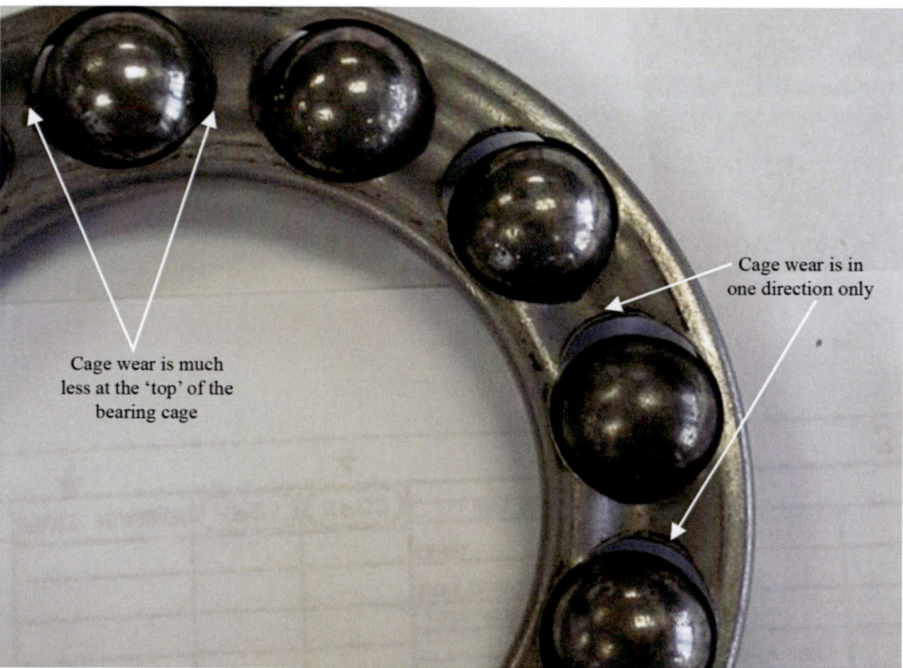

FIGURE 7.15. Wear in a traditional 'cage' bearing separator that occurs as a consequence of mechanical loads on blade bearing balls. Pitch-wise total pressure variations induce the mechanical loads.

FIGURE 7.16. Sheard and Rhoden's (2012) bearing separator concept.

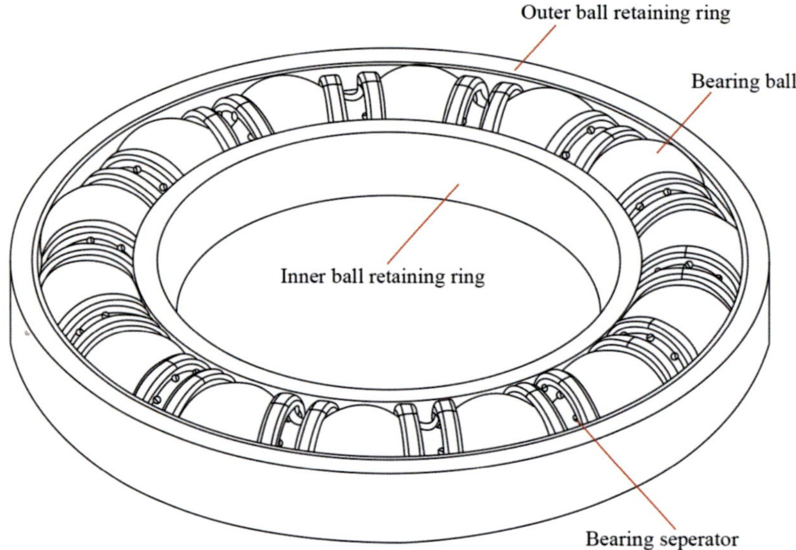

FIGURE 7.17. A blade bearing, fitted with Sheard and Rhoden's (2012) bearing separator.

each separator that then transports lubricant into the bearing ball to bearing separator contact zone.

Sheard and Rhoden's (2012) bearing separator design has demonstrated good in-service reliability, withstanding the imposed loads that arise with the cyclic variation in total pressure due to the uneven pressure distribution emanating from the inlet. An investigation after 22,000 in-service hours indicated that approximately 50 per cent of bearing life had been consumed. The bearing itself was still well packed with lubrication that showed good persistence in all contact areas within the bearing. The bearing balls, separator and raceway showed no visible signs of damage.

CONCLUSIONS

I studied an inlet box in its five stays original form and an improved configuration with eight stays plus eight inlet guide vanes, designed to pre-swirl the flow before discharge onto a fan stage. The two inlet box simulations enabled me to identify handed flow effects in the original configuration which are still present in the improved configuration in the form of small, but detectable tangential speed variations. The corollary is that for complex inlet assembly geometries, particularly for higher speed applications, there is a need to account for the cumulative flow effects in order to avoid discharged flow pitch-wise non-uniformities. We can best meet this objective through some form of computational method which is able to resolve the three-dimensional flow-field. For the case under study, the fan's stall margin in its original configuration is adequate to accommodate the additional loading associated with the improved configurations' additional stays and inlet guide vanes. Nevertheless, the fan remains susceptible to erosion effects.

The analysis of the improved eight stays plus eight inlet guide vanes rotor-stator simulations over the eight pitch-wise sectors enabled the identification of a 2 per cent cyclic variation in pressure due to the uneven pressure distribution emanating from the inlet box. Although the aerodynamic impact is small, there is a mechanical consequence for the studied variable pitch in motion fan. The resultant mechanical loads on the blade-bearing result in the bearing ball separator's rapid wear and consequently, premature blade bearing failure. I used Sheard and Rhoden's (2012) novel bearing separator design that has proven capable of withstanding the imposed mechanical loads and thus eliminating the need to reduce further the cyclic variations in total pressure. After 22,000 hours' in-service operation, I estimated that the blade bearings fitted with the novel bearing separator had approximately 50 per cent bearing life remaining. As such, the novel bearing separator design facilitates an estimated five year bearing life, self-consistent with the major service intervals of variable pitch in motion fans in induced-draft application.

REFERENCES

Bianchi, S., Corsini, A. and Sheard, A.G. (2012), 'Installed Aeroacoustic Performance of Cooling Axial Fans Fitted with End-plates', *Noise Control Engineering Journal*, vol. 60, pp. 519–527.

Corsini, A., Marchegiani, A., Minotti, S. and Sheard, A.G. (2011), 'Numerical Investigation on the Aerodynamic Influence of Eroded Leading-edge Geometry on Boiler Fan Performance', *Proceedings of the 9th European Conference on Turbomachinery, Fluid Dynamics & Thermodynamics,* 21–25 March, Istanbul, Turkey, pp. 879–892.

Corsini, A., Marchegiani, A., Rispoli, F., Sheard, A.G. and Venturini, P. (2012a), 'Predicting Blade Leading Edge Erosion in an Axial Induced Draft Fan', *Transactions of the ASME, Journal of Engineering for Gas Turbines & Power,* vol. 134, paper no. 042601, pp. 1–9.

Corsini, A., Rispoli, F., Sheard, A.G. and Venturini, P. (2012b), 'Numerical Simulation of Coal Fly-ash Erosion in an Induced Draft Fan', *Proceedings of the 57th American Society of Mechanical Engineers Turbine and Aeroengine Congress,* 11–15 June, Copenhagen, Denmark, paper no. GT2012-69048.

Galpin, P.F. and Raithby, G.D. (1986), 'Numerical Solution of Problems in Incompressible Fluid Flow: Treatment of the Temperature-velocity Coupling', *Journal of Numerical Heat Transfer*, vol. 10(2), pp. 105–129.

Greenzweig, J.T., Henry, R. and Holm, T. (2011), 'The Effect on Inlet Flow Distortion on the Performance of Centrifugal Fans for Utility Applications', *Proceedings of the 56th American Society of Mechanical Engineers Turbine and Aeroengine Congress,* 6–10 June, Vancouver, Canada, paper no. GT2011-45422.

Guédel, A. (2006), 'Prediction of the Noise Installation Effect Induced by a Bend at the Inlet of an Axial Flow Fan', *Noise Control Engineering Journal*, vol. 54(1), pp. 21–26.

Pascu, M., Miclea, M., Epple, P., Delgado A. and Durst, F. (2009), 'Analytical and Numerical Investigation of the Optimum Pressure Distribution along a Low-pressure Axial Fan Blade', *Proceedings of the Institution of Mechanical Engineers, Part C: Journal of Mechanical Engineering Science*, vol. 223(3), pp. 643–657.

Perkinson, R.H., Soule, M.C. and Holl, H.J. (2012), 'Separator for Bearing Assemblies with Cyclic Loads', patent no. US 8,167,501 B2, 1 May.

Roslyng, O. (1984), 'Installation Effect on Axial Flow Fan Caused by Swirl and Non-uniform Velocity Distribution', *Proceedings of the Institution of Mechanical Engineers (IMechE) Conference 1984-4 on the Installation Effects in Ducted Fan Systems*, 1–2 May, London, UK, paper no. C114/84, pp. 21–28.

Sheard, A.G. and Rhoden, A. (2012), 'Variable Pitch Fans', patent no. GB 2,485,634 B, 12 September.

Staiger, M. and Stetter, H. (1993), 'Aerodynamic Response of Axial Fan Bladings to Non-uniform Inlet Flow Fields', *Institution of Mechanical Engineers Seminar on Installation Effects in Fan Systems*, 28 September, London, UK.

TASCflow (2003), *TASCflow Documentation*, ANSYS Canada Ltd, Waterloo, Ontario, Canada.

Teixeira, J.A., Tourlidakis, A., Ivey, P.C., Sheard, A.G., Molin, J.A. and Kinghorn, I.R. (2003), 'Computational Analysis of an Industrial Axial-flow Fan Including Installation Effects', *Proceedings of the IMechE Advances of CFD in Fluid Machinery Design Seminar,* 18 November, London, UK.

Chapter 8

The Conceptual Design of High-pressure Reversible Axial Tunnel Ventilation Fans

A.G. Sheard and K. Daneshkhah

ABSTRACT

Tunnel ventilation fans, classically, must have the ability to both supply and extract air from a tunnel system, with the operator's choice dependent on the tunnel ventilation system's operating mode most appropriate at any given point in time. Consequently, tunnel ventilation fans must incorporate a reversible aerodynamic design which limits the maximum fan pressure rise. This chapter presents three high-pressure reversible fan concepts. These comprise a two-stage counter-rotating fan, a single-stage high-speed fan and a two-stage fan with a single motor and impeller on each end of the motor shaft. We consider the relative merits of each concept. The third concept offers the most compact fan, transform, silencer and damper package size. We discuss the mechanical design challenges that occur with a two-stage fan with a single motor and impeller on each end of the motor shaft. We then present and consider a selected motor bearing arrangement and casing design for maintainability. Finally, we present both prototype fan and full scale package aerodynamic and acoustic performance, before discussing the challenges presented by high-temperature certification in accordance with the Euro Norm EN 12101-3: 2012 requirements.

This chapter is a revised and extended version of Sheard, A.G. and Daneshkhah, K. (2012), 'The Conceptual Design of High Pressure Reversible Axial Tunnel Ventilation Fans'. *Advances in Acoustics and Vibration*, vol. 2012, article ID 562309, pp. 1–11.

NOMENCLATURE

Latin letters

BPF	[Hz]	Blade passage frequency
D_t	[m]	Blade tip diameter
f	[Hz]	Frequency
L_W	[dB]	Sound power level
p	[Pa]	Static pressure
P	[kW]	Power
p_p	[Pa]	Pressure pulse
p_t	[Pa]	Total pressure
V	[m^3/s]	Volume flow rate

Greek letters

η_m	[%]	Minimum efficiency
η_M	[%]	Peak efficiency
ρ		Air density
σ		Solidity

Acronyms

FCP	Factory production control
PWM	Pulse width modulation

INTRODUCTION

The current limit on the pressure developing capability of single-stage reversible tunnel ventilation fan technology is 2,000 Pa. As tunnel length increases, the system resistance that occurs with a required flow can exceed this 2,000 Pa limit. A second factor that escalates the required tunnel ventilation fan pressure developing capability is increasing train speeds in modern metro systems. Travelling in a tunnel, a train induces a positive pressure pulse ahead and a negative pressure pulse behind. When the train passes a ventilation shaft, a tunnel ventilation fan within the shaft will experience first a positive and then a negative pressure pulse.

Sheard and Corsini (2012) considered pressure pulse magnitudes in tunnel ventilation systems and concluded that the historic norm is 300 Pa. However, they identified a third factor which increases the required tunnel ventilation fan pressure development capability. The introduction of platform screen doors has increased the size of pressure pulses to typically 700 Pa and in combination with high train speeds as high as 1,000 Pa. Within the context of a historic system resistance of typically 1,200 to 1,500 Pa, an increase in the pressure pulse magnitude from 300 to 700 Pa can take the total required pressure beyond the capability of a single reversible fan stage.

The above requirements imply a level of complexity in the design process that is beyond the historic norm within the air movement and control community. Empirical design methods are no longer appropriate, with tunnel ventilation fan designers increasingly moving from empirical fan to industrial turbomachinery design methodologies. Wallis (1961) and Daly (1985) exemplify the traditional empirical approach to tunnel ventilation fan design and describe a trial-and-error approach that relies on the designer's experience.

More recently, tunnel ventilation fan designers have started to utilise computational fluid dynamics analyses at the beginning of the design process. Vad *et al.* (2001) and Vad (2001) have proposed the application of computational methods in the fan design process. They have developed a family of high-performance swept fans for mine ventilation by feeding-back the three-dimensional design criterion with computed aerodynamic data which provides insight into the size and nature of rotor secondary flows.

Lee *et al.* (2008) recently applied an inverse approach to the design of cooling fans for electronic appliances. This included the combined use of a 'design of experiments' step and computational fluid dynamics to explore the space available for design solutions. Thus, Lee *et al.* (2008) transferred methodologies that researchers had originally developed for the aerospace community into the air movement and control community. Daneshkhah and Sheard (2013) are part of the air movement and control community and also utilised turbomachinery design methodologies. They developed an experimental design approach in combination with the use of computational fluid dynamics to optimise a tunnel ventilation jet fan blade's design. They based the design methodologies on Horlock and Denton's (2005) design practices which are reliant on computational fluid dynamics to develop appropriate three-dimensional blade sections.

Another issue of importance to the contemporary design of tunnel ventilation fans is that standards for the fans' acoustic performance during both routine and emergency operations have become progressively more stringent. During routine night-time operations, tunnel ventilation system noise at the exhaust portals is typically limited to 50 dBA to avoid disturbing people who live in the vicinity. During emergency operations, operators typically limit noise to 85 dBA to allow emergency-services personnel to communicate effectively.

Borello *et al.* (2012) studied the practicality of using an unsteady computational fluid dynamic code to assist in the assessment of tunnel ventilation fan acoustic characteristics. Borello *et al.* (2012) concluded that more aggressive aerodynamic designs produce higher acoustic emissions, and consequently, acoustic optimisation was as important an aspect of a design methodology as aerodynamic and mechanical optimisation.

When responding to the constraints imposed upon them, tunnel ventilation fan designers require a methodology that provides an optimum balance between acoustic, aerodynamic and mechanical requirements during both emergency operations at elevated temperature and during routine operation. In order to be effective, the tunnel ventilation fan designer must combine finite-element analysis (for mechanical analysis) and computational fluid dynamics (for aerodynamic and aero-acoustic analysis) to provide a *virtual prototyping* design methodology that replaces traditional test and evaluation methods in fan development.

Although virtual-prototyping techniques are presently uncommon in the air movement and control community, others have used such techniques extensively.

Pratt (1994) and Jasnoch *et al.* (1994) have studied virtual-prototyping techniques. Both researchers independently concluded that they provide cross-functional evaluations of competing objectives. They enable issues that others have previously considered 'downstream issues' to be considered during the initial design stages. By developing virtual prototypes as 'digital mock-ups', the process of virtual prototyping reduces the need to build physical prototypes and facilitates identifying design problems early.

Sheard *et al.* (2009) documented a case study of a virtual-prototyping approach application in developing a new tunnel ventilation fan range. They used virtual-prototypes to characterise the new generation tunnel ventilation fans' aerodynamic and aero-acoustic performance. Sheard *et al.*'s (2009) work focused exclusively on single-stage fan development. In this chapter we expanded this work and used it to design three tunnel ventilation fan concepts. These include a two-stage counter-rotating fan, a high-speed single-stage fan and a two-stage fan with a single motor and impeller on each end of the motor shaft.

In this chapter, we consider the mechanical challenges that occur with three tunnel ventilation fan concepts. We present the aerodynamic performance of the three fan concepts and discuss the mechanical design challenges. The chapter then moves on to present the fan concept realisation that resulted in the most compact ventilation package: a two-stage fan with a single motor and impeller on each end of the motor shaft. Following a description of the realised design, we compare the laboratory performance of a prototype and first production fan, transform, silencer and damper package to that which we predicted during the design process. The chapter closes with a review of the challenges that may occur with high-temperature certification of large tunnel ventilation fans in accordance with the Euro Norm EN 12101-3: 2012 requirements, followed by our conclusions and recommendations.

FAN DESIGN REQUIREMENTS

We developed the reversible fan concepts presented in this chapter using the Marmaray project specification as an example of a 'next generation' tunnel ventilation system that requires a reversible fan beyond the current single-stage state of the art. Sheard *et al.* (2013) described the Marmaray tunnel system design and observed that Parsons Brinckerhoff developed a specification for its tunnel ventilation fans, Table 8.1. The specification includes a minimum fan efficiency of 70 per cent. Although modest by uni-directional fan standards, an efficiency of 70 per cent is close to the state of the art for a conventional single-stage reversible fan, and therefore a specified efficiency of 70 per cent, and its associated power of 710 kW at the design flow and pressure, represents a challenge for tunnel ventilation fan designers.

The acoustic specification and requirement for EN 12101-3 certification also have an implication for the tunnel ventilation fans' pressure development capability. Any silencer has an associated pressure loss, so we must add this to the pressure that the fan generates. The Euro Norm EN 12101-3 requires tunnel ventilation fans to be

Table 8.1. *Marmaray fan, transform, silencer, damper package specification.*

Static pressure p (Pa)	3,400
Pressure pulse p_p (Pa)	+ / – 500
Volume flow rate V (m^3/s)	144
Power P (kW)	710
Minimum efficiency η_m (%)	70
dBA (tunnel side)	85 ten metres from vent shaft
dBA (atmospheric side)	85 ten metres from the vent shaft
EN 12101-3 certification	250° C for one hour
Reversibility	100% reversible
Air density	1.38 kg/m^3

capable of clearing smoke in the event of a tunnel fire. The clearance between aluminium blades and the steel casing within which they run must increase at ambient conditions to ensure that it does not close completely at elevated temperatures during a tunnel fire. Increasing blade tip-to-casing clearance reduces the tunnel ventilation fans' efficiency, as well as their pressure development capability.

The specification (Table 8.1) is for the fan, transform, damper and silencer package and includes a 3,400 Pa duty point pressure. This is a pressure rise across the package, and therefore after accounting for the pressure loss through the silencers. Additionally, the package must achieve its specification at ambient temperature, with the associated increased blade tip-to-casing clearance needed to enable the fan to operate at elevated temperatures during a tunnel fire.

Although not explicit in the Parsons Brinckerhoff specification, the ventilation plant room's size dictated the overall size of the fan, transform, damper and silencer package. Despite the constraint on the ventilation plant room size, it was necessary to retain the ability to undertake routine fan maintenance within the plant room's confines.

TUNNEL VENTILATION FAN CONCEPTUAL DESIGN

Two fans, counter rotating, are particularly effective when 100 per cent reversibility is required. The reason is that the second fan removes the exit flow swirl from the first. This results in axial flow exiting the second stage and therefore an inherently more efficient aerodynamic solution than is possible with a single-stage fan. Simply changing both fans' rotation direction achieves reversibility. The counter-rotating two-stage fans' inherent aerodynamic efficiency translates into achievable pressure, which is typically 2.5 times the pressure that one of the two fans operating in isolation can achieve. However, two fans counter rotating result in unsteady flow from the first fan entering the second. This increases noise production in the second. Consequently, silencers must be correspondingly larger for a fixed acoustic specification.

When acoustic considerations make a two-stage counter-rotating fan-based package potentially too large in a specific application, a second option available to the tunnel ventilation fan designer is to move beyond current mechanical limits for single-stage reversible fans. The choice of aluminium as the blade material limits today's tunnel ventilation fan technology. Aluminium enables the designer to design fans with a tip speed of up to 175 m/s. Within the context of the required specification, Table 8.1, this would equate to a 2.24 metre diameter fan running at 1,500 rpm. A reversible 2.24 metre fan running at 1,500 rpm has the potential to generate 3,000 Pa pressure rise; however, operators would not normally consider this as the selection would result in the fan operating far from its peak aerodynamic efficiency. In effect, it would be impossible to achieve the target 70 per cent minimum fan efficiency.

Despite the above reservation about fan efficiency, fan pressure rise increases in proportion to the ratio of fan diameter squared. As such, a 12 per cent increase in fan diameter from 2.24 metres to 2.5 metres will increase fan pressure developing capability by 24 per cent, increasing potential pressure from 3,000 Pa to 3,700 Pa. This increase in fan diameter would necessitate an increase in blade tip speed from 175 m/s to 196 m/s. In practice, improving aluminium blade design could not increase in tip speed. Sheard *et al.* (2009) presented fan blade mechanical design optimisation and described a computational-based methodology for minimising blade stress for a given aerodynamic design point. In effect, Sheard *et al.* (2009) maximised aerodynamic performance within the blade material's mechanical limits. Therefore, only moving away from aluminium to a higher strength-to-weight material, with titanium as the preferred choice within the aerospace industry, can achieve an additional increase in blade tip speed. Carbon fibre blades are not an option as a consequence of the EN 12101-3 high-temperature emergency requirement.

Although a move from aluminium to titanium has the potential to achieve a higher pressure within a single-stage reversible fan, there is an acoustic consequence with running at an increased tip speed. A 2.50 metre diameter fan running at 1,500 revolutions per minute will generate approximately 4 dB more noise than a 2.24 metre diameter fan running at the same speed. Consequently, silencers must be correspondingly larger for a fixed acoustic specification.

A third concept is to remain within the aluminum blades' mechanical limits, return to a lower speed two-stage fan, but this time using a single motor with an impeller on both ends. As the fan motor separates the two impellers, the solution is relatively quiet compared to either a two-stage counter-rotating or high-speed single-stage solution.

The above results in three high-pressure reversible tunnel ventilation fan concepts:

- a two-stage counter-rotating fan;
- a single-stage high-speed fan; and
- a two-stage double-ended motor fan.

Each is a viable option, with its own strengths and weaknesses, Table 8.2.

Table 8.2. *Relative merits of the three fan system concepts.*

	Concept 1 (Counter rotating)	Concept 2 (High speed)	Concept 3 (Double-ended motor)
Static pressure p	2.5	2.25	1.8
Peak efficiency η_M	+8%	+2%	+5 %
Noise	15 dB noisier	10 dB noisier	8 dB noisier
Cost	2.0 a single fan	2.5 a single fan	2.0 a single fan
Size	2.0 the length of a single fan	1.1 the length of a single fan	1.3 the length of a single fan
Technical risk	Low	High	Medium

Note 1: All comparisons are relative to that of a single-stage fan of the same diameter.

Note 2: Peak efficiency is relative to that of a current state of the art single-stage reversible fan with a peak efficiency of 70 per cent.

Although assessment of technical risk is subjective, we can, at least, broadly characterise the technical risks of each concept. The two-stage counter-rotating fan concept is a low technical risk, as it is the standard solution when pressure exceeds that which a single stage can achieve. The concept has a good in-service reliability record.

The single-stage high-speed fan concept is high technical risk, as it involves moving from a traditional tunnel ventilation fan blade material to a material which engineers usually associate with the aerospace industry. Although the move is not inevitably problematic, new supply chain development and the mechanical validation of titanium blades would be time consuming, with problems emerging during the mechanical validation process. The nature and extent of the mechanical design problems are difficult to quantify due to the application's novelty and therefore constitute a high technical risk.

The two-stage double-ended motor option is a medium technical risk, as it remains within existing mechanical limits; however, as a consequence of the novel double-ended configuration, will not fall within existing EN 12101-3 certification. EN 12101-3 certification of a range of double-ended motor fans implicitly requires large fan high-temperature testing in accordance with EN 12101-3 requirements. Large fan high-temperature tests are challenging and therefore constitute a significant, but quantifiable technical risk.

Concept 1: counter-rotating fan

We estimated the aerodynamic performance of a two-stage counter-rotating fan that will achieve the specification in Table 8.1 by scaling the smaller single-stage fans' performance using Daly's method (1985), Figure 8.1. The selection is good, as a consequence of the required duty point falling close to the fan characteristic's peak efficiency point. The two-stage counter-rotating configuration's inherent aerodynamic efficiency results in an efficiency that comfortably exceeds the specified 70 per cent minimum.

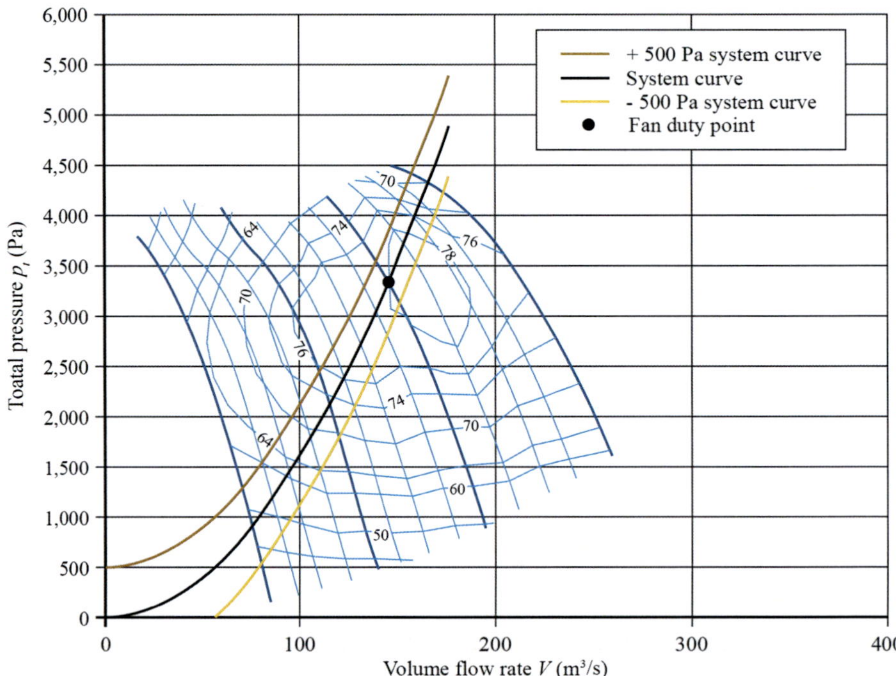

FIGURE 8.1. The characteristics of a 2.5 metre diameter, 6-pole two-stage counter-rotating fan, with an estimated 78 per cent efficiency at the Marmaray duty point.

Concept 2: high-speed single-stage fan

We estimated the single-stage high-speed fan's aerodynamic performance that will achieve the specification in Table 8.1 by scaling a lower speed fan's performance, Figure 8.2. The selection is also good, as a consequence of the required duty point falling close to fan characteristic's peak efficiency point. The exit swirl from a single stage results in a lower aerodynamic efficiency than that of the two-stage counter-rotating selection. Consequently, the single-stage high-speed fan has a lower predicted efficiency of 71 per cent. The uncertainty that we see with scaling fan performance data is typically no better than +/– 1 per cent and so a laboratory performance test would need to validate this fan selection's ability to achieve the minimum 70 per cent efficiency.

Concept 3: double-ended fan

We estimated the two-stage double-ended motor fan's aerodynamic performance that will achieve the specification in Table 8.1 by scaling the single-stage fan's performance, Figure 8.3. The selection is also good, as a consequence of the required duty point falling close to fan characteristic's peak efficiency point. The results include an assumption about the inter-stage flow straighteners' efficiency,

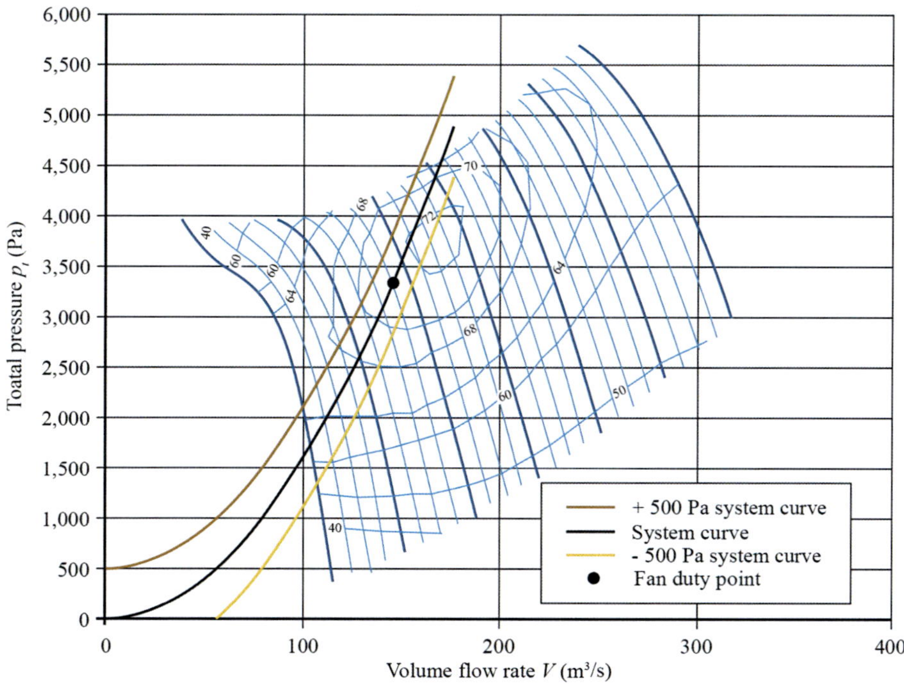

FIGURE 8.2. The characteristics of a 2.5 metre diameter, 4-pole single-stage high-speed fan, with an estimated 71 per cent efficiency at the Marmaray duty point.

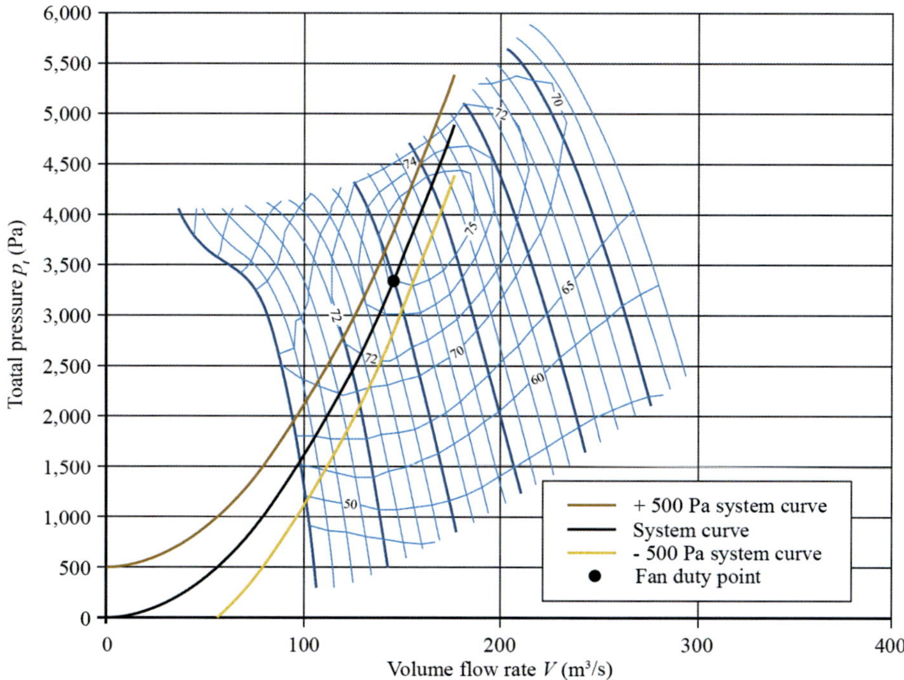

FIGURE 8.3. The characteristics of a 2.8 metre diameter, 6-pole two-stage double-ended motor fan, with an estimated 75 per cent efficiency at the Marmaray duty point.

with the assumption resulting in a predicted overall fan efficiency of 75 per cent. We estimated uncertainty with the inter-stage flow straighteners' actual performance at 3 per cent, and consequently, a potential minimum 72 per cent fan efficiency. As a consequence of the uncertainty in fan efficiency, a laboratory test is necessary to validate this selection.

Tunnel ventilation fan concept

Operators can use any of the above mentioned three fan concepts as a basis for a tunnel ventilation system fan, transform, silencer and damper package. Sheard *et al.* (2013) considered the relative merits of the three fan concepts and concluded that the package's capital and through-life cost based on any one of the three fan concepts was similar. However, the package volume was significantly smaller for a package based on the third fan concept.

Package size has a direct impact on the underground plant room's required size for tunnel ventilation installation, and consequently, a direct impact on the plant room's cost. Because of the third fan concept's relatively small package size, a two-stage double-ended motor fan is the most suitable for a high-pressure reversible tunnel ventilation fan.

FAN DESIGN REALISATION

Mine ventilation applications have utilised two-stage double-ended motor axial fans since the 1950s, Figure 8.4. Space is restricted in a mine environment, and hence the two-stage double-ended motor concept has historically proven practical in mine ventilation systems. However, mine ventilation applications generally use the

FIGURE 8.4. A 1959 vintage Woods of Colchester J-Range double-ended mine ventilation fan.

two-stage double-ended motor fans as booster fans, which are therefore classically up to 1.6 metres in diameter and up to 100 kW power. Sheard *et al.* (2013) observed that the Marmaray tunnel ventilation fan selection required a 2.8 metre fan with a 710 kW motor. This represented a significant product extension beyond the historic range of mine ventilation fans.

The mechanical challenge which the product range extension posed was primarily confined to the motor bearing arrangement. The impellers remained within historic limits, and therefore did not present a mechanical challenge. Reversible tunnel ventilation fans can operate in either direction. As such, aerodynamic loads associated with the pressure across the fan can also be in either direction. Second, the motor rotor and twin impeller also impose a significant gravitational load on the motor bearings. The loads imposed upon motor bearings require accommodation in both vertical and horizontal arrangements.

Historically, tunnel ventilation fan manufacturers have specified motors with deep-groove ball bearings. As fan power and size have increased, there has been a recent requirement to move to angular-contact bearings for vertical installations, as Sheard and Jones (2011) describe. A move to single large motors with an impeller on each end increases bearing loads significantly beyond the deep-groove ball bearing's load carrying capability in horizontal application or an angular-contact ball bearing in vertical application.

The bearing arrangement, which we adopted for horizontal application, Figure 8.5, utilised roller-bearings at each end of the motor shaft. The roller-bearings were

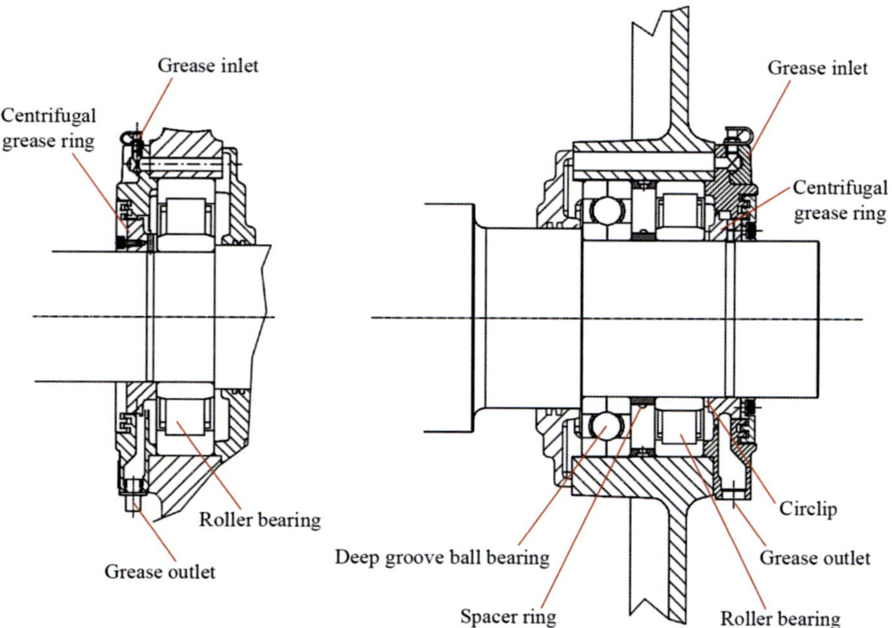

FIGURE 8.5. Fan motor bearing configuration for the horizontal two-stage double-ended motor fan.

free to float, with a single angular contact ball bearing accommodating the axial load. The bearing arrangement which we adopted for vertical application, Figure 8.6, utilised a pair of angular contact ball bearings to accommodate the gravitational and aerodynamic loads. The aerodynamic loads either added to or reduced the gravitational load, depending on the flow direction through the fan. The aerodynamic loads are significantly lower than the gravitational load, and consequently, although the vertical bearing load's magnitude changes as the fan changes from forward to reverse operation, it is always in the same direction. We completed the vertical bearing arrangement with a floating deep-groove ball bearing.

The tunnel ventilation system design drives the requirement to install tunnel ventilation fans either horizontally or vertically. The requirement has only minimal impact on the fans' aerodynamic performance, and no impact on the fan impellers' mechanical design. However, the fan casing and motor mounting arrangement are significantly different for vertical or horizontal installation.

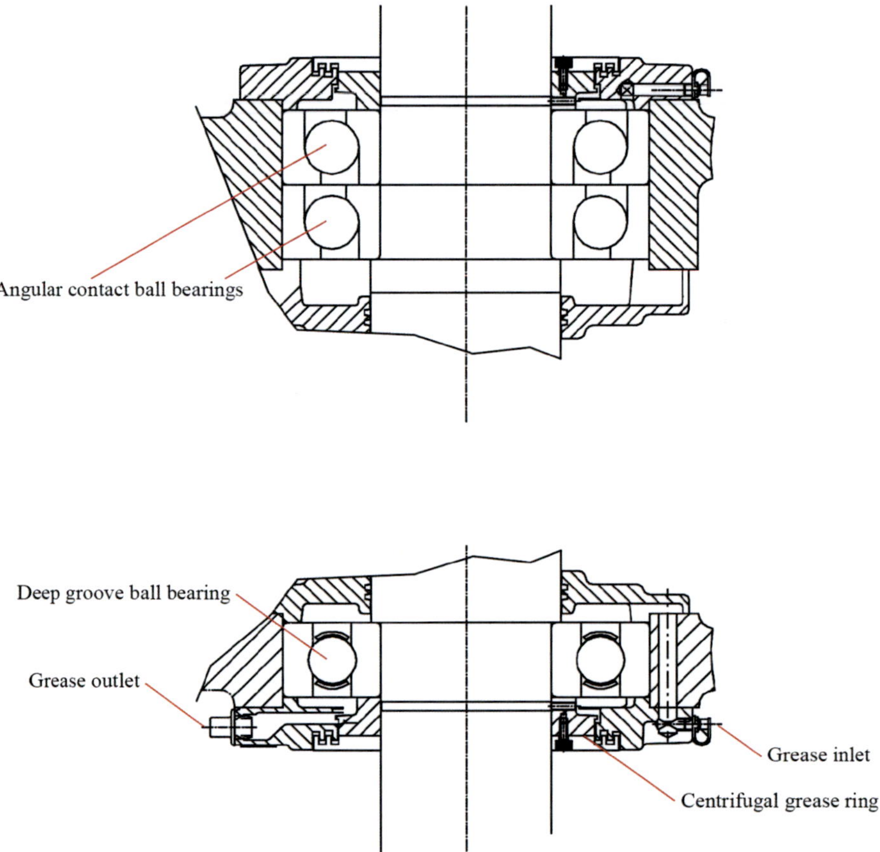

FIGURE 8.6. Fan motor bearing configuration for the vertical two-stage double-ended motor fan.

When installed horizontally, the preference is a foot mount motor arrangement, Figure 8.7, as it allows mechanical loads to pass through the casing and into the fan's mounting frame. Foot mounted motor design within tunnel ventilation fans is well established and consequently, a mechanically functional casing design is routine. Fan size and the twin impeller concept result in design for maintainability became the greater challenge. The Marmaray tunnel ventilation fan package location within a plant room restricted access for maintenance, and consequently, the team had to design the fan casing to facilitate maintenance within a confined environment, as Sheard *et al.* (2013) describe. The casing design is split to allow casing removal in segments, leaving the motor and impellers accessible. Lifting point locations in the plant room and special tooling design for casing segment, impeller and motor removal were all considerations during the fan mechanical design.

The preference was a flange mounted motor arrangement when installing vertically, Figure 8.8. The motor mounts inside a drum, with mechanical loads transferring through flow-straighteners to a heavy casing that then mounts onto a base-frame via four mounting arms. As with the horizontal casing design, the casing is segmented to allow maintenance access to the two impellers and the motor's top.

A requirement to fit combined deep-groove, roller and angular-contact bearings, plus the requirement to include multiple casing split lines both had an

FIGURE 8.7. A computer aided design model of the horizontal 2.8 metre two-stage double-ended motor fan designed for the Marmaray project.

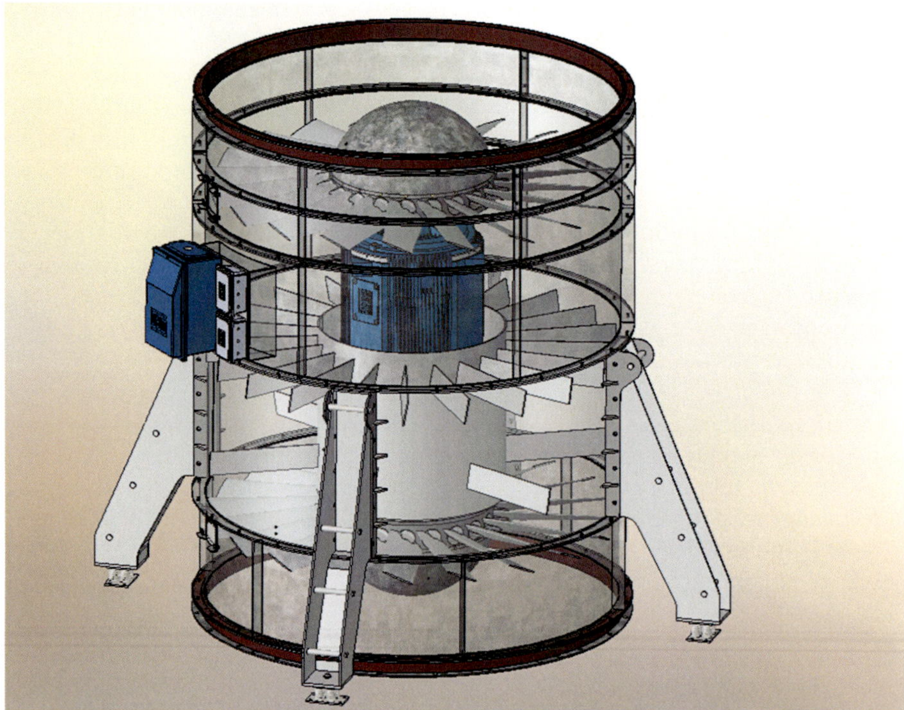

FIGURE 8.8. A computer aided design model of the vertical 2.8 metre two-stage double-ended motor fan designed for the Marmaray project.

implication for the two-stage double-ended motor fans' EN 12101-3 certification. The certifying authority considers a change in bearing arrangement and casing split line inclusion as fundamental changes that require validation via third-party independent testing. Sheard and Jones (2012a) describe this process.

Prototype laboratory performance

Following the Marmaray fan mechanical design completion, we built an 1,800 mm scaled prototype of the full size design, Table 8.3. We tested the prototype in a ducted system in accordance with ISO 5801: 2007 requirements (ISO 5801, 2007). We adopted Mattern *et al.*'s recommendations (2012) for calculating the fan pressure rise. We obtained experimental results with both impellers fitted and with one impeller fitted. Knowing the prototype's performance with both one and two impellers fitted enabled us to empirically establish the scaling rules which we needed to predict twin impeller fan aerodynamic performance from single impeller performance.

The agreement was good between the prototype's aerodynamic performance from the single impeller data when subject to the scaling rules and the performance data which we obtained from the prototype when testing with both impellers, Figure 8.9.

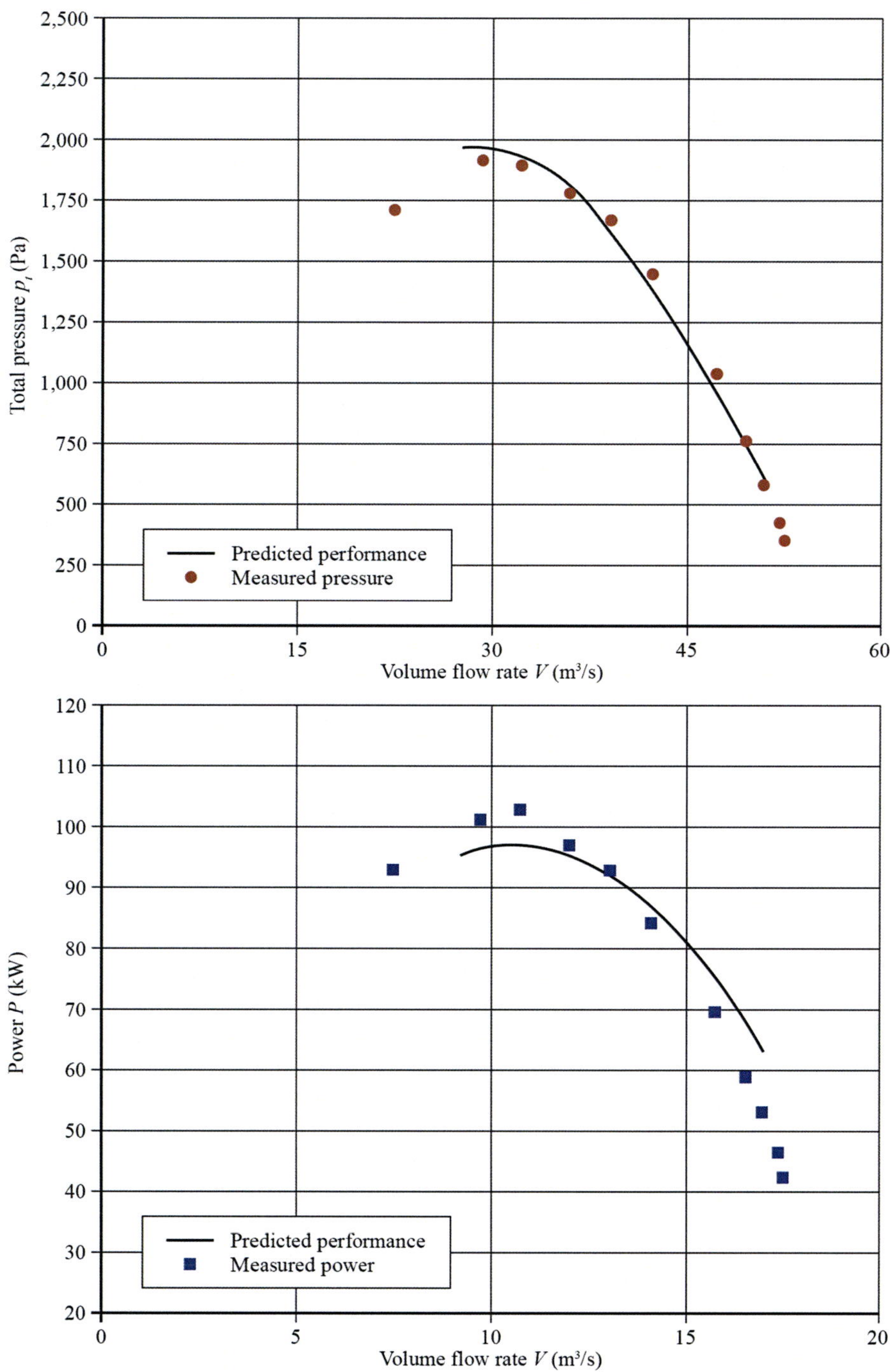

FIGURE 8.9. Aerodynamic performance of a 1.8 metre prototype two-stage double-ended motor fan.

This gave confidence in both the accuracy and conservatism of the assumptions when deriving the scaling rules.

We measured the prototype's acoustic performance with both single and twin-impellers, Figure 8.10, in accordance with ISO 10302 requirements (ISO 10302, 1996) (equivalent to British Standard BS 848-2: 2000). This enabled us to empirically establish the spectrum correction factors needed to predict twin impeller acoustic performance from single impeller performance.

Although the prototype fan had been manufactured with a casing arrangement that would facilitate the optimisation of the motor mounting arrangement and inter-stage flow straighteners, we considered the results for the 'as designed' geometry

Table 8.3. *Prototype and full size double-ended motor tunnel ventilation fan specification.*

	Prototype fan	Full size fan
Blade tip diameter D_t (mm)	1,800	2,800
Hub ratio	0.44	0.40
Solidity σ	0.5	0.5
Blade number	16	20
Design speed (rpm)	990	890
Blade passing frequency (Hz)	264	297

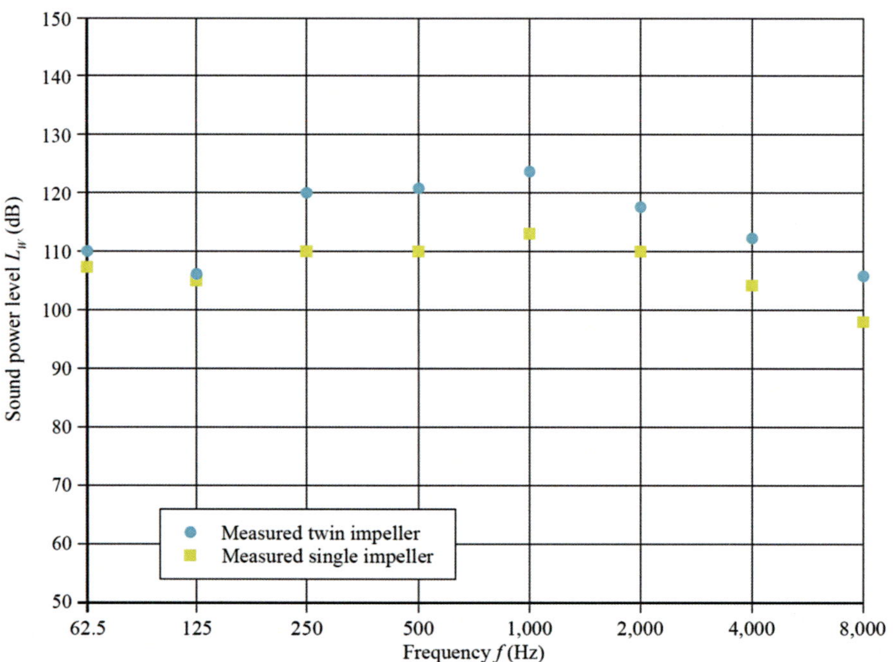

FIGURE 8.10. Acoustic performance of a 1.8 metre prototype two-stage double-ended motor fan, tested with one and two impellers fitted.

good enough to make further optimisation unnecessary. We undertook some experimental work to establish the sensitivity of both aerodynamic and acoustic performance to those changes in geometry that might occur as a consequence of manufacturing tolerances. Additionally, we evaluated the impact of inlet flow distortion on both aerodynamic and acoustic performance. Following the experimental campaign, we deemed the mechanical design both aerodynamically and acoustically fit for purpose. This initiated the manufacturing of the first production unit.

Package performance

Laboratory testing of the prototype fan was a fan-only test which we used to verify the fan performance. Package performance verification necessitated us to test a production fan with production transforms, silencers and dampers. The package size, Figure 8.11, necessitated an outdoor test which introduced some additional uncertainty into the measured aerodynamic and acoustic performance. Testing on a still day minimised the effects of cross-winds on measured aerodynamic performance and background noise. We again measured aerodynamic performance in accordance with ISO 5801 requirements (ISO 5801, 2007) and acoustic performance in accordance with ISO 10302 requirements (ISO 10302, 1996).

The package's aerodynamic performance, Figure 8.12, was in good agreement with the scaled performance prediction. The experimental performance data exhibited a slightly steeper curve than the scaled data, indicating that the off-design performance was not as well predicted as the design point performance. Given the

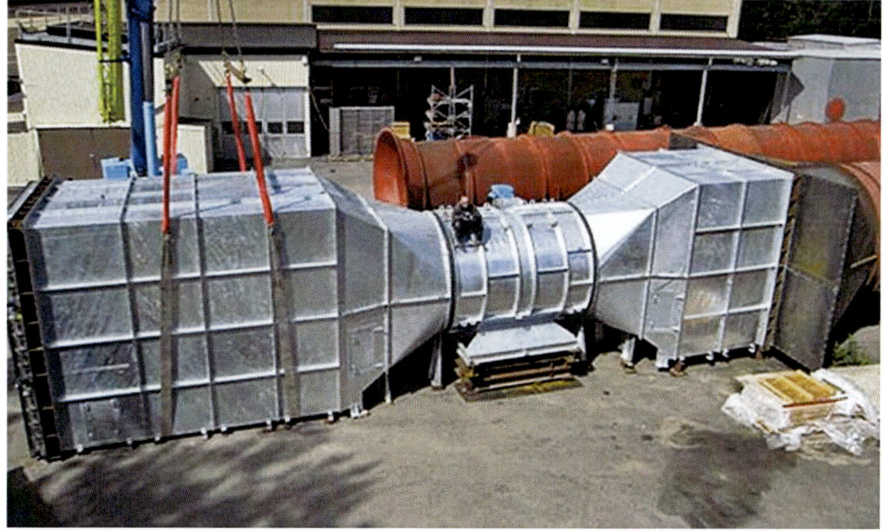

FIGURE 8.11. The first production Marmaray 2.8 metre two-stage double-ended motor fan, transform, silencer and damper package set up prior to its aerodynamic and acoustic test.

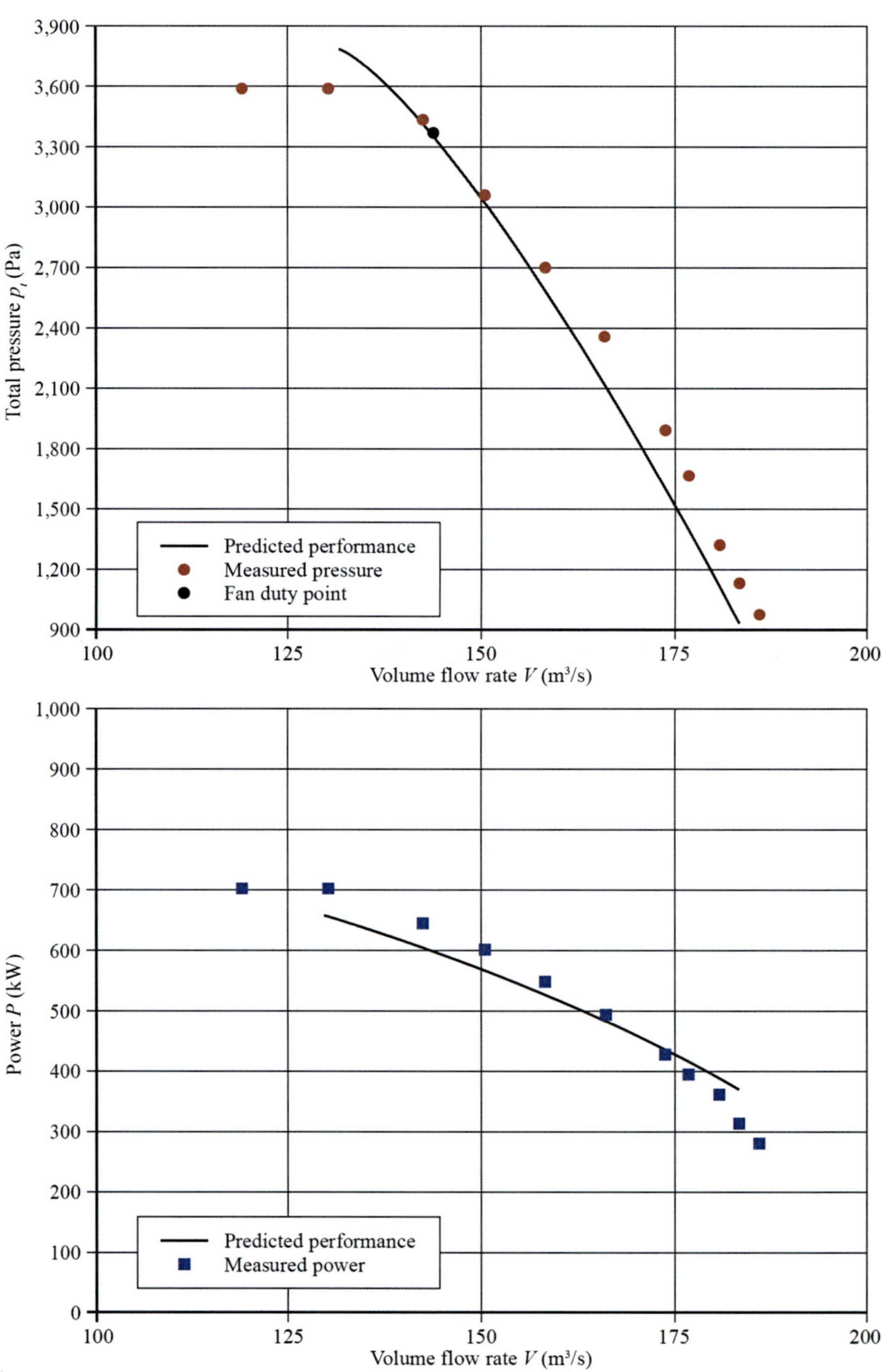

FIGURE 8.12. Predicted and measured Marmaray project package aerodynamic performance.

assumptions when scaling the two-stage fan data from single-stage data, plus the assumptions regarding transition, silencer and damper aerodynamic loss, we judged the correlation between scaled and experimentally measured performance as good.

Experimentally measured acoustic data for the full size fan also matched the predicted data well, Figure 8.13. There was some discrepancy between predicted and measured spectrum; however, the overall measured outlet noise level of 135 dB compared well with the predicted value of 136 dB. The discrepancy in the low-frequency band suggests that the spectrum correction factor obtained from the measurement on the prototype fan are not accurate, and therefore further refinement would be necessary to enable the designer to make a more accurate prediction of the twin impeller fans' acoustic spectrum using the single-stage source data. However, despite this reservation about the spectrum correction factors, we predicted the overall sound levels within 1 dB of the measured value. With more than 1 dB uncertainty in the measured value, we can conclude that the prediction of overall sound level is correct within the uncertainty of the measurement.

We combined experimentally measured acoustic data for the full size fan, silencer insertion losses and system effect to predict the ventilation package tunnel and atmospheric side sound level, Table 8.4. The predicted ventilation package tunnel and atmospheric side sound levels were 81 dB and 82 dB respectively, which were within the 85 dB specification, and therefore, we considered them as good.

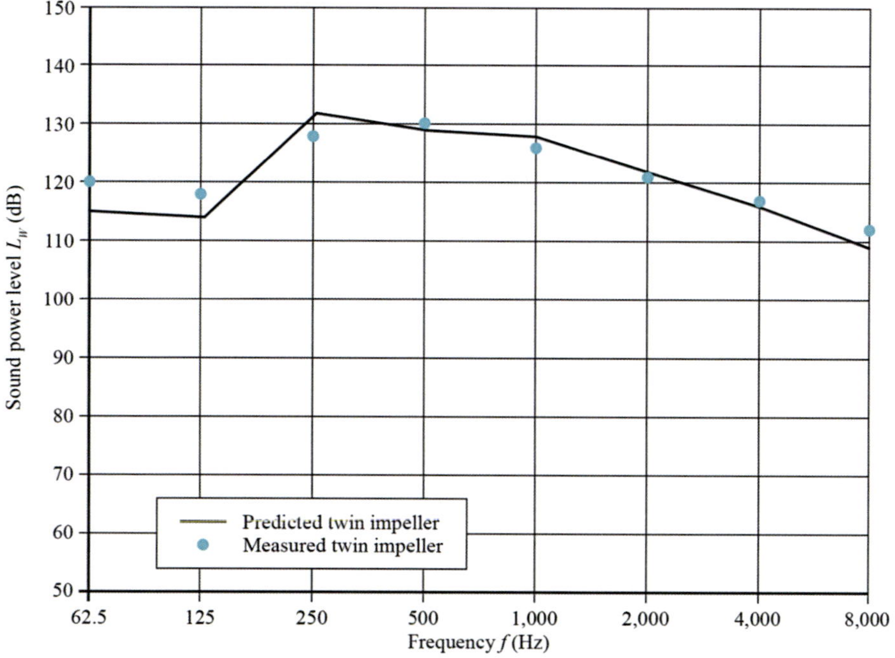

FIGURE 8.13. Predicted and measured Marmaray project package outlet acoustic performance.

Table 8.4. *System acoustic calculation at the tunnel and atmospheric side.*

		63Hz	125Hz	250Hz	500Hz	1KHz	2KHz	4KHz	8KHz
Tunnel side									
Fan sound power level L_W		123	120	130	130	127	121	120	114
System losses		0	0	0	0	0	0	0	0
Room effect		-3	-5	-7	-9	-10	-11	-11	-12
Attenuator insertion loss		-16	-32	-45	-50	-50	-50	-38	-27
Resultant L_WA (85 dBA spec)	**81**	104	83	78	71	67	60	71	75
Atmospheric side									
Fan sound power level L_W		123	120	130	130	127	121	120	114
System losses		-2	-3	-4	-5	-5	-5	-6	-6
Atmos effect		-14	-14	-14	-13	-13	-13	-13	-13
Attenuator insertion loss		-8	-20	-33	-40	-50	-48	-25	-15
Resultant (85 dBA spec)	**82.3**	99	83	79	72	59	55	76	80

EN 12101-3 CERTIFICATION

Although the specification, Table 8.1, only required EN 12101-3 high-temperature certification at 250°C for one hour, the pre-existing Fläkt Woods EN 12101-3 certification was at 300°C for two hours. Designing and testing at 300°C for two hours enabled us to extend the existing range certification. This required less testing and documentation than creating an entirely new 250°C test for a one hour range.

Within the European Union it is a legal requirement for tunnel ventilation fans to comply with Euro Norm EN 12101-3. The Marmaray project is part of Istanbul's infrastructure upgrade programme. The capital of Turkey, Istanbul is not part of the European Union. As such, it is not a legal requirement for fans in the Marmaray project to comply with EN 12101-3 requirements. However, Turkey aspires to join the European Union and an aspect influencing the final decision is Turkey's ability to implement Euro Norms. This is why the tunnel ventilation fans in the Marmaray project complied with EN 12101-3.

The Euro Norm EN 12101-3 defines how to approve 'powered smoke and heat control' fans. For the purpose of type approval, Sheard and Jones (2012a) observed that it is not necessary to test every fan size in a product range. Initially testing the following will significantly reduce testing:

- the fan with the most highly stressed impeller, and the fans with impellers in which the individual stress in any component, weld or fastening is the highest, as appropriate, if not the same;
- fans with motors mounted in an enclosure which restricts the cooling; this is the fan with the smallest free area of the motor enclosure or the smallest section of exit or entry airway for cooling air;
- at least two fan sizes at their highest rotational speed;
- the fan with the smallest motor frame size to be used, except for fans where the impeller is not mounted on the motor shaft and the motors are out of the airstream in ambient air and heat transfer from the fan or the fan construction does not affect the motor's cooling;
- if the highest impeller stress levels are determined by geometric similarity conditions, sufficient sizes of fans to ensure that the impeller diameters of the range are from 0.80 to 1.26 of those tested;
- if the highest impeller stress levels are determined by calculation, sufficient sizes of fans to ensure that the impeller diameters of the range are from 0.63 to 1.26 of those tested.

The Euro Norm EN 12101-3 was introduced in 2002, and has the full designation EN 12101-3: 2002. The working group responsible for EN 12101-3: 2002 has recently been drafting a revision. Sheard and Jones (2012b) considered the revision and noted that the primary changes address Factory Production Control (FPC) of motors and the approval of variable frequency drives.

The potential impacts of a variable frequency drive on a three-phase motor's winding insulation and bearings are significant. That is why the revision provides information on fan testing which will operate with a variable frequency drive. To approve a fan for operation via a variable frequency drive at both ambient and at a high temperature, one can install a test fan and operate it with the electrical supply provided from a pulse-width modulation (PWM) variable frequency drive. We can assume that a fan is able to withstand the same maximum peak to peak voltage values and voltage rise rate at the motor terminals in an installation as it did during the test, independent of the type or brand of PWM variable frequency drive that the operator uses. Alternatively, one may use a fan that has only been approved from direct-on-line testing with a PWM drive if one installs it with line chokes or filters and the motor power is suitably derated.

If an operator only intends to use a fan driven by a PWM variable frequency drive at ambient temperature, then one may install a fan tested direct-on-line with a variable frequency drive, provided that one by-passes the drive during an emergency smoke and heat exhausting operation.

The revision to EN 12101-3 provides detailed information on how a motor manufacturer's Factory Production Control (FPC) system can demonstrate on-going motor compliance for installation within a smoke control fan, evaluated by an independent body according to the following rules:

- The motor manufacturer selects a body for inspection and, if necessary, a body for tests.
- Only the selected inspection body performs the motor manufacturer's factory production control assessment and writes an inspection report in English.
- The motor manufacturer can then transmit the inspection report to all fan manufacturing customers, or directly to a customer's notified test authority for reasons of confidentiality.
- If the motor manufacturer changes or wants to change some design element or construction, he or she must inform the selected test laboratory which will perform, if required by the rules indicated in the standard, additional test(s). Then the motor manufacturer provides the test report (also written in English) to the inspection body so that it can note these change in the inspection report.

The revision of EN 12101-3 also requires fan manufacturers to have documented procedures which specify how to deal with non-complying products, corrective actions and the handling, storage and product packaging (with suitable storage areas preventing damage or deterioration). The revision then emphasises that the fan manufacturer's Factory Production Control system must address the EN 12101-3 product specific requirements, in effect introducing an EN 12101-3 specific factory production control system. Tables provide details of materials and parameters that one must check and information which one must record. The revision provides guidance for independent Notified Body assessors on what is required from an initial factory and Factory Production Control inspection. It also provides additional guidance on requirements for the continuous surveillance of a suitable factory production control system.

As part of the Marmaray project, we high-temperature tested a 2.8 metre diameter tunnel ventilation fan with a variable speed drive at 300°C for two hours in accordance with the revised EN 12101-3 requirements. To the best of our knowledge, this 2.8 metre diameter fan is the largest diameter fan that anyone has tested in accordance with EN 12101-3 requirements. Sheard and Jones (2011) reported a 1,050 kW fan as the highest power fan tested in accordance with EN 12101-3: 2002 requirements. Sheard and Jones (2011) tested that fan with a variable speed drive, and included a 1,050 kW motor from a motor manufacturer with a factory production control system. That test is also compliant with the revised version requirements of EN 12101-3.

We combined the 2.8 metre Marmaray project tunnel ventilation fan high-temperature test with other high-temperature tests which Sheard and Jones (2012a) previously undertook to certify a range of two-stage double-ended motor fans.

CONCLUSIONS

The current state of the art in tunnel ventilation fan design limits the maximum pressure rise that an operator can practically achieve in a single reversible fan stage to 2,000 Pa. The reversibility requirement results in the necessity for symmetrical fan blade aerofoil sections. Also, one must reduce guide vanes to flow-straighteners if the fan is to produce the same flow and pressure in both directions. It is necessary for tunnel ventilation fans to produce the same flow and pressure in both supply and extract modes to enable the tunnel ventilation fans to accommodate the varied requirements for routine ventilation and to keep escape routes clear in the event of a fire.

As tunnels become longer and deeper, the pressure drop that occurs with the flow necessary to ventilate the tunnel can rise above the 2,000 Pa limit of a current state of the art single reversible fan stage. In such cases, the traditional solution is to offer a two-stage fan with counter-rotating stages. A disadvantage of the two-stage counter-rotating fan is that it produces 15 dB more noise than one of the fans operating in isolation. Consequently, silencers must be larger for the same overall acoustic emissions at the silencer inlet and outlet.

In a situation where minimum overall fan, transform, silencer and damper package size is desirable, the two-stage counter-rotating fan may not be the optimum fan concept. In such circumstances, the tunnel ventilation fan designer has two further fan concepts available. These include a high-speed single-stage fan or a two-stage double-ended motor fan. Each represents a valid technical solution, and has the potential to achieve a specified duty point. The three fan concepts are the:

- two-stage counter-rotating fan;
- single-stage high-speed fan; and
- two-stage double-ended motor fan.

The two-stage double-ended motor fan concept resulted in the minimum package size, thus we selected it as the preferred concept. We primarily encountered two mechanical design challenges regarding the two-stage double-ended motor fan. First, there was the issue of whether the motor bearing system design could accommodate the imposed aerodynamic and gravitational loads in both horizontal and vertical configurations. Second, was the issue of designing a casing that operators could dismantle within a plant room's confines where engineers installed the fans and their associated package.

Having successfully addressed the mechanical design challenges, we measured the prototype fan's aerodynamic and acoustic performance. We then measured the full size package aerodynamic performance and full fan acoustic performance. Both aerodynamic and acoustic results were in good agreement with the original performance prediction scaled from single-stage fan data. Mechanical validation ended with a successful 300°C test for two hours of the full size fan, in accordance with EN 12101-3 requirements.

REFERENCES

ISO 10302 (1996), *Fans for General Purposes. Methods of Noise Testing. Airborne Noise Emitted by Small Air-moving Devices.*

ISO 5801 (2007), *Industrial Fans – Performance Testing Using Standardized Airways.*

Borello, D., Corsini, A., Rispoli, F. and Sheard, A.G. (2012), 'Investigation on the Unsteady Aerodynamics of an Industrial Fan', *Proceedings of the Fan 2012 Conference*, Senlis, France, 18–20 April.

Daly, B.B. (1985), *Woods Practical Guide to Fan Engineering*, Woods of Colchester Ltd, Colchester, UK.

Daneshkhah, K. and Sheard, A.G. (2013), 'A Parametric Study of Reversible Jet-fan Blades Aerodynamic Performance', *Transactions of the ASME, Journal of Engineering for Gas Turbines and Power*, vol. 135(2), paper no. 024503, pp. 1–4.

Horlock, J.H. and Denton, J.D. (2005), 'A Review of Some Early Design Practice using Computational Fluid Dynamics and a Current Perspective', *Transactions of the ASME, Journal of Turbomachinery*, vol. 127, pp. 5–13.

Jasnoch, U., Kress, H. and Rix, J. (1994), 'Towards a Virtual Prototyping Environment', in Rix, J. (Ed.), *Virtual Prototyping: Virtual Environments and the Product Design Process*, Chapman and Hall, London, UK, pp. 173–183.

Lee, K.Y., Choi, Y.S., Kim, Y.L. and Yun, J.H. (2008), 'Design of Axial Fan using Inverse Design Method', *Journal of Mechanical Science and Technology*, vol. 22, pp. 1883–1888.

Mattern, P., Sieber, S., Cantrak, D., Frohlig, F., Caglar, Ş. and Gabi, M. (2012), 'Investigations on the Swirl Flow Caused by Axial Fans – a Contribution to the Revision of ISO 5801', *Proceedings of the Fan 2012 Conference*, Senlis, France, 18–20 April.

Pratt, M.J. (1994), 'Virtual Prototypes and Product Models in Mechanical Engineering', in Rix, J. (Ed.), *Virtual Prototyping: Virtual Environments and the Product Design Process*, Chapman and Hall, London, UK, pp. 113–128.

Sheard, A.G. and Corsini, A. (2012), 'The Mechanical Impact of Aerodynamic Stall on Tunnel Ventilation Fans', *International Journal of Rotating Machinery*, vol. 2012, paper no. 402763, pp. 1–12.

Sheard, A.G. and Jones, N.M. (2011), 'The Development of a Fan Range to Meet Increasingly Demanding Safety Criteria Inside Tunnels', *Proceedings of the 14th Australasian Tunnelling Conference: Development of Underground Space*, Sky City, Auckland, New Zealand, 8–9 March, pp. 305–318.

Sheard, A.G. and Jones, N.M. (2012a), 'Powered Smoke and Heat Exhaust Ventilators: the Impact of EN 12101-3 and ISO 21927-3', *Tunnelling and Underground Space Technology*, vol. 28, pp. 174–182.

Sheard, A.G. and Jones, N.M. (2012b), 'An Initial Assessment of the Changes that will Appear in a Forthcoming (2012) Revision of EN 12101-3', *Proceedings of the Fan 2012 Conference*, Senlis, France, 18–20 April.

Sheard, A.G., Corsini, A., Minotti, S. and Sciulli, F. (2009), 'The Role of Computational Methods in the Development of an Aero-acoustic Design Methodology: Application in a Family of Large Industrial Fans', *Proceedings of the 14th International Conference on Modelling Fluid Flow Technologies*, Budapest, Hungary, 9–12 September, pp. 71–79.

Sheard, A.G., Daneshkhah, K. and Corsini, A. (2013), 'Fan Conceptual Design as Applied to the Marmaray Tunnel Ventilation System', *Proceedings of the 58th American Society of Mechanical Engineers Gas Turbine and Aeroengine Congress*, San Antonio, TX, USA, 3–7 June, paper no. GT2013-94548.

Vad, J. (2001), 'Incorporation of Forward Blade Sweep in the Non-free Vortex Design Method of Axial Flow Turbomachinery Rotors', *Periodica Polytechnica, Mechanical Engineering*, vol. 45(2), pp. 217–237.

Vad, J., Bencze, F., Corsini, A. and Rispoli, F. (2001), 'Non Free Vortex Flow Effects in an Axial Flow Rotor', *Periodica Polytechnica, Mechanical Engineering*, vol. 45(2), pp. 201–216.

Wallis, R.A. (1961), *Axial Flow Fans: Design and Practice*, George Newnes Ltd, London, UK.

Chapter 9

Predicting Blade Leading Edge Erosion in an Axial Induced Draft Fan

A. Corsini, A. Marchegiani, F. Rispoli, A.G. Sheard and P. Venturini

ABSTRACT

Induced draft fans extract coal-fired boiler combustion products, including particles of unburnt coal and ash. As a consequence of the particles, the axial fan blades' leading edges are subject to erosion. Erosion results in a loss of the blade leading edge aerodynamic profile and a reduction of blade chord and effective camber that together degrade aerodynamic performance. In this chapter we present an experimental study that demonstrates that whilst the degradation of aerodynamic performance begins gradually, it collapses as blade erosion reaches a critical limit. We complement the experimental study with a numerical study on the evolution of blade leading edge erosion patterns. We calculate particle trajectories using a Reynolds-averaged Navier–Stokes solver coupled with a trajectory predicting solver based on an original finite element interpolation scheme. The numerical study clarifies the influence of flow structure, initial blade geometry, particle size and concentration on erosion pattern.

This chapter is a revised and extended version of Corsini, A., Marchegiani, A., Rispoli, F., Sheard, A.G. and Venturini, P. (2012), 'Predicting Blade Leading Edge Erosion in an Axial Induced Draft Fan', *Transactions of the ASME, Journal of Engineering for Gas Turbines & Power*, vol. 134, paper no. 042601, pp. 1–9.

NOMENCLATURE

Latin letters

c	[mm]	Blade chord
C_D		Particle drag coefficient
c_μ		Numerical coefficient
D_{hc}		Hub to casing diameter ratio
d_p		Particle diameter
D_t	[mm]	Blade tip diameter
E_R		Erosion rate
f_α		Erosion angular function
f_r	[rpm]	Fan design speed
g		Acceleration of gravity
h_b		Blade height fraction
I_f		Impact frequency
k		Turbulent kinetic energy
p	[Pa]	Static pressure
p_t	[Pa]	Total pressure
t		Time abscissa
u		Fluid velocity vector
V	[m^3/s]	Volume flow rate
v		Particle velocity vector
v_{bc}		Particle velocity vector in the centre of the cell
x		Axial Cartesian coordinate
y^+		Normalised wall distance

Greek letters

α		Impact angle
β		Exchange coefficient
δ	[deg]	Stagger angle
ε		Turbulent kinetic dissipation rate
θ	[deg]	Chamber angle
ρ		Density
σ		Solidity
$\sigma(t)$		Standard deviation of the particle distribution in the cloud
χ	[% span]	Blade tip clearance
τ		Time scale
τ_L		Lagrangian time scale

Acronyms

BBO	Basset-Boussinesque-Oseen
FEM	Finite element method
LE	Leading edge
MG	Multi-grid
MPI	Message passage interface
PCT	Particle cloud tracking

PDF	Probability density function
PFS	Pan Fläkt Single
PG	Petrov-Galerkin
PS	Pressure side
RANS	Reynolds-averaged Navier–Stokes
SS	Suction side
TE	Trailing edge

Subscript and symbols

bc	Boundary cell
f	Fluid
m	Cloud's centre of mass
p	Particle
$\langle\ \rangle$	Ensemble average

INTRODUCTION

When installed as induced draft fans in coal-fired power plant boiler applications, erosion degrades both axial and centrifugal fan performance. Particle laden flows characterise the operating environment as a consequence of both solid and molten particle transport. These particles are an unwanted by-product of the combustion process. The combustion process produces ash that comprises those parts of the heavy oil or synthetic fuel that will not burn. In coal-fired boilers, degradation of the ball-mills that grind coal into a powder before it enters the boiler can result in coal particles too large to fully burn. The result is unburnt coal particles leaving the boiler, with a combination of ash and unburnt coal particles larger than 20 μm which typically drive the erosive process (Kurz and Brun, 2001; Hamed *et al.*, 2006).

Researchers have correlated the erosion to a degradation of aerodynamic performance, driven by the deterioration of the original blade profiles (Grant and Tabakoff, 1975). There are three factors that result in the degradation of aerodynamic performance. First, the blades' ability to turn the flow diminishes because of the reduced blade chord. Second, the local erosion at the leading edge aggravates the tendency of boundary layers to separate. Separated boundary layers reduce a fan's stable operating range. Finally, aerodynamic losses increase as a consequence of changes to the blade's leading edge, trailing edge, thickness and blade tip-to-casing clearance (Kurz and Brun, 2001).

A range of factors drives the erosive process. The physical properties and geometry of particles is important as some particles are intrinsically more erosive than others. The fan's gas-path geometry and its ducting system and fan blade geometry are important as they influence where particles will impact the fan blades. The material from which the fan blades are manufactured is important as some materials resist erosion more effectively than others. Finally, the quality of the coal used in coal-fired boilers is important as the silica content of low-grade coal can be significant. Silica particles are highly erosive (Richardson *et al.*, 1979).

The erosion pattern one observes when inspecting fan blades removed from service are a function of impact velocity that is in turn a function of blade tip speed (Hamed *et al.*, 2006). Some scholars have studied erosion of high-pressure compressors (Sallee *et al.*, 1975; Balan and Tabakoff, 1984) and fans (Ghenaiet *et al.*, 2004). They concluded that blade erosion occurred primarily over the outer 50 per cent of the blade span. Page *et al.* (1979) studied induced draft fan blade erosion. They concluded that erosion rate was proportional to particle velocity raised to a power of between 1.99 and 5.00. The actual value was dependent on the percentage of silica and quartz in the fly-ash passing through the fan.

When studying the effect of sand ingestion, Ghenaiet *et al.* (2004) concluded that the resultant erosion increased blade tip-to-casing clearance and reduced the blade chord. This increase in blade tip-to-casing clearance and reduction of the blade chord resulted in a ten per cent drop in fan efficiency and pressure developing capability. They concluded that this drop was primarily a consequence of two factors. First, the blade leading edge profile was destroyed. Second, there was a loss of the blade profile over the pressure surface in the leading edge region. Sugano *et al.* (1982) reported similar results when studying the erosive effects of fly-ash on axial induced-draft fan blades.

A practical problem that researchers face when studying erosion is that any experimental based study is both difficult to control and expensive to undertake. An experimental study is difficult to control as there are practical problems associated with accounting for all the parameters that govern the erosive process. Despite the difficulty of using an experimental study to characterise the erosive process, inspecting eroded blades can provide useful qualitative information. It is possible to characterise the erosive particles that have resulted in the in-service erosion. It is also possible to know accurately the number of hours in-service blades operated, and the operating point at which they operated. This is not enough to provide an insight into the erosive process itself. However, it does place qualitative data into a context. This contextualisation provides some insight into the severity of the erosive process.

We may conclude that the study of blades removed from service can provide some useful qualitative data on the effects of the erosive process. However, it provides no useful quantitative data on the physical mechanisms underlying that erosive process. Consequently, we may conclude that a numerical simulation of a fan's blade-to-blade flow-field has the potential assist in elucidating these underlying physical mechanisms. A numerical simulation offers the potential to predict particle trajectories and therefore the impact frequency of particles over the blade surface. In turn, one may use this prediction of impact frequency to predict erosion rate.

Hussein and Tabakoff (1974) pioneered particle trajectory simulations through rotating and stationary axial turbomachinery. They complemented their computational work with an experimental programme to determine particle rebound factors. Elfeki and Tabakoff (1987) demonstrated that in centrifugal compressors, particle trajectories are difficult to predict. This difficulty is a consequence of the complexity of flow-field features and the effect of centrifugal force on features. Ghenaiet (2005) conducted a similar numerical simulation to that of Elfeki and Tabakoff (1987). Ghenaiet (2005) studied the effect of sand ingestion on a small radial compressor and a large cooling fan. Despite these scholars undoubted success at elucidating some aspects of the erosive process, the prediction of particle trajectories and the resultant blade erosion remains challenging. There are many factors that influence the erosion process, and these factors are prone to both complex and non-liner interaction. Therefore, predicting erosion using a numerical simulation remains challenging.

This chapter presents an erosion process model based on a simulation of the three-dimensional blade-to-blade flow-field. This simulation is coupled with an erosion model that predicts particle trajectories, and their impact on the blade surface. We then move on to predict the erosive effect of particles impacting the blade surface, and compare the resulting predicted erosion with the erosion observed on blades removed from service. We conducted our numerical simulation using a three-dimensional Reynolds-averaged Navier–Stokes (RANS) based solver. We adopted a parallel multi-grid scheme developed for an in-house finite element method (FEM) solver (Borello *et al.*, 2003), utilising Kirk *et al.'s* (2006) C++ libMesh libraries. We based the finite element method formulation on a stabilised Petrov-Galerkin (PG) scheme originally developed by Corsini *et al.* (2006).

We used the numerical simulation to predict the blade-to-blade flow-field, an essential precursor to the prediction of particle trajectories. To predict the trajectory of particles with mass, we utilised a Lagrangian particle cloud tracking (PCT) model, which Baxter (1989) first proposed. Other scholars, notably Wang (1990), Litchford and Jeng (1991), Baxter and Smith (1993) and Jain (1995) have subsequently developed the model. Predicting erosion requires tracking a large number of particles if the resulting prediction is to be statistically significant (Kær, 2001). Tracking large numbers of individual particles is possible, but computationally intensive. The particle cloud tracking model reduces the computational effort needed to track large numbers of particles. It is used to predict the trajectory of a relatively small number of cloud cores. One then uses the flow-field around each cloud core to model the dispersion of individual particles around the cloud core.

In the research presented in this chapter, we implemented the particle cloud tracking model using an original finite element (FE) based tracking scheme. We then used this tracking scheme to predict the trajectory of cloud cores followed by the dispersion of individual particles around the cloud core (Venturini *et al.*, 2010). The resultant tracking scheme was able to predict both the frequency with which particles impacted on different regions of the fan blade, and the angle with which particles impacted. In combination, we used impact frequency and angle to predict erosion rate, which we were then able to compare with the eroded blades that had operated in an erosive environment.

BOILER FAN DESCRIPTION

We undertook this research using a 26 blade fan with C4 profile blades, Table 9.1. The studied fan is equipped with a mechanism for varying the blade pitch-angle whilst rotating at its design speed, Figure 9.1. The industrial fan community refers to

Table 9.1. *The studied fan constitutes one fan from a family of fans intended for induced-draft application in coal-fired power plant. The authors selected this example as typical of the family and used it as a basis for numerical simulation of the flow-field, particle impact frequency and erosion rate.*

Rotor data		
Blade number	26	
Hub to casing diameter ratio D_{hc}	0.7	
Blade tip diameter D_t (mm)	1,767	
Blade tip clearance χ (% span)	6	
Fan design speed f_r (rpm)	890	
Blade material	Stainless steel	
Blade geometry	**Hub**	**Tip**
Blade chord c (mm)	487	446
Solidity σ	1.48	0.96
Stagger angle δ (degrees)	38	60
Camber angle θ (degrees)	45	36

FIGURE 9.1. A cut-away view of the studied fan illustrating the variable pitch in motion blades, their inlet and outlet guide vanes. This family of fans is configured with an inlet box that turns the flow through 90 degrees. The flow into the inlet box is vertically down and the flow out is horizontal.

any fan fitted with this mechanism as a variable pitch in motion (VPIM) fan. Design engineers originally developed variable pitch in motion fans in the 1950s for application with coal-fired boilers. The coal's variable calorific value results in a variation in boiler pressure. Maintaining boiler pressure within safe limits requires a boiler's forced- and induced-draft fans to have the ability to respond rapidly to changing flow-rate demand. The size and, therefore, inertia of these forced- and induced-draft fans was high, thus the fan speed could not change quickly enough to follow demand. Consequently, the engineers developed constant speed variable pitch in motion fans. Despite the mechanical complexity of variable pitch in motion fans, they still find wide-spread application in thermal power stations' coal-fired boiler systems. The cost of high-power variable-speed drives remains high. Consequently, fixed-pitch variable-speed forced- and induced-draft fans are not desirable.

In-service erosion

The studied fan was typical of a forced- or induced-draft fan used in coal-fired boiler applications. It was coded PFS, an abbreviation of Pan Fläkt Single were pan and fläkt are respectively the Swedish words for boiler and fan. In the research reported in this chapter, we installed each fan as one of a pair of induced draft fans operating in parallel to effectively double volume flow rate. We fired the boiler with low-quality coal with a high mineral content. Further, the mineral content resulted in wearing the ball-mills that grind the coal before it blows into the boiler. The result was that the ball-mills produced large coal particles that did not burn completely in the boiler. Therefore, in addition to the fly-ash that formed from the coal's mineral content, the flue gas also included unburnt carbon particles. Consequently, the induced draft fans were subjected to the effects of erosive particles comprising of both fly-ash and carbon.

Flue gas left the boiler economizer at approximately 300°C, entered the induced-draft fan at approximately 165°C and left at approximately 195°C due to compressibility effects through the fan. A temperature of 165°C to 195°C significantly increases the erosive effect of particles when compared to the same particles at atmospheric temperature. We were able to qualitatively assess the impact of erosive particles on the studied fan by inspecting blades after 9,000 hours' in-service operation, Figure 9.2. The blade leading edge profile was compromised and outlet guide vanes had lost almost half their chord. There was also evidence of erosion on the blade pressure surface in the hub region. Blade suction surface erosion was probably a consequence of particles entering the blade-to-blade passage at high negative incidence angles. We observed an eroded region of the blade suction surface at approximately mid-span. We hypothesise that this mid-span erosion may be a consequence of particles impacting the suction's surface downstream of the blade leading edge over the first 30 per cent of blade chord.

After inspecting the blades, we measured the leading edge profile, Figure 9.3. The leading edge wear is distinctive, with material wearing from both the blade's pressure and suction surface. This wear pattern had developed over 9,000 in-service

FIGURE 9.2. The authors assessed the impact of erosive particles on the studied fan after 9,000 hours' in-service operation. The blade leading edge profile was compromised and outlet guide vanes had lost almost half their chord, left. There is also evidence of erosion on the blade pressure surface in the hub region, right.

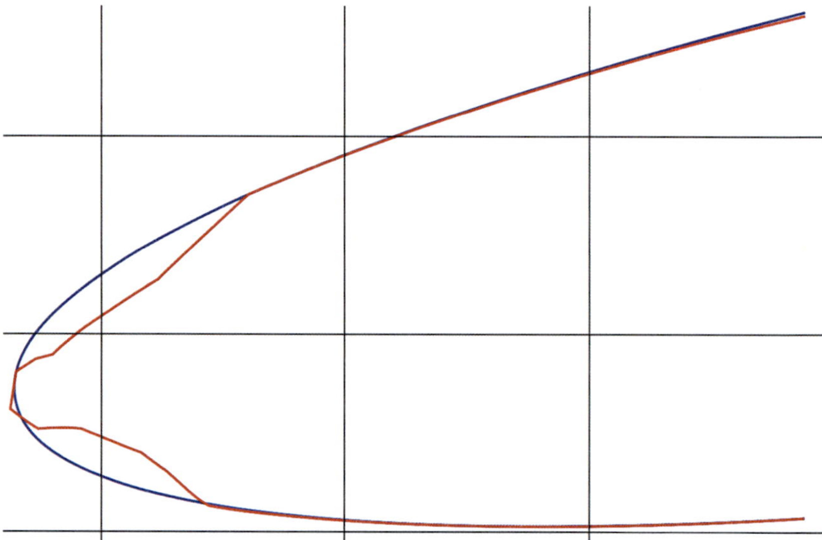

FIGURE 9.3. A comparison of the blade leading edge profile as built, shown in blue and after 9,000 hours' in-service operation, shown in red.

operating hours. The wear effectively destroyed the blade leading edge profile. Estimating the leading edge profile's degradation impact on fan performance was challenging. However, we were able to identify the overall effect of erosion on fan performance. We removed the induced-draft fans from service when the electric motors driving the fans tripped on high current. This represented approximately a two-thirds increase in the necessary electrical power to drive the fan. Consequently, we may conclude that the reduced fan efficiency that occurred over 9,000 hours of in-service operation resulted in the two thirds increase in electrical power.

METHODOLOGY

We may calculate the trajectory of a single solid particle by following Baxter's (1989) method, solving a reduced particle motion equation, the Basset-Boussinesque-Oseen (BBO) equation. This equation accounts for inertia, drag and buoyancy forces on a particle. Following Baxter's (1989) method, we assumed that the gas-to-solid particle density ratio was below 10^{-3} and therefore the Basset-Boussinesque-Oseen equation reduces to:

$$\rho_p \frac{d\mathrm{v}}{dt} = -\frac{3}{4d_p}\rho_f C_D(\mathrm{v}-\mathrm{u})\,|\,\mathrm{v}-\mathrm{u}\,| + (\rho_p - \rho_f)\mathrm{g} \tag{1}$$

where u and v indicate the gas and solid particle velocities respectively; ρ_p and ρ_f are the particle and fluid physical densities respectively, and d_p is the particle diameter. Following Crowe *et al.* (1996) and Kær's (2001) methods, we expressed the drag coefficient C_D as a particle Reynolds number-based modification of the Stokes drag coefficient.

Particle cloud tracking model

We adopted Baxter's (1989) particle cloud tracking model that we used to derive the equation of motion of a group of particles. We may calculate the starting position of the group of particles by ensemble-averaging the Basset-Boussinesque-Oseen equation:

$$\frac{d\langle \mathrm{v}(t)\rangle}{dt} = \langle \beta(t)\rangle \cdot [\langle \mathrm{u}(t)\rangle - \langle \mathrm{v}(t)\rangle] + \left(1 - \frac{\rho_f}{\rho_p}\right)\mathrm{g} \tag{2.1}$$

where the ensemble average of the exchange coefficient β reads as:

$$\langle \beta(t)\rangle = \frac{3}{4d_p} C_D \frac{\rho_f}{\rho_p}\,|\,\langle \mathrm{u}(t)\rangle - \langle \mathrm{v}(t)\rangle\,| \tag{2.2}$$

and $\langle v(t)\rangle$ is the average cloud velocity. We computed the average cloud position $\langle x_p(t)\rangle$, defined as the particle cloud's centre of mass, by time-integrating Equation (2.1):

$$\langle x_p(t)\rangle = \int_0^t \langle v(t)\rangle dt + \langle x_p(t=0)\rangle \tag{3}$$

We assume particle distribution around $\langle x_p(t)\rangle$ to be Gaussian:

$$PDF(x;t) = \frac{1}{(2\pi)^{\frac{1}{2}}\sigma(t)} \exp\left[-\frac{1}{2}\left(\frac{x-\langle x\rangle}{\sigma(t)}\right)^2\right] \tag{4}$$

where σ(t) is the standard deviation of the distribution. The standard deviation is a function of particle-turbulence interaction. We evaluated the particle-turbulence interaction by assuming the Markovian approximation:

$$\sigma^2(t) = 2\langle v^2\rangle\left[\frac{t}{\tau_L} - \frac{1}{\tau_L^2}\left(1-e^{-\frac{t}{\tau_L}}\right)\right] \tag{5.1}$$

$$\tau_L = \max(\tau_f, \tau_m) \tag{5.2}$$

where τ_L is the Lagrangian time scale, defined as the maximum difference between the turbulence and the particle motion time scales, and:

$$\langle v^2\rangle = \langle u^2\rangle\left(1-e^{-\frac{\tau_f}{\tau_p}}\right) = \frac{2}{3}k\left(1-e^{-\frac{\tau_f}{\tau_p}}\right) \tag{5.3}$$

is the fluctuating component of the particle velocity as induced by the particle-turbulence interaction. The time scale definitions are:

$$\tau_f = \frac{C_\mu^{\frac{3}{4}}k}{0.817\varepsilon} \tag{6.1}$$

and

$$\tau_m = \left\langle \frac{1}{\beta(t)} \right\rangle \tag{6.2}$$

Furthermore, the ensemble averaged velocity in the cloud volume reads as:

$$\langle u(t)\rangle = \iiint_{x=\pm\infty} u(x;t)\cdot PDF(x;t)dx \tag{7}$$

where *PDF* indicates the probability density function of the particle distance from the cloud centre.

Cloud tracking algorithm

We decomposed the computational domain into hexahedral finite elements. This decomposition enabled us to determine the integral of Equation (7) as a summation of the contributions of each element within the cloud. At each time step, the particle cloud tracking model requires us to compute the cloud's average position and radius, and the elements within the cloud volume. The practical approach we adopted to undertake this computation was to use Venturini's (2010) finite element approach. This finite element approach maps each element within the computational grid onto its logical image using an iso-parametric scheme. This mapping standardises the search process and makes it independent of the elements' actual geometry.

At each new time step we evaluate the position of elements relative to the cloud centre to determine if an element remains within the cloud. Using a pre-processing computation, one can identify all elements surrounding a given element at the cloud centre (Venturini, 2010; Venturini *et al.*, 2010). Once one has identified the elements surrounding the element at the cloud centre, one may extend the search process to find the elements around each of the previously identified elements. The search logic proceeds towards the cloud's periphery until it identifies every element within the cloud. Using the flow-field parameters associated with each element, one may then compute the trajectory of particles from one time step to the next.

Erosion model

The original particle cloud tracking formulation is based on a hypothesis: rigid body motion of clouds composed of particles moving at the velocity of the cloud's centre of mass. This hypothesis assumes that there is no influence associated with the non-uniform velocity distribution both within the cloud and when the cloud impacts on a solid boundary. Therefore, the original formulation includes significant simplifying assumptions. In order to partially account for wall-induced velocity effects, we have adopted Venturini's (2010) developed form of the particle cloud tracking model. We assume that particles within a cloud that are impacting on a solid surface keep the same drift velocity as the cloud's centre. Using this assumption, we are able to derive particle velocity vector in the centre of the cell (v_{bc}) as:

$$|v_{bc}| = |u_{bc}|\left(1 - \frac{|v_{drift}|}{|u_m|}\right). \tag{8}$$

Here, we define u_{bc} as the average velocity vector at the centre of a boundary cell, $v_{drift} = u_m - v_m$ is the drift velocity vector at the cloud's centre, u_m is the flow's velocity vector at the cloud centre, and v_m is the velocity vector of the cloud centre of mass, Figure 9.4.

Among the factors influencing the erosion of ductile materials, the energy that an impacting particle transfers to the target plays a critical role. According to Tabakoff *et al.* (1979), the angle of impact α onto the target surface also influences erosion. Tabakoff *et al.* (1979) experimentally measured the erosion rate of 304 grade stainless steel, Figure 9.5. In order to account for anisotropy of particle impact angle, we characterised impact angle using a polynomial function.

Numerical technique

In this chapter we report the results of a computational analysis using a Reynolds-averaged Navier–Stokes (RANS) simulation. We used the same numerical procedure as Borello *et al.* (2003), incorporating a topology-free low-Reynolds number formulation incorporating Kirk *et al.*'s (2006). libMesh libraries. We modelled the incompressible three-dimensional turbulent flow in the rotating frame of reference with a two-equation closure (Craft *et al.*, 1996). The two-equation closure was able to cope with non-isotropic and non-equilibrium turbulence effects without a significant increase in the required computational effort. The numerical procedure utilised a parallel multi-grid (MG) numerical scheme embedded in a finite element method (FEM) code.

The finite element method formulation is based on a stabilised Petrov-Galerkin (PG) method. This method is modified for application to three-dimensional equal- and mixed-order spaces of approximation (Corsini *et al.*, 2006). The stabilised Petrov-Galerkin (PG) scheme controls the instability that affects the advective-diffusive incompressible flow, and the momentum's reaction and turbulent scale equations. The turbulent scale equations relate to Coriolis acceleration (Corsini *et al.*, 2003). We used equal-order Q1-Q1 and mixed-order Q2-Q1 interpolation for primary turbulent scale equations and constrained secondary variables, implicitly eliminating the undesirable pressure-checker boarding effects. We performed the computational analysis using a hybrid full linear multi-grid accelerator running on an overlapping parallel solver (Borello *et al.*, 2001).

We parallelised an iterative Krylov method when smoothing and solving using the adopted multi-grid finite element method scheme with an original additive domain decomposition algorithm. This algorithm managed the message passing operations using the Message Passing Interface (MPI) libraries, a standardised message-passing system. We used the preconditioned GMRes (5) and GMRes (50) algorithms respectively as a smoother and as the core solver (Richardson *et al.*, 1979).

Fan modelling

We undertook the blade-to-blade flow-field and particle transport numerical simulations using their own dedicated mesh. In both cases the computational mesh

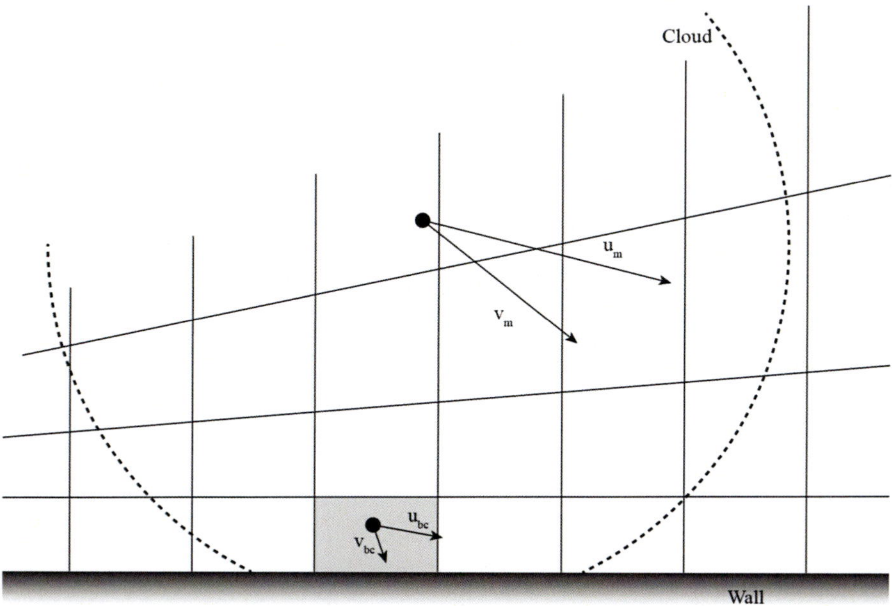

FIGURE 9.4. The particle cloud tracking scheme for characterising the cloud velocity distribution and consequently, the velocity of particles and angle (α) to a solid wall at impact.

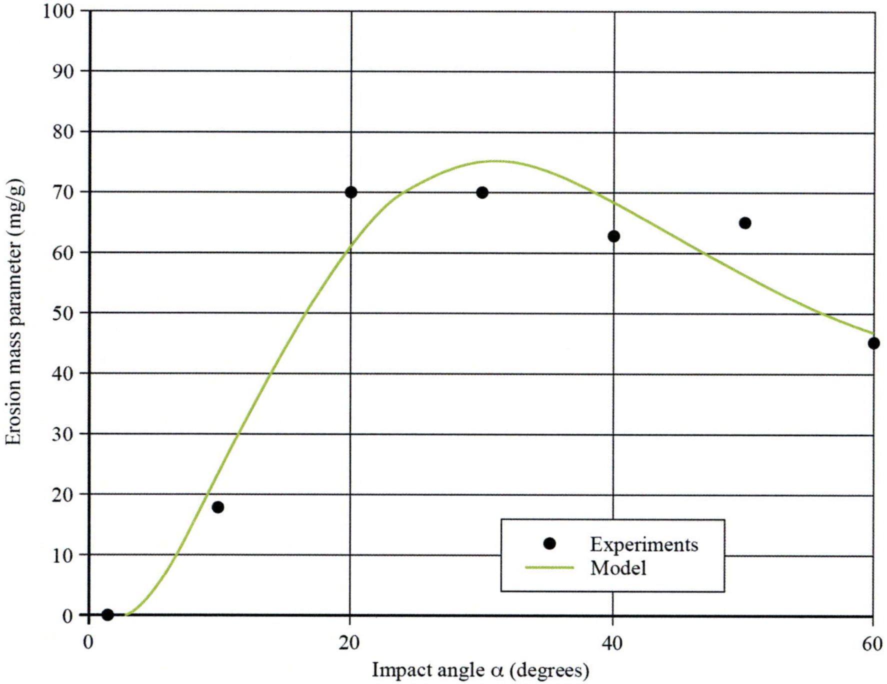

FIGURE 9.5. The experimentally derived erosion rate of 304 grade stainless steel measured by Tabakoff *et al.* (1979). The change of erosion rate with impact angle (α) illustrates the importance of predicting impact angle accurately to a prediction of erosion rate.

distributed 16 per cent of all elements upstream of the blade leading edge, 50 per cent in the blade passage and 34 per cent downstream of the blade trailing edge. We modelled the blade tip-to-casing gap with 55 grid nodes across the gap. In both cases the meshes were clustered toward solid boundaries, with the ratio of minimum grid spacing on solid walls to mid-span blade chord set at 7×10^{-4}. The adopted grid refinement toward the solid surfaces controlled the normalised wall distance, δ^+ value to about one at the first row of nodes.

We undertook the blade-to-blade flow-field numerical simulation using a tetrahedral computational mesh, Figure 9.6. We constructed the mesh using a non-orthogonal body fitted coordinate system with fully unstructured topology. The mesh consists of approximately 4.4 million tetrahedral elements, Table 9.2.

Table 9.2. *Computational mesh data for the flow-field and particle transport numerical simulations.*

	Flow-field	Particle transport
Element type	Tetrahedral	Hexahedral
Nodes	786,768	922,383
Cells	4,384,173	880,488

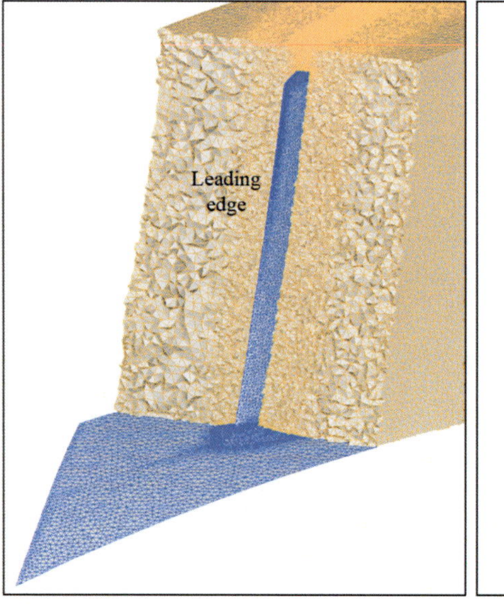

FIGURE 9.6. The tetrahedral computational mesh used for the numerical simulation of the fan blade-to-blade flow-field, left and the hexahedral computational mesh used for the numerical simulation of the particle motion, right.

We undertook the particle transport numerical simulation using a hexahedral computational mesh, Figure 9.6. We solved the Lagrangian particle transport equation with a non-orthogonal body-fitted coordinate system and a block-structured h-type topology. We constructed the mesh by merging a mesh in the blade-to-blade region with a second mesh embedded in the blade tip-to-casing gap. The mesh consists of approximately 880,000 hexahedral elements, Table 9.2.

Boundary condition and particle seeding

We defined the boundary conditions following Sheard *et al.*'s (2009) method. We defined inflow turbulence following Corsini and Rispoli's (2004) method. We determined the number of particles and their initial positions along the blade span using Ghenaiet's (2009) concentration profile, Figure 9.7. This concentration profile features a peak particle concentration at the blade mid-span, a concentration profile consistent with the erosion that we observed from blades removed after in-service operation in an erosive environment. The concentration profile had an average value of 750 milligrams of particles per cubic metre, assuming and average particle size 25 microns.

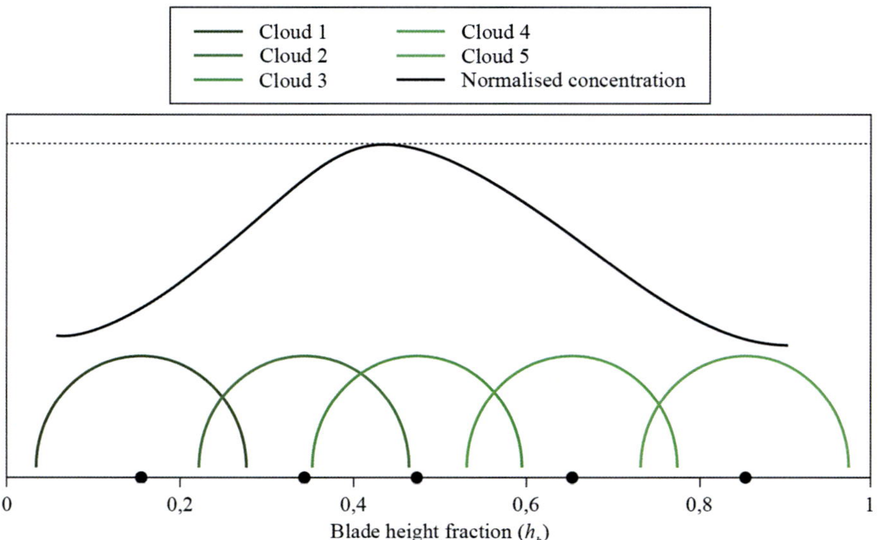

FIGURE 9.7. The span-wise distribution of clouds, and the resulting span-wise particle concentration profile.

RESULTS

Overall fan performance

We established the studied fan performance experimentally in accordance with ISO 5801 requirements (2007), Figure 9.8. We numerically predicted the studied fan's performance with a blade angle of 60 degrees, Figure 9.9. We computed fan performance using both a standard k-ε turbulence model and a nonlinear cubic k-ε turbulence model, described more fully by Corsini *et al.* (2011). The nonlinear cubic k-ε turbulence model resulted in a more accurate prediction of fan performance off-design, therefore we used it in all subsequent numerical simulations. The fan design point corresponded to a flow rate of 375 m^3/s. However, in practice the fan operated in-service at the off-design operating point corresponding to a flow rate of 425 m^3/s. Therefore, we computed all subsequent numerical results with a blade angle of 60 degrees at a flow rate of 425 m^3/s.

We may gain an insight into the flow-field features through the blade-to-blade passage by studying iso-surfaces of relative total pressure coloured using static pressure contours, Figure 9.10. The blade suction surface is characterised by boundary layer flow that remains attached to the blade with no evidence of separated flow regions. The blade wake is distorted around the mid-span as a consequence of the higher aerodynamic loading. In the blade tip region the blade tip-to-casing leakage vortex is evident, after its inception at approximately 25 per cent blade chord

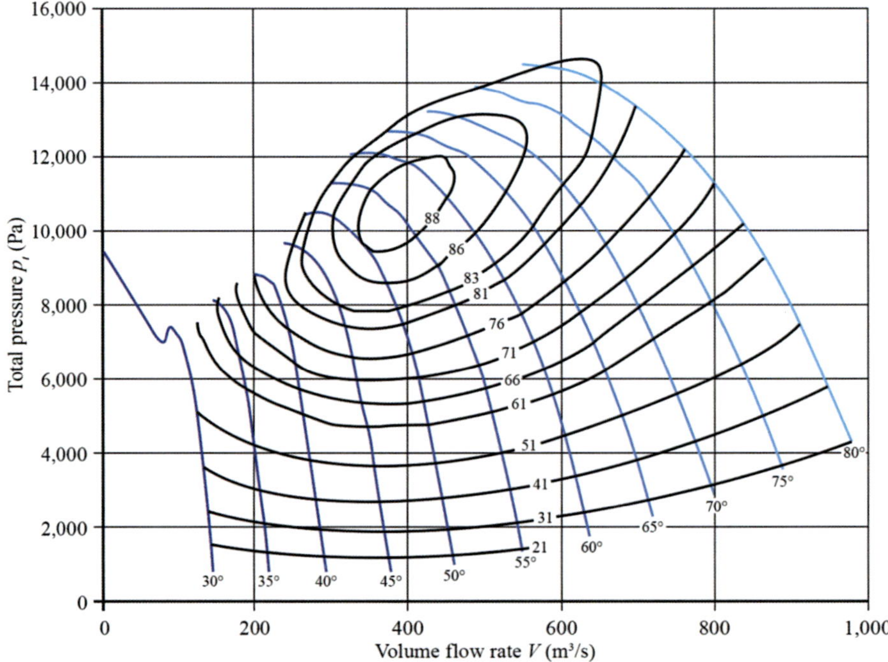

FIGURE 9.8. The performance characteristics of the studied fan measured over a range of blade tip stagger angles from 30 to 80 degrees in accordance with ISO 5801:2007 requirements (2007).

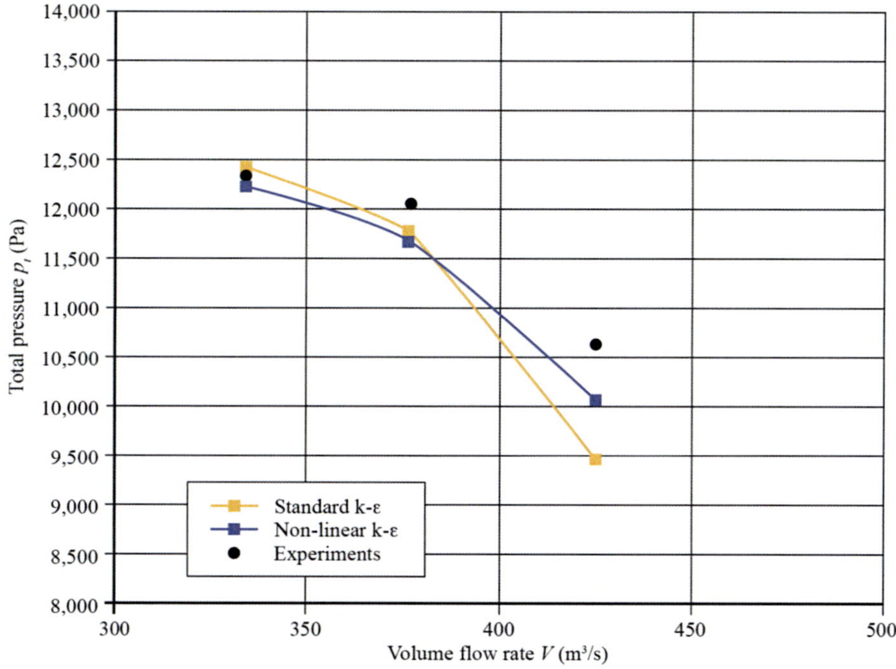

FIGURE 9.9. A comparison of fan performance when measured and when numerically predicted using a tetrahedral computational mesh with first a standard k-ε and second a non-linear k-ε turbulence model. The authors used a 60 degree blade angle. The non-linear k-ε turbulence model results in a better off-design prediction of fan performance.

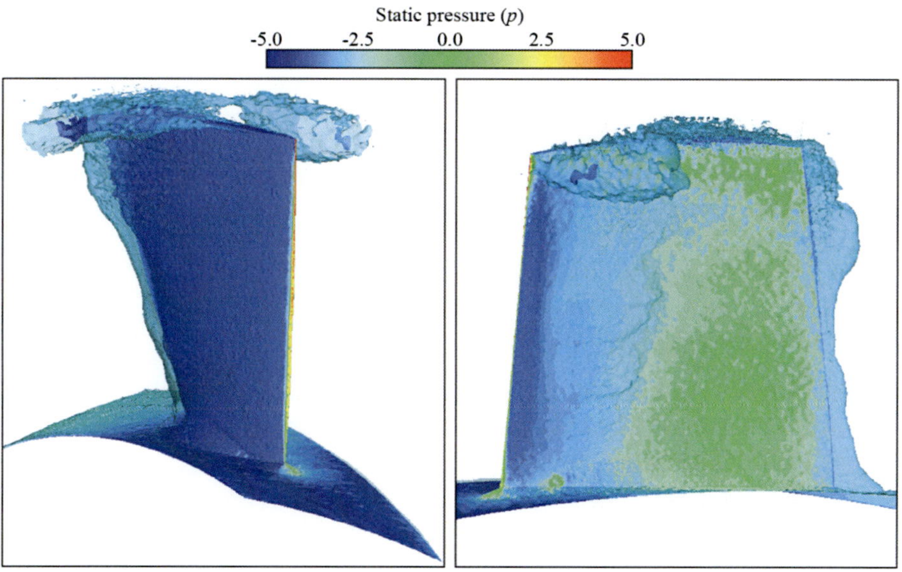

FIGURE 9.10. Iso-surfaces of relative total pressure coloured using static pressure contours over the blade surface numerically predicted using a tetrahedral computational mesh and a non-linear k-ε turbulence model. Results are shown for the studied fan blade suction's surface, left and pressure surface, right.

from the leading edge. As the leakage vortex is transported towards the blade trailing edge, it also migrates across the blade passage.

We may gain a further insight into the flow-field features through the blade-to-blade passage by studying three-dimensional streamlines through the blade-to-blade passage, Figure 9.11. The three-dimensional streamlines illustrate that the flow is essentially attached with no evidence of separated flow regions. There is also no evidence of secondary flow features distorting the three-dimensional streamlines. There is one exception to this observation. There is a distortion of the three-dimensional streamlines at the trailing edge in the hub region. We may attribute this distortion to the over-turning of wake flow driven by the blade-to-blade passage vortex. If we study the blade pressure surface, it is apparent that the three-dimensional streamlines migrate radially outward. We may attribute this radial migration to a leading edge separation bubble that we may characterise as a negative feature of the flow-field.

Particle impact and erosion

The agreement between measured and predicted fan performance was reasonable. Therefore, we may assume that the blade-to-blade flow-field and its associated secondary flow features were also reasonably well predicted. Thus, one may use the numerical simulation in conjunction with the cloud particle tracking model to predict particle trajectory.

Consider the predicted impact frequency I_f distributions over the blade pressure surface and suction surface, Figure 9.12. We define impact frequency as the number of impacts per square metre per gram of ingested particles. The impact frequency distributions are broadly similar to the erosion patters that we observed over blades removed following in-service operation in an erosive environment. The leading edge region is subject to the highest impact frequency. When we compare the impact frequency distributions over the suction surface with those over the pressure surface, it is evident that the impact frequency is highest over the pressure surface. In the blade pressure surface's near-hub region there is a peak region of impact frequency approximately 25 per cent blade chord downstream of the blade leading edge. We may associate this region with a secondary flow features in the blade root region that direct particles towards the blade root.

As we previously mentioned, erosion rate is a function of both impact velocity and angle. Consequently, although we expect some similarity between a prediction of impact frequency and erosion rate, the predicted erosion rate is distinct from the predicted impact frequency. We calculated erosion rate using Tabakoff *et al.*'s (1979) experimentally derived model coefficients for 304 grade stainless steel. The blade was actually manufactured from nodular iron; however, Tabakoff *et al.* (1979) had not characterised nodular iron. The 304 grade stainless steel data was the closest available material data. Table 9.3 presents the resultant prediction of overall erosion rate for the blade leading edge, tip and surfaces.

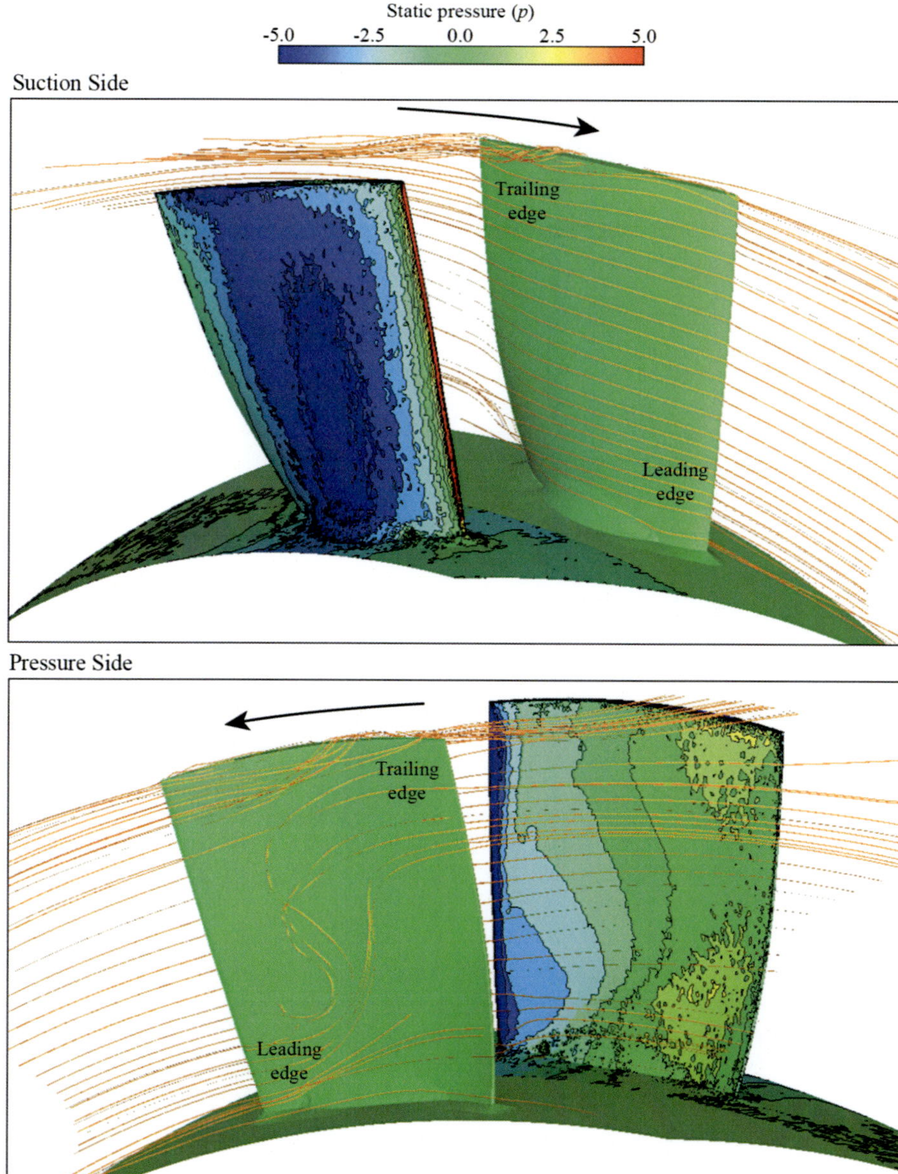

FIGURE 9.11. Three-dimensional streamlines through the blade-to-blade passage, with the blade surfaces coloured using static pressure contours. Both were numerically predicted using a tetrahedral computational mesh and a non-linear k-ε turbulence model. Results are shown for the studied fan blade suction's surface, top and pressure surface, bottom.

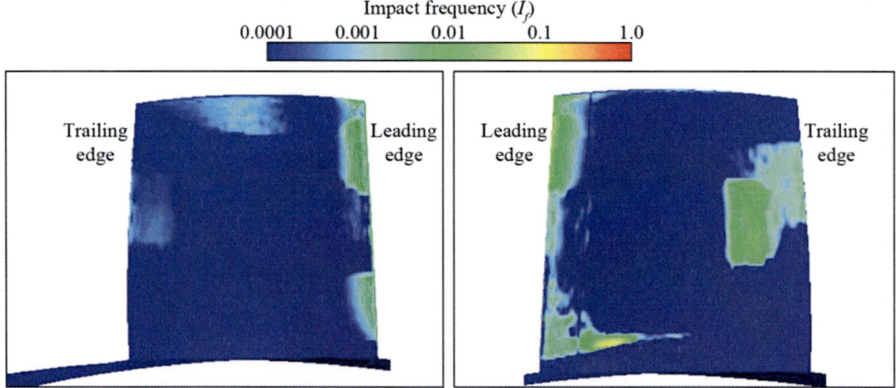

FIGURE 9.12. Predicted impact frequency I_f distributions over the blade pressure surface, left and suction surface, right. The authors define impact frequency as the number of impacts per square metre per gram of ingested particles.

Table 9.3. *The blade erosion at different blade locations after one hour of particle ingestion, predicted using the numerical simulation.*

Eroded areas	Erosion (mg)
Leading edge	36.460
Blade tip	2.549
Blade surfaces	61.570

Consider the predicted erosion rate E_R distributions over the blade pressure surface and suction surface, Figure 9.13. We define erosion rate as the milligrams of material eroded per square metre per gram of ingested particle per second. The predicted erosion rate distributions indicate that three factors dominate the erosive process. These are:

- the incidence angle at which the flow enters the blade-to-blade passage;
- the flow deviation induced by the work distribution along the blade span; and,
- the blade speed.

The incidence angle at which the flow enters the blade-to-blade passage correlates with erosion rate at the blade leading edge. When we study the leading edge erosion it is apparent that the blade pressure surface is more prone to erosion than the suction surface. Additionally, the leading edge erosion rate becomes progressively higher as we move from the blade hub to tip. At the blade mid-span there is a peak erosion rate region on the pressure surface near the blade trailing edge. This region correlates with the blade's work distribution and the deviation of particle trajectories from the flow streamlines as a consequence of their inertia.

In order to provide additional insight into the erosion process, we compared numerically predicted and observed in-service erosion rate E_R distributions, Figure 9.14. The comparison is inevitably qualitative, and therefore we must be cautions

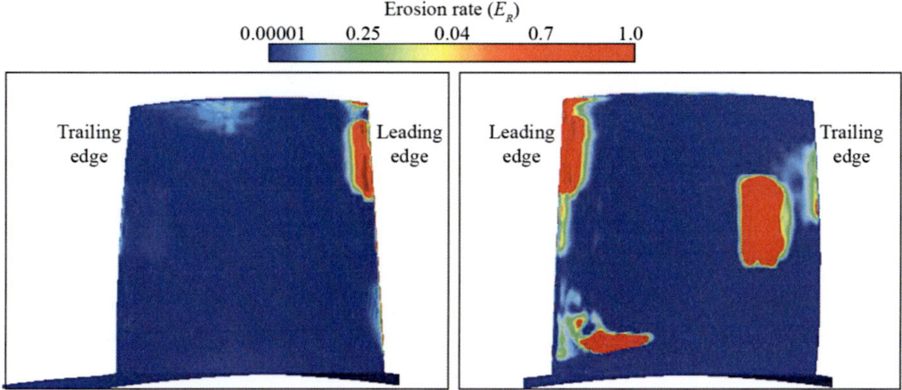

FIGURE 9.13. Predicted erosion rate E_R distributions over the blade pressure surface, left and suction surface, right. The authors define erosion rate as the milligrams of material eroded per square metre per gram of ingested particle per second.

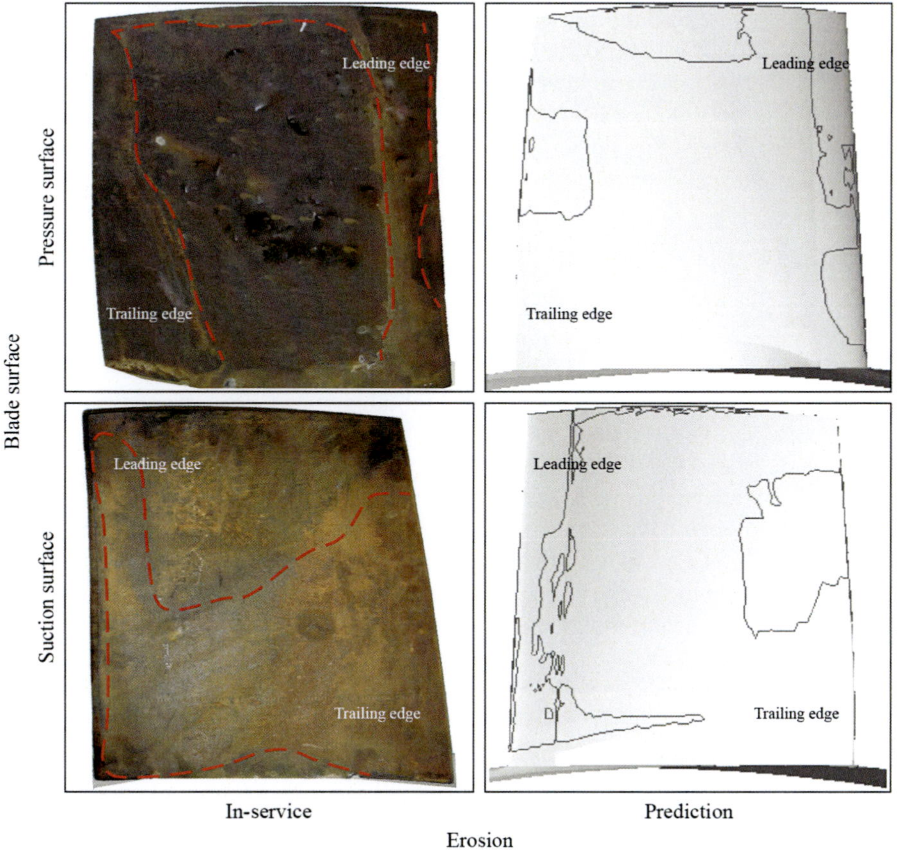

FIGURE 9.14. A comparison of erosion rate E_R distributions over the blade surfaces, numerically predicted and as observed in-service after 9,000 hours' operation. The authors considered the agreement between predicted and in-service erosion rate to be reasonable.

when drawing conclusions for the comparison. Despite this caveat, we may consider the agreement between the numerically predicted and in-service erosion patterns reasonable. The blade suction surface is characterised by the presence eroded regions around the blade's periphery and not the blade's centre. In contrast, the pressure surface of the blade is characterised by more intense erosion at the blade's centre. From this difference we may hypothesis that curvature and rotation of the blade have a significant impact on particle trajectory and the resulting erosion rate distribution. As Page *et al.* (1979) concluded, erosion rate is proportional to particle velocity raised to a power of between 1.99 and 5.00. Therefore, we expected the erosion patterns to indicate that the blade tip leading edge was subject to the highest erosion rates, a phenomena confirmed by both numerically predicted and observed in-service erosion rate distributions.

CONCLUSIONS

This chapter presents a numerical investigation of the erosion process through the blading of an induced-draft fan. We conducted a numerical simulation of the unburnt coal and fly-ash driven erosion process. We conducted the numerical simulation with finite element based in-house solvers that we used to predict the fan flow-field. We then used a Lagrangian cloud particle tracking model to predict particle trajectories. We then used the resultant prediction of particle impact frequency, velocity and angle as input to an empirical erosion model to predict erosion rate.

We conducted the fan flow-field's numerical simulation at an operating point typical of that associated with the fan's actual in-service operation. At this operating point the fan flow-field had no separated flow regions through the blade-to-blade passage. There was some evidence of a separation bubble on the blade pressure surface immediately downstream of the blade leading edge. However, although this separation bubble resulted in distortion of the near-wall fluid streamlines, it did not have a significant effect on the overall blade-to-blade flow-field or fan performance.

The prediction of overall fan performance was in reasonable agreement with experimentally measured fan performance. Therefore, we concluded that the blade-to-blade flow-field and its associated features were reasonably predicted and that one could use the numerical simulation as an input to the cloud particle tracking model. We used the cloud particle tracking model to predict the impact frequency and particle velocity and angle of impact. We then used this to predict the erosion rate. We compared the resulting prediction of erosion patterns over the blade suction and pressure surfaces with the erosion patterns over blades removed from in-service operation in an erosive environment. We concluded that the agreement between the numerically predicted and observed in-service erosion patterns was reasonable.

The numerical simulation results facilitated an analysis of the erosion process. We concluded that three factors dominate the erosion process:

- the incidence angle at which the flow enters the blade-to-blade passage;
- the flow deviation induced by the work distribution along the blade span; and,
- the blade speed.

This insight into the erosion process enabled us to characterise the erosion regions associated with the blade leading edge, suction surface and pressure surface. The blade leading edge was subject to increasing erosion with increasing blade span. Only the inner ten per cent of the leading edge in the blade root region was unaffected by erosion. We also identified a region of the blade pressure surface near the blade trailing edge at mid-span that was susceptible to erosion. This region was subject to erosion as a consequence of the deviation of heavier particles form the flow streamlines. The region most susceptible to erosion was the blade tip leading edge region as a consequence of the relatively high blade tip speed in this region.

REFERENCES

ISO 5801:2007 (2007), *Industrial Fans: Performance Testing Using Standardised Airways.*

Balan, C. and Tabakoff, W. (1984), 'Axial Flow Compressor Performance Deterioration', *Proceedings of the AIAA.SAE/ASME 20th Joint Propulsion Conference*, Cincinnati, OH, USA, 11–13 June, paper no. AIAA-84-1208.

Baxter, L.L. (1989), 'Turbulent Transport of Particles', PhD thesis, Brigham Young University, Provo, UT, USA.

Baxter, L.L. and Smith, P.J. (1993), 'Turbulent Dispersion of Particles: the STP Model', *Energy and Fuels*, vol. 7(6), pp. 852–859.

Borello, D., Corsini, A. and Rispoli, F. (2003), 'A Finite Element Overlapping Scheme for Turbomachinery Flows on Parallel Platforms, *Computational Fluids*, vol. 32(7), pp. 1017–1047.

Borello, D., Borrelli, P., Quagliata, E. and Rispoli, F. (2001), 'A Multi-grid Additive and Distributive Parallel Algorithm for FEM Turbomachinery CFD', *Proceedings of the European Congress on Computational Methods in Applied Sciences (ECCOMAS CFD 2001)*, Swansea, UK, 4–7 September.

Corsini, A. and Rispoli, F. (2004), 'Using Sweep to Extend Stall-free Operational Range in Axial Fan Rotors', *Proceedings of the IMechE Part A, Journal of Power and Energy*, vol. 218, pp. 129–139.

Corsini, A., Rispoli, F., and Santoriello, A. (2003), 'A High Order Petrov-Galerkin Stabilized Finite Element Method for Incompressible RANS in Presence of Strong Reaction Effects', *Proceedings of CMFF-03, the Conference on Modelling Fluid Flow*, Budapest, Hungary, 3–6 September.

Corsini, A., Rispoli, F., Santoriello, A. and Tezduyar, T.E. (2006), 'Improved Discontinuity-capturing Finite Element Techniques for Reaction Effects in Turbulence Computation', *Computational Mechanics*, vol. 38, pp. 356–364.

Corsini, A., Marchegiani, A., Minotti, S. and Sheard, A.G. (2011), 'Numerical Investigation on the Aerodynamic Influence of Eroded Leading-edge Geometry on Boiler Fan Performance', *Proceedings of the 9th European Conference on Turbomachinery, Fluid Dynamics & Thermodynamics*, Istanbul, Turkey, 21–25 March, pp. 879–892.

Craft, T.J., Launder, B.E. and Suga, K. (1996), 'Development and Application of a Cubic Eddy-viscosity Model of Turbulence', *International Journal of Heat and Fluid Flow*, vol. 17(2), pp. 108-115.

Crowe, C.T., Troutt, T.R. and Chung, J.N. (1996), 'Numerical Models of Two-phase Turbulent Flows', *Fluid Mechanics*, vol. 28, pp. 11–43.

Elfeki, S. and Tabakoff, W. (1987), 'Erosion Study of Radial Flow Compressor with Splitters', *Transactions of the ASME, Journal of Turbomachinery*, vol. 109(1), pp. 62–69.

Ghenaiet, A. (2005), 'Numerical Simulations of Flow and Particle Dynamics within a Centrifugal Turbomachine', *Proceedings of the Institution of Mechanical Engineers International Conference on Compressors and their Systems*, London, UK, 4–7 September, paper no. C639-652, pp. 497–506.

Ghenaiet, A. (2009), 'Numerical Study of Sand Ingestion through a Ventilating System', *Proceedings of the World Congress on Engineering 2009*, London, UK, 1–3 July.

Ghenaiet, A., Tan, S.C. and Elder, R.L. (2004), 'Experimental Investigation of Axial Fan Erosion and Performance Degradation', *Proceedings of the IMechE Part A, Journal of Power and Energy*, vol. 218(6), pp. 437–450.

Grant, G. and Tabakoff, W. (1975), 'Erosion Prediction in Turbomachinery Resulting from Environmental Solid Particles', *Journal of Aircraft*, vol. 12(5), pp. 471–478.

Hamed, A., Tabakoff, W. and Wenglarz, R. (2006), 'Erosion and Deposition in Turbomachinery', *Journal of Propulsion and Power*, vol. 22, pp. 350–360.

Hussein, M.F. and Tabakoff, W. (1974), 'Computation and Plotting of Solid Particle Flow in Rotating Cascades', *Computational Fluids*, vol. 2(1), pp. 1–15.

Jain, S. (1995), 'Three-dimensional Simulation of Turbulent Particle Dispersion', PhD thesis, University of Utah, Salt Lake City, UT, USA.

Kær, S.K. (2001), 'Numerical Investigation of Ash Deposition in Straw-fired Furnaces', PhD thesis, Aalborg University, Aalborg, Denmark.

Kirk, B.S., Peterson, J.W., Stogner, R.H. and Carey, G.F. (2006), 'libMesh: A C++ Library for Parallel Adaptive Mesh Refinement/Coarsening Simulations', *Engineering With Computers*, vol. 22, pp. 237–254.

Kurz, R. and Brun, K. (2001), 'Degradation in Gas Turbine Systems', *Transactions of the ASME, Journal of Engineering for Gas Turbines and Power*, vol. 123, pp. 70–77.

Litchford, R.J. and Jeng, S.M. (1991), 'Efficient Statistical Transport Model for Turbulent Particle Dispersion in Sprays', *AIAA Journal*, vol. 29, pp. 1443–1451.

Page, A.L., Elseewi, A.A. and Straughan, I.R. (1979), 'Physical and Chemical Properties of Fly Ash from Coal-fired Power Plants with Reference to Environmental Impacts', *Residue Review*, vol. 71, pp. 83–120.

Richardson, J.H., Sallee, G.P. and Smakula, F.K. (1979), 'Causes of High Pressure Compressor Deterioration in Service', *Proceedings of the AIAA / SAE / ASME 15th Joint Propulsion Conference*, Las Vegas, NV, USA, 18–20 June, paper no. AIAA 79-1234.

Sallee, G.P., Kruckenburg, H.D. and Toomey, E.H. (1975), 'Analysis of Turbofan Engine Performance Deterioration and Proposed Follow-on tests', Report NASA-CR-134769.

Sheard, A.G., Corsini. A., Minotti, S. and Sciulli, F. (2009), 'The Role of Computational Methods in the Development of an Aero-acoustic Design Methodology: Application in a Family of Large Industrial Fans', *Proceedings of the 14th International Conference on Modelling Fluid Flow Technologies*, Budapest, Hungary, 9–12 September, pp. 71–79.

Sugano, H., Yamaguchi, N. and Taguchi, S. (1982), 'A Study on the Ash Erosion of Axial Induced Draft Fans of Coal-fired Boilers', *Mitsubishi Heavy Industries Technical Review*, vol. V19(N1), pp. 12–22.

Tabakoff, W., Kotwal, R. and Hamed, A. (1979), 'Erosion Study of Different Materials Affected by Coal Ash Particles', *Wear*, vol. 52(1), pp. 161–173.

Venturini, P. (2010), 'Modelling of Particle Wall-deposition in Two Phase Gas-solid Flows', PhD thesis, Sapienza University of Rome, Rome, Italy.

Venturini, P., Borello, D., Iossa, C., Lentini, D. and Rispoli F. (2010), 'Modelling of Multiphase Combustion and Deposit Formation and Deposit Formation in a Biomass-fed Boiler', *Energy*, vol. 35(7), pp. 3008–3021.

Wang, L.P. (1990) 'On the Dispersion of Heavy Particles by Turbulent Motion', PhD thesis, Washington State University, Pullman, WA, USA.

Chapter 10

Numerical Simulation of Coal Fly-ash Erosion in an Induced Draft Fan

A. Corsini, F. Rispoli, A.G. Sheard and P. Venturini

ABSTRACT

Induced draft fans extract coal-fired boiler exhaust gases in the form of a two-phase flow with a dispersed solid phase comprising unburnt coal and fly-ash. Consequently, fan blades are subject to erosion causing wear of the blade leading edge, trailing edge, suction and pressure surfaces. Erosion results in a loss of blade material, a reduction of blade chord and effective camber that collectively degrade fan aerodynamic performance. This chapter presents a numerical study of the erosive processes at play within the blade-to-blade passage of an induced draft fan. We first used a Reynolds-averaged Navier–Stokes code to predict the blade-to-blade flow-field. We then conducted a numerical simulation of particle trajectory through the blade-to-blade passage using a particle cloud tracking technique that models drifting near wall velocity and incorporates an algebraic erosion model. The numerical study clarifies a fan operating points' influence on the resultant erosion patterns over the blade suction and pressure surfaces. The study specifically focuses on elucidating the role particle size and mass distribution play on erosion patterns. We analysed a fly-ash sample taken from an induced draft fan that had operated in a highly erosive environment, characterising the particles within the fly-ash and used that characterisation as an input to our cloud particle tracking model. The resulting prediction of erosion patterns over the blade suction and pressure surfaces provides an insight into the effect of particle size and mass.

This chapter is a revised and extended version of Corsini, A., Rispoli, F., Sheard, A.G. and Venturini, P. (2013), 'Numerical Simulation of Coal Fly-ash Erosion in an Induced Draft Fan', *Transactions of the ASME, Journal of Fluids Engineering*, vol. 135, paper no. 081303, pp. 1–12.

NOMENCLATURE

Latin letters

a		Length of the surface
b		Width of the surface
c	[mm]	Blade chord
c_μ		Numerical coefficient
C_D		Particle drag coefficient
d		Diameter
D_{hc}		Hub to casing diameter ratio
D_t	[mm]	Blade tip diameter
$<d>$	[μm]	Average size
E_r		Erosion rate
f_α		Erosion angular function
g		Acceleration of gravity
h_e		Thickness of the eroded leading edge
I_f		Impact frequency
k		Turbulent kinetic energy
K_c, K_1, K_2, K_{12}		Erosion model coefficients
l_ε		Dissipation length scale
m	[%]	Mass
m^+		Normalised particle mass
$\dot{m}^+$		Normalised mass flow
P	[kW]	Power
p	[Pa]	Static pressure
$\dot{p}^+$		Normalised particle flow
t		Time
u		Fluid velocity vector
v		Particle velocity vector
V_{er}		Volume of eroded material
x		Axial Cartesian coordinate
y^+		Normalised wall distance

Greek letters

α		Impact angle
β		Exchange coefficient
γ	[deg]	Stagger angle
ε		Turbulent kinetic dissipation rate
θ	[deg]	Camber angle
μ		Mean position of the cloud
ρ		Density
ρ_s		Density of surface material
σ		Solidity
$\sigma(t)$		Variance of the particle distribution in the cloud
τ_L		Lagrangian time scale
χ	[% span]	Tip clearance

Subscript and symbols

bc	Boundary cell
f	Fluid
m	Cloud's centre of mass
p	Particle
< >	Ensemble average

Acronyms

FEM	Finite element method
MG	Multi grid
MPI	Message passage interface
PCT	Particle cloud tracking
PDE	Partial differential equation
PDF	Probability density function
PFS	Pan Fläkt single
PG	Petrov-Galerkin
PSPG	Pressure-stabilising / Petrov-Galerkin
RANS	Reynolds-averaged Navier–Stokes
SP	Stagnation point
SUPG	Streamline-upwind / Petrov-Galerkin

INTRODUCTION

Particle laden flows are a common feature of many industrial processes. The air movement fans used in these processes are inevitably subject to both fouling and erosion. The induced draft fans in coal-fired boiler applications classically suffer deterioration in-service as a consequence of fan blade erosion. The exhaust gas passing through the induced draft fans is erosive as a consequence of both solid and molten particles within the gas. These particles are comprised of unburnt coal particles and fly-ash formed from those elements within the coal that does not burn. Both unburnt coal and fly-ash particles form in the boiler during the combustion process (Kurz and Brun, 2001; Hamed *et al.*, 2006).

Scholars have previously correlated fan and compressor blade erosion to the degradation of aerodynamic performance (Grant and Tabakoff, 1975). A consequence of erosion is a degradation of aerodynamic performance. However, when erosion progresses beyond a critical limit it can lead to a structural failure. Therefore, there is interest within the air movement and control community in better understanding the physical phenomena underpinning the erosive process. Those scholars who have studied the erosive process have concluded that three limiting phenomena characterise fan performance degradation as they are subject to erosion. First, the blades' ability to turn the flow diminishes, partly as a consequence of the chord reduction. Second, erosion of the blade leading edge destroys the leading edge profile resulting in a tendency of boundary layers to separate. Boundary layer separation impacts the location and extent of laminar-to-turbulence boundary layer transition that then reduces the fan's stable operating range. Finally, aerodynamic loss mechanisms increase in intensity, driven by the change in blade leading edge, trailing edge, blade thickness and blade tip-to-casing clearance (Kurz and Brun, 2001).

Grant and Tabakoff (1975), Richardson *et al.* (1979) and Okita *et al.* (2012) and others have studied the factors that underpin the erosive process. These scholars have concluded that significant factors driving the erosive process are the physical properties and geometry of particles. Sallee *et al.* (1975) and Balan and Tabakoff (1984) studied the erosive process within compressors and Ghenaiet *et al.* (2004) studied it within fans. These scholars chose to use the military specification MIL-E-5007E to characterise the physical properties and geometry of particles. This standard characterises quartz sand particles over the size range 75 to 1,000 μm which they used during their study of fan and compressor blade erosion due to sand ingestion. These scholars concluded that erosion occurred over the outer 50 per cent of the blade span.

Ghenaiet *et al.* (2004) studied the performance of an eroded gas turbine compressor. They concluded that blade leading edge erosion resulted in an increase in blade tip-to-casing clearance and a reduction in blade chord at the tip. The impact on compressor performance was a 7 per cent reduction in efficiency and 10 per cent reduction in pressure rise coefficient. This reduction occurred after 9 hours' operation in a desert environment. Sugano *et al.* (1982) studied the erosion of induced draft fans in coal-fired boiler applications. They characterised the influence of the particle size distribution associated with different types of coal. Kurz and Brun (2001) and Hamed *et al.* (2006) studied the influence of particle size on the erosive process. They concluded that particles larger than 20 μm drive the erosive process. Bons *et al.* (2001) studied the erosion of a land-based gas turbine compressor. They concluded that the erosive process was primarily driven by particles smaller than 30 μm.

Characterising the particle sizes most influential in the erosive process and the impact of erosion on fan and compressor performance is a useful contribution to the community. However, to capitalise on this insight, compressor and fan designers require a deeper level of insight in the effects of erosion on fan and compressor performance. For a specific distribution and concentration of particle sizes a designer must be able to predict the resultant erosion patterns over specific blade geometry. This will facilitate a prediction of the rate of performance degradation with time. It will also provide an input into a design process aimed at developing more erosion tolerant blade designs. In practice, it is challenging to use experimental techniques to study the erosion process. Reproducing the in-service environment is particularly difficult, and consequently the applicability of any experimental results is questionable. Assessing blades removed after in-service operation also provides only qualitative data. The size, distribution and concentration of the particles responsible for the observed erosion patterns are difficult to measure accurately. Consequently, although the in-service environment is reproduced, it is still not possible to draw quantitative conclusions that can be of direct use to the fan or compressor designer.

The difficulty of experimentally characterising the erosive process has resulted in scholars focusing on developing numerical techniques to predict erosion. Starting in the early 1970s, researchers began to develop numerical techniques to simulate particle trajectories and resulting fan and compressor blade erosion. Hussein and Tabakoff (1974) and Elfeki and Tabakoff (1987) pioneered particle trajectory simulations through axial and radial turbomachinery. They focused on modelling the influence of inertial forces and forces induced by rotation on particles with mass. Ghenaiet (2005, 2009) and Suzuki and Yamamoto (2011) have reviewed the numerical studies focused on modelling the erosive processes associated with sand ingestion. Despite the undoubted progress in characterising the physical phenomena associated with erosion, predicting blade erosion patterns remains challenging.

This chapter presents a method for numerically predicting particle trajectories, their impact velocity and angle on blade surfaces and the resultant erosion. The prediction uses as input a distribution of particle size and mass typical of the particles produced during the combustion process in a coal-fired boiler. The numerical method is based on an Euler-Lagrangian approach. The blade-to-blade flow-field is modelled using a three-dimensional Reynolds-averaged Navier–Stokes (RANS) finite element (FEM) solver originally developed by Borello *et al.* (2003). We model particle trajectory, impact velocity and angle using a particle cloud tracking (PTC) approach originally developed by Baxter (1989). We predict erosion using an empirical material wear model originally developed by Tabakoff *et al.* (1979). The numerical procedure provides fan and compressor designers with a method they can use to establish the erosive consequences of a specific particle distribution. They may use this insight to inform the design process. The objective is to facilitate the design of more erosion tolerant blades that retain their aerodynamic performance more effectively when eroded.

CHARACTERISATION OF THE PARTICLE LADEN FLOW

The studied fan is equipped with a mechanism for varying the blade pitch-angle whilst rotating at its design speed, Figure 10.1. The industrial fan community refers to any fan fitted with this mechanism as a variable pitch in motion (VPIM) fan. Design engineers originally developed variable pitch in motion fans in the 1950s for application with coal-fired boilers. Despite the mechanical complexity of variable pitch in motion fans, they still find wide-spread application in the coal-fired boilers systems of thermal power stations. The cost of high-power variable-speed drives remains high. Consequently, the industry does not favour fixed-pitch variable-speed forced- and induced-draft fans.

The studied fan is installed as part of a coal-fired boiler system in induced draft arrangement. We were able to obtain a sample of fly-ash from the fan during a scheduled maintenance outage. When we analysed this sample we found that the particles were divided into two size groups. The first size group contained particles ranging in size from less than 1 μm to 45 μm, Figure 10.2 and Table 10.1. The second group contained particles ranging in size from 45 μm to 150 μm, Figure 10.3 and Table 10.1. This division of particle size is self-consistent with Seggiani *et al.* (2000) and Pandian's (2004) findings. The first group comprises particles that primarily consist of silica compounds. The second group comprises particles that are

FIGURE 10.1. A cut-away view of the studied fan illustrating the variable pitch in motion blades, their inlet and outlet guide vanes. This family of fans is configured with an inlet box that turns the flow through 90 degrees. The flow into the inlet box is vertically down and the flow out is horizontal.

Table 10.1. *The distribution of silicon (Si) compound particles measured in a particle sample taken from an induced draft fan installed in a coal-fired power plant. The authors used particle size distribution in the numerical simulation of particle motion, impact frequency and erosion rate. The Stokes number is a dimensionless number corresponding to the behaviour of particles suspended in a fluid flow. It is defined as the ratio of the characteristic time of a particle to a characteristic time of the flow.*

Particle size (μm)	Mass fraction	Stokes number
0.75	2.006×10^{-6}	2.26×10^{-2}
1.75	1.875×10^{-5}	1.23×10^{-1}
3.75	2.342×10^{-5}	5.65×10^{-1}
7.50	3.127×10^{-5}	2.26×10
15.00	2.032×10^{-5}	9.05×10
32.50	2.692×10^{-5}	4.25×10^{1}
52.25	2.113×10^{-1}	1.10×10^{2}
67.50	2.569×10^{-1}	1.83×10^{2}
82.50	2.663×10^{-1}	2.74×10^{2}
97.50	4.303×10^{-2}	3.82×10^{2}
112.50	9.655×10^{-2}	5.09×10^{2}
135.00	1.258×10^{-1}	7.33×10^{2}

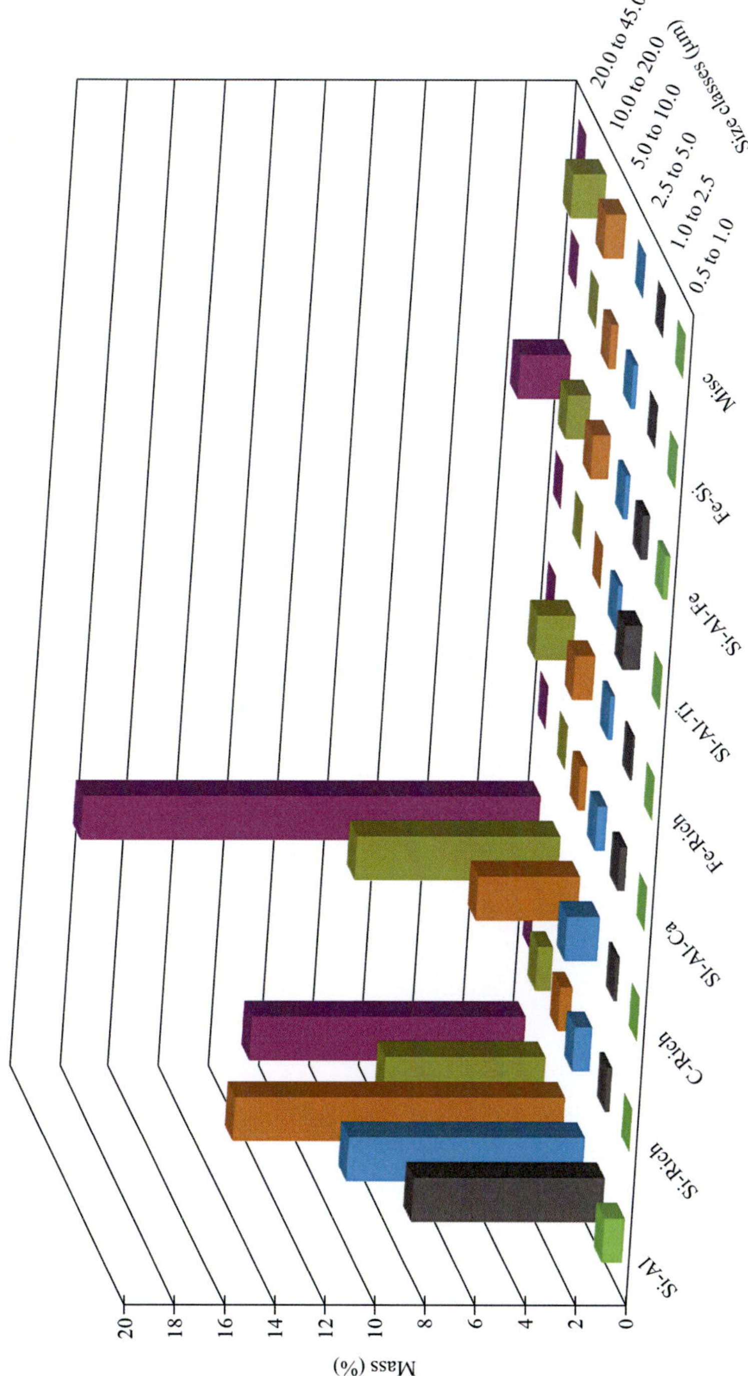

FIGURE 10.2. An analysis of the fine particles within a dust sample taken from an induced draft fan installed in a typical coal-fired power plant. The analysis characterises the size, composition and mass distribution of particles in the dust sample.

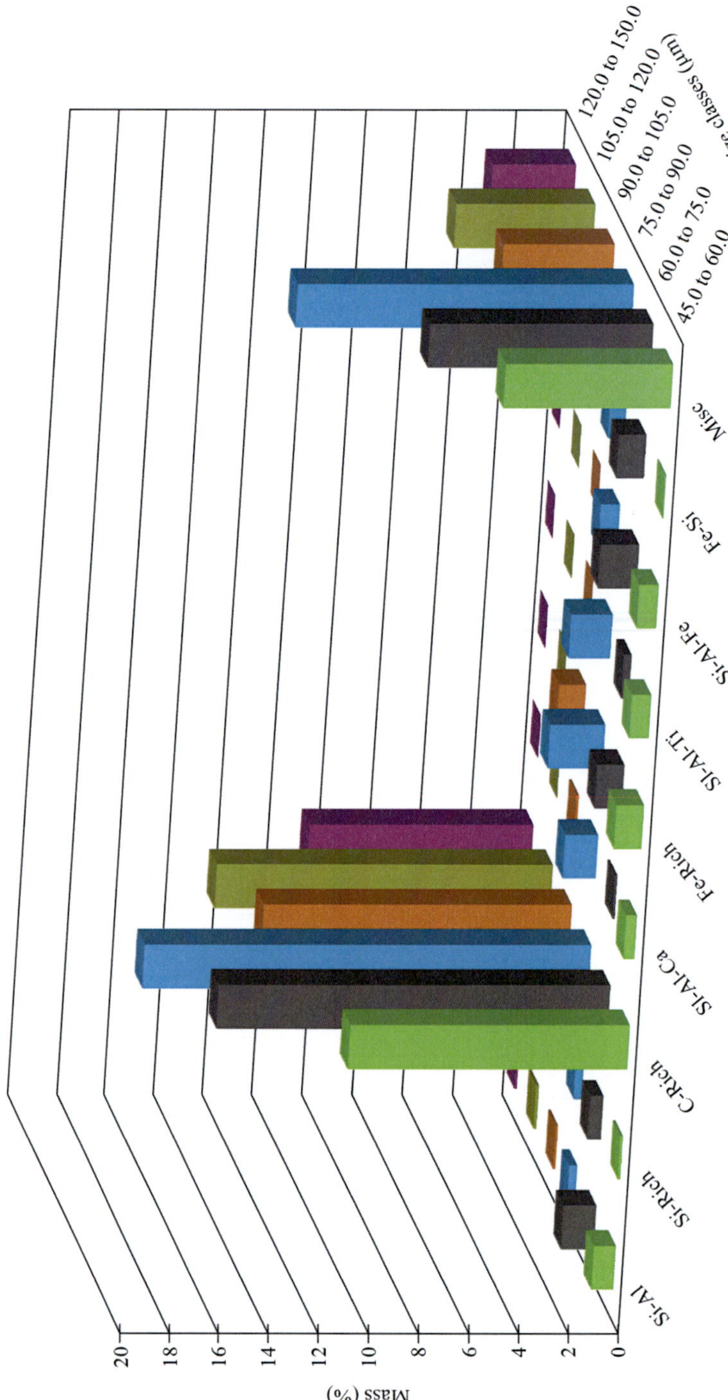

FIGURE 10.3. An analysis of the coarse particles within a dust sample taken from an induced draft fan installed in a typical coal-fired power plant. The analysis characterises the size, composition and mass distribution of particles in the dust sample.

primarily carbon-rich. The silica compounds form as a consequence of the mineral content present in the coal burnt in the boiler. The carbon-rich particles form as a consequence of coal particles that are too large to burn in the boiler.

Kleis and Kulu (2008) studied the effect of particles consisting of silica compounds and those that are carbon-rich. They concluded that the silica particles are harder than the coal-rich particles and are primarily responsible for erosion. The carbon-rich particles are primarily responsible for fouling. As the carbon-rich particles are not erosive, we chose to simplify our modelling of particle distribution by modelling only the silica compound particles. We were also able to take an air sample whilst the studied fan was operating from which we estimated that the particle concentration was approximately 64 mg/m^3. Our analysis of particles indicated that the particle density was approximately 2,000 kg/m^3. Pandian (2004) studied particle density, concluding that it could vary from 1,600 to 3,100 kg/m^3 and therefore we concluded that our measured particle density was typical of coal-fired boiler fly-ash.

NUMERICAL METHODOLOGY

In this chapter we report the results of a computational analysis that we undertook using a Reynolds-averaged Navier–Stokes (RANS) simulation. We used the same numerical procedure as Borello *et al.* (2003) using a topology-free low-Reynolds number formulation incorporating Kirk *et al.*'s (2006) libMesh libraries. We modelled the incompressible three-dimensional turbulent flow in the rotating frame of reference with a two-equation closure (Craft *et al.*, 1996), which was able to cope with non-isotropic and non-equilibrium turbulence effects without a significant increase in the required computational effort. The numerical procedure utilised a parallel multi-grid (MG) numerical scheme embedded in a finite element method (FEM) code.

Corsini and Rispoli (2005) and Corsini *et al.* (2009, 2012b) previously validated this computational approach. Corsini *et al.* (2009) studied the flow-field through a compressor cascade. Corsini and Rispoli (2005) studied the flow-field through a high-pressure axial flow. Corsini *et al.* (2012b) studied the flow-field induced in the blade tip-to-casing gap by a blade-tip end-plate. These researchers concluded that the computational approach was able to predict fan or compressor performance with reasonable accuracy. It was also able to model effectively the flow-field features within the blade-to-blade passage.

The finite element method formulation is based on a stabilised Petrov-Galerkin (PG) method. We modified this method for application to three-dimensional equal- and mixed-order spaces of approximation (Corsini *et al.*, 2006). The stabilised Petrov-Galerkin (PG) scheme controls the instability that affects the advective-diffusive incompressible flow, and the momentum's reaction and turbulent scale equations. The turbulent scale equations relate to Coriolis acceleration (Corsini *et al.*, 2003). We used equal-order Q1-Q1 and mixed-order Q2-Q1 interpolation for primary turbulent scale equations and constrained secondary variables, implicitly eliminating the undesirable pressure-checker boarding effects. We performed the computational analysis using the hybrid full linear multi-grid accelerator running on an overlapping parallel solver (Borello *et al.*, 2001).

We parallelised an iterative Krylov method when smoothing and solving using the adopted multi-grid finite element method scheme with an original additive domain decomposition algorithm. This algorithm managed the message passing operations using the Message Passing Interface (MPI) libraries, a standardised message-passing system. We used the preconditioned GMRes (5) and GMRes (50) algorithms respectively as a smoother and as the core solver (Richardson *et al.*, 1979).

Particle solver

If one accepts that particle concentration is low, we may assume that the flow-field affects particle motion, but particles do not affect the flow-field. If we accept this assumption then we may use a prediction of the blade-to-blade flow-field as an input into a particle tracking algorithm without modelling the particles when simulating the flow-field. This simplifying assumption reduces the computational effort needed to predict particle motion. However, predicting the trajectory of individual particles still required a computational effort beyond that available within the air movement and control community. A way to reduce the required computational effort required to predict particle motion is the particle cloud tracking (PCT) approach, which Baxter (1989) originally developed. Wang (1990), Litchford and Jeng (1991), Baxter and Smith (1993) and Kær (2001) further developed it. Venturini *et al.* (2010) and Corsini *et al.* (2012a) successfully validated it.

The particle cloud tracking approach is able to reduce the required computational effort as it does not attempt to predict the motion of single particles. The approach is based on predicting particle clouds. We may express the cloud's mean position (μ) at a given time as:

$$\mu = (x_p) = \int_0^t (v)d + (x_p 0) \tag{1}$$

This equation of motion for the cloud particles is an ensemble-averaged version of the Basset-Boussinesque-Oseen equation, which accounts for inertia, drag and buoyancy forces on a particle. Following Baxter's (1989) method, we assumed that the gas-to-solid particle density ratio was below 10^{-3} and therefore the Basset-Boussinesque-Oseen equation reduces to:

$$\frac{d\langle v\rangle}{dt} = \langle\beta\rangle \cdot [\langle u\rangle - \langle v\rangle] + \left(1 - \frac{\rho_f}{\rho_p}\right) g \tag{2}$$

where:

$$\langle\beta\rangle = \frac{3}{4d_p} C_D \frac{\rho_f}{\rho_p} \,|\, \langle u \rangle - \langle v \rangle \,| \tag{3}$$

The cloud particle tracking modelling approach assumes that particle position distribution within a cloud is Gaussian. We further assume that the cloud size varies in time as a function of the blade-to-blade flow-field turbulence, Figure 10.4,

$$PDF(\mathrm{x}) = \frac{1}{(2\pi)^{\frac{1}{2}}\sigma(t)} \exp\left[-\frac{1}{2}\left(\frac{\mathrm{x} - \langle \mathrm{x} \rangle}{\sigma^{(t)}}\right)\right] \tag{4}$$

where $\sigma(\tau)$ is the standard deviation of the distribution. The standard deviation is a function of particle-turbulence interaction that drives the turbulent dispersion of particles. Lecrivain and Hampel (2012) concluded that the turbulent dispersion of particles is proportional to the time scale of turbulence that partially accounts for the anisotropic flow-field. We evaluated the particle-turbulence interaction by assuming the Markovian approximation:

$$\sigma(t)^2 = 2\langle \mathrm{v}^2 \rangle \left[\frac{t}{\tau_L} - \frac{1}{\tau_L^2}\left(1 - e^{-\frac{t}{\tau_L}}\right)\right] \tag{5}$$

where:

$$\tau_L = \max(\tau_f, \tau_p) \tag{6a}$$

where τ_L is the Lagrangian time scale, defined as the maximum difference between the turbulence and the particle motion time scales, and:

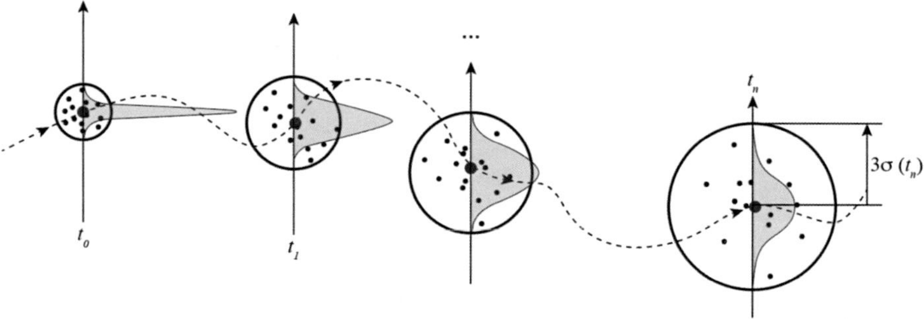

FIGURE 10.4. The particle cloud tracking (PCT) modelling approach assumes that particle position distribution within a cloud is Gaussian, and that the cloud size varies in time as a function of the blade-to-blade flow-field turbulence.

$$\langle \acute{v}^2 \rangle = \langle \acute{u}^2 \rangle \left(1 - e^{-\frac{\tau_f}{\tau_p}}\right) = \frac{2}{3} k \left(1 - e^{-\frac{\tau_f}{\tau_p}}\right) \tag{6b}$$

is the fluctuating component of the particle velocity as induced by the particle-turbulence interaction (Smith, 1993). The time scale definitions are (Baxter, 1989):

$$\tau_f = \frac{C_\mu^{\frac{3}{4}} k}{0.817\varepsilon} \tag{6c}$$

and

$$\tau_p - \beta^{-1} \tag{6d}$$

We assume the cloud radius to be three standard deviations and that each cloud consists of perfectly spherical particles with the same chemical and physical characteristics. When considering particle interaction with a solid wall, we assume that all particles within a cloud stop interacting with the solid wall after the cloud centre has interacted. We also assume that particles impacting a solid wall rebound elastically when the clouds centre reaches the wall. Finally, we assume that impacting particles bounce off solid walls and then remain within the cloud. Therefore, the particles may impact again the same or other solid walls.

Erosion model

The original particle cloud tracking formulation is based on a hypothesis: rigid body motion of clouds comprised of particles moving at the velocity of the cloud's centre of mass. This hypothesis assumes that there is no influence associated with the non-uniform velocity distribution both within the cloud and when the cloud impacts on a solid boundary. Therefore, the original formulation includes significant simplifying assumptions. In order to partially account for wall-induced velocity effects, we have adopted Venturini's (2010) and Borello *et al.*'s (2013) developed form of the particle cloud tracking model. We assume that particles within a cloud that is impacting on a solid surface keep the same drift velocity at the cloud's centre. Using this assumption we are able to derive the actual velocity of impact of each particle v_{bc} as:

$$|v_{bc}| = |u_{bc}| \left(1 - \frac{|v_{drift}|}{|u_m|}\right) \tag{7}$$

Here, we define u_{bc} as the average velocity vector at the centre of a boundary cell, $v_{drift} = u_m - v_m$ is the drift velocity vector at the cloud's centre, u_m is the velocity vector of the flow at the cloud centre and v_m is the velocity vector of the cloud centre of mass, Figure 10.5.

As a cloud approaches a wall, we may compute the local velocity of particle using Equation (7). The necessary local particle velocity is one parameter to calculate the erosion that occurs as a consequence of particles impacting a solid wall. In addition to local particle velocity, erosion is a function of impact angle, α, the angle between a particle's trajectory and the direction normal to the target surface. Tabakoff *et al.* (1979), Oka *et al.* (1997) and Pandian (2004) have studied the erosion of ductile materials. Their experimental investigations indicated that particles are at their most erosive when impacting a solid wall at between 20 and 30 degrees. Tabakoff *et al.* (1979) developed an empirical model to characterise erosion rate (E_R) per unit mass of impacting particles. We define erosion rate as the milligrams of material eroded per square metre per gram of ingested particles per second. We may describe erosion rate as:

$$E_R = K_1 f_\alpha |v_{bc}|^2 \cos^2\alpha (1 - R_T^2) + f_i \tag{8}$$

with:

$$R_T = 1 - 0.0061 |v_{bc}| \sin \alpha \tag{9a}$$

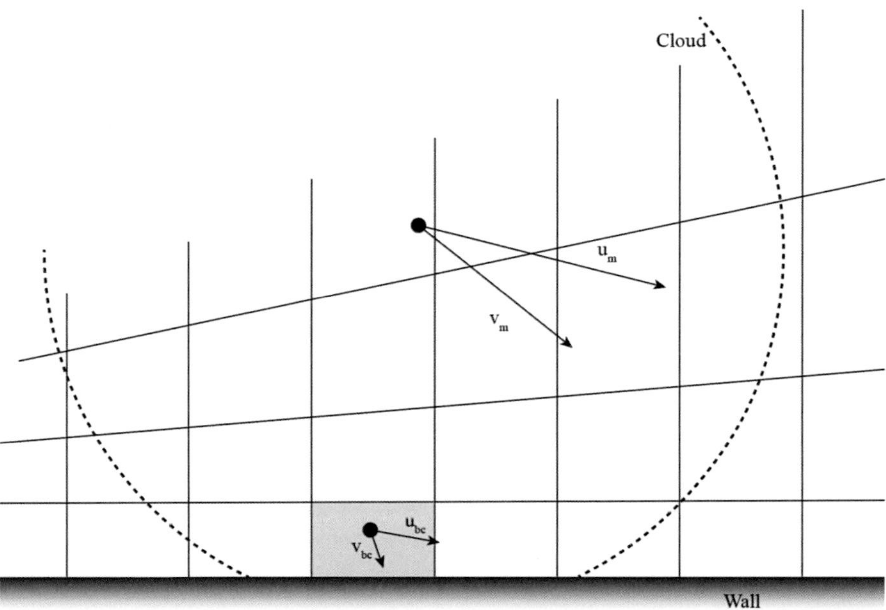

FIGURE 10.5. The particle cloud tracking scheme for characterising the cloud velocity distribution and consequently, the velocity of particles and angle (α) to a solid wall at impact.

$$f_\alpha = \left\{1 + K_C\left[K_{12}\sin\left(\frac{90\alpha}{\alpha_0}\right)\right]\right\}^2 \tag{9b}$$

$$f_\alpha = K_2(|v_{bc}|\sin\alpha)^4 \tag{9c}$$

Tabakoff *et al.* (1979) experimentally established values of the material specific coefficients for 304 grade stainless steel, Table 10.2. The studied fan's blades were actually manufactured from nodular iron, however Tabakoff *et al.* (1979) had not characterised nodular iron. The 304 grade stainless steel data was the closest available material data.

THE STUDIED FAN

We undertook the research presented in this chapter through the study of a 26 blade fan with C4 profile blades, Table 10.3. This fan is typical of those used in induced-draft application with a coal-fired power plant. The studied fan incorporates un-swept blades designed with a free vortex radial work distribution. The blade design utilises a C4 profile family. When numerically modelling the studied fan, we chose a blade tip stagger angle of 60 degrees, which is the point at which the fan achieves its peak efficiency. We selected this peak efficiency operating point as it is typical of the operating point at which the studied fan operates in-service.

Computational domain

Following Corsini *et al.*'s method (2012a), we undertook the blade-to-blade flow-field and particle transport numerical simulations using their own dedicated mesh. In both cases the computational mesh distributed 16 per cent of all elements upstream of the blade leading edge, 50 per cent in the blade passage and 34 per cent downstream of the blade trailing edge. We modelled the blade tip-to-casing gap with 55 grid nodes across the gap. In both cases the meshes were clustered toward solid boundaries, with the ratio of minimum grid spacing on solid walls to mid-span blade chord set at 7×10^{-4}. The adopted grid refinement toward the solid surfaces controlled the normalised wall distance, δ^+ value to about one at the first row of nodes.

We evaluated two tetrahedral computational meshes for the blade-to-blade flow-field numerical simulation. We constructed both meshes using a non-orthogonal body fitted coordinate system with fully unstructured topology. Mesh one consisted of approximately 4.5 million and mesh two had 4.3 million tetrahedral elements, Table 10.4. Mesh one resulted in the best agreement between predicted fan power at the peak efficiency operating point and experimentally measured fan power. Therefore, we used mesh one for all blade-to-blade flow-field numerical simulations, Figure 10.6.

Table 10.2. *Erosion model coefficients (Tabakoff et al., 1979).*

K_C	Impact angle $\alpha \leq 3\,\alpha_0$	1
	Impact angle $\alpha > 3\,\alpha_0$	0
K_1		1.505101×10^{-6}
K_{12}		0.296077
K_2		5.0×10^{-12}

Table 10.3. *The studied fan constitutes one fan from a family of fans intended for induced-draft application in a coal-fired power plant. The authors selected this example as typical of the family and used it as a basis for numerical simulation of the flow-field, particle impact frequency and erosion rate.*

Rotor data		
Blade number		26
Hub to casing diameter ratio D_{hc}		0.7
Blade tip diameter D_t (mm)		1,767
Blade tip clearance χ (% span)		6
Fan design speed (rpm)		890
Blade material		Nodular iron
Blade geometry	**Hub**	**Tip**
Blade chord c (mm)	487	446
Solidity σ	1.48	0.96
Stagger angle γ (degrees)	38	60
Camber angle θ (degrees)	45	36

Table 10.4. *Computational mesh data for the blade-to-blade flow-field numerical simulations and a comparison of the studied fan's experimentally measured and numerically predicted absorbed power at its peak efficiency operating point.*

	Mesh one	Mesh two	Experiments (Corsini *et al.*, 2012a)
Nodes	7.87×10^5	7×10^5	—
Elements	4.5×10^6	4.3×10^6	—
	Tetrahedral	Tetrahedral	
Power P (kW)	4,961.2	4,970.8	4,897.7
Error (%)	1.3	1.4	—

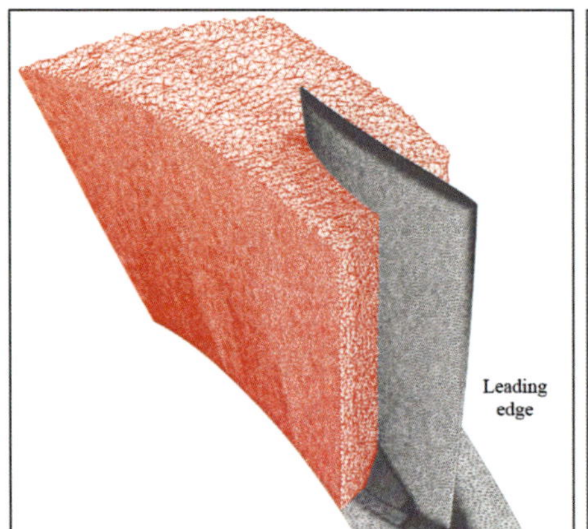

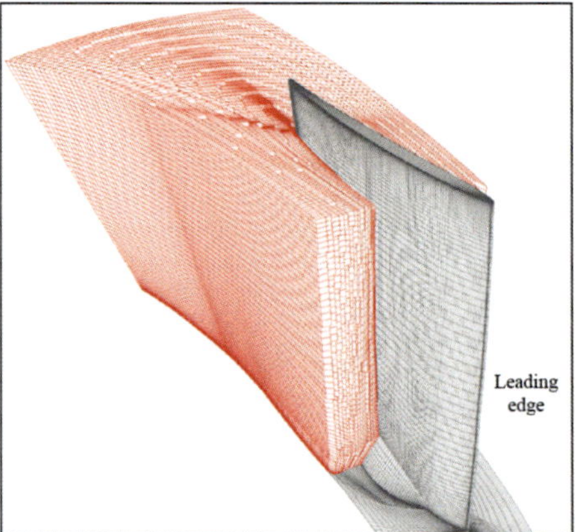

FIGURE 10.6. The tetrahedral computational mesh used for the numerical simulation of the fan blade-to-blade flow-field, left and the hexahedral computational mesh used for the numerical simulation of the particle motion, right.

We undertook the particle transport numerical simulation using a hexahedral computational mesh, Figure 10.6. We solved the Lagrangian particle transport equation with a non-orthogonal body-fitted coordinate system and a block-structured h-type topology. We constructed the mesh by merging a mesh in the blade-to-blade region with a second mesh embedded in the blade tip-to-casing gap. The mesh consisted of approximately 880,000 hexahedral elements.

Boundary and initial conditions

We defined the boundary conditions following Sheard *et al.*'s (2009) method. We defined inflow turbulence following Corsini and Rispoli's (2004) method. We defined the boundary conditions according to Corsini and Rispoli's (2004) recommendations. They studied the performance of ducted high-solidity fans. The Dirichlet conditions for the relative velocity components were imposed at the inflow section, half a mid-span chord upstream of the leading edge, as obtained from a flow simulation in an annular passage of identical hub-to-casing diameter ratio that included an upstream spinner cone. The turbulence intensity profile was a near-uniform six per cent across the annulus, increasing to ten per cent near the casing. We used the turbulent kinetic energy's pitch-wise profile to calculate the radial distribution of dissipation based on a dissipation length scale (l_ε) of one per cent of blade mid-span pitch. We completed the boundary conditions by assuming periodicity both up- and down-stream of the fan blades.

The particle cloud tracking (PCT) modelling approach requires as an input parameter the number and mass of particles. We estimated the distribution of particle number and mass from the analysis of fine and coarse particles, Figure 10.7. We normalised particle characteristics against those of a 0.75 μm particle to facilitate particle modelling in the numerical simulation, Table 10.5. We estimated the number of seeded particles assuming the particle load equals 64.1 milligrams of particles per cubic metre and a fan volume flow rate of 380 metres cubed per second. The fan volume flow rate is 380 metres cubed per second at the fan's peak efficiency operating point.

Following Tabakoff *et al.*'s method (1979), we modelled erosion without accounting for the actual shape of particles or the chemical composition. We used the same approach as Corsini *et al.* (2012a) to determine the initial particle positions along the blade span. Particle clouds entered the computational domain at five span-wise positions, with each cloud comprising one million individual particles. We then scaled the resulting particle distribution to match that actual particle concentration estimated from the assumed particle load and fan volume flow rate.

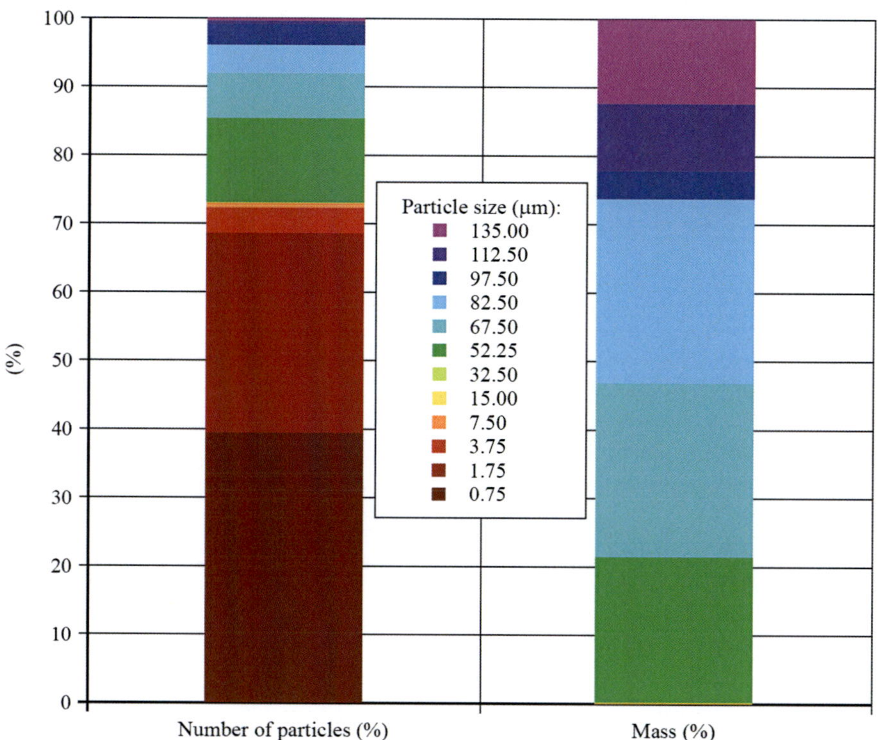

FIGURE 10.7. The particle cloud tracking (PCT) modelling approach requires as an input parameter the number and mass of particles. We estimated the distribution of particle number and mass from the analysis of fine and coarse particles within a dust sample taken from an induced draft fan installed in a typical coal-fired power plant.

Table 10.5. *Particle characteristics normalised against those of a 0.75 μm particle to facilitate the modelling of particles in the numerical simulation. Normalised particle mass is defined as particle mass divided by the mass of a 0.75 μm particle. Normalised mass flow is defined as particle mass flow rate divided by the mass flow rate of a 0.75 μm particle. Normalised particle flow is defined as the number of particles divided by the number of 0.75 μm particles.*

Average size (μm)	Normalised particle mass	Normalised mass flow	Normalised particle flow
0.75	1.00	1.00	1.00
1.75	1.27×10^{1}	9.35	7.36×10^{-1}
3.75	1.25×10^{2}	1.17×10^{1}	9.34×10^{-2}
7.5	1.00×10^{3}	1.56×10^{1}	1.56×10^{-2}
15	8.00×10^{3}	1.01×10^{1}	1.27×10^{-3}
32.5	8.14×10^{4}	1.34×10^{1}	1.65×10^{-4}
52.25	3.38×10^{5}	1.05×10^{5}	3.12×10^{-1}
67.5	7.29×10^{5}	1.28×10^{5}	1.76×10^{-1}
82.5	1.33×10^{6}	1.33×10^{5}	9.98×10^{-2}
97.5	6.86×10^{6}	2.14×10^{4}	3.13×10^{-3}
112.5	3.37×10^{6}	4.81×10^{4}	1.43×10^{-2}
135	5.83×10^{6}	6.27×10^{4}	1.08×10^{-2}

RESULTS

Corsini *et al.* (2011) conducted a numerical study into the aerodynamic impact of an eroded blade leading edge. Their focus was on the aerodynamic consequence of erosion rather than the erosive process itself. Despite this caveat, Corsini *et al.* (2011) did conclude that regions of separated flow and vortexes play an important role in the erosive process. Both affect the velocity and angle with which particles impact and therefore have a primary effect on the resulting pattern of erosion over both the blade pressure and suction surfaces. When studying the same fan that is the subject of the research reported in this chapter, Corsini *et al.* (2011) predicted the flow-field through the blade-to-blade passage. They studied three-dimensional streamlines, with the blade surfaces coloured using static pressure contours, Figure 10.8. When one studies the three-dimension streamlines and static pressure contours, a separated flow region is evident on the pressure surface in the hub region near the blade leading edge. The separated flow region results in near-hub fluid centrifuging up the blade pressure surface, having a primary impact of the resulting pressure surface erosion pattern.

We may characterise the impact of near-hub fluid centrifuging up the blade pressure surface by numerically predicted particle cloud trajectories, Figure 10.9. To differentiate between particle trajectories and flow-field streamlines, we presented particle trajectories as a series of white spheres and flow-field streamlines as orange lines. Particles were driven up the blade pressure surface until the separated flow reattached, following which the particle trajectories became almost axial and then

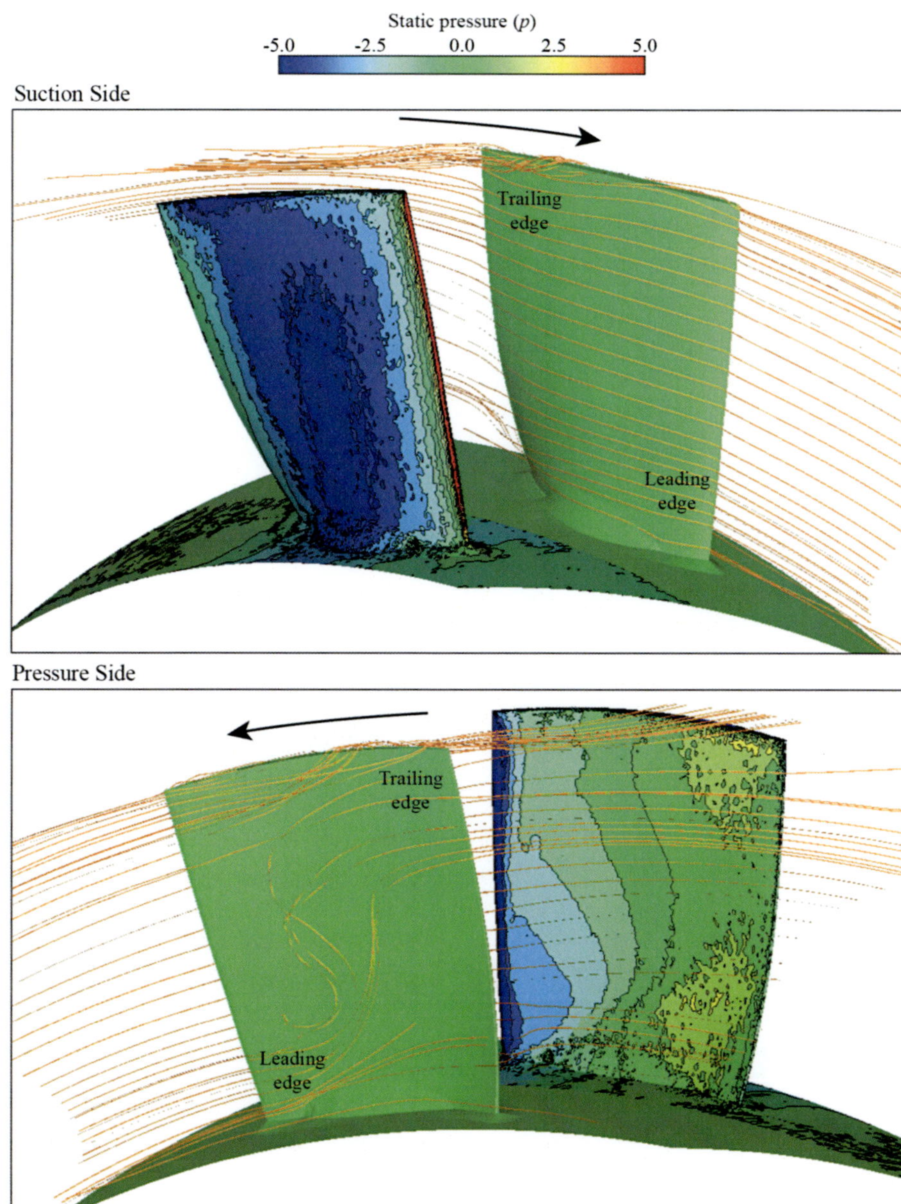

FIGURE 10.8. Three-dimensional streamlines through the blade-to-blade passage, with the blade surfaces coloured using static pressure contours. Corsini *et al.* (2011) numerically predicted both using a tetrahedral computational mesh. Results are shown for the studied fan blade suction's surface, top and pressure surface, bottom.

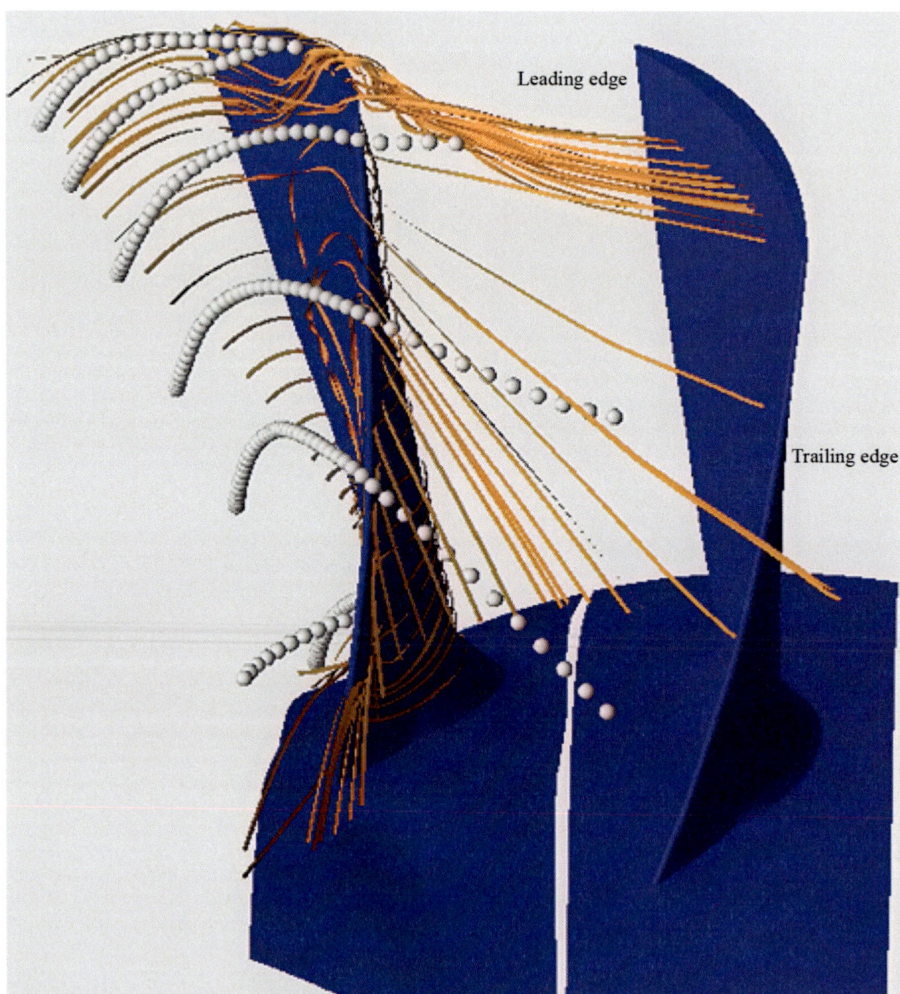

FIGURE 10.9. Numerically predicted particle cloud trajectories, shown as spheres and flow-field streamlines, shown as orange lines illustrating the extent to which particle trajectories are influenced by secondary flow features in the blade-to-blade passage.

impacted on the blade trailing edge. In the blade tip region the particle trajectories drew into the blade tip-to-casing leakage vortex. We may expect that these two phenomenon would result in elevated erosion rates over the blade trailing edge in the tip region and generally through the blade tip-to-casing gap.

Blade erosion by particle size classes

Predicting erosion rate is a two-step process. First, we use the predicted blade-to-blade flow-field prediction as input to our particle cloud tracking algorithm.

Second, we use Tabakoff *et al.*'s (1979) empirical model to predict erosion rate (E_R) per unit mass of impacting particles. However, the resulting prediction of erosion rate is dependent on particle size. The larger a particle, the more significantly its trajectory is influenced by centrifugal force as it passes through the blade-to-blade passage. We may characterise the effect of particle size by predicting the trajectory of different size particles, Figure 10.10. In this numerical simulation, particles from 0.75 μm to 15.00 μm pass over the blade pressure surface whilst particles from 32.50 μm to 135 μm pass over the blade suction surface. The effect of centrifugal force on particles becomes significant for the particles over 30.00 μm. The particles over 100.00 μm are effectively driven radially outward onto the fan casing before they enter the blade-to-blade passage. This is a reason why larger particles play relatively little part in blade pressure and suction surface erosion.

As previously mentioned, predicting erosion rate is a two-step process. First, we use the predicted blade-to-blade flow-field prediction as input to our particle cloud tracking algorithm that we use to predict the particles' impact frequency. We may use a prediction of impact frequency to characterise the effect of particle size on impact frequency. Therefore, we used the particle cloud tracking algorithm to predict the impact frequency over both the blade pressure and suction surface for particles of 0.75, 15.00 and 52.50 μm diameter, Figure 10.11, and 82.50, 112.50 and 135.00 μm diameters, Figure 10.12. These six particle sizes represent over 70 per cent of the total mass of all particles.

The predicted impact frequency over both the blade pressure and suction surface is a strong function of particle size. It is apparent that particles below 50 μm diameter impact on every part of the blade pressure and suction surface. Therefore,

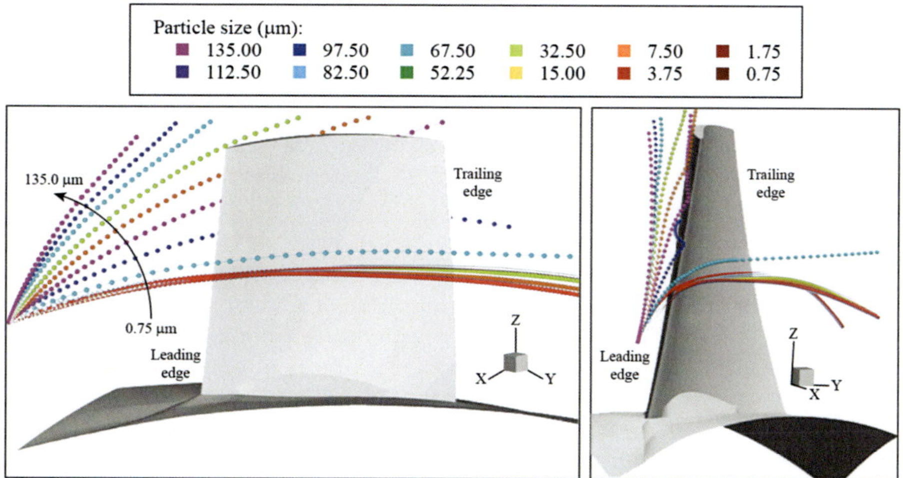

FIGURE 10.10. The numerically predicted particle cloud trajectories for each size of particle. The distribution of trajectories indicates that fine particles introduced at the leading edge migrate over the blade pressure surface at a near-constant span-wise plane. In contrast, coarse particles migrate both to the blade suction surface and blade tip region as they pass through the blade-to-blade passage.

we may regard these smaller particles as primary drivers of the erosive process. The smallest 0.75 and 15.00 μm particles result in impact frequency patterns that are essentially similar over both the blade pressure and suction surface, Figure 10.11. This indicates that these smaller diameter particles are driven primarily by the direction of fluid-flow and not by inertial forces. Therefore, their motion is strongly affected by regions of secondary flow and vortices. As we would expect, we see evidence of high impact frequency at the blade pressure surface leading edge associated with the separated flow in this region. We also see evidence of high impact frequency at the blade pressure surface trailing edge as a consequence of the hub separation.

In the blade tip region the 0.75 and 15.00 μm diameter particle impact frequencies are higher over the blade suction surface than the blade pressure surface. We may associate these smaller diameter particles with the blade tip-to-casing leakage vortex as it is only the smaller particles that are able to follow the flow-field streamlines within the vortex. They are transported by the flow-field from the blade leading edge to the suction's surface trailing edge where they impact on the blade surface. As particle size increases to 52.50, 82.50, 112.50 and finally, 135.00 μm diameter, the impact frequency is increasingly concentrated over the outer 50 per cent of the blade-span. This concentration is a direct consequence of the centrifugal force acting on the particles. By the time the particles reach 135.00 μm diameter, they impact only over the outer 40 per cent of the blade pressure surface and a small proportion of the blade suction surface. Therefore, we may conclude that the contribution of larger particles to the erosive process reduces as they are driven onto the fan casing.

We may illustrate the overall effect of the distribution of particle sizes that we used in this research by considering the resultant impact frequency distributions. Consider the blade pressure surface, Figure 10.13. There are two regions of the blade pressure surface area (A) and suction surface area (F) that particle impact does not affect. However, particle impact affects the majority of both the blade pressure and suction surface. The pressure surface region most affected by particle impact is the outer 20 per cent of the leading edge blade span, area (B). There is a second region of the blade pressure surface near the blade tip at approximately 66 per cent blade span that is affected by particle impact, area (C). We may associate this second region with an effect of the blade tip-to-casing leakage vortex. There are two other pressure surface regions affected by particle impact, a region close to the trailing edge, area (D), and a region close to the hub, area (E). There is a separated flow region on the pressure surface in the hub region near the blade leading edge. This results in near-hub fluid centrifuging up the blade pressure surface. We may associate area (D) and area (E) with the effect of the separated flow region on the pressure surface in the hub region.

Consider the blade suction surface, Figure 10.13. In contrast to the pressure surface particle, the impact is most concentrated in the blade tip region, area (G). There are also elevated particle impact levels along the entire blade-span at the leading edge, area (H), and in a near-hub region close to the blade trailing edge, area (I). The mechanisms responsible for particle impact in these three regions will be the

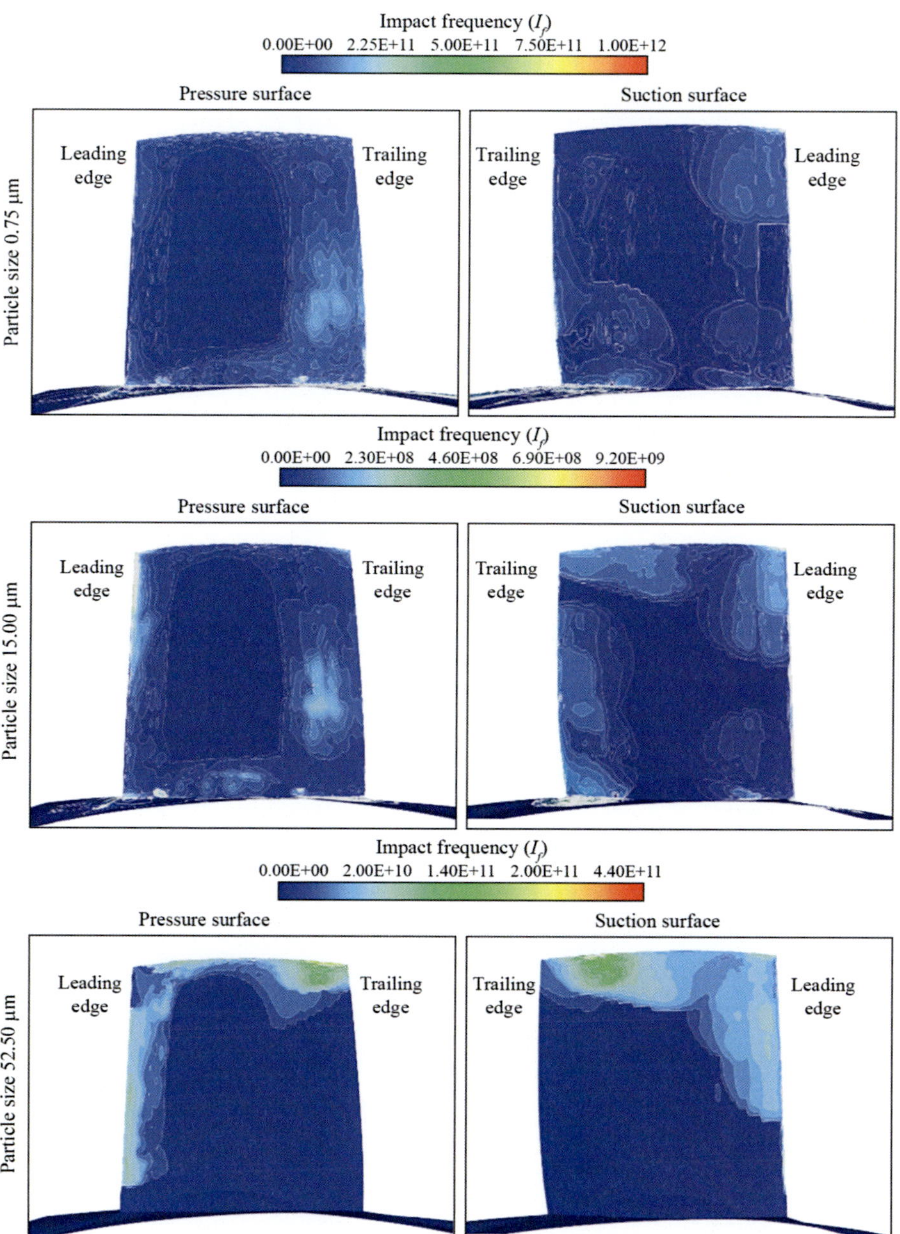

FIGURE 10.11. Predicted Impact frequency I_f distributions over the blade pressure surface, left and suction surface, right for 0.75 μm, 15.00 μm and 52.50 μm diameter particles. Impact frequency is defined as the number of impacts per square metre per gram of ingested particles.

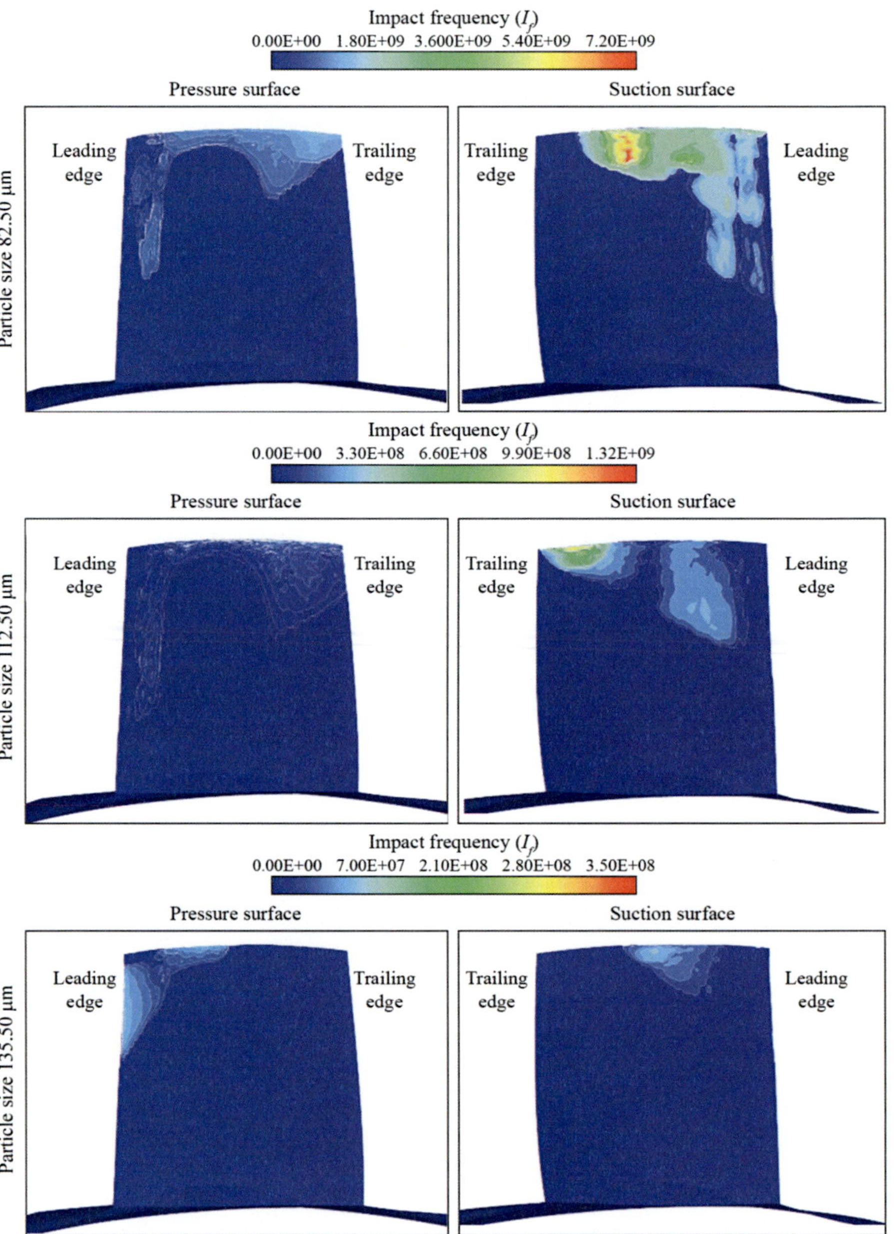

FIGURE 10.12. Predicted Impact frequency I_f distributions over the blade pressure surface, left and suction surface, right for 82.50 µm, 112.50 µm and 135.00 µm diameter particles. Impact frequency is defined as the number of impacts per square metre per gram of ingested particles.

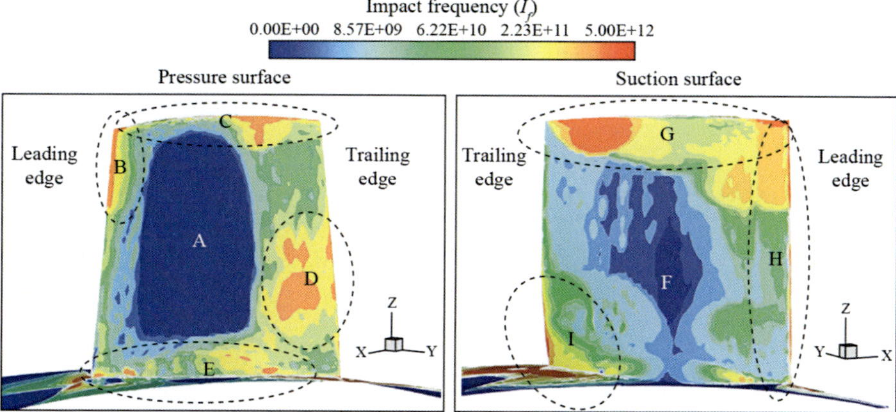

FIGURE 10.13. Impact frequency I_f regions predicted over the blade pressure surface, left and suction surface, right. The authors derived the regions from the predictions of impact frequency for each particle size, combined in proportion to the in-service distribution of particle size.

same as those at play over the pressure surface. We may associate the elevated particle impact levels in the blade tip region, and particularly the concentration of impact towards the trailing edge, with an effect of the blade tip-to-casing leakage vortex. We may associate the elevated particle impact levels along the blade leading edge with the particles impacting the blade stagnation line. The elevated particle impact level in the near-hub region close to the blade trailing edge is also probably a consequence of a secondary flow-feature in the hub region.

We used predicted particle impact frequency as an input into our erosion model to predict erosion rate distributions over the blade pressure and suction surface. As with particle impact frequency, we predicted erosion rate for 0.75 μm, 15.0 μm, 52.5 μm, 82.5 μm, 112.5 μm and 135.0 μm diameter particles, Figures 10.14 and 10.15. Erosion rate is related to impact frequency and therefore we would expect predicted impact frequency distributions and erosion rate distributions to show some similarity. However, as erosion rate is a function of impact frequency, impact velocity and impact angle predicted erosion rate distributions are not the same as predicted impact frequency distributions.

Consider the erosion rate distributions for the smaller particles, Figure 10.14. The 0.75 μm diameter particles are primarily responsible for erosion in the blade trailing edge region and in the near-hub region. These small diameter particles are less affected by inertia effect, moving more closely with the fluid-flow streamlines. Consequently, they follow fluid-flow streamlines through secondary flow features. In contrast to the 0.75 μm particles, the 15.0 μm diameter particles result in an erosion rate distribution that is more concentrated over the blade leading edge specifically and the blade tip region generally. This trend continues with the move from 15.0 μm diamter particles to the 52.5 μm diameter particles. The increasing effect of inertia on the particles drives this trend.

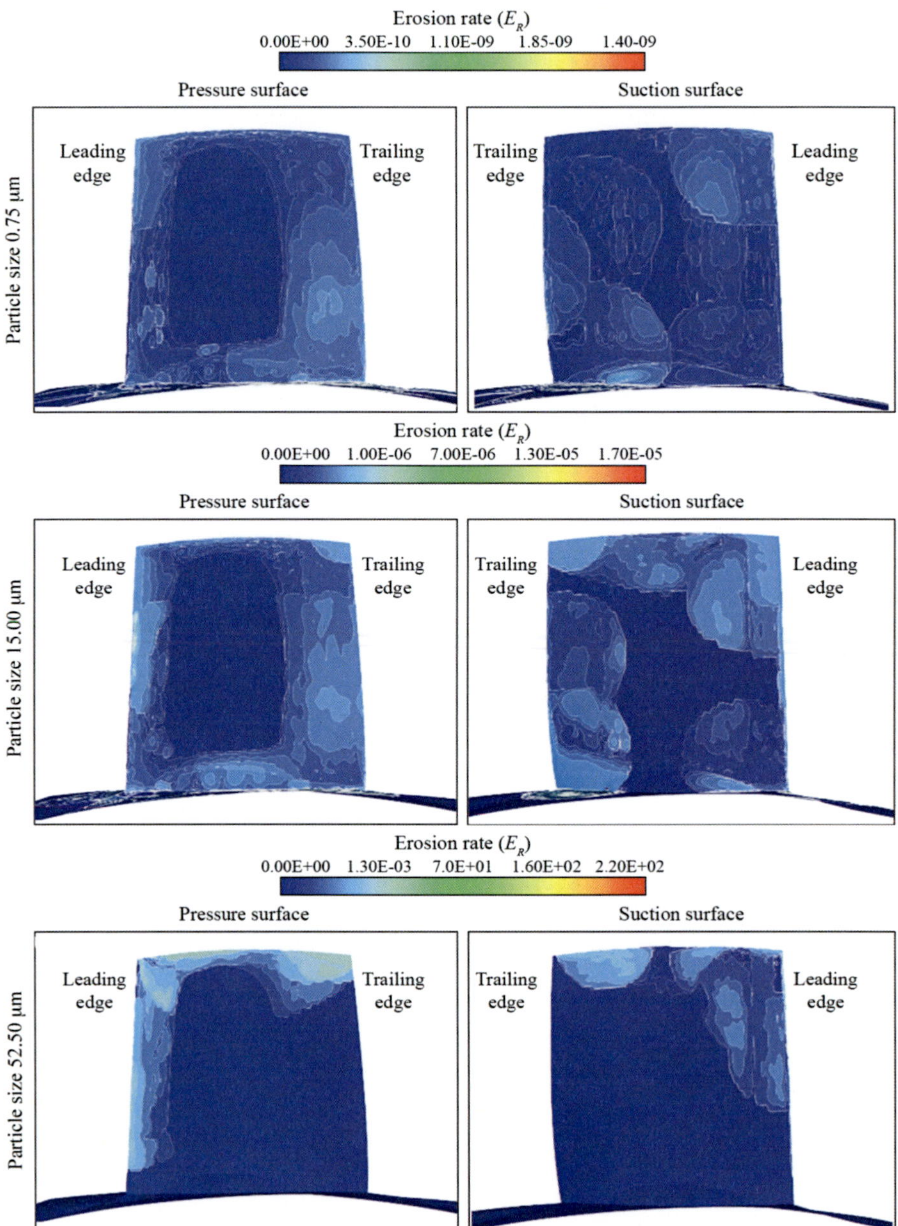

FIGURE 10.14. Predicted erosion rate E_R distributions over the blade pressure surface, left and suction surface, right for 0.75 μm, 15.0 μm and 52.5 μm diameter particles. Erosion rate is defined as the milligrams of material eroded per square metre per gram of ingested particles per second.

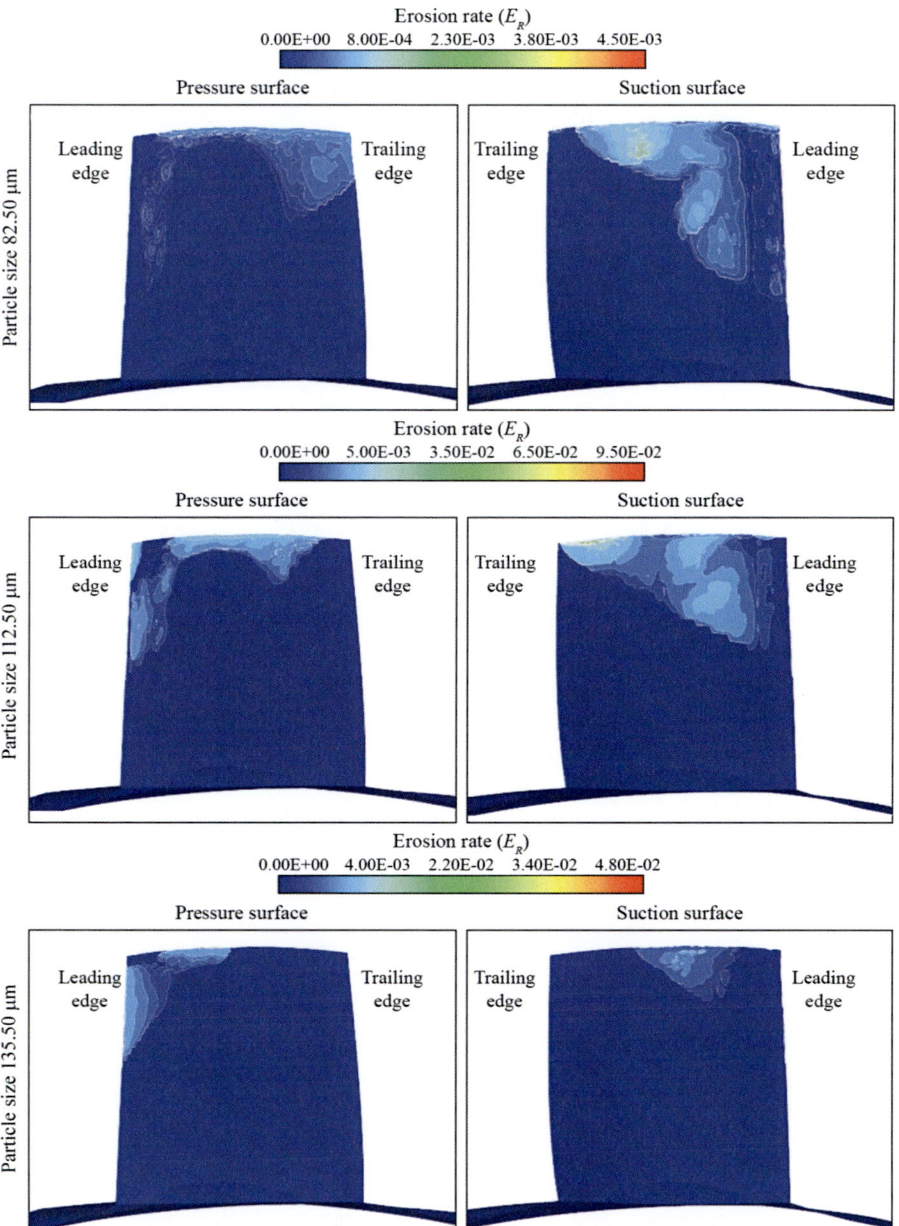

FIGURE 10.15. Predicted erosion rate E_R distributions over the blade pressure surface, left and suction surface, right for 82.5 μm, 112.5 μm and 135.0 μm diameter particles. Erosion rate is defined as the milligrams of material eroded per square metre per gram of ingested particles per second.

Consider the erosion rate distribution for the larger particles, Figure 10.15. As particle size increases from 82.5 μm to 112.5 μm and finally 135.0 μm, the erosion rate reduces. Large particles are not less erosive than small particles. Quite the contrary, a larger particle impacting a point with the same velocity and angle of impact as a smaller particle will result in a higher erosion rate. The reducing contribution to erosion rate with increasing particle size is a consequence of inertial effects dominating the trajectory of these larger particles. Centrifugal force results in larger particles driving outward and onto the fan casing. Once on the fan casing, they play no role in the erosion of either the blade pressure or suction surface.

We may illustrate the overall effect of the distribution of particle sizes used in this research by considering the resultant erosion rate distributions, Figure 10.16. We may clarify the link between erosion rate distributions and impact frequency distributions by considering impact frequency distributions, Figure 10.13.

Consider the impact frequency distribution over the blade pressure surface, Figure 10.13. The pressure surface region most affected by particle impact is the outer 20 per cent of the leading edge blade span, area (B). Consider the erosion rate distribution over the blade pressure surface, Figure 10.16. The high impact frequency in area (B) does not result in a similar peak in erosion rate. In contrast to the impact frequency, the blade leading edge erosion rate is characterised by an elevated erosion rate region at the blade mid-span, area (L). The impact frequency distribution in the hub region, area (E) is characterised by small regions of elevated impact frequency, increasing towards the blade trailing edge. In contrast, the erosion rate in the same region is elevated at the blade leading edge and reduces towards the trailing edge.

Consider the impact frequency distribution over the blade suction surface, Figure 10.13. There are elevated particle impact levels in a near-hub region close to

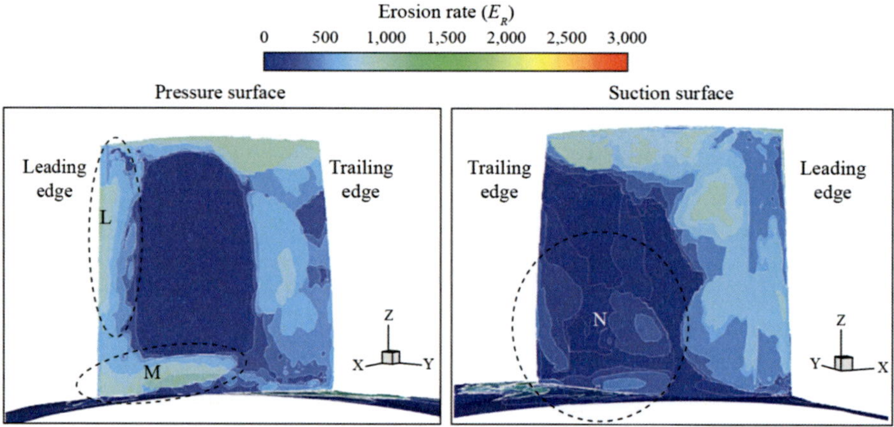

FIGURE 10.16. Erosion rate E_R regions predicted over the blade pressure surface, left and suction surface, right. The regions are derived from the predictions of erosion rate for each particle size, combined in proportion to the in-service distribution of particle size.

the blade trailing edge, area (I). Consider the erosion rate distribution over the blade suction surface, Figure 10.16. The high impact frequency in area (N) does not result in a similar peak in erosion rate. Therefore, we may conclude that the regions of high erosion rate, area (L), (M) and (N), are not associated with high impact frequency. In contrast, the high impact frequency regions, area (C), (D) and (G) do correlate with regions of high erosion rate.

LEADING EDGE EROSION BY PARTICLE SIZE CLASSES

The erosion rate distributions over the blade pressure and suction surface represent the overall effect of the distribution of particle sizes that we used in the research. It is this overall erosion rate distribution that is needed to identify those blade areas most prone to erosion when subjected to a specific combination of particles sizes. We ultimately use these overall erosion rate distributions to estimate the loss of blade mass with time. Industrial fan manufacturers typically specify the maximum allowable erosion in critical blade regions and the maximum allowable loss of blade weight due to erosion. Therefore, it is helpful to be able to predict both.

The erosion of an axial fan blade has an impact on fan performance. Erosion of all blade regions effect performance adversely. However, one region has a particularly detrimental effect on fan performance. Erosion of the blade leading edge results in a loss of blade chord and camber. In combination, the loss of blade chord and camber result in reduced fan pressure developing capacity. Additionally, destruction of the blade leading edge geometry results in reduced fan efficiency. Therefore, we have a particular interest in the erosive process in the blade leading edge region.

To facilitate further insight into the blade leading edge erosive process we calculated the contribution to mass removed of each particle size. The total mass removed is always 100 per cent, with the per cent split between the particles illustrating the relative importance of each particle size. We calculated the contribution to mass removed of each particle size at four blade span-wise sections, S1, S2, S3 and S4, Figure 10.17. Having previously predicted erosion rate (E_R), we may express the mass (m) of material removed by erosion from surface of length a and width b as:

$$m = E_R \cdot ab \tag{10}$$

We may also express the mass of material removed as:

$$m = \rho_s V_{er} = \rho_s \cdot abh_e \tag{11}$$

where V_{er} is the volume of eroded material, ρ_s is the density of surface material and h_e is the thickness of the eroded volume. By combining Equations (10) and (11), we may express the thickness of the eroded leading edge as:

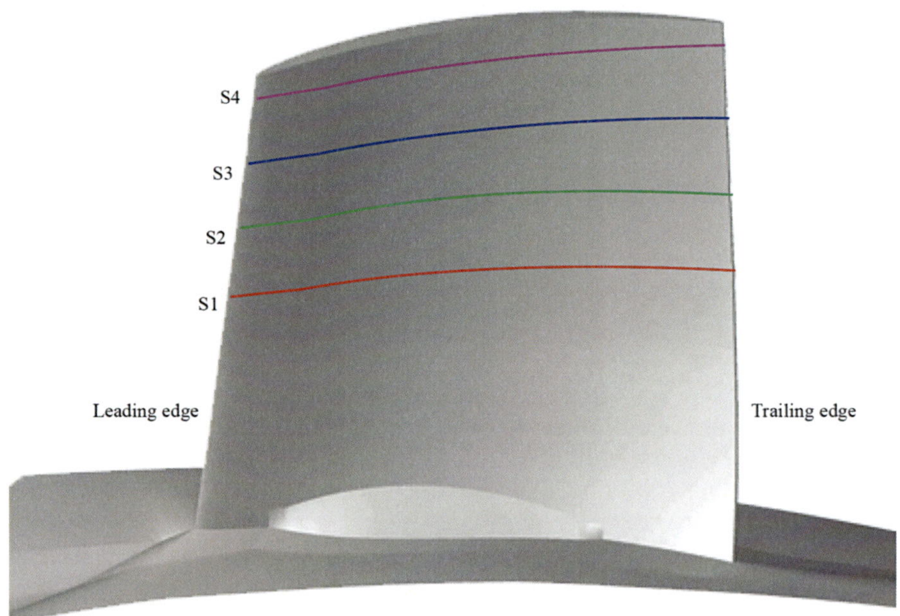

FIGURE 10.17. The four blade span-wise sections, S1, S2, S3 and S4 used for an analysis of numerically predicted mass removal due to particle erosion. An inspection of blades removed from service indicated that the blade leading edge region was most susceptible to the effect of erosive particles and therefore our analysis focused on the blade leading edge region.

$$h = \frac{ER}{\rho_s} \tag{12}$$

When presenting the thickness of the eroded leading edge, we use a normalised angular coordinate system, defined along the blade span-wise sections, Figure 10.18. This normalised angular coordinate system facilitates the presentation of erosion against angular position. A positive angular position refers to a position on the blade pressure surface and a negative angular position refers to a position on the blade suction surface.

Consider the split between the mass removed by each particle size at the blade span-wise section S1 located at approximately the blade mid-span, Figure 10.19. The 52.5 μm diameter particles make the largest contribution to erosion for angular positions between –8 to –1 degrees. This particle size accounts for between 60 and 90 per cent of all erosion over the angular position range. As we approach the stagnation point, an angular position of zero, the contribution to erosion of the 52.5 μm diameter particles drops to 10 per cent, and then rapidly rises to over 95 per cent. In contrast, the 67.5 μm diameter particles contribution to erosion rises to 90 per cent at the stagnation point. When we study the combined contribution to erosion of the 52.5 μm and 67.5 μm diameter particles, it is apparent that these two particle

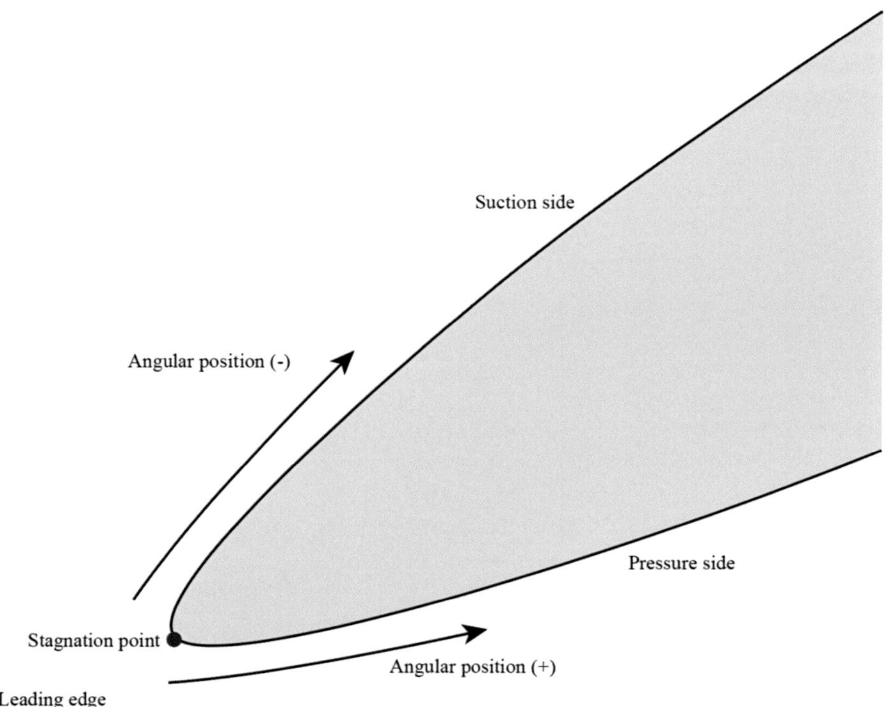

FIGURE 10.18. The normalised angular coordinate system defined along the blade span-wise sections to facilitate the analysis of numerically predicted mass removal due to particle erosion.

sizes are responsible for almost 100 per cent of the erosion at the blade leading edge. The only exception is a relatively minor 15 per cent contribution of the 32.5 μm diameter particles for angular positions between the stagnation point and +1 degrees.

Consider the split between the mass removed by each particle size at the blade span-wise sections S2, S3 and S4, Figures 10.20, 10.21 and 10.22. With increasing blade-span, the blade-span contribution to erosion of the 67.5 μm size particles decreases. This trend is particularly evident at section S3 where erosion is almost completely dominated by 52.5 μm size particles. Section S4 erosion is also dominated by the 52.5 μm size particles, with minor contributions to erosion from the 67.5 μm, 82.5 μm and 87.5 μm size particles. The observation that larger particles are less involved in the erosive process is self-consistent with their susceptibility to inertial effects. By centrifuging onto the fan casing they no longer result in blade erosion. There is one anomaly in the section S4 split between mass removed by each particle size. The 112.5 μm size particles account for almost 100 per cent of the erosion between and angular position of +0.5 and +1 degrees. Despite our caveat about the 112.5 μm size particles between an angular position of +0.5 and +1 degrees, the erosive process is primarily driven by particles between 32.5 μm and 67.5 μm.

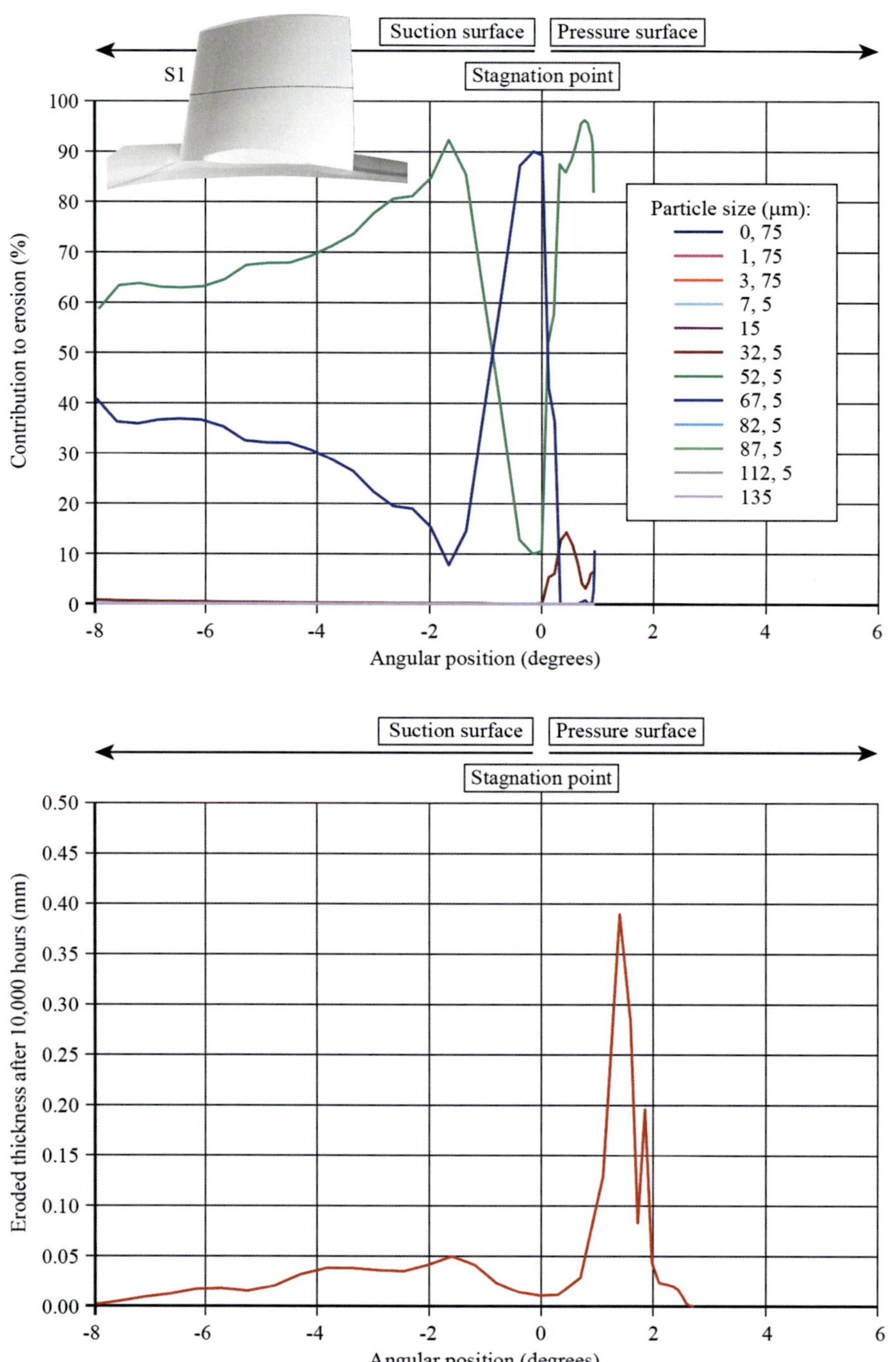

FIGURE 10.19. Predicted eroded thickness of the blade leading edge at section S1 and the contribution to the erosion of each size of particle. The prediction of eroded thickness is for 10,000 operating hours.

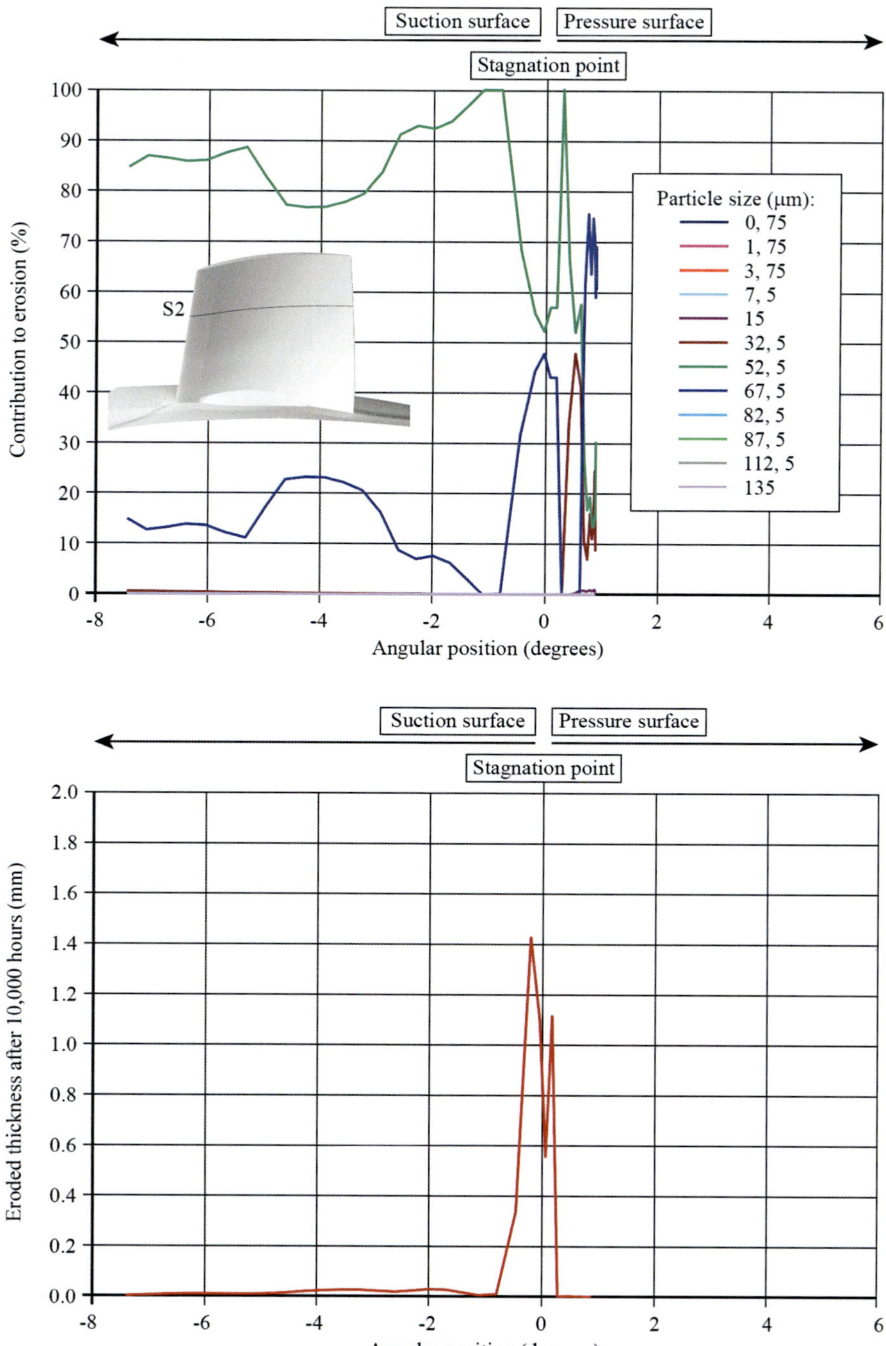

FIGURE 10.20. Predicted eroded thickness of the blade leading edge at section S2 and the contribution to the erosion of each size of particle. The prediction of eroded thickness is for 10,000 operating hours.

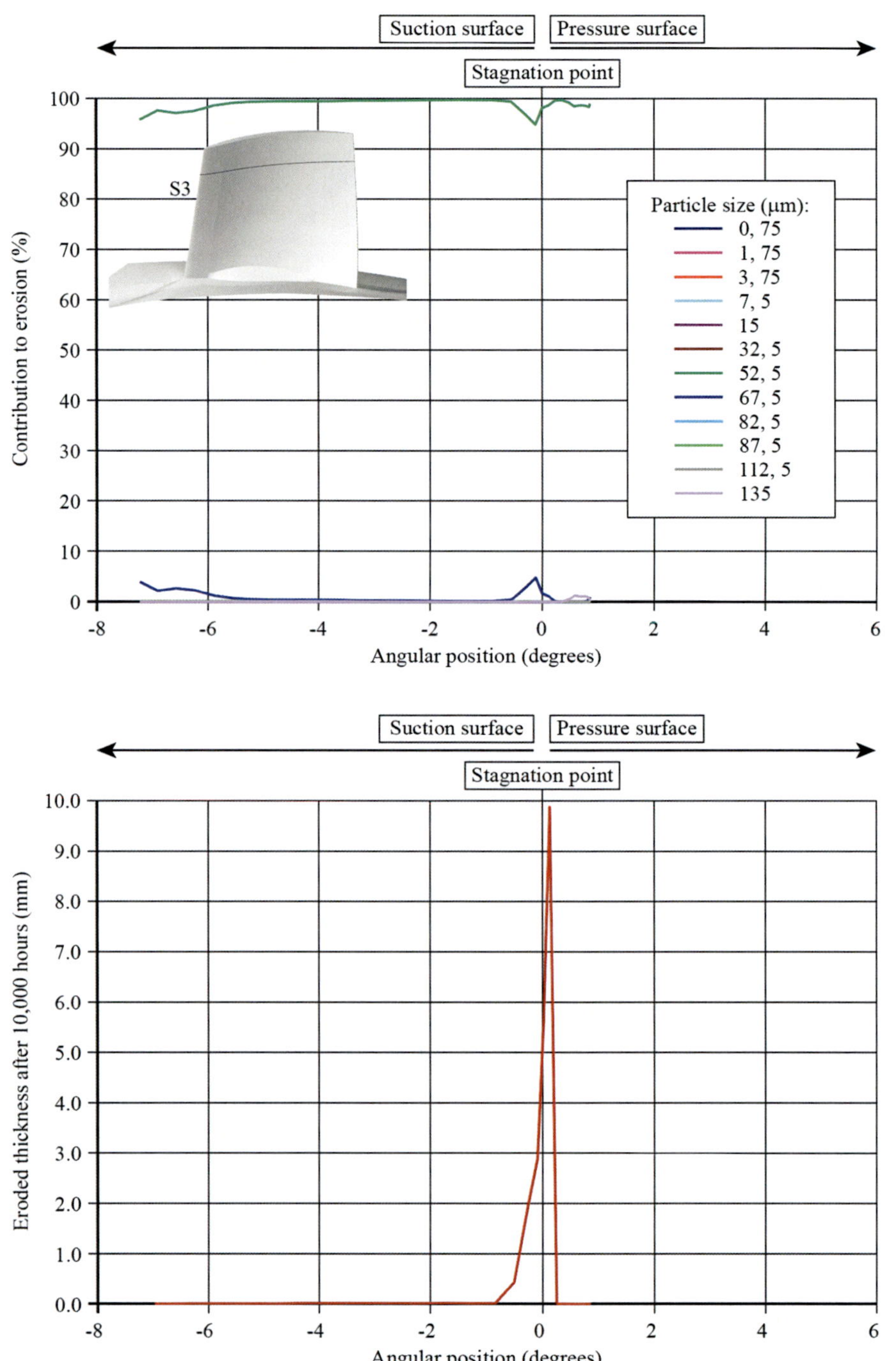

FIGURE 10.21. Predicted eroded thickness of the blade leading edge at section S3 and the contribution to the erosion of each size of particle. The prediction of eroded thickness is for 10,000 operating hours.

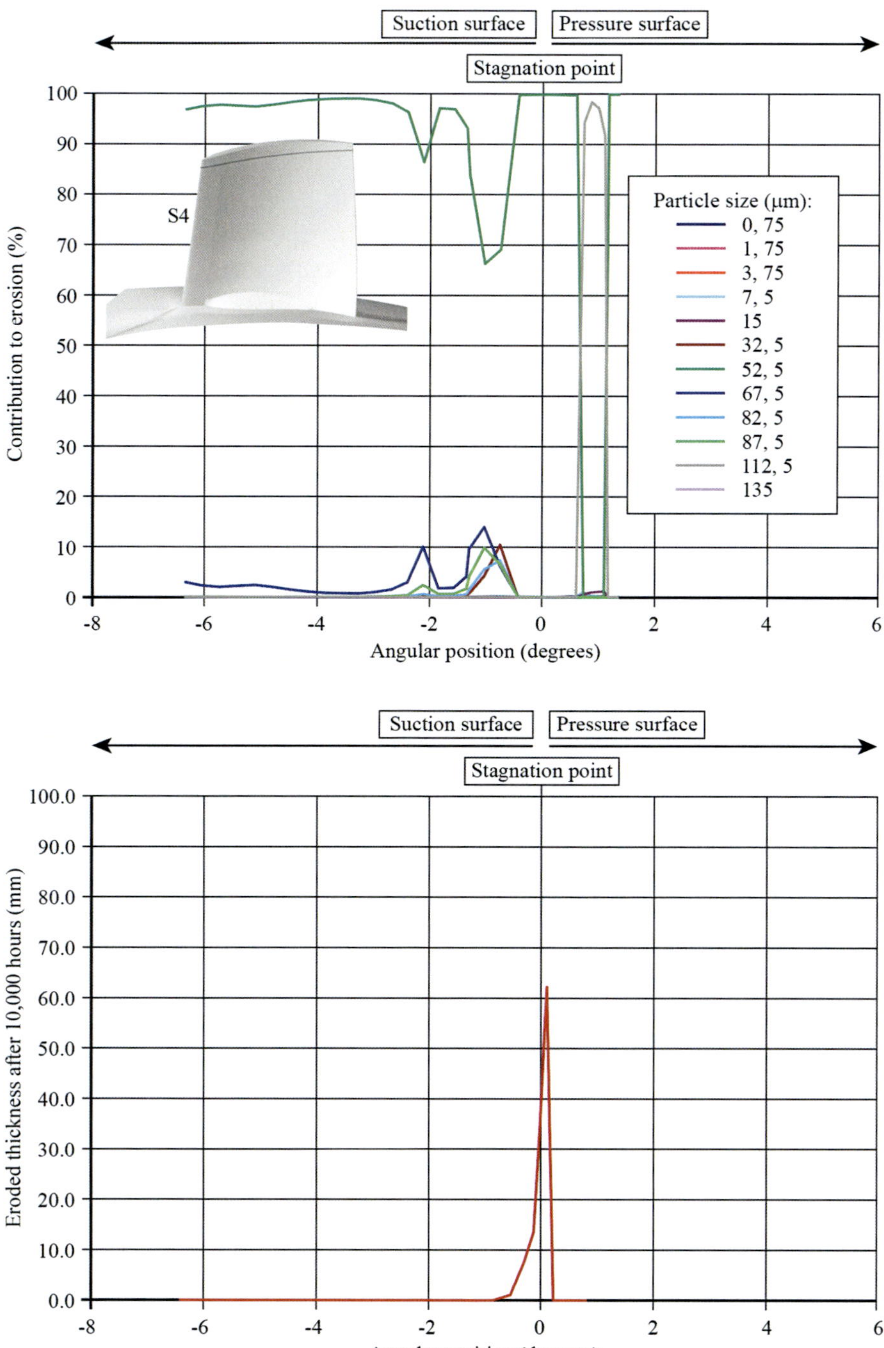

FIGURE 10.22. Predicted eroded thickness of the blade leading edge at section S4 and the contribution to the erosion of each size of particle. The prediction of eroded thickness is for 10,000 operating hours.

Consider the eroded thickness of blade leading edge material at the blade span-wise sections S1, S2, S3 and S4, Figures 10.19, 10.20, 10.21 and 10.22. The change in eroded thickness with angular position is similar at each section. However, the thickness or eroded material increases as we move from the blade mid-span (section S1) to the blade tip (section S4). We may attribute this increase to the increasing velocity of blade sections with increasing blade radius. This observation is self-consistent with Page *et al.*'s (1979) conclusions. They studied the erosion of induced draft fan blades. Page *et al.* (1979) concluded that erosion rate was proportional to particle velocity raised to a power of between 1.99 and 5.00. The actual value was dependent on the percentage of silica and quartz in the fly-ash passing through the fan. When we study the eroded thickness at each of the studied sections, we see a peak at section S2. Therefore, we may conclude that section S2 represents a peak combination of particle velocity and number of erosive particles. At section S3 and S4 the eroded thickness reduces indicating that the proportion of particles centrifuged onto the fan casing has a greater effect than the increasing blade velocity.

Blade deterioration prediction

We undertook the research reported in this chapter following the unscheduled shutdown of a coal-fired power plant incorporating the studied induced draft fans. The power plant was shut down after 10,382 in-service hours as the induced draft fan performance had fallen so far that the power plant boiler was no longer able to operate at its design duty point. Therefore, we may define the loss of blade mass due to erosion after 10,382 in-service hours as 100 per cent of the actual blade life.

We measured the weight of the blades removed from in-service operation after 10,382 hours. We then used our predicted erosion rate over the blade pressure and suction surface to predict the time it would take to erode the same mass of material from a blade. We accomplished the prediction of time to erode the same mass of material from a blade first, by predicting the average erosion rate. Our predicted average erosion rate was approximately 4.74×10^{-8} kilograms of material eroded per square metre per gram of ingested particles per second. Following Smith's (1993) method, we used this average erosion rate to predict time to erode the same mass of blade material as 11,628, Table 10.6.

We considered the comparison between measure in-service time and predicted time reasonable, with an error between the two of approximately 11.5 per cent. It is noteworthy that the prediction of erosion rate does not account for the effect of changing blade geometry over time with blade wear. The predicted number of hours to remove 100 per cent of the eroded mass was higher than the measured in-service hours. It is possible that if the change in blade geometry with erosion had been modelled, the resulting prediction of time to erode 100 per cent of the eroded mass would have been closer to the in-service measurement. An alternative reason for the discrepancy may simply be that our estimate of particle size distribution was correct for the sample we analysed, but that sample may not have been representative of the average particle size distribution over the 10,382 in-service operating hours. Lastly,

Table 10.6. *Blade deterioration rate when subject to erosion, with 100 per cent wear defined as the wear of blades removed from service after 10,382 in-service hours. These blades were removed from service as the fan was no longer able to operate at its design point. The authors considered the agreement between predicted and measured 100 per cent blade life reasonable.*

Blade wear (%)	Predicted blade life (hours)	Measured blade life (hours)
25%	2,907	—
50%	5,814	—
100%	11,628	10,382
150%	17,442	—

erosion is a function of temperature and we did not account for the variation in temperature of the gas passing through the induced draft fan. We installed this particular induced draft fan in a coal-fired power plant boiler system that operated in a region prone to extreme temperatures. In summer the ambient temperature peaked at 50°C and in winter at –40°C. This variation in ambient temperature may have resulted in a variation in temperature of the gas passing through the induced draft fan significantly enough to affect erosion rate.

We extended our analysis of blade wear to include a prediction of the time it takes to erode 25, 50 and 150 per cent of the measured eroded mass after 10,382 in-service operational hours, Table 10.6. The predicted time to erode 50 per cent of the measured eroded mass is 5,814 hours. Although somewhat subjective, we estimated that the ability of the induced draft fan to operate was relatively unaffected by erosion up to 50 per cent of the measured eroded mass. After 50 per cent, the erosion impact became progressively more problematic. If we chose to define the time taken to erode 50 per cent of the blade mass as the time at which the blades should be replaced with new blades, the predicted time of 5,814 hours is significant. Coal-fired power plants typically schedule one maintenance interval per year. With approximately 8,800 hours a year, that implies that blade life must be at least 8,800 hours to operate from one scheduled maintenance outage to the next. An actual in-service life of 5,814 hours is significantly less than that required. This indicates that a more erosion tolerant blade design is required.

CONCLUSIONS

This chapter reported on a numerical investigation into the erosive processes at play over the blade pressure and suction surface of an induced draft fan. We used a finite element based Reynolds-averaged Navier–Stokes solvers to model the blade-to-blade flow-field. We used a Lagrangian particle tracking solver to predict particle trajectories. We used particle impact frequency, velocity, angle and the eroded surface physical properties as input to an erosion model.

We took a sample of particles from an induced draft fan operating in an erosive environment and measured the distribution of particle size. We characterised the distribution of particle size and then went on to use this characterisation to study the effect of particle size. We found that particle size had a significant impact on the resulting erosion patterns over the blade pressure and suction surface. We conclude that changing particle size results in erosion over different blade pressure and suction surface regions.

The blade pressure surface proved to be most susceptible to erosion in the blade tip and leading edge regions and over a region close to the trailing edge at approximately blade mid-span. Further, there is a region in the centre of the blade pressure surface that was unaffected by erosion. The blade suction surface featured erosion around the blade tip and in the leading edge region. We calculated the average erosion rate over the entire pressure and suction surface. We used this average erosion rate to estimate the time it took to erode the same mass of blade material eroded from blades removed from in-service operation. The error between predicted and measured time was approximately 11.5 per cent. The discrepancy between the two may be a consequence of three factors. First, we did not account for the change in blade geometry as the erosion progressed. Second, we modelled the particle size distribution we measured from one particle sample. This distribution may not have been representative of the average distribution over the fan's in-service life. Third, erosion is a function of temperature and we did not account for any variation of temperature with time. Despite the above caveats, we considered an 11.5 per cent discrepancy between predicted and measured time to be reasonable.

We analysed the erosive effect of different size particles, concluding that particle of diameter 32.5 μm to 67.5 μm diameter were the most erosive in this context. Erosion of the blade leading edge has the most impact on fan performance. The combination of reduced blade chord and reduced blade camber results in a reduction in fan pressure developing capability. Destroying the blade leading edge geometry also results in reduced fan efficiency. We may conclude that the prediction of blade life may assist fan designers when estimating the change in blade life with changing particle size distribution or concentration. It may also be useful when estimating the change in erosion with changes in blade geometry. As such, the erosion rate predictions may assist during the optimisation of blade geometry, helping the fan designer minimise the susceptibility of a blade design to erosion.

REFERENCES

Balan, C. and Tabakoff, W. (1984), 'Axial Compressor Performance Deterioration', *Proceedings of the AIAA / SAE / ASME20th Joint Propulsion Conference*, Cincinnati, OH, USA, 11–13 June, paper no. AIAA-84-1208.

Baxter, L.L. (1989), 'Turbulent Transport of Particles', PhD thesis, Brigham Young University, Provo, UT, USA.

Baxter, L.L. and Smith, P.J. (1993), 'Turbulent Dispersion of Particles: the STP Model', *Energy and Fuels*, vol. 7(6), pp. 852–859.

Bons, J.P., Taylor, R.J., McClain, S.T. and Rivir, R. (2001), 'The Many Manifestations of Turbine Surface Roughness', *Transactions of the ASME, Journal of Turbomachinery*, vol. 123(4), pp. 739–748.

Borello, D., Borrelli, P., Quagliata, E. and Rispoli, F. (2001), 'A Multi-grid Additive and Distributive Parallel Algorithm for FEM Turbomachinery CFD', *Proceedings of the European Congress on Computational Methods in Applied Sciences (ECCOMAS CFD 2001)*, Swansea, UK, 4–7 September.

Borello, D., Corsini, A. and Rispoli, F. (2003), 'A Finite Element Overlapping Scheme for Turbomachinery Flows on Parallel Platforms', *Computational Fluids*, vol. 32, pp. 1017–1047.

Borello, D., Venturini, P., Rispoli, F. and Saavedra, G.Z.R. (2013), 'Prediction of Multiphase Combustion and Ash Deposition within a Biomass Furnace', *Applied Energy*, vol. 101, pp. 413–422.

Corsini, A. and Rispoli, F. (2004), 'Using Sweep to Extend Stall-free Operational Range in Axial Fan Rotors', *Proceedings of the IMechE Part A, Journal of Power and Energy*, vol. 218, pp. 129–139.

Corsini, A. and Rispoli, F. (2005), 'Flow Analyses in a High-pressure Axial Ventilation Fan with a Non-linear Eddy-viscosity Closure', *International Journal of Heat and Fluid Flow*, vol. 26, pp. 349–361.

Corsini, A., Rispoli, F., and Santoriello, A. (2003), 'A High Order Petrov-Galerkin Stabilized Finite Element Method for Incompressible RANS in Presence of Strong Reaction Effects', *Proceedings of CMFF-03, the Conference on Modelling Fluid Flow*, Budapest, Hungary, 3–6 September.

Corsini, A., Rispoli, F., Santoriello, A. and Tezduyar, T. (2006), 'Improved Discontinuity-capturing Finite Element Techniques for Reaction Effects in Turbulence Computation', *Computational Mechanics*, vol. 38, pp. 356–364.

Corsini, A., Santoriello, A., Tezduyar, T.E., Menichini, F. and Rispoli, F. (2009), 'A Multiscale Finite Element Formulation with Discontinuity Capturing for Turbulence Models with Dominant Reaction Like Terms', *Transactions of the ASME, Journal of Applied Mechanics*, vol. 76(2), paper no. 021211, pp. 1–8.

Corsini, A., Marchegiani, A., Minotti, S. and Sheard, A.G. (2011), 'Numerical Investigation on the Aerodynamic Influence of Eroded Leading-edge Geometry on Boiler Fan Performance', *Proceedings of the 9th European Conference on Turbomachinery, Fluid Dynamics & Thermodynamics*, Istanbul, Turkey, 21–25 March, pp. 879–892.

Corsini, A., Marchegiani, A., Rispoli, F., Sheard, A.G. and Venturini, P. (2012a), 'Predicting Blade Leading Edge Erosion in an Axial Induced Draft Fan', *Transactions of the ASME, Journal of Engineering for Gas Turbines & Power*, vol. 134, paper no. 042601, pp. 1–9.

Corsini, A., Rispoli, F., Sheard, A.G. and Tezduyar, T.E. (2012b), 'Computational Analysis of Noise Reduction Devices in Axial Fans with Stabilized Finite Element Formulations', *Computational Mechanics*, vol. 50, pp. 695–705.

Craft, T.J., Launder, B.E. and Suga, K. (1996), 'Development and Application of a Cubic Eddy-viscosity Model of Turbulence', *International Journal of Heat and Fluid Flow*, vol. 17(2), pp. 108–115.

Elfeki, S. and Tabakoff, W. (1987), 'Erosion Study of Radial Flow Compressor with Splitters', *Transactions of the ASME, Journal of Turbomachinery*, vol. 109(1), pp. 62–69.

Ghenaiet, A. (2005), 'Numerical Simulations of Flow and Particle Dynamics within a Centrifugal Turbomachine', *Proceedings of the Institution of Mechanical Engineers*

International Conference on Compressors and their Systems, London, UK, 4–7 September, paper no. C639-652, pp. 497–506.

Ghenaiet, A. (2009), 'Numerical Study of Sand Ingestion through a Ventilating System', *Proceedings of the World Congress on Engineering 2009*, London, UK, 1–3 July.

Ghenaiet, A., Tan, S.C. and Elder, R.L. (2004), 'Experimental Investigation of Axial Fan Erosion and Performance Degradation', *Proceedings of the IMechE Part A, Journal of Power and Energy*, vol. 218(6), pp. 437–450.

Grant, G. and Tabakoff, W. (1975), 'Erosion Prediction in Turbomachinery Resulting from Environmental Solid Particles', *Journal of Aircraft*, vol. 12(5), pp. 471–478.

Hamed, A., Tabakoff, W. and Wenglarz, R. (2006), 'Erosion and Deposition in Turbomachinery', *Journal of Propulsion and Power*, vol. 22, pp. 350–360.

Hussein, M.F. and Tabakoff, W. (1974), 'Computation and Plotting of Solid Particle Flow in Rotating Cascades', *Computational Fluids*, vol. 2(1), pp. 1–15.

Kær, S.K. (2001), 'Numerical Investigation of Ash Deposition in Straw-fired Furnaces', PhD thesis, Aalborg University, Aalborg, Denmark.

Kirk, B.S., Peterson, J.W., Stogner, R.H. and Carey, G.F. (2006), 'LibMesh: a C++ Library for Parallel Adaptive Mesh Refinement/Coarsening Simulations', *Engineering with Computers*, vol. 22, pp. 237–254.

Kleis, I. and Kulu, P. (2008), *Solid Particle Erosion: Occurrence, Prediction and Control*, Springer, London, UK.

Kurz, R. and Brun, K. (2001), 'Degradation in Gas Turbine Systems', *Transactions of the ASME, Journal of Engineering for Gas Turbines and Power*, vol. 123, pp. 70–77.

Lecrivain, G. and Hampel, U. (2012), 'Influence of the Lagrangian Integral Time Scale Estimation in the Near Wall Region on Particle Deposition', *Transactions of the ASME. Journal of Fluids Engineering*, vol. 134(7), paper no. 074502, pp. 1–6.

Litchford, R.J. and Jeng, S.M. (1991), 'Efficient Statistical Transport Model for Turbulent Particle Dispersion in Sprays', *AIAA Journal*, vol. 29, pp. 1443–1451.

Oka, Y.I., Ohnogi, H., Hosokawa, T. and Matsumura, M. (1997), 'The Impact Angle Dependence of Erosion Damage Caused by Solid Particle Impact', *Wear*, vols 203–204, pp. 573–579.

Okita, R., Zhang, Y., McLaury, B.S. and Shirazi, S.A. (2012), 'Experimental and Computational Investigations to Evaluate the Effects of Fluid Viscosity and Particle Size on Erosion Damage', *Transactions of the ASME, Journal of Fluids Engineering*, vol. 134(6), pp. 1–13, paper no. 061301.

Page, A.L., Elseewi, A.A. and Straughan, I.R. (1979), 'Physical and Chemical Properties of Fly Ash from Coal-fired Power Plants with Reference to Environmental Impacts', *Residue Review*, vol. 71, pp. 83–120.

Pandian, N.S. (2004), 'Fly Ash Characterization with Reference to Geotechnical Applications', *Journal of the Indian Institute of Science*, vol. 84, pp. 189–216.

Richardson, J.H., Sallee, G.P. and Smakula, F.K. (1979), 'Causes of High Pressure Compressor Deterioration in Service', *Proceedings of the AIAA / SAE / ASME 15th Joint Propulsion Conference*, Las Vegas, NV, USA, 18–20 June, paper no. AIAA 79-1234.

Sallee, G.P., Kruckenburg, H.D. and Toomey, E.H. (1975), 'Analysis of Turbofan Engine Performance Deterioration and Proposed Follow-On Tests', Report NASA-CR-134769.

Seggiani, M., Bardi, A. and Vitolo, S. (2000), 'Prediction of Fly-ash Size Distribution: A Correlation between the Char Transition Radius and Coal Properties', *Fuel*, vol. 79, pp. 999–1002.

Sheard, A.G., Corsini. A., Minotti, S. and Sciulli, F. (2009), 'The Role of Computational Methods in the Development of an Aero-acoustic Design Methodology: Application in a Family of Large Industrial Fans', *Proceedings of the 14th International Conference on Modelling Fluid Flow Technologies*, Budapest, Hungary, 9–12 September, pp. 71–79.

Smith, P.J. (1993), 'Three-dimensional Turbulent Particle Dispersion Submodel Development', Department of Energy, Pittsburgh Energy Technology Center, Pittsburgh, PA, Report DOE/PC/90094-T5.

Sugano, H., Yamaguchi, N. and Taguchi, S. (1982), 'A Study on the Ash Erosion of Axial Induced Draft Fans of Coal-fired Boilers', *Mitsubishi Heavy Industries Technical Review*, vol. V19(N1), pp. 12-22.

Suzuki, M. and Yamamoto, M. (2011), 'Numerical Simulation of Sand Erosion Phenomena in a Single-stage Axial Compressor', *Journal of Fluid Science and Technology*, vol. 6, pp. 98–113.

Tabakoff, W., Kotwal R. and Hamed, A. (1979), 'Erosion Study of Different Materials Affected by Coal Ash Particles', *Wear*, vol. 52(1), pp. 161–173.

Venturini, P. (2010), 'Modelling of Particle Wall-deposition in Two Phase Gas-solid Flows', PhD thesis, Sapienza University of Rome, Rome, Italy.

Venturini, P., Borello, D., Iossa, C., Lentini, D. and Rispoli F. (2010), 'Modelling of Multiphase Combustion and Deposit Formation and Deposit Formation in a Biomass-fed Boiler', *Energy*, vol. 35(7), pp. 3008–3021.

Wang, L.P. (1990) 'On the Dispersion of Heavy Particles by Turbulent Motion', PhD thesis, Washington State University, Pullman, WA, USA.

Chapter 11

Predicting the Performance of an Industrial Centrifugal Fan Incorporating Cambered Plate Impeller Blades

L. Cardillo, A. Corsini, G. Delibra, F. Rispoli and A.G. Sheard

ABSTRACT

Application of computational methods to industrial fan design processes has progressed steadily over the past decade. The reducing cost of the computer hardware upon which codes run has brought it within reach of the air movement and control community. However, the cost of commercially available codes remains high. Open-source codes provide an alternative. Scholars have used the finite volume open-source code OpenFOAM to predict the performance of industrial centrifugal fans incorporating impeller blades constructed from cambered plate, but not by designers working within the air movement and control community. This chapter presents a modelling approach which we developed for application as part of an industrial fan manufacturer's order related design process. We compare numerical performance predictions with experimental results both at peak pressure and at peak efficiency operating points, concluding that predicted and experimentally measured performance agree within the uncertainty of the measurements. As a further possible application of the developed modelling approach, we used the simulated flow-field as input to a particle tracking model to predict erosion patterns over the impeller. The predicted erosions patterns were not a good match with those which we observed over impellers removed from service after operation in an erosive environment. We concluded that the developed steady modelling approach was adequate to predict overall fan performance, but not to predict particle trajectories and therefore erosion patterns. We recommend an unsteady modelling approach for predicting particle trajectories.

This chapter is a revised and extended version of Cardillo, L., Corsini, A., Delibra, G., Rispoli, F. and Sheard, A.G. (2014), 'Predicting the Performance of an Industrial Centrifugal Fan Incorporating Cambered Plate Impeller Blades', *Periodica Polytechnique, Mechanical Engineering*, vol. 58(1), pp. 15–25.

NOMENCLATURE

Latin letters

b_i	[mm]	Impeller blade width
b_v	[mm]	Volute width
D_{in}	[mm]	Impeller inlet diameter
D_{out}	[mm]	Impeller outlet diameter
D_t	[mm]	Blade tip diameter
D_v	[mm]	Volute outlet diameter
D_2	[m]	Size range
d_p	[μm]	Particle diameter
E_R		Erosion rate
f		Erosion angular function
k	[m²/s²]	Turbulent kinetic energy
K_c, K_1, K_2, K_{12}		Erosion model coefficients
p	[Pa]	Static pressure
p_t	[Pa]	Total pressure
Δp_t	[kPa]	Total pressure rise
U_{tip}	[m/s]	Tip blade velocity
u		Flow velocity
u^+		Normalised velocity
u_{bc}		Flow velocity vector in the centre of the cell
u_m		Cloud's centre of mass rotation velocity
V	[m³/s]	Volume flow rate
v		Particle velocity
v_{bc}		Particle velocity vector in the centre of the cell
V_c		Cell volume
v_{drift}		Drift velocity vector at the cloud centre $v_{drift} = u_m - v_m$
v_m		Cloud's centre of mass absolute velocity
y^+	[-]	Normalised wall-distance
w^+		Normalised relative velocity magnitude

Greek letters

α		Impact angle
ε	[m²/s³]	Dissipation of k
η_{pt}	[-]	Efficiency based on total pressure rise
λ_P		Power coefficient
ρ_p	[kg/m³]	Particle density
φ	[-]	Normalised axial velocity, $\varphi = U_{ax}/U_{bulk}$
ω		Rotational speed

Acronyms

AMI	Arbitrary mesh interface
CFD	Computational fluid dynamics
EXP	Experimental
PE	Peak efficiency
PP	Peak pressure

INTRODUCTION

The industrial centrifugal fans used in cement, steel and power applications are typically tailored to the applications' specific flow and pressure requirements (Godichon and Sheard, 2009). Although aerodynamic performance has historically been scaled from previous laboratory test data, the impeller, shaft and housing mechanical design, bearing selection and rotor-dynamic analysis are invariably unique. A consequence of the need for a unique design for each fan has resulted in a historic focus on automating the processes associated with mechanical design and rotor-dynamic analysis (Godichon and Sheard, 2009). Design methods that embed previously established limits into a computer code are coupled with parametric three-dimensional computer-aided design (CAD) models that have associated two-dimensional drawing packs. The result are design methods that engineers can use to deliver a full set of manufacturing drawings for a centrifugal industrial fan with less than one man week of engineering time.

A historic focus on the automation and control of industrial centrifugal fan order related engineering and production of manufacturing drawings has helped to ensure that in-service failures are rare. However, this focus ultimately limits the resultant design's aerodynamic efficiency. Scaling and interpolating fan performance using methods developed in the 1950s (Eck, 1973; Daly, 1978) results in reliably predicting aerodynamic performance, but inevitably also sub-optimal performance.

The European Union Regulation 327 sets minimum Fan and Motor Efficiency Grades (FMEGs) for industrial fans. The 2013 minimum FMEGs have resulted in approximately 33 per cent of fan sales before 1 January, 2013, now illegal within Europe as a consequence of their low efficiency. The European Union will raise minimum FMEGs on 1 January, 2015. In the USA, the Department of Energy has been monitoring the regulatory activity within Europe. On 1 February, 2013, the U.S. federal government published a framework document in the *Federal Register* outlining the intended approach to fan regulation that aims to eliminate inefficient industrial fans within the USA by 2019. The air movement and control community widely anticipates that the Department of Energy will adopt the same approach as the European Union, increasing the minimum allowable fan and motor efficiency within three years of introducing the initial 2019 targets.

With today's regulatory framework it is reasonable to assume that the practice of scaling aerodynamic performance from historic test data will become progressively less acceptable. The resultant fan efficiency will be increasingly unlikely to meet the minimum allowable fan efficiency. The air movement and control community will be driven to apply the same computational methods that are commonplace within the aerospace community, optimising fan performance within the mechanical constraints imposed by each specific application.

Over the last two decades, scholars working within the air movement and control community have developed computational fluid dynamic techniques specifically focused on centrifugal fans. Zhang *et al.* (1996) computed the three-dimensional viscous flow in a backwards swept centrifugal fan impeller's blade passage at its design point by solving the Reynolds-averaged Navier–Stokes (RANS) equations using standard eddy viscosity turbulence closure. Thakur *et al.* (2002) studied a centrifugal blower and developed a quasi-steady rotor-stator modelling technique. Khelladi *et al.* (2005) studied the impeller-diffuser unsteady interactions by using an unsteady computational method.

A reason for the difficulty associated with accurately predicting centrifugal industrial fan performance is that their impellers are typically fabricated from cambered plate. A consequence of using cambered plate impeller blades is that engineers inevitably associate impellers with regions of separated flow. This makes accurately predicting fan performance using computational methods challenging. However, recent research has focused on elucidating the flow-field features associated with interaction between the impeller and diffuser (Vasudeva Karanth and Yagnesh Sharma, 2009), interaction between the impeller-scroll-inlet ring (Zhao *et al.*, 2005) and the clearance flow impact on overall fan performance (Lee, 2010; Lee *et al.*, 2011). The research objective has been to systematically characterise the flow-field in poorly conditioned regions of separated flow, thus providing a basis for modelling the actual flow-field.

In this chapter we present a method for predicting an industrial centrifugal fan's performance when the fan impeller incorporates cambered plate blades using the unstructured open-source finite volume solver OpenFOAM (Weller *et al.*, 1998; Jasak, 2010). We model the time-varying flow at a fan impeller-volute interface using the 'frozen rotor' approach that researchers have shown results in accurate predictions of centrifugal fan performance (Seo *et al.*, 2003; Lee *et al.*, 2011). We describe how we connected the impeller and volute grids using an arbitrary mesh interface (AMI) (Petit *et al.*, 2009). We conducted the research on a typical example of an industrial centrifugal fan incorporating an impeller with cambered plate blades.

We then explain how we verified the chosen modelling approach's effectiveness by comparing computational results with experimental results at both peak efficiency and peak pressure operating points. Agreement between the two was within the measured data's uncertainty, and consequently, the computational method was effective in the application. We then explain how we analysed secondary flow structures to provide an insight into the flow-field physics, and then go on to use the flow-field prediction as input to a particle tracking model. The particle tracking model predicts the trajectory of particles as they pass through the impeller. The purpose of predicting particle trajectories was to facilitate a prediction of erosion patterns over the impeller. Our aim was to identify those areas of the impeller most prone to erosion. Insight into the location of those areas most prone to erosion is potentially useful. As a designer optimises an impeller's geometry, it is helpful to have an insight into the erosive consequences of any change in geometry. An insight into the impact of a change in geometry on the susceptibility of an impeller design to erosion facilitates an optimisation process that both maximises efficiency whilst minimising the resultant impeller's susceptibility to erosion.

FAN DESCRIPTION

The studied fan is part of a family of fans intended for process industry application with a pressure developing requirement of up to 20 kPa. The family includes impellers with backward-swept cambered plate blades that are the subject of the reported research, Table 11.1. A double-inlet with the inlet plenum incorporating 90 degree bends and a double impeller characterises the chosen fan, Figure 11.1 and Table 11.2. We experimentally derived the studied fan's aero-dynamic performance in accordance with ISO 5801 (2007) requirements.

MODELLING APPROACH

An issue when modelling centrifugal fans is that the geometry does not allow for common techniques that engineers use when modelling axial turbomachinery. For example, engineers usually apply periodic boundary conditions, which simulate the flow-field in just one blade passage. When modelling a centrifugal fan it is necessary to model the entire impeller in order to account for the coupling between the impeller and the asymmetric volute casing within which it is housed.

The 'frozen rotor' approach is a reliable and accurate modelling technique for predicting the flow-field for an entire centrifugal fan impeller when in an asymmetric volute casing (Launder and Sharma, 1974; Ferziger and Peric, 2002). This approach neglects the unsteady coupling between the impeller and volute casing. The

Table 11.1. *Fan range description.*

Blade tip diameter D_t	3 to 4 metres
Performance standard	ISO 5801 – ISO 13347
Volume flow rate V	Up 750 m^3/s
Total pressure rise Δp_t	Up to 50 kPa
Rotational speed ω	900–1,000 rpm
Blade sections	Cambered plate

Table 11.2. *The studied fan geometry.*

Impeller inlet diameter D_{in}	1,804 mm
Impeller outlet diameter D_{out}	3,440 mm
Volute outlet diameter D_v	5,600 mm
Impeller blade width b_i	400 mm
Volute width b_v	200 mm
Impeller blade count	11
Rotational speed ω	900 rpm

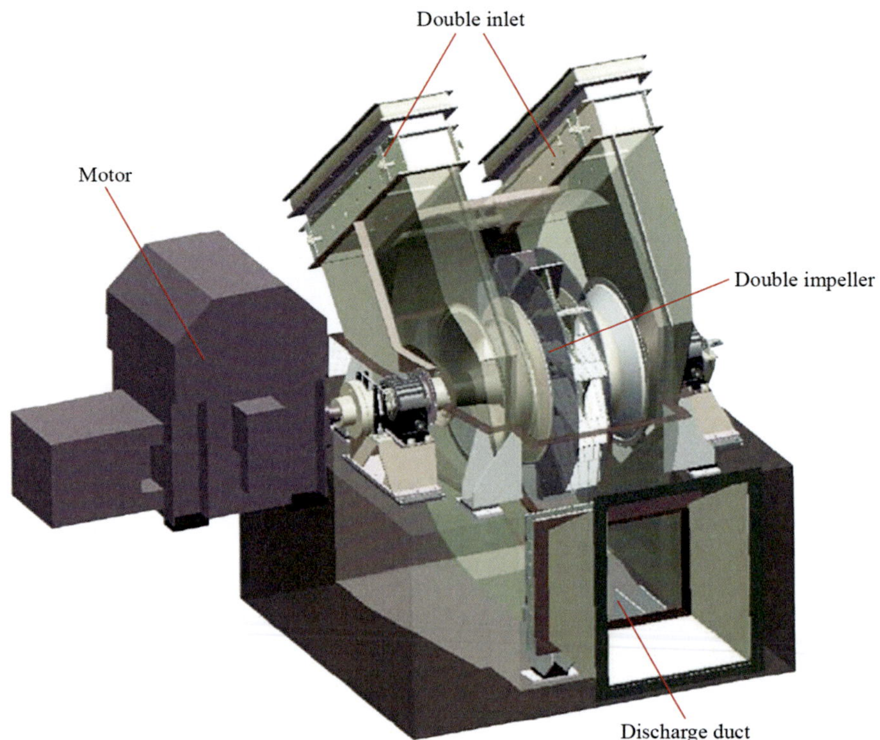

FIGURE 11.1. An example of a centrifugal fan with double inlet and cambered plate blades from the studied family of industrial fans.

impeller's position is fixed during the simulation relative to the volute casing, with the resulting computation at steady-state. We can account for the impeller motion relative to the volute casing by using a multiple frame methodology (Ferziger and Peric, 2002). A multiple frame methodology enables one to account for the Coriolis and centrifugal forces within the impeller. We explicitly computed Coriolis forces within the Navier-Stokes momentum equation and applied centrifugal forces after we completed the steady-state computation (Denton and Dawes, 1999).

We spatially resolved the coupling between the impeller and volute casing computational domains over the impeller's periphery using an arbitrary mesh interface (AMI) (Petit *et al.*, 2009). We solved the incompressible Reynolds-averaged Navier–Stokes equations using the finite volume method and a standard k-ε turbulence model with the synthetic wall-treatment available in OpenFOAM (Launder and Sharma, 1974).

We carried out simulations with OpenFOAM (Weller *et al.*, 1998; Jasak, 2010) using the built-in solver *MFRSimpleFOAM*. We carried out computations using a second order upstream interpolation for a convective kinetics divergence scheme in order to diminish the numerical diffusivity. We solved the linearised system of equations by adopting an ILU preconditioned semi-iterative conjugate gradient linear solver, in combination with a SIMPLE segregation scheme and OpenFOAM's *smoothSolver* for all other equations. In all cases the convergence tolerance was set to 10^{-8} for all the computed quantities. We performed the computations on the computer cluster Matrix of Cineca (Rome) over 48 processors. The initial fan flow-field's computation at the peak efficiency duty point required 48 hours of computational time. We then used the peak efficiency duty point solution as a starting point for the fan flow-field's peak pressure computation that required a further eight hours of computational time.

Fan geometry, boundary conditions and grid sensitivity analysis

An impeller with both a double inlet and double impeller characterises the studied fan, Figure 11.1. To minimise the required computational effort for the simulation, we modelled only one half of the impeller to exploit the impeller's symmetry and effectively reduce the necessary computational effort to predict the flow-field by half. We modelled the resultant half-fan in three parts: the inlet, the impeller and the volute casing, Figure 11.2. We coupled these three regions using AMI interfaces.

FIGURE 11.2. The modelling technique adopted modelled one half of the double inlet – double width impeller. The computational domain was split into three parts: an inlet region (shown in purple), an impeller region (shown in black) and a volute casing region (shown in grey).

We assessed simulation sensitivity to grid density with a power coefficient as a convergence parameter, Table 11.3. The comparative analysis between grids focused on the inlet and impeller blade passage grids. We increased the blade-to-blade grid density within the impeller from an initial coarse level of 50,000 cells to a fine level of 300,000 cells with both hub-to-shroud and blade-to-blade refinement. The simulation results appeared unaffected by the blade-to-blade grid density, and therefore we used the coarse blade-to-blade passage grid.

Coupling the aforementioned three regions of the mesh (inlet, impeller and volute) by means of AMI interface, we were able to produce a high-quality hexahedral mesh without the usual constraints of hexahedral meshes and kept the total count of cells around 6.1 million. Figure 11.3 shows the details of the mesh, whilst Table 11.4 provides an assessment of the grid quality and resolution that provides data for the mesh quality indicators aspect ratio, minimum included angle, volume ratio, skewness and cell number for the three different grid regions. The same Table 11.4 includes y+ values onto the solid surfaces to provide quantitative data of the mesh resolution. Exploiting the AMI modelling approach, we were able to produce a good mesh using only hexahedral cells. We estimate from previous experience that an unstructured tetrahedral mesh would require approximately 20 million cells, increasing the computational effort because of both more cells and the necessary iterations to reach a converged solution.

We set the inflow boundary conditions to mimic an inflow distortion which is produced because of the inlet plenum geometry. We derived the distorted inflow from previous simulations that included the inlet plenum that we used to complement Greenzweig *et al.*'s (2011) systematic experimental study. The inflow velocity distribution illustrates the presence of a large flow distortion, with both a separated flow region and highly asymmetric flow into the impeller, Figure 11.4. This flow distortion occurs because of the 90 degree bend and the limited space available for the inlet plenum in practical application. We chose an average inflow turbulence level of ten per cent and set a dissipation length scale at ten per cent of mean blade chord. We used a standard wall function at the solid walls and imposed zero-gradient boundary conditions on the diffuser outflow.

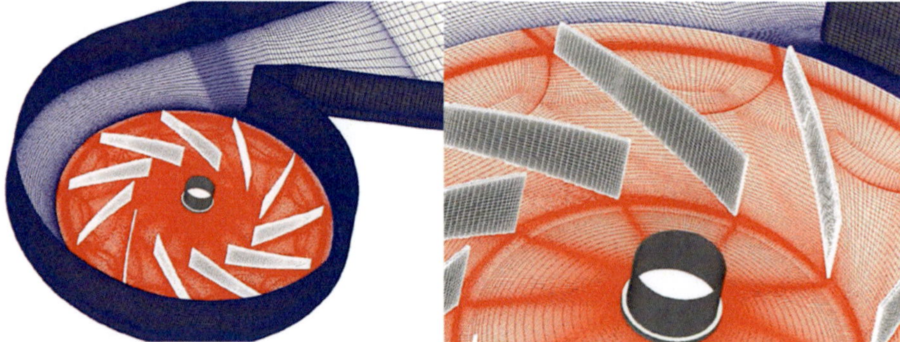

FIGURE 11.3. The computational mesh for the studied fan illustrating mesh density over the fan impeller and scroll, left, with further details of the impeller, right.

Table 11.3. *Grid sensitivity analysis.*

	Coarse	Fine	Exp $\phi = 0.047$
Power coefficient λ_P	0.0451	0.0453	0.0459

Table 11.4. *Numerical simulation mesh quality indicators and non-dimensional wall distance (y^+) values at the solid walls.*

	Impeller — number of cells: 3.3M		
	Min.	Average	Max.
Aspect ratio	1.0	5	32.8
Minimum included angle	20.3	65.3	90
Volume ratio	1.0	1.4	22.5
Skewness	0.0	0.13	0.67
	Volute — number of cells: 1.0M		
	Min.	Average	Max.
Aspect ratio	1.0	5.2	116.7
Minimum included angle	17.1	75.4	90
Volume ratio	1.0	1.3	2.5
Skewness	0.0	0.05	0.64
	Inlet — number of cells: 1.8M		
	Min.	Average	Max.
Aspect ratio	1.0	4.4	33.3
Minimum included angle	17.6	76.2	89.7
Volume ratio	1.02	1.47	6.65
Skewness	0.0	0.06	0.69
	y^+ values on solid walls		
	Min.	Average	Max.
Blades	2	90	290
Hub	7	118	118
Shroud	4	72	146
Volute	8	94	165

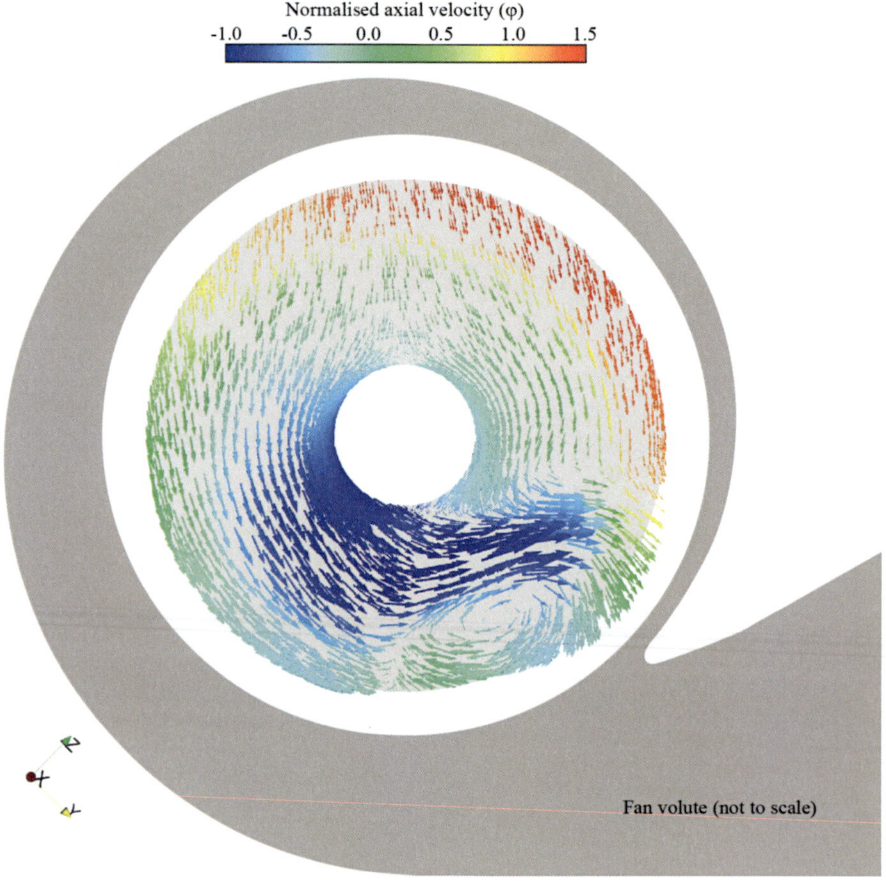

FIGURE 11.4. Inflow velocity profiles coloured with normalised axial velocity (φ). The volute is not shown to scale.

Particle transport and dispersion, and erosion modelling

We may predict particle trajectory via one of two approaches. The first is single particle tracking, computing the motion of each particle (Baxter, 1989). The second is particle cloud tracking, modelling the motion of a cluster of particles by computing the motion of the cloud centre (Corsini *et al.*, 2013b). The former approach requires the Basset-Boussinesq-Oseen particle equation of motion to be solved at run-time using the calculated flow-field parameters at each particle's current location (Sommerfeld *et al.*, 2008). In contrast, the latter approach involves tracking a statistical representation of clusters of particles.

Baxter (1989) first proposed particle cloud tracking and then scholars working with both the air movement and control community and the aerospace community developed and applied it (Wang, 1990; Litchford and Jeng, 1991; Jain, 1995; Kær, 2001; Borello *et al.*, 2013; Corsini *et al.*, 2013a). Although single particle tracking is the more accurate modelling approach, the computational effort required is between one and two orders of magnitude larger than that required for particle cloud tracking. It is the relatively low computational effort associated with particle cloud tracking that makes it attractive to scholars working with the air movement and control community (Venturini *et al.*, 2010; Corsini *et al.*, 2012; Borello *et al.*, 2013; Corsini *et al.*, 2013a).

The cloud equation of motion is the ensemble averaged version of the Basset-Boussinesq-Oseen particle equation, reduced in accordance to the application's constraints (Baxter, 1989). We assume that particle distribution within a cloud is Gaussian, and the cloud size depends on the flow's turbulent characteristics. In Baxter's (1989) original particle cloud tracking method, each particle moved with the mean cloud velocity. Assuming that each particle moves with the mean cloud velocity strongly affects the erosion prediction and therefore we have modified the particle cloud tracking method. We assume that the ratio between the drift and the flow velocity at the cloud's mean position is constant within a cloud. Accordingly, we compute the particle local velocity at a boundary cell as:

$$|\mathbf{v}_{\mathrm{bc}}| = |\mathbf{u}_{\mathrm{bc}}|\left(1 - \frac{|\mathbf{v}_{\mathrm{drift}}|}{|\mathbf{u}_{\mathrm{m}}|}\right) \tag{1}$$

where $\mathbf{u}_{\mathrm{bc}}$, $\mathbf{v}_{\mathrm{bc}}$ are respectively the flow and particle velocity vectors in the centre of the cell, $\mathbf{v}_{\mathrm{drift}} = \mathbf{u}_{\mathrm{m}} - \mathbf{v}_{\mathrm{m}}$ is the drift velocity vector at the cloud's centre and $\mathbf{u}_{\mathrm{m}}$ and $\mathbf{v}_{\mathrm{m}}$ are respectively the flow and cloud velocity vectors at the cloud's centre. One may find a full presentation of our modified particle cloud tracking method in Borello *et al.* (2013), Corsini *et al.* (2012, 2013a), Venturini *et al.* (2010) and Venturini (2010).

An ability to predict particle trajectory is an essential pre-cursor to the prediction of erosion, but it is not itself an erosion prediction method. Erosion is a complex phenomenon dependent on particle impact velocity, particle impact angle, particle shape, and material properties of both particles and the target. The complexity of the erosive process results in erosion models being by necessity semi-empirical. Tabackoff *et al.* (1979) developed one of the most widely used semi-empirical erosion models which predicts the erosion rate (E_R) per unit mass of impacting particles (mg/g). Tabackoff *et al.* (1979) expressed erosion rate as:

$$E_R = K_1 f_\alpha |\mathbf{v}_{\mathrm{bc}}|^2 \cos^2\alpha (1 - R_T^2) + f_i \tag{2}$$

being:

$$R_T = 1 - 0.0061|\mathbf{v}_{\mathrm{bc}}|\sin\alpha$$

$$f_\alpha = \{1 + K_C[K_{12}\sin(90\alpha/\alpha_0)]\}^2$$

$$f_i = K_2(|\mathbf{v}_{\mathrm{bc}}|\sin\alpha)^4$$

Table 11.5. *Erosion model coefficients (Tabakoff* et al., *1979).*

K_C $\alpha \leq 3\alpha_0$	1
$\alpha > 3\alpha_0$	0
K_1	$1.505101 \cdot 10^{-06}$
K_{12}	$2.96077 \cdot 10^{-01}$
K_2	$5.0 \cdot 10^{-12}$

where α is the impact angle, K_1, K_2, K_{12} and K_C are functions of material properties of particles and target surface. We used Tabackoff *et al.*'s (1979) model in this study, and although the fan was manufactured from mild steel, we assumed it was manufactured from stainless steel. We made this assumption as stainless steel was the closest material to that from which the fan was actually manufactured for which Tabackoff *et al.* (1979) had derived material coefficients, Table 11.5. Also following Tabackoff *et al.*'s (1979) recommendation, we assumed that particles impacting with an impact angle, α_0, of less than 30 degrees had no erosive effect.

RESULTS

Flow-field

The reported research focuses on developing and validating a modelling approach that will enable the reliable and routine use of the open-source solver OpenFOAM to predict industrial centrifugal fan performance with cambered plate impeller blades. End users of the studied class of industrial fan generally specify a design and a test-block operating point. Engineers typically select the design point close to the fans' peak efficiency operating point to minimise fan running costs. They generally select the test-block point close to the fans' peak pressure operating point to minimise fan size for a given application in order to minimise capital cost. Therefore, we predicted the studied fan's performance at two operating points, peak pressure (PP) corresponding to a volume flow rate of 110 m^3/s and the peak efficiency (PE) corresponding to a volume flow rate of 221 m^3/s.

We selected the peak pressure operating point at the very peak of the studied fan's characteristic, therefore approaching stall. In practical application fan designers would not select this class of fan with a peak pressure operating point at the fan characteristic's peak. They would use a margin on pressure and would choose the peak pressure operating point at a higher flow rate than that which we adopted in the reported research. We selected the peak pressure operating point at the fan characteristic's very peak to predict fan performance when stall is incipient. This is a demanding task. Consequently, any modelling approach that can predict fan performance when stall is incipient is inherently robust.

We first established the studied fan's performance experimentally in accordance with ISO 5801 (2007) requirements, Figure 11.5. The developed modelling approach was able to predict fan performance well, Figure 11.5. We under predicted the

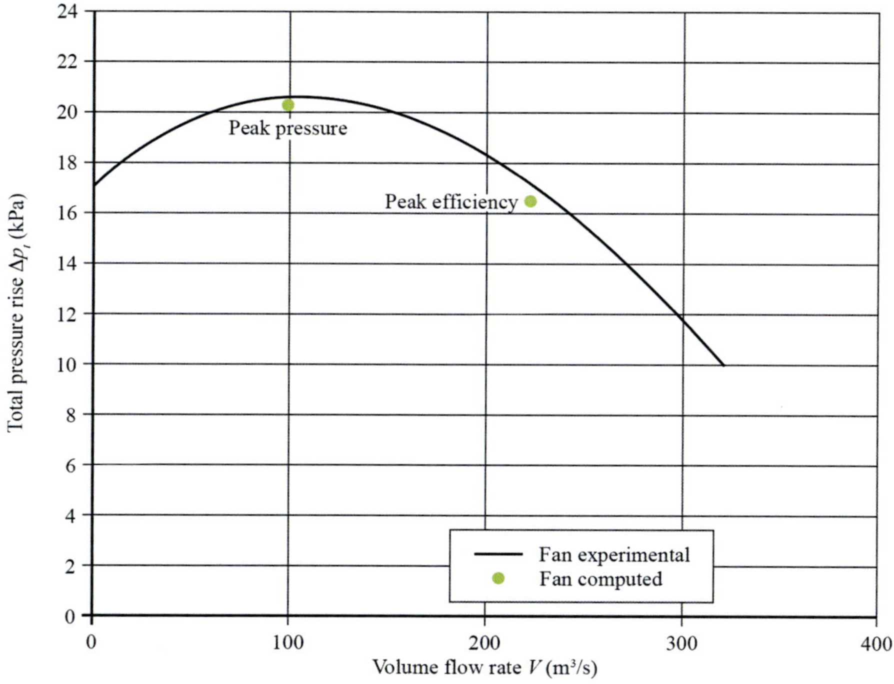

FIGURE 11.5. Pressure-volume characteristic curve for the studied fan illustrating the position of the peak efficiency and peak pressure operating points on the fan characteristic.

calculated peak pressure operating point total pressure by 0.6 per cent and the peak efficiency operating point by 1.5 per cent. We over predicted the calculated peak pressure operating point efficiency by 3.0 per cent and the peak efficiency operating point by 1.0 per cent, Table 11.6. We estimated the measured data's uncertainty as +/−1.0 per cent.

Although we predicted the peak pressure operating point total pressure within the measured data's uncertainty, we under predicted both the peak pressure and peak efficiency operating point total pressures. The under prediction appeared to correlate with volume flow rate. This trend of under predicting total pressure is probably a consequence of the adopted modelling approach which neglected the unsteady interaction between the impeller and volute. It is this interaction that generates artificial diffusivity, and therefore the adopted computational grid, numerical method

Table 11.6. *A comparison of experimental and computed total pressure rise and efficiency.*

Duty point	**V [m³/s]**		**Δp_t [kPa]**			η_{pt}	
		EXP	CFD	Δ	EXP	CFD	Δ
Peak pressure	110	20.78	20.46	−1.5%	75%	78%	+3%
Peak efficiency	221	17.00	16.89	−0.6%	82%	81%	−1%

and turbulence model did not include the necessary mechanisms to model artificial diffusivity. Artificial diffusivity plays a role in calculating the complex flow-field associated with centrifugal fan impellers.

Despite under predicting the total pressure, we considered the prediction of the peak pressure operating point's efficiency was in good agreement with the measured data. We selected the peak pressure operating point at the very peak of the fan's characteristic where stall is incipient. The ability of the developed modelling approach to predict total pressure and efficiency to within 1.5 per cent and 3.0 per cent respectively when stall is incipient confirms the effectiveness of the adopted modelling approach. The modelling approach was able to adequately compute the circumferential blade-fluid interaction and the resultant torque on the impeller without accounting for the unsteady impeller-volute interaction. Consequently, we may consider the adopted modelling approach as both robust enough to facilitate a converged solution when stall is incipient, and to provide a reasonable degree of accuracy with minimal computational effort.

We may assess the adopted modelling approach's effectiveness by considering the relative velocity streamlines through the impeller. These provide an insight into the flow-field's three-dimensionality around the impeller. To aid in interpreting the data, we have labelled each impeller blade, Figure 11.6. We show the relative velocity streamlines for both the peak pressure and peak efficiency operating points, Figure 11.7.

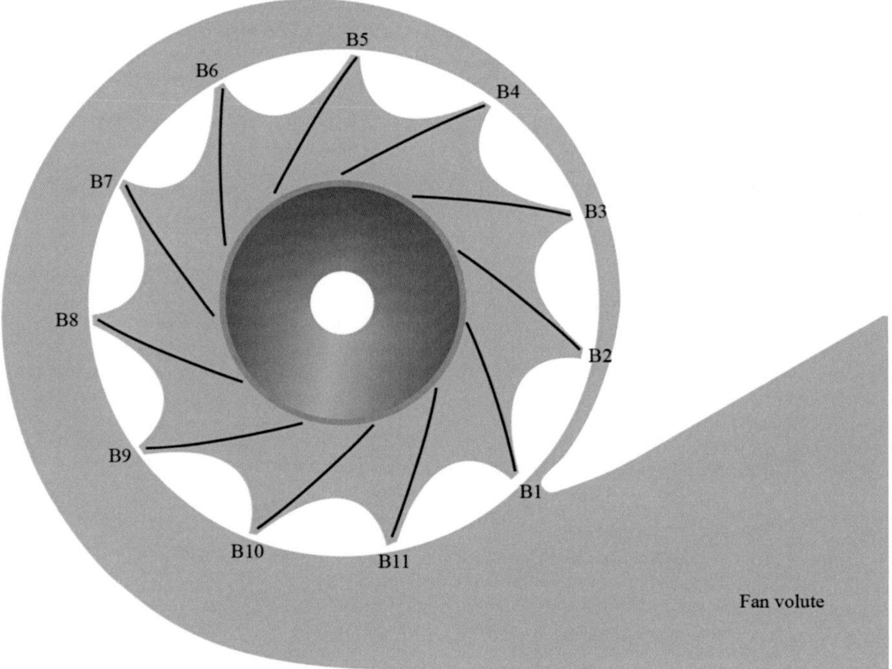

FIGURE 11.6. Blade reference labels B1 to B11 used to identify the position of each blade relative to the fan volute.

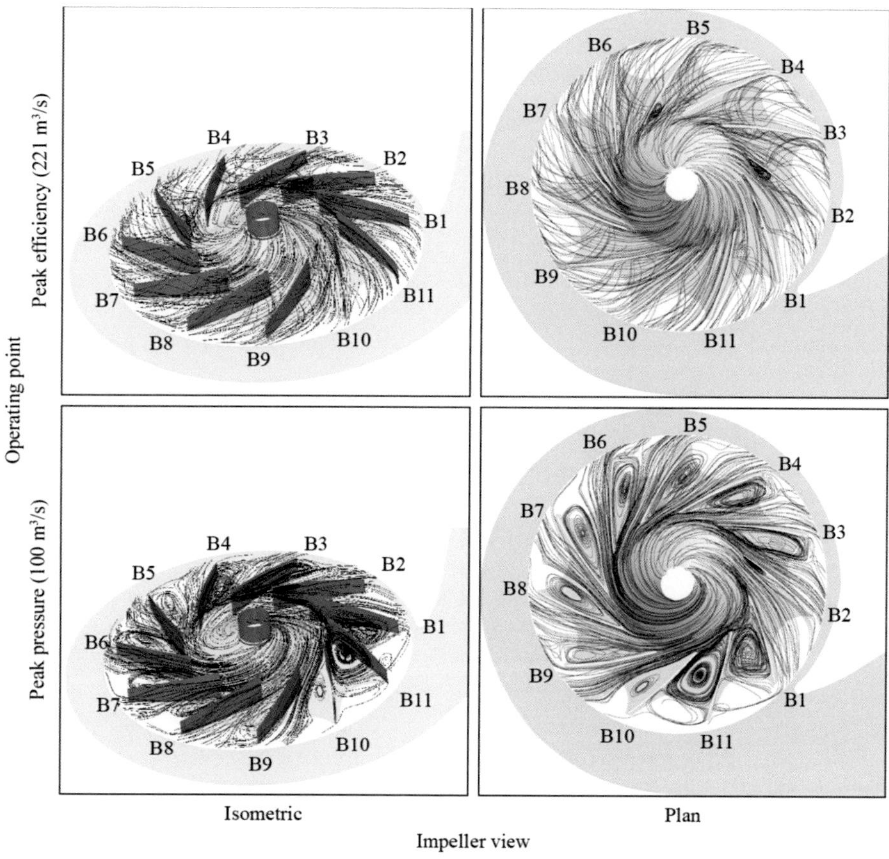

FIGURE 11.7. Three-dimensional relative velocity streamlines through the impeller for peak efficiency (top) and peak pressure (bottom) operating points.

The relative velocity streamlines at the peak efficiency operating point indicate the presence of a leading edge separation from blades B1 to B5, immediately after the junction of the blade and shroud. The blade shroud junction has resulted in the flow's non-optimal alignment onto the blades' leading edge which then results in a separated flow region within each blade-to-blade passage. In contrast, the relative velocity streamlines at the peak pressure operating point indicate that the fan flow-field is adapting to the reduced flow-rate associated with incipient stall. The relative velocity streamlines indicate that separated flow is present between most of the impeller blades. The only blade-to-blade passage not affected by the separation occurring in other passages is the passage between impeller blades B1 and B2. The relative velocity streamlines indicate that all other blade-to-blade passages are either partly stalled with the blade passage between blade B10 and B11 having a double separation that indicates it is completely stalled.

Relative velocity magnitude may provide a further insight into the flow-field's complexity. To aid in interpreting the data, we have defined two impeller reference sections, Figure 11.8. We show the relative velocity magnitudes for both the peak pressure and peak efficiency operating points, Figure 11.9.

The relative velocity magnitude contours at the peak pressure operating point indicate that each impeller blade-to-blade passage is associated with separated flow. The only exception is the passage between blade B1 and B2 immediately after the volute throat. The double separation in the blade passage B11 to B1 is clearly evident, with a double separation also establishing itself in the blade passage B10 to B11. The relative velocity magnitude contours indicate that the extent of the separated flow is greater at the reference section S1 than S2. This is to be expected as the reference section S1 is closer to the impeller's back-plate and therefore the boundary layer flow on the back-plate more heavily influences it.

In contrast to the peak pressure operating point, we do not associate the peak efficiency operating point with relative velocity magnitudes that indicate the presence of separated flow. When we studied the relative velocity magnitudes for blades B1 to B5, there was some evidence of recirculation at the blade's leading edge. This recirculation may be indicative of a non-optimal blade leading edge profile, a problem that is difficult to address when using cambered plate blades.

We may achieve a final assessment of the adopted modelling approach's effectiveness by considering the secondary flow structure in the discharge volute. The secondary flow structures provide an insight into the flow-field's three-dimensionality leaving the impeller, Figure 11.10. The secondary flow structures for both the peak pressure and peak efficiency operating points indicate that the flow through the discharge volute is highly three-dimensional. At the peak pressure operating point the flow is able to remain attached through the diffuser, with a

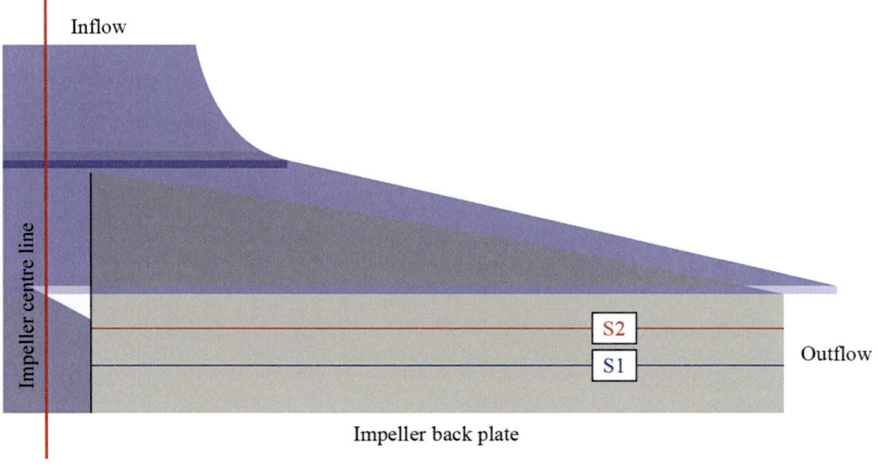

FIGURE 11.8. The location of impeller reference sections S1 and S2 used to identify the proximity of each section to the impeller back-plate.

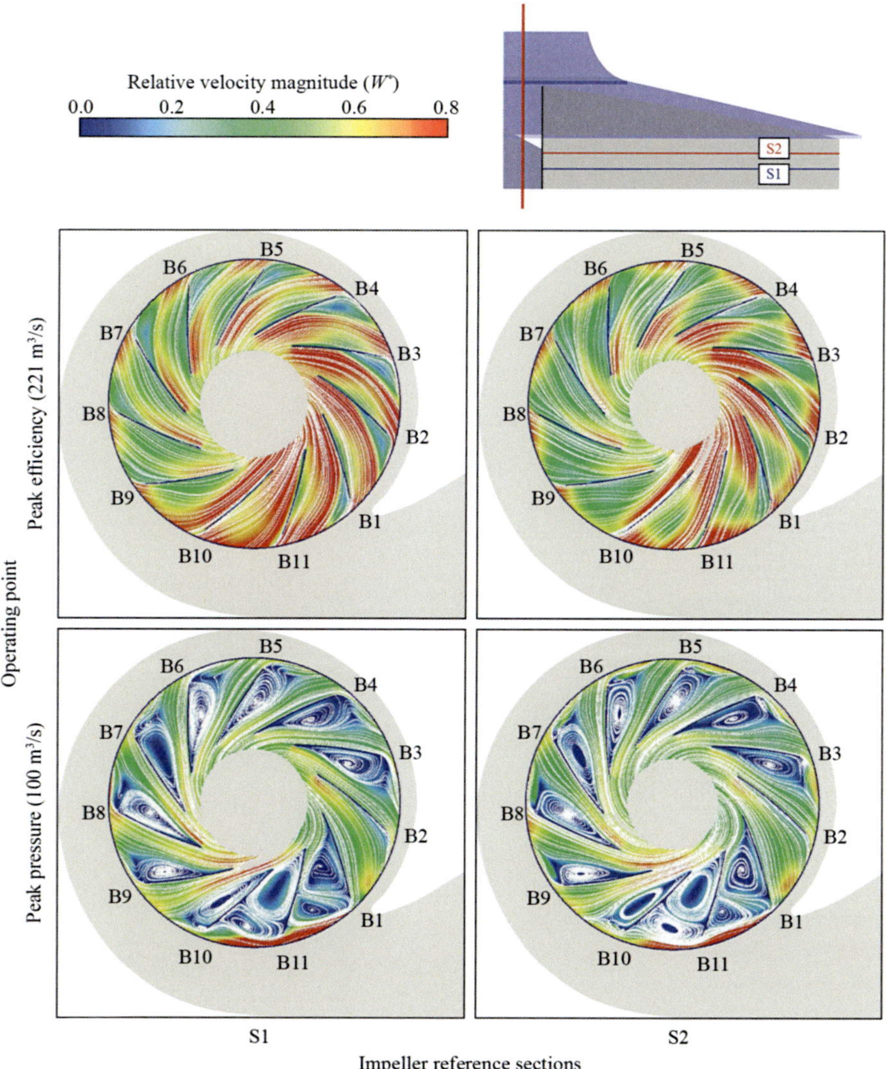

FIGURE 11.9. Contours of relative velocity magnitude (w^+) through the impeller with streamlines for peak efficiency (top) and peak pressure (bottom) operating points.

relatively weak secondary flow system establishing itself through the discharge volute. In contrast, a relatively stronger secondary flow characterises the peak efficiency operating point and features through both the diffuser and the entire discharge volute. A separated flow region that establishes itself through the diffuser is clearly evident, indicating that the diffuser angle may be larger than optimal at the peak efficiency operating point.

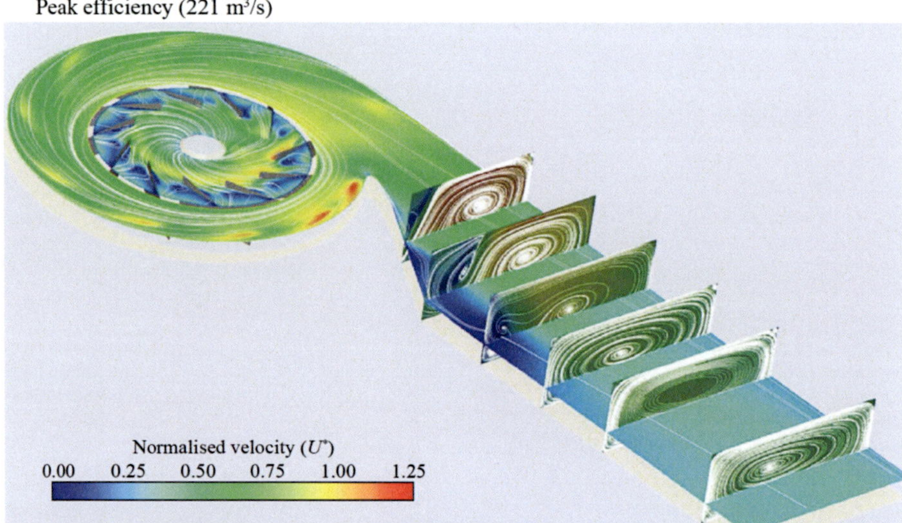

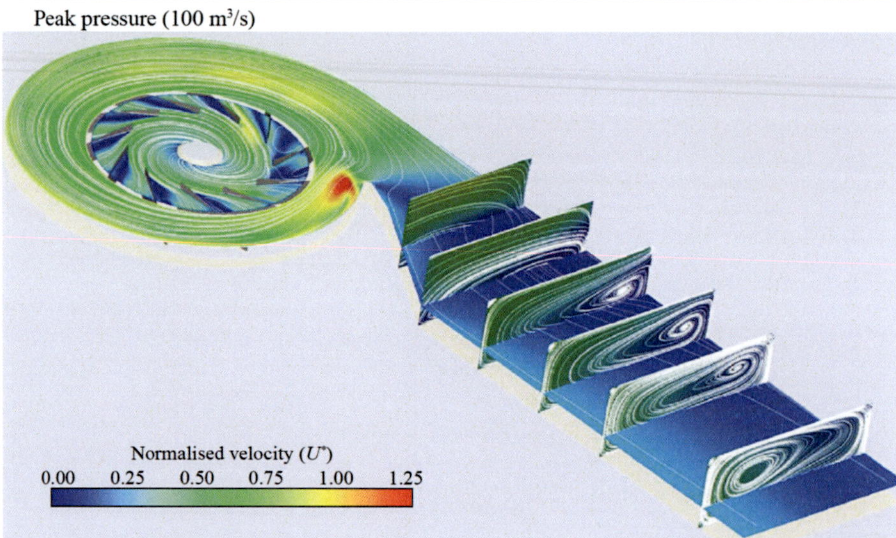

FIGURE 11.10. Secondary flow feature visualisation using contours of normalised velocity (u^{+}) through the fan volute and discharge duct for peak efficiency (top) and peak pressure (bottom) operating points.

Particle erosion

To facilitate a study of the erosive process we consider only the impeller, which is the component most susceptible to the erosive effect of particles passing through the fan. We undertook particle cloud tracking simulations using the predicted flow-field for the peak efficiency operating point. We chose to use 55 particle clouds, each containing 50 million 5 μm diameter spherical non-rotating and non-reacting particles, Table 11.7. The particle clouds were evenly distributed across the impeller inlet, Figure 11.11.

The particle clouds enter a blade passage transported by the flow-field and approach the blade leading edge, Figure 11.12. As they pass through the blade-to-blade passages, the combined effect of centrifugal force and secondary flow structures drives the particle cloud trajectories. The impeller's rotation induces the centrifugal force. The blade geometry induces secondary flows. The combined effect of these forces on particle cloud trajectories is that as they move through the blade-

Table 11.7. *Main characteristics of the simulated particles.*

Particle density, ρ_p	*Particle diameter, d_p*	*Initial velocity*
2,000 kg/m^3	5 μm	Local flow velocity

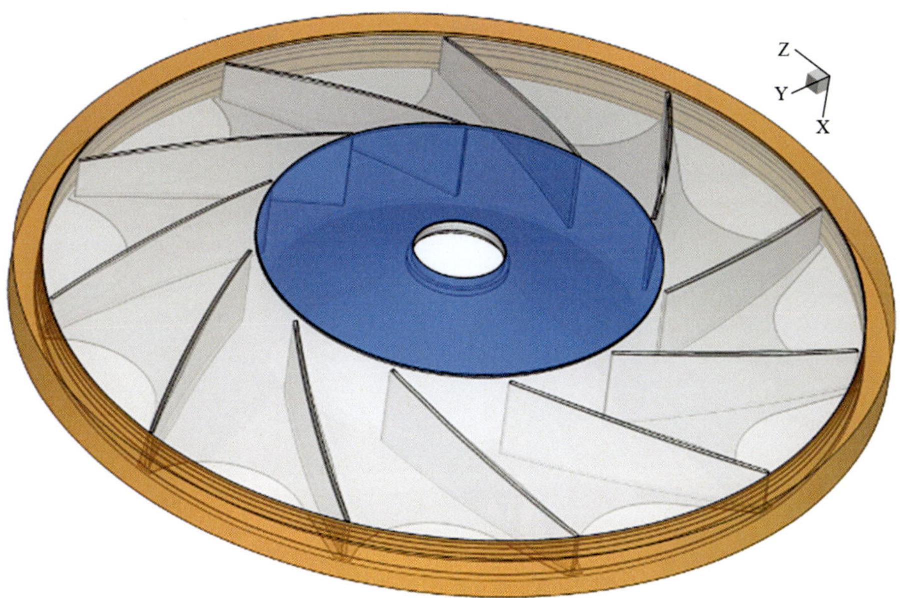

FIGURE 11.11. The computational domain for particle transport and erosion. The particle clouds enter the impeller at the fan inlet, shown in blue, and exit at the periphery of the impeller, shown in orange.

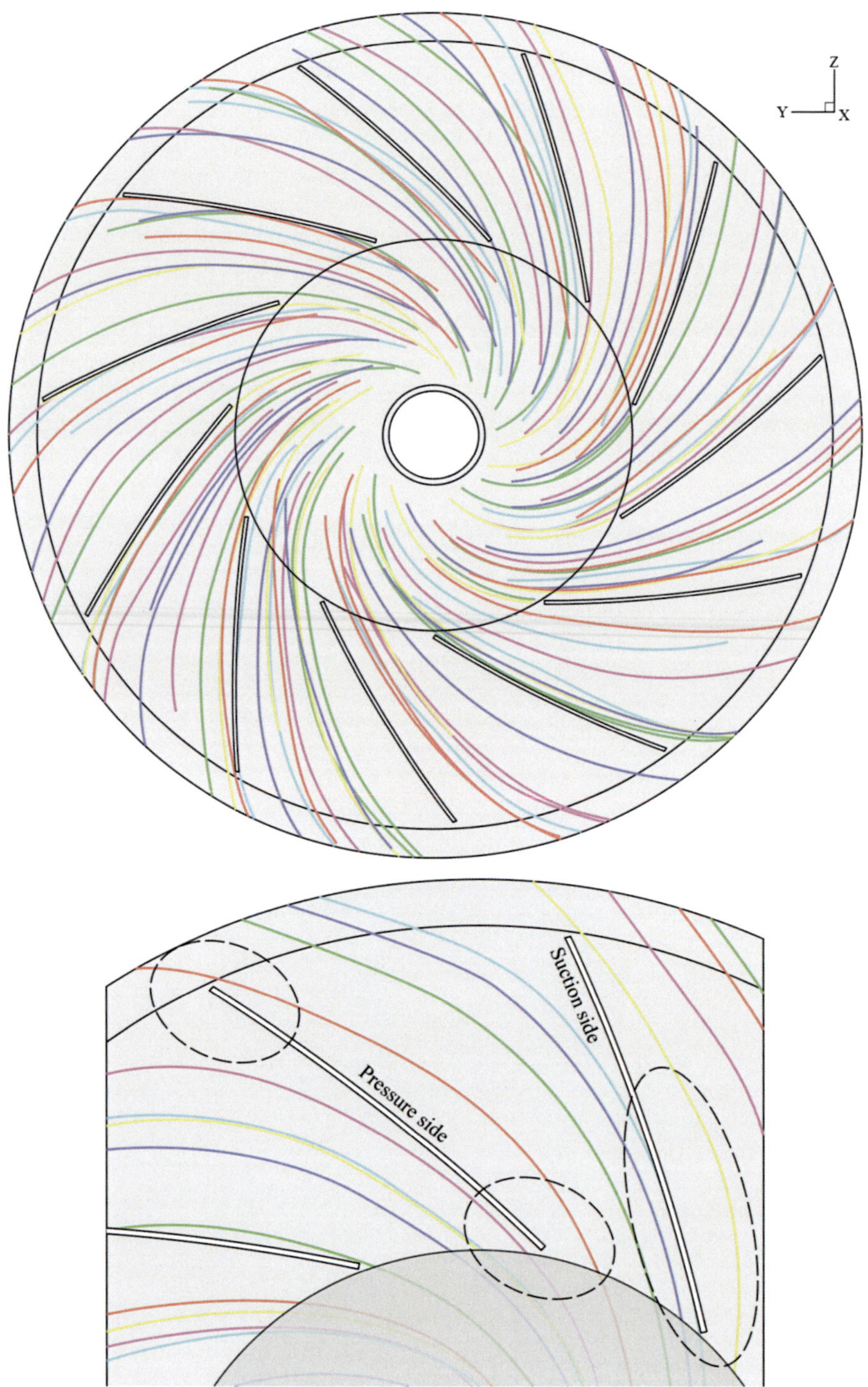

FIGURE 11.12. Cloud trajectories used in the numerical simulation (top), and blade passage details (below) for the peak efficiency operating point. The ellipses indicate the zones where trajectories are closer to blade leading and trailing edges.

to-blade passages, they are deflected away from the blade leading edge region. An effect of this deflection is to drive the particle clouds towards the blade pressure surface as they approach the trailing edge, Figure 11.12. Consequently, the particle cloud trajectories result in a high particle impact frequency over the blade pressure surface at the trailing edge. A prediction of high impact frequency is indicative of a high erosion rate.

We may normalise the erosion rate calculated using Equation two using the maximum predicted value of erosion rate to facilitate interpretation. We present erosion rates for the top and bottom of the impeller, Figure 11.13, and the top of the impeller and blades, Figure 11.14. The prediction of normalised erosion rate indicates that it is not uniform, particularly over the impeller's rear face where erosion is concentrated over the conical surface opposite the fan inlet and in the transition region between the impeller's bottom and the blades. The erosion patterns over the blades are not uniform from blade passage to blade passage. This non uniformity occurs as a consequence of the asymmetry of the flow within the impeller which is driven by the interaction between the impeller and the volute within which it is housed.

The prediction of erosion rate has been successful as the predicted erosion rate is not uniform and may be correlated with known features of the flow-field. However, the predicted erosion rate distribution over the impeller's top and rear face and the blades is not in good agreement with the distributions we observed when we remove impellers from service in an erosive application. A possible reason for the difference between predicted erosion rate and observed erosion patterns is the limitations of the chosen computational technique which is a steady technique. An unsteady technique would enable a more complete simulation of the flow-field physics by considering the unsteady coupling between the impeller and its housing.

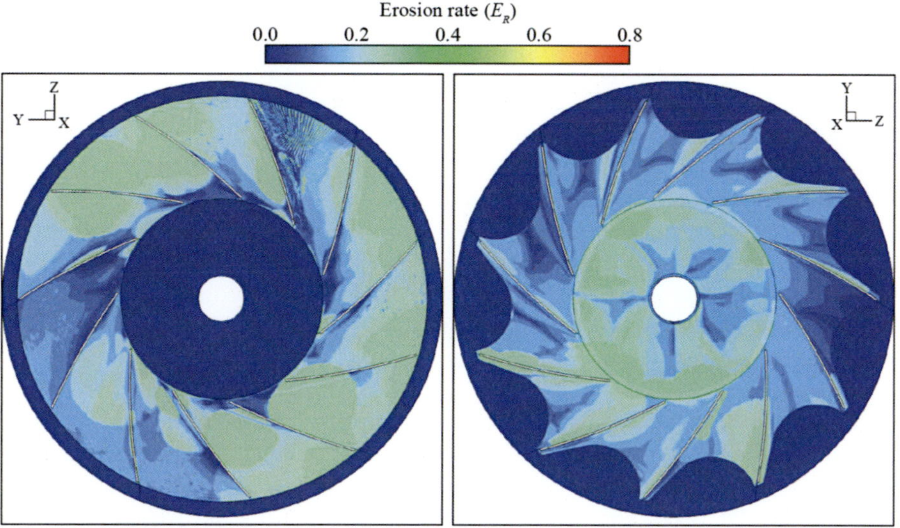

FIGURE 11.13. Normalised erosion rate (E_R) over the top of the impeller, left and bottom of the impeller, right for the peak efficiency operating point.

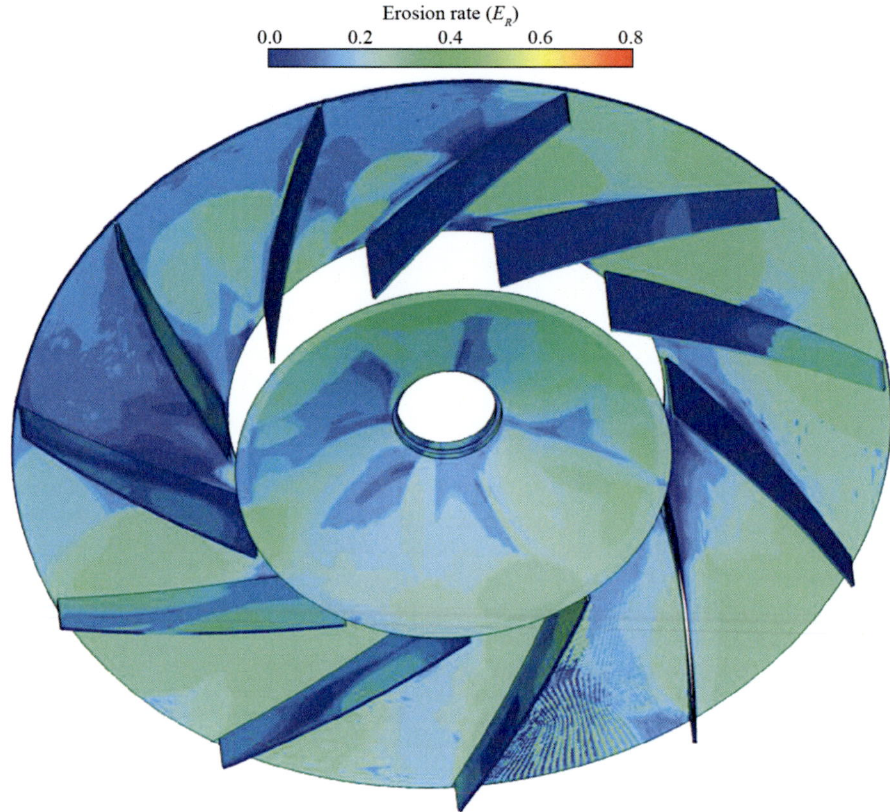

FIGURE 11.14. Normalised erosion rate (E_R) over the impeller blades for the peak efficiency operating point.

Modelling the unsteady coupling between the impeller and its housing is necessary to reproduce the evolution of flow-field features in the blade-to-blade passages, and the interaction of those flow-field features with the fan's discharge volute.

The difference between erosion rate when predicted with the steady computational technique used in the programme of work reported in this chapter and observed over impellers removed from service is significant. This difference indicates that modelling the unsteady coupling between the impeller and its housing may be critical when predicting the trajectory of erosive particles, and therefore on the resultant prediction of erosion rate patterns. A disadvantage of unsteady computational techniques is that they require approximately one order of magnitude more computational effort than an equivalent steady technique. This increase in computational effort takes unsteady techniques beyond that practical within the air movement and control community at the time of writing. Despite this caveat, scholars working with the air movement and control community do have access to high-performance computer clusters and therefore are able to develop unsteady computational techniques. Therefore, a subject for further research is the application of an unsteady computational technique to predict the flow-field, followed by an assessment of the accuracy of subsequent erosion rate pattern predictions.

CONCLUSIONS

The use of open-source software is an alternative to commercially available software that has the potential to provide the air movement and control community with a cost effective mean of optimising the performance of industrial centrifugal fans that incorporate cambered impeller blades. The reported research in this chapter presents the development and validation of a computational technique that we implemented in the open-source solver OpenFOAM. The computational technique characterises the studied fan using a 'frozen' rotor approach and an arbitrary mesh interface. The developed technique predicted the studied fan's pressure rise and efficiency reasonably accurately at a peak pressure operating point when stall is incipient. This success occurred despite the frozen rotor modelling assumption that neglected unsteady interaction between the impeller and its volute casing and the use of a basic turbulence model. An analysis of the flow-field within the impeller, volute casing and discharge duct indicated that the computational technique provides a credible prediction of the fan's secondary flow-field. Secondary flow structure predictions are possible in spite of the presence of extensive separated flow regions, particularly at the peak pressure operating point.

The success of the developed computational technique when implemented in the open-source solver OpenFOAM provides the air movement and control community with a cost effective method for optimising industrial centrifugal fan performance when the fan incorporates an impeller with cambered plate blades. The regulatory environment facing the air movement and control community has made minimum fan and motor efficiency grades a legally binding requirement within Europe, and will do so in the USA by 2019. This regulatory environment will result in the majority of industrial centrifugal fans with cambered plate impeller blades not meeting planning minimum efficiency levels, and therefore it is imperative that designers optimise the performance of their portfolio of industrial centrifugal fans that incorporate impellers with cambered plate blades. The low cost, robustness and accuracy of the developed computational technique when implemented in the open-source solver OpenFOAM provides a practical way for designers to engage in the optimisation process.

We used the predicted fan flow-field as an input to a particle cloud tracking model that predicted the trajectory of particles through the fan. One may use a prediction of particle trajectories to predict impact frequency of particles over the fan's front and rear face and the blade pressure and suction surface. In turn, one may use this prediction of impact frequency to predict erosion rate distributions. The predicted erosion rate patterns indicated that 5 μm particles resulted in the highest erosion rates over the impeller back plate over the conical section opposite the fan inlet. Despite success in the prediction of erosion rate patterns, the resultant patterns are not in good agreement with those we observed over impellers removed from service in erosive applications. The difference indicates that the steady computational approach may be adequate to predict fan performance, but not the trajectory of particles. We conclude that a more accurate prediction of particle trajectories requires an unsteady computational technique that models the coupling between the impeller and its housing.

REFERENCES

ISO 5801 (2007), *Industrial Fans – Performance Testing Using Standardized Airways.*

Baxter, L.L. (1989), 'Turbulent Transport of Particles', PhD thesis, Brigham Young University, Provo, UT, USA.

Borello, D., Venturini, P., Rispoli, F. and Saavedra G.Z.R. (2013), 'Prediction of Multiphase Combustion and Ash Deposition within a Biomass Furnace', *Applied Energy*, vol. 101, pp. 413–422.

Corsini, A., Marchegiani, A., Rispoli, F., Sheard, A.G. and Venturini, P. (2012), 'Predicting Blade Leading Edge Erosion in an Axial Induced Draft Fan', *Transactions of the ASME, Journal of Engineering for Gas Turbines & Power*, vol. 134, paper no. 042601, pp. 1–9.

Corsini, A., Rispoli, F., Sheard, A.G. and Venturini, P. (2013a), 'Numerical Simulation of Coal Fly-ash Erosion in an Induced Draft Fan', *Transactions of the ASME, Journal of Fluids Engineering*, vol. 135, paper no. 081303, pp. 1–12.

Corsini, A., Delibra, G. and Sheard, A.G. (2013b), 'A Critical Review of Computational Methods and Their Application in Industrial Fan Design', *International Scholarly Research Network, Mechanical Engineering*, vol. 2013, article ID 625175, 20 pages, http://dx.doi.org/10.1155/2013/625175.

Daly, B.B. (1978), *Woods Practical Guide to Fan Engineering*, Woods of Colchester, Colchester, UK.

Denton, J.D. and Dawes, W.N. (1999), 'Computational Fluid Dynamics for Turbomachinery Design', *Proceedings of the IMechE Part C, Journal of Mechanical Engineering Science*, vol. 213, pp. 107–124.

Eck, B. (1973), *Fans: Design and Operation of Centrifugal, Axial-flow, and Cross-flow Fans*, Pergamon Press, Oxford, UK.

Ferziger, J.H. and Peric, M. (2002), *Computational Methods for Fluid Dynamics*, Springer, London, UK.

Godichon, A.F.E. and Sheard, A.G. (2009), 'Customer-oriented Design', *International Cement Review*, October, pp. 95–99.

Greenzweig, J., Henry, R. and Holm, T. (2011), 'The Effect of Inlet Flow Distortion on the Performance of Centrifugal Fans for Utility Applications', *Proceedings of the 56th American Society of Mechanical Engineers Turbine and Aeroengine Congress*, Vancouver, Canada, 6–10 June, paper no. GT2011-45422.

Jain, S. (1995), 'Three-dimensional Simulation of Turbulent Particle Dispersion', PhD thesis, University of Utah, Salt Lake City, UT, USA.

Jasak, H. (2010), 'OpenFOAM: a Year in Review', *Proceedings of the Fifth OpenFOAM Workshop*, Gothenburg, Sweden, 21–24 June.

Kær, S.K. (2001), 'Numerical Investigation of Ash Deposition in Straw-fired Furnaces', PhD thesis, Aalborg University, Aalborg, Denmark.

Khelladi, S., Koudiri, S., Bakir, F. and Rey, R (2005), 'Flow Study in the Impeller-diffuser Interface of a Vaned Centrifugal Fan', *Transactions of the ASME, Journal of Fluids Engineering*, vol. 127(3), pp. 495–502.

Launder, B.E. and Sharma, B.R. (1974) 'Application of the Energy-dissipation Model of Turbulence to the Calculation of Flow Near a Spinning Disc', *Letters in Heat and Mass Transfer*, vol. 1(2), pp. 131–137.

Lee, Y.T. (2010), 'Impact of Fan Gap Flow on the Centrifugal Impeller Aerodynamics', *Transactions of the ASME, Journal of Fluids Engineering*, vol. 132, pp. 1–7, paper no. 091103.

Lee, Y.T., Ahuja, V., Hosagandi, A., Slipper, M.E., Mulvihill, L.P., Birkbeck, R. and Coleman, R. (2011), 'Impeller Design of a Centrifugal Fan with Blade Optimization', *International Journal of Rotating Machinery*, vol. 2011, paper no. 537824, pp. 1–16.

Litchford, R.J. and Jeng, S.M. (1991), 'Efficient Statistical Transport Model for Turbulent Particle Dispersion in Sprays', *AIAA Journal*, vol. 29, pp. 1443–1451.

Petit, P., Page, M., Beaudoin, M. and Nilsson, H. (2009), 'The ERCOFTAC Centrifugal Pump OpenFOAM Case Study', *Proceedings of the 3rd IAHR International Meeting of the Workgroup on Cavitation and Dynamic Problems in Hydraulic Machinery and Systems*, Brno, Czech Republic, 14–16 October.

Seo, S.J., Kim, K.Y. and Kang, S.H. (2003), 'Calculations of Three-dimensional Viscous Flow in a Multi-blade Centrifugal Fan by Modelling Blade Forces', *Proceedings of the IMechE Part A, Journal of Power and Energy*, vol. 217, pp. 287–297.

Sommerfeld, M., van Wachem, B. and Oliemans, R. (2008), *Best Practice Guidelines for Computational Fluid Dynamics of Dispersed Multiphase Flows*, European Research Community on Flow, Turbulence and Combustion (ERCOFTAC), Brussels, Belgium.

Tabakoff, W., Kotwal, R. and Hamed, A. (1979), 'Erosion Study of Different Materials Affected by Coal Ash Particles', *Wear*, vol. 52(1), pp. 161–173.

Thakur, S., Lin, W. and Wright, J. (2002), 'Prediction of Flow in Centrifugal Blower using Quasi-steady Rotor-stator Models', *Journal of Engineering Mechanics*, vol. 128, pp. 1039–1049.

Vasudeva Karanth, K. and Yagnesh Sharma, N. (2009), 'CFD Analysis on the Effect of Radial Gap on Impeller-diffuser Flow Interaction as Well as on the Flow Characteristics of a Centrifugal Fan', *International Journal of Rotating Machinery*, vol. 2009, paper no. 293508, pp. 1–8.

Venturini, P. (2010), 'Modelling of Particle Wall-deposition in Two Phase Gas-solid Flows', PhD thesis, Sapienza University of Rome, Rome, Italy.

Venturini, P., Borello, D., Iossa, C., Lentini, D. and Rispoli F. (2010), 'Modelling of Multiphase Combustion and Deposit Formation in a Biomass-fed Boiler', *Energy*, vol. 35(7), pp. 3008–3021.

Wang, L.P. (1990) 'On the Dispersion of Heavy Particles by Turbulent Motion', PhD thesis, Washington State University, Pullman, WA, USA.

Weller, H.G., Tabor, G., Jasak, H. and Fureby, C. (1998), 'A Tensorial Approach to Continuum Mechanics Using Object-oriented Technique', *Computers in Physics*, vol. 12(6), pp. 620–631.

Zhang, M.J., Pomfret, M.J. and Wong, C.M. (1996), 'Three-dimensional Viscous Flow Simulation in a Backswept Centrifugal Impeller at the Design Point', *Computers & Fluids*, vol. 25(5), pp. 497–507.

Zhao, Y., Song, L., Wenqi, H., Weixiong, W., Dongtao, H. and Zhichi, Z. (2005), 'Numerical Simulation of Flow Field for a Whole Centrifugal Fan and Analysis of the Effects of Blade Inlet Angle and Impeller Gap', *HVAC&R Research*, vol. 11(2), pp. 263–283.

Chapter 12

Simulation of Particle-laden Flows in a Large Centrifugal Fan for Erosion Prediction

L. Cardillo, A. Corsini, G. Delibra, F. Rispoli, A.G. Sheard and P. Venturini

ABSTRACT

Industrial fans utilised in power generation, cement and steel applications must operate as part of a process that produces erosive particles. Over time these erosive particles erode centrifugal fan impeller blades, changing the blade profile and consequently, degrading fan performance. To replace the eroded impellers, operators must shut down the process. If one must replace an impeller between scheduled maintenance intervals, the costs associated with lost production become significant. Consequently, the air movement and control community is interested in predicting the erosion, and ultimately, a fan impeller's in-service life when operating in an erosive environment. Fan designers face challenges when attempting to predict impeller erosion. Centrifugal fan impeller blades are routinely constructed from cambered plate with backward sweep blades that are inevitably associated with regions of separated flow. This separated flow is highly three-dimensional making difficult an accurate prediction of the blade-to-blade flow-field. Assuming that one can accurately predict this three-dimensional blade-to-blade flow-field, one must then go on to simulate the erosive particles' trajectory.

This chapter builds on the work of other scholars who have developed a computational approach that predicts the performance of a centrifugal fan incorporating cambered plate impeller blades. We report an unsteady numerical analysis using the finite volume open-source code OpenFOAM. The analysis uses a moving mesh technique, based on arbitrary mesh interface (AMI) technology. We solved Reynolds-averaged Navier–Stokes (RANS) equations for incompressible flow with a non-linear first order turbulence closure. We modelled particle transport and dispersion using a Lagrangian approach coupled with a particle cloud tracking (PCT) model. A study of the effect of particle size facilitates identifying critical

This chapter is a revised and extended version of Cardillo, L., Corsini, A., Delibra, G., Sheard, A.G. and Venturini, P. (2014), 'Simulation of Particle-laden Flows in a Large Centrifugal fan for Erosion Prediction', *Proceedings of the 59th American Society of Mechanical Engineers Turbine and Aeroengine Congress*, Dusseldorf, Germany, 16–20 June, paper no. GT2014-25865.

regions on the impeller blades that are most prone to erosion for each combination of particle sizes. Identifying the most critical regions thus provides a basis for modifying overall impeller and individual blade geometry in an effort to reduce susceptibility to erosion. This increases in-service life, and consequently the time between maintenance intervals.

NOMENCLATURE

Latin letters

b_i	[mm]	Impeller blade width
b_v	[mm]	Volute width
C	[Nm]	Torque
C_D		Particle drag coefficient
d_p	[μm]	Particle diameter
D_{in}	[mm]	Impeller inlet diameter
D_{out}	[mm]	Impeller outlet diameter
D_v	[mm]	Volute outlet diameter
D_t	[mm]	Blade tip diameter
E_R		Erosion rate
f_α		Erosion angular function
f_r	[rpm]	Rotational frequency
g	[m²/s]	Acceleration of gravity
h_o		Impeller outlet height
I_f		Impact frequency
L_t	[m]	Turbulence length scale
k	[m²/s²]	Turbulent kinetic energy
K_c, K_1, K_2, K_{12}		Erosion model coefficients
Δp_t	[kPa]	Total pressure rise
Re	[-]	Reynolds number: $Re = V_{tip}D_t / \nu = 39 \times 10^6$
Stk		Stokes number
t	[s]	Time
T		Integral time scale
T_L		Lagrangian integrated time scale
T_{mE}		Eulerian integral time scale
u, v, w	[m/s]	Rotation, absolute and relative velocity
u_{bc}		Boundary cell rotation velocity
u_m		Cloud's centre of mass rotation velocity
u_τ		Friction velocity
V	[m³/s]	Volume flow rate
V_{tip}	[m/s]	Tip velocity
v_{bc}		Boundary cell absolute velocity
v_{drift}		Drift velocity vector at the cloud centre: $v_{drift} = u_m - v_m$
v_m		Cloud's centre of mass absolute velocity
$\bar{v}$	[%]	Mean volume
w^+		Relative velocity magnitude

x, y, z		Cartesian coordinates
y_n	[m]	Distance from the wall
y^+	[-]	Non dimensional wall distance: $y^+ = y_n u_\tau / v$

Greek letters

α		Impact angle
β		Exchange coefficient
ε		Turbulent kinetic dissipation rate
η	[%]	Efficiency
ν	[m^2/s]	Kinematic viscosity
ν_t	[m^2/s]	Turbulent kinematic viscosity
ρ_f	[kg/m^3]	Fluid density
ρ_p	[kg/m^3]	Particle density
$\sigma(t)$		Standard deviation of the particle position
τ_L		Lagrangian time scale
τ_f		Fluid time scale
τ_p		Particle time scale
ω	[1/s]	Angular velocity

Acronyms

AMI	Arbitrary mesh interface
CFD	Computational fluid dynamics
LE	Leading edge
PCT	Particle cloud tracking
PDF	Probability density function
QUICK	Quadratic upstream interpolation for a convective kinetics
RANS	Reynolds-averaged Navier–Stokes
TE	Trailing edge
TI	Turbulent intensity
U-RANS	Unsteady Reynolds-averaged Navier–Stokes

INTRODUCTION

Centrifugal fans are applied into a wide range of industrial applications. The smallest are used to cool electronic equipment whilst the largest classically serve cement, steel and power applications. Each application has its own unique challenges, requiring application specific knowledge for a trouble-free in-service life. When considering large centrifugal fan applications, a factor that challenges reliable in-service operation is the erosive effect of the two-phase flow passing through the fans. Engineers associate cement, steel and power applications with the presence of solid particles which the process generates.

Because of the relatively high velocities through a centrifugal fan, the solid particles in the gas stream impact with high velocity on both static and rotating components. Impacting particles erode the components, with the loss of material resulting in a change in blade geometry with an associated loss in fan efficiency and pressure developing capability. Over time the effect of erosion is to degrade fan performance and ultimately limit in-service life. Consequently, the air movement and control community has an interest in minimising erosion rates through optimising fan geometry.

One may use a numerical simulation of the blade-to-blade flow-field through a fan to predict the static pressure field and velocities. One may then go on to use that blade-to-blade flow-field prediction to predict the trajectory of particles' with mass as they pass through the fan. In combination with laboratory data which helps researchers relate particle mass, velocity and angle of impact to erosion rate, numerical simulations can provide an estimate of erosion patterns over impeller surfaces.

Corsini *et al.* (2012, 2013a) have previously reported a computational method to simulate axial fan blade erosion. The axial fans were variable pitch in motion fans which Corsini *et al.* (2012, 2013a) applied in an induced draft application removing combustion products from a boiler burning low-grade coal. The erosion rate predictions were in reasonable agreement with qualitative data. Corsini *et al.* (2012, 2013a) obtained qualitative blade erosion pattern data when they inspected the fan blades during scheduled maintenance. The research reported in this chapter applies the developed numerical technique for application with an axial fan in an induced draft power generation application to a centrifugal fan in a cement application, Figure 12.1.

There are few reports in the open literature on centrifugal fan erosion. There are some experimental and numerical studies primarily focused on centrifugal fans used in coal-fired power plants (Rugang, 2000; Mei and Xing, 2007; Hu *et al.*, 2008). Whilst a welcome contribution, these studies are aimed primarily at understanding the erosive process, not on providing fan designers with the necessary insight to optimise overall impeller or individual blade geometry to minimise susceptibility to the erosive effect of different size particle combinations. Predicting the effect of different particle size combinations is important as particle samples are typically available to the designer prior to undertaking the fan design. Consequently, they know the particle distribution and concentration, and if a suitable method is available, may use it to predict the effect on erosion rate and therefore, service life.

DESCRIPTION OF THE TEST CASE

The studied fan is part of a family of fans intended for process industry application with a pressure developing requirement of up to 20 kPa, Table 12.1. The family includes impellers with backward-swept cambered plate blades that are the subject of the reported research, Figure 12.2. A double-inlet with the inlet plenum incorporating 90 degree bends and a double impeller characterises the chosen fan, Figure 12.1 and Table 12.2. We experimentally derived the studied fan's aero-dynamic performance in accordance with ISO 5801 requirements (ISO 5801, 2007). We conducted the studied fan's numerical analysis at its peak efficiency operating point with a 221m^3/s flow rate, a 17 kPa pressure and 82 per cent efficiency.

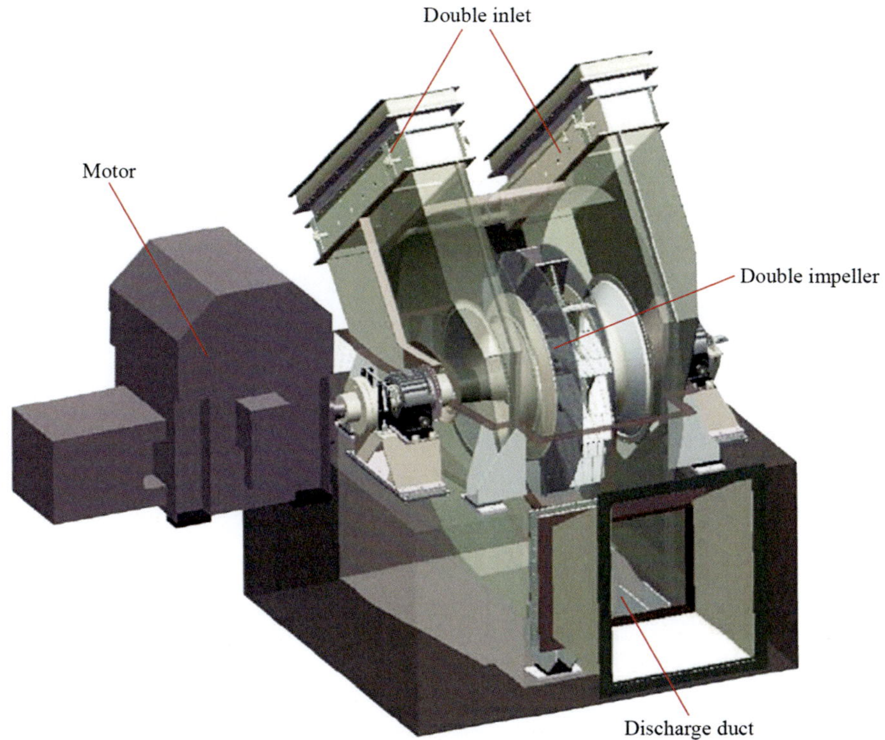

FIGURE 12.1. An example of a centrifugal fan with double inlet and cambered plate blades from the studied family of industrial fans.

Table 12.1. *The studied family of fans operating range.*

Blade tip diameter D_t	3 to 4 metres
Performance standard	ISO 5801 – ISO 13347
Volume flow rate V	Up 750 m^3/s
Total pressure rise Δp_t	Up to 50 kPa
Rotational speed ω	900–1,000 rpm
Blade sections	Cambered plate

Table 12.2. *The studied fan geometry.*

Impeller inlet diameter D_{in}	1,804 mm
Impeller outlet diameter D_{out}	3,440 mm
Volute outlet diameter D_v	5,600 mm
Impeller blade width b_i	400 mm
Volute width b_v	200 mm
Impeller blade count	11
Rotational speed ω	900 rpm

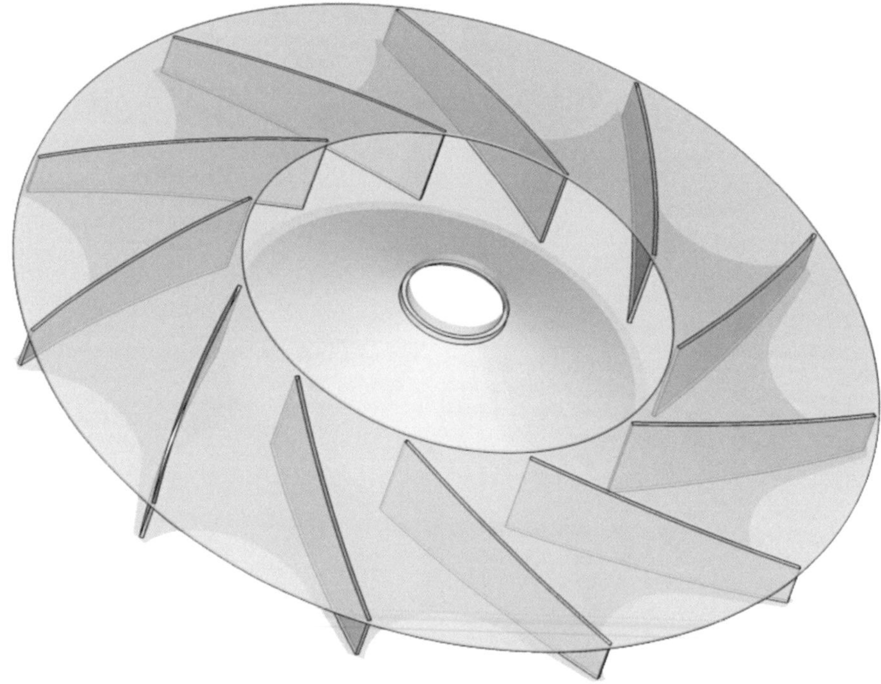

FIGURE 12.2. The studied fan centrifugal impeller with backward-swept cambered plate blades.

NUMERICAL METHODOLOGY

When modelling centrifugal fans, one cannot reduce the computational effort by modelling a single blade passage with imposed periodic boundary conditions. Centrifugal fan simulations must be full rotor simulations to account for the interaction between the impeller and housing within which it is situated. As a consequence of housing geometry, the flow-field through any single blade-to-blade passage is inherently periodic and consequently, requires more computational effort than an axial fan single blade passage flow-field simulation. Additionally, the flow-field through a centrifugal fan impeller is highly three-dimensional and if the blades are rolled flat plate, separated flow regions typically characterise the blade-to-blade flow-field. Therefore, the flow-field through a centrifugal fan is inherently both periodic and unsteady, making an unsteady prediction of the flow-field necessary to adequately model the flow-field physics.

Fluid phase U-RANS solver

Within the air movement and control community, fan designers favour the 'frozen rotor' approach as it is able to predict the overall fan performance characteristics with good accuracy (Corsini *et al.*, 2013b). The primary advantage of the frozen rotor approach is that the impeller-to-housing coupling does not require a moving mesh. The relative position of the two rows is fixed in time, with relative motion managed using different frames of reference and adding Coriolis and centrifugal forces to the momentum equation in the rotating frame of reference. The frozen rotor approach is able to account for a non-uniform circumferential distribution of velocity and pressure. This leads to a more realistic representation of wake mixing, and consequently, a more accurate simulation of the blade-to-blade flow-field.

Although the air movement and control community has embraced widely the frozen rotor approach, this does not mean that the approach is without drawbacks. Adding Coriolis and centrifugal forces to the momentum equation in the rotating frame of reference constitutes an approximation of the flow-field physics that limits the resulting simulations' accuracy. Cardillo *et al.* (2014) predicted the studied fan's pressure rise and efficiency using the frozen rotor approach. The resultant erosion patterns were not in good agreement with those observed over impellers removed from service in erosive applications. Cardillo *et al.* (2014) concluded that the difference indicates that a steady computational approach is not adequate to predict the particles' trajectory.

To accurately simulate centrifugal fan flow-field physics it is necessary to undertake a fully unsteady coupling of the impeller and housing. This requires a moving mesh that can increase the required computational effort by up to an order of magnitude compared to a frozen rotor simulation. The increase in computational effort is a consequence of the required time to account for the interaction between the impeller and housing. Despite the increase in computational effort, we chose to use a moving mesh in the research reported in this chapter. We utilised an open-source finite volume code OpenFOAM 2.2.x to solve unsteady Reynolds-averaged Navier–Stokes (U-RANS) incompressible equations (Weller *et al.*, 1998).

We spatially resolved the coupling between the impeller and volute casing computational domains over the impeller's periphery using an arbitrary mesh interface (AMI) (Farrell and Maddison, 2011). We solved the incompressible Reynolds-averaged Navier–Stokes equations using the finite volume method and a standard k-ε turbulence model with the synthetic wall-treatment available in OpenFOAM (Launder and Sharma, 1974).

We carried out computations using an unsteady Reynolds-averaged Navier–Stokes solver with second order quadratic upstream interpolation for a convective kinetics (QUICK) divergence scheme and a GAMG solver for pressure. We solved other equations using OpenFOAM's *smoothSolver* and conjugate gradient solver for the inner and final iterations of the PIMPLE algorithm respectively. In all cases, we set the convergence tolerance to 10^{-7} for all the computed quantities. We carried out simulations on the Fermi HPC system of Cineca over 264 processors. Full convergence required 150 hours of computational time, corresponding to 2.2 revolutions of the impeller starting from initial conditions. The flow-field's initial stabilisation required 0.7 of the 2.2 revolutions.

Solid phase solver

We carried out the numerical prediction of particle transport and erosion process by using Lagrangian computation of particle dispersion either via single particle tracking or via particle cloud tracking (Corsini *et al.*, 2013a). Single particle tracking involves computing single particle trajectories at run-time using the calculated flow-field parameters at the particle's current location. In contrast, particle cloud tracking involves tracking a statistical representation of an entire particle cloud, only computing the cloud's centre trajectory (Baxter, 1989; Wang, 1990; Litchford and Jeng, 1991; Baxter and Smith, 1993; Jain, 1995; Kær, 2001). Single particle tracking is the more accurate approach. However, the computational effort required for single particle tracking is between one and two orders of magnitude larger than that required for particle cloud tracking. For this reason, the air movement and control community favour particle cloud tracking (Venturini *et al.*, 2010; Corsini *et al.*, 2012, 2013b; Borello *et al.*, 2013).

The cloud equation of motion is the ensemble average version of the Basset-Boussinesq-Oseen equation that is Newton's second law, reduced in accordance to the application's constraints (Baxter, 1989):

$$\frac{d\langle \mathbf{v}\rangle}{dt} = \langle\beta\rangle\cdot[\langle \mathbf{u}\rangle - \langle \mathbf{v}\rangle] + \left(1 - \frac{\rho_f}{\rho_p}\right)\mathbf{g} \tag{1}$$

where $\langle\ \rangle$ denotes the ensemble average, and $\langle\beta\rangle$ reads:

$$\langle\beta\rangle = \frac{3}{4d_p} C_D \frac{\rho_f}{\rho_p} |\langle \mathrm{u}\rangle - \langle \mathrm{v}\rangle| \tag{2}$$

In the particle cloud tracking model, we assume the particle distribution within each cloud to be Gaussian and described by a probability density function *PDF*:

$$PDF(\mathbf{x}) = \frac{1}{(2\pi)^{1/2}\sigma}\exp\left[-\frac{1}{2}\left(\frac{\mathbf{x} - \langle \mathbf{x}\rangle}{\sigma}\right)^2\right] \tag{3}$$

In Equation (3), $\sigma(t)$ represents the square root of the particle position's variance and accounts for the particles' turbulent dispersion:

$$\sigma^2 = 2\langle \mathbf{v}'^2\rangle\left[\frac{t}{\tau_L} - \frac{1}{\tau_L^2}\left(1 - e^{-t/\tau_L}\right)\right] \tag{4}$$

where:

$$\tau_L = \max(\tau_f, \tau_p) \tag{5}$$

is the Lagrangian integral time scale, the maximum between the particle motion and the turbulence time scale, and:

$$\langle \acute{\mathbf{v}}^2 \rangle = \langle \acute{\mathbf{u}}^2 \rangle \left(1 - e^{-\tau_f/\tau_p}\right) = \frac{2}{3} k \left(1 - e^{-\tau_f/\tau_p}\right) \tag{6}$$

is the cloud radius that varies in time according to the turbulent properties, that we assume it to be equal to 3σ.

Erosion modelling

Erosion rate is modelled as a function of the particles' impact velocity, of the impact angle and the material properties of both the particles and the target surface. Tabakoff *et al.* (1979) developed an empirical correlation for predicting the erosion per unit mass of impacting particles (E_R, in mg/g) of different materials:

$$E_R = K_1 f_\alpha |\mathbf{v}_{bc}|^2 \cos^2\alpha (1 - R_T^2) + f_i \tag{7}$$

where:

$$R_T = 1 - 0.0061|\mathbf{v}_{bc}|\sin\alpha$$

$$f_\alpha = \{1 + K_C[K_{12}\sin(90\alpha/\alpha_0)]\}^2$$

$$f_i = K_2(|\mathbf{v}_{bc}|\sin\alpha)^4$$

where α is the impact angle, K_1, K_2, K_{12} and K_C are functions of material properties of particles and target surface. We used Tabakoff *et al.*'s (1979) model in this study, and although the fan was manufactured from mild steel, we assumed it was manufactured from stainless steel. We made this assumption as stainless steel was the closest material to that from which the fan was actually manufactured for which Tabakoff *et al.* (1979) had derived material coefficients, Table 12.3. Also following Tabakoff *et al.*'s (1979) recommendation we assumed that particles impacting with an impact angle, α_0, of less than 30 degrees had no erosive effect.

Baxter's (1989) original particle cloud tracking formulation assumes that all particles within a cloud move with the cloud velocity, thus the Baxter (1989) model

Table 12.3. *Erosion model coefficients (Tabakoff* et al., *1979).*

K_C $\alpha \le 3\,\alpha_0$	1
$\alpha > 3\,\alpha_0$	0
K_1	$1.505101 \cdot 10^{-06}$
K_{12}	$2.96077 \cdot 10^{-01}$
K_2	$5.0 \cdot 10^{-12}$

did not consider the effect of velocity non-uniformities close to a solid wall. One may improve the accuracy of the particle cloud tracking approach by accounting for the presence of a solid wall. One may achieve this by modelling the distribution of the particle velocities according to the fluid velocity defect (Corsini *et al.*, 2012; Borello, 2013). We assume that the ratio between the drift and the flow velocity at the cloud's mean position is constant within a cloud. Accordingly, we computed the particle local velocity at a boundary cell which read as:

$$\|\mathbf{v}_{bc}\| = \|\mathbf{u}_{bc}\| \left(1 - \frac{|\mathbf{v}_{drift}\|}{|\,\mathbf{u}_{m}|} \right) \tag{8}$$

where $\mathbf{u}_{bc}$, $\mathbf{v}_{bc}$ were respectively the flow and particle velocity vectors in the cell's centre, $\mathbf{v}_{drift} = \mathbf{u}_m - \mathbf{v}_m$ was the drift velocity vector at the cloud's centre and $\mathbf{u}_m$ and $\mathbf{v}_m$ were respectively the flow and cloud velocity vectors at the cloud centre. One can find a full presentation of our modified particle cloud tracking method in Borello *et al.* (2013), Corsini *et al.* (2012, 2013b), Venturini *et al.* (2010) and Venturini (2010).

Numerical grid and boundary conditions

Fluid phase

In the research reported in this chapter we used the same boundary conditions as Cardillo *et al.* (2014), Table 12.4. We derived the inflow velocity profile from Cardillo *et al.*'s (2014) simulation that accounted for the presence of the inlet plenum. We modelled only one half of the impeller to exploit the impeller's symmetry and effectively reduce the necessary computational effort to predict the flow-field by half. We modelled the resultant half-fan in three parts: the inlet, the impeller and the volute casing connecting them with an arbitrary mesh interface, Figure 12.3. We used an inflow average turbulence intensity of ten per cent whilst the dissipation length scale was set to ten per cent of the mean blade chord.

We conducted a grid sensitivity analysis using pressure rise, lift and drag on the rotor blade as convergence parameters. We undertook the grid sensitivity analysis using Cardillo *et al.*'s (2014) computational grid. The computational grid required for a mesh independent solution incorporated 6.2 million hexahedra, Figure 12.4. We used approximately one million hexahedra to mesh the volute, 3.5 million for the impeller and 1.7 million for the inlet region, Table 12.5.

Solid phase solver

We used Corsini *et al.*'s (2012, 2013a) method to predict particle cloud trajectories. In the research reported in this chapter, we used 99 clouds uniformly distributed across the fan inlet. Each cloud contains 5×10^7 spherical particles, having the same size and properties. Cloud initial positions may affect the erosion

Table 12.4. *Numerical model boundary conditions.*

Inflow	Velocity profile (Cardillo *et al.*, 2014), TI = 5% , L_t=0.1c
Outflow	Zero gradient
Volute, bellmouth	No-slip conditions, wall functions for k, ε, ν_t
Impeller	v=ωR, wall functions for k, ε, ν_t

Table 12.5. *Numerical simulation mesh quality indicators and non-dimensional wall distance (y^+) values at the solid walls.*

		Impeller	
	Min.	Average	Max.
Aspect ratio	1.0	5	32.8
Minimum included angle	20.3	65.3	90
Volume ratio	1.0	1.4	22.5
Skewness	0.0	0.13	0.67
		Volute	
	Min.	Average	Max.
Aspect ratio	1.0	5.2	116.7
Minimum included angle	17.1	75.4	90
Volume ratio	1.0	1.3	2.5
Skewness	0.0	0.05	0.64
		Inlet	
	Min.	Average	Max.
Aspect ratio	1.0	4.4	33.3
Minimum included angle	17.6	76.2	89.7
Volume ratio	1.02	1.47	6.65
Skewness	0.0	0.06	0.69
		y^+	
	Min.	Average	Max.
Blades	2.0	90.6	290.9
Hub	7.5	118.0	117.9
Shroud	4.5	72.0	146.8
Volute	8.0	94.6	165.4

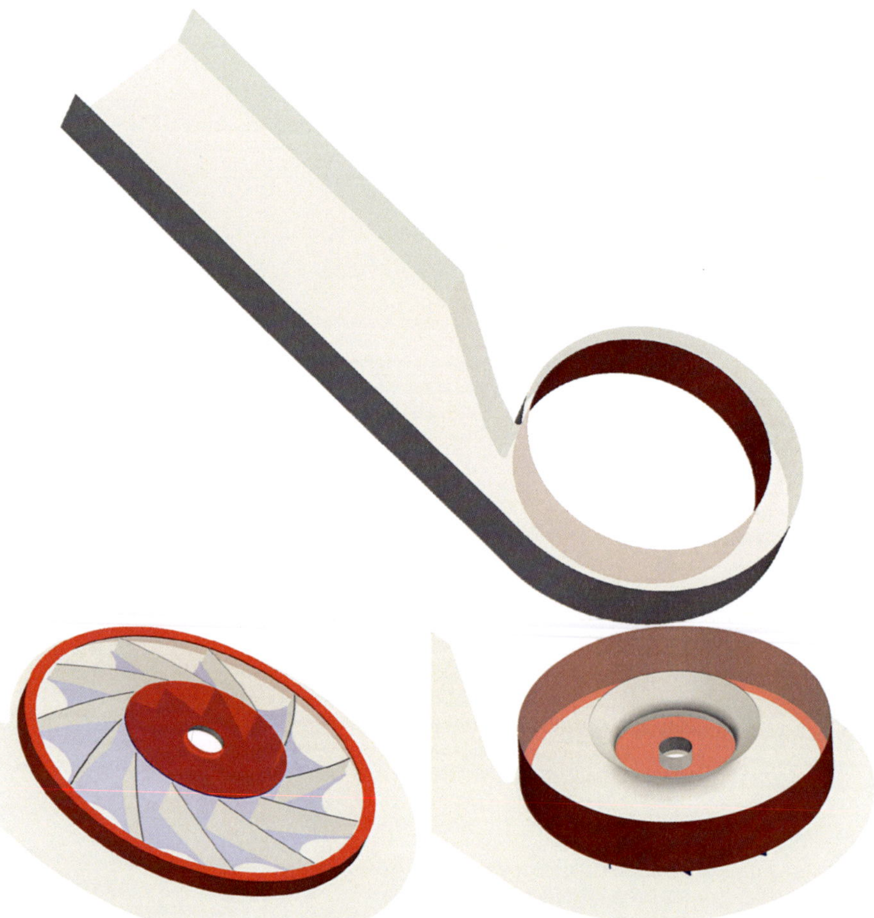

FIGURE 12.3. Details of the computational domain, with arbitrary mesh interface (AMI) surfaces in red. The computational domain is split into three regions: the volute, top; the impeller, bottom left; the inlet, bottom right.

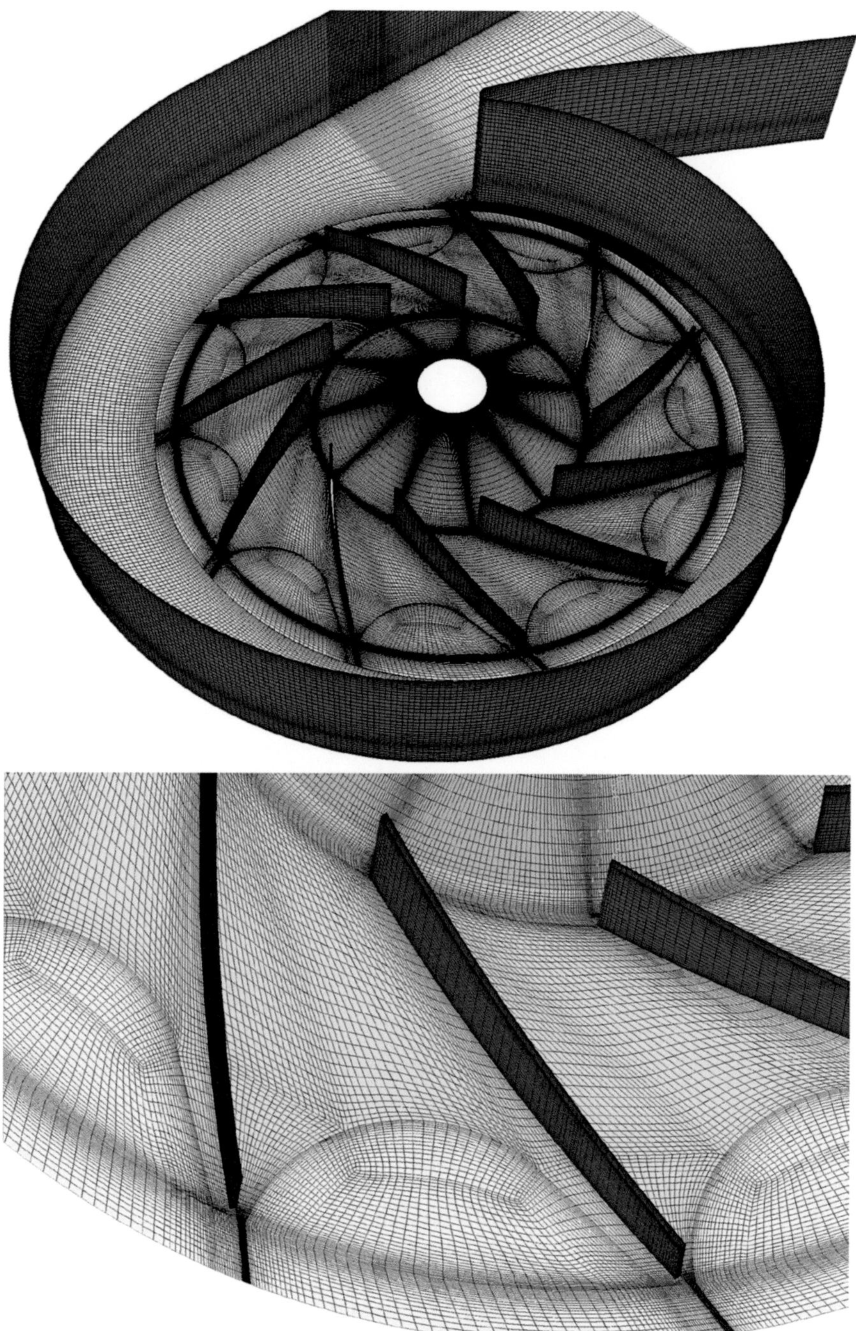

FIGURE 12.4. The computational mesh for the studied fan illustrating mesh density over the fan impeller and scroll, top, with further details of the impeller, bottom.

patterns. To minimise the effect of initial cloud position, we injected nine clouds per blade passage, covering the whole vane inlet.

The centrifugal fan that is the subject of the reported research is installed in a cement production facility. An analysis of the particle size distribution of a sample of particles from a similar fan installed in a cement facility indicates that there are two types of particle present, fine particles and course particles, Figure 12.5 and Figure 12.6 respectively. The fine particles range from 0.1 μm to 20.0 μm, having an average diameter of 5.0 μm. The coarse particles range from 160 μm to 760 μm, having an average diameter of 400 μm. An analysis of the particles' chemical composition indicated that the finer particles are the most erosive. Therefore, in the present study we predict the trajectory of only the fine particle. To simplify the modelling approach, we assumed that all fine particles had a diameter of 5.0 μm. A final simplifying assumption was to neglect the erosion's effect on the impeller front and back plate, modelling erosion of the blade-to-blade passages only. In-service experience indicates that it is blade erosion that limits impeller life, and therefore, we considered it reasonable to reduce the computational domain to the impeller only, Figure 12.7.

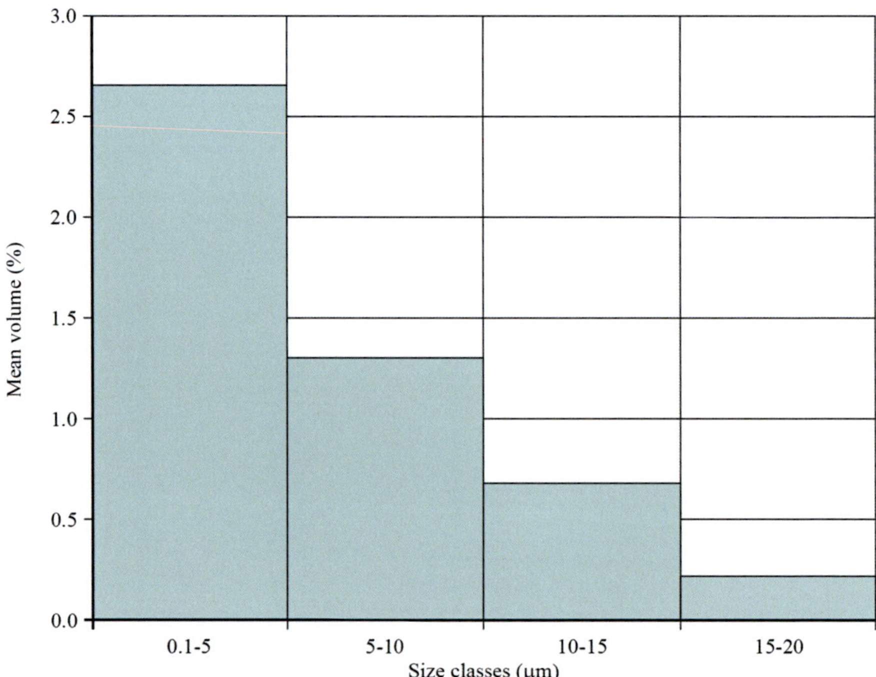

FIGURE 12.5. An analysis of the diameter of fine particles taken from a fan installed in a cement production facility. The average diameter of fine particles was 5.0 μm.

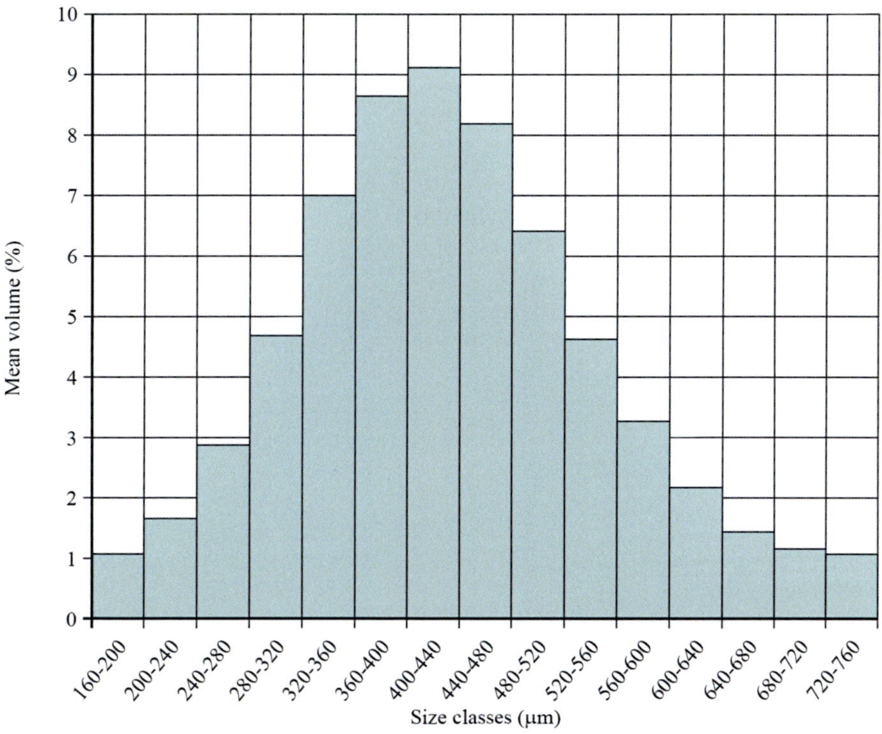

FIGURE 12.6. An analysis of the diameter of course particles taken from a fan installed in a cement production facility. The average diameter of course particles was 400 μm.

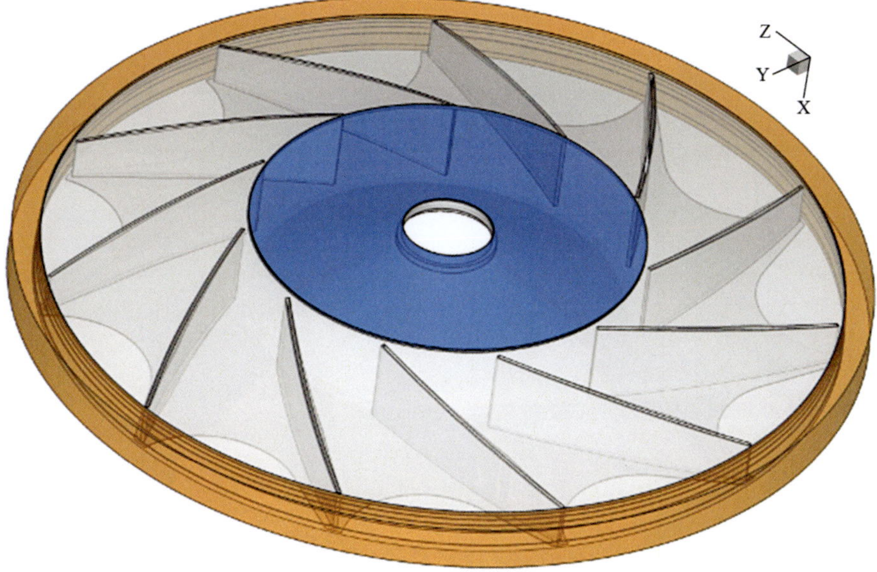

FIGURE 12.7. The computational domain for particle transport and erosion. The particle clouds enter the impeller at the fan inlet, shown in blue, and exit at the impeller's periphery, shown in orange.

RESULTS

Pressure rise and efficiency

The performance of the studied fan was first established experimentally in accordance with ISO 5801 requirements (2007). The international standard ISO 5801 allows an uncertainty in the prediction of fan efficiency of up to 3.2 per cent. The difference between the experimentally measured and predicted fan efficiency was 3.4 per cent, Table 12.6. When predicting fan efficiency using an unsteady simulation, one derives the resultant efficiency from a prediction of torque. As we use an unsteady simulation, the predicted torque is periodic, Figure 12.8. The converged unsteady simulation predicts torque with amplitude that is within one per cent of the mean torque value. This relatively low amplitude gives confidence in the accuracy of the converged solution. However, when one studies the oscillations in predicted torque, there are differences in the amplitude of oscillation with time. These differences could partly explain the 3.4 per cent difference between experimentally measured and predicted efficiency. Therefore, we considered 3.4 per cent to be close enough to the allowable uncertainty in ISO 5801 as acceptable.

Unsteady flow prediction

We may illustrate the impact of using an unsteady, as opposed to a frozen rotor or steady rotor-stator coupling model through considering the impeller-volute, Figure 12.9. We plotted velocity magnitude contours with streamlines at four different times. Each time corresponds one fourth of the impeller period of revolution and therefore every 90 degrees. Consider a fixed region in space, for example, the space between vanes B8 and B9 and at time t_0. Velocity contours do not change with time and therefore we may conclude that each vane is subjected to the same fluid dynamics as it rotates.

If this is the case, then Cardillo *et al.*'s (2014) steady state frozen rotor simulation will accurately predict fan pressure rise and efficiency. However, Cardillo *et al.* (2014) concluded that an unsteady solution was needed to reproduce the evolution of flow-field features within the blade-to-blade passage. This was a consequence of the need to account for the interaction between the blade-to-blade flow-field and the fan's discharge volute. This interaction is critical as it has a fundamental effect of particle trajectory through the blade-to-blade passage, and

Table 12.6. *A comparison of the studied fan's experimentally measured and numerically predicted performance at its peak efficiency operating point.*

	Total pressure rise Δp_t	Efficiency η
CFD	16.4 kPa	78.4 %
experiments	16.8 kPa	81.8 %

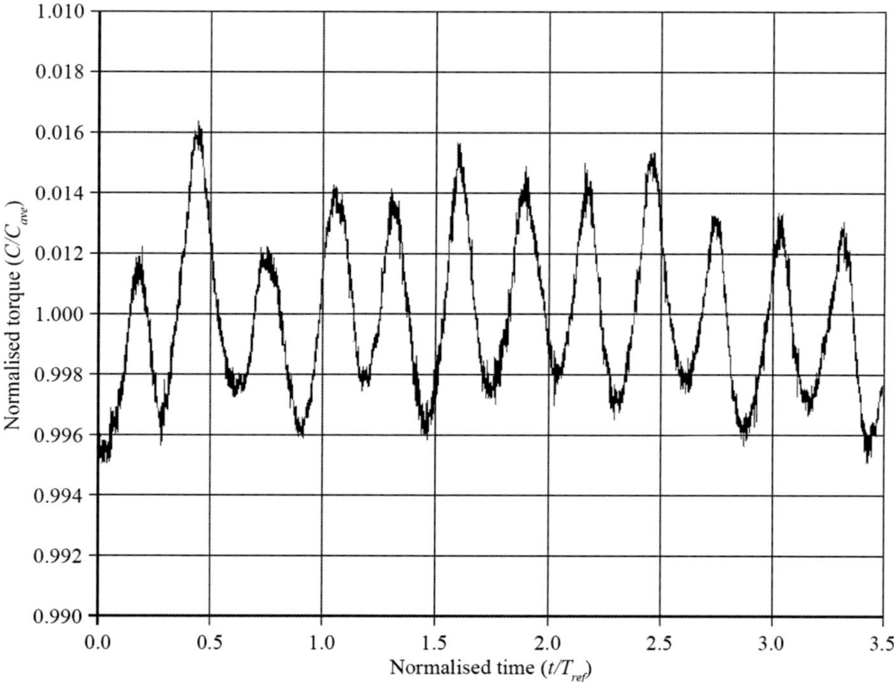

FIGURE 12.8. Normalised torque, predicted using an unsteady Reynolds-averaged Navier–Stokes numerical simulation illustrating the prediction's periodic nature.

therefore on the resultant prediction of erosion patterns through the blade-to-blade passage.

We may illustrate the importance of an unsteady simulation by considering velocity contours and streamlines at three time-steps corresponding to 10 degree revolutions of the impeller, Figure 12.10. As the impeller moves from one time step to the next, the impeller blades move relative to the volute's tongue. As the leading edge of each blade passes the volute's tongue, it passes through a region of low-velocity fluid. This induces a low-velocity fluid core at the leading edge of each blade pressure surface. The size of this low-velocity core changes as the impeller rotates in the volute and deforms a high-velocity jet that releases from the the blade suction surface's leading edge. This unsteady interaction changes the evolution of the blade-to-blade flow-field. We do not predict the unsteady interaction between the volute's tongue and the blade-to-blade flow-field using the frozen rotor approach. Consequently, an unsteady approach is needed to predict the blade-to-blade flow-field with sufficient accuracy to facilitate an adequate prediction of particle trajectory.

We may illustrate the significance of the unsteady impeller-volute interaction by studying the predicted variation of velocity with time at four difference axial immersions for a single point near the volute's tongue, Figure 12.11. The predicted

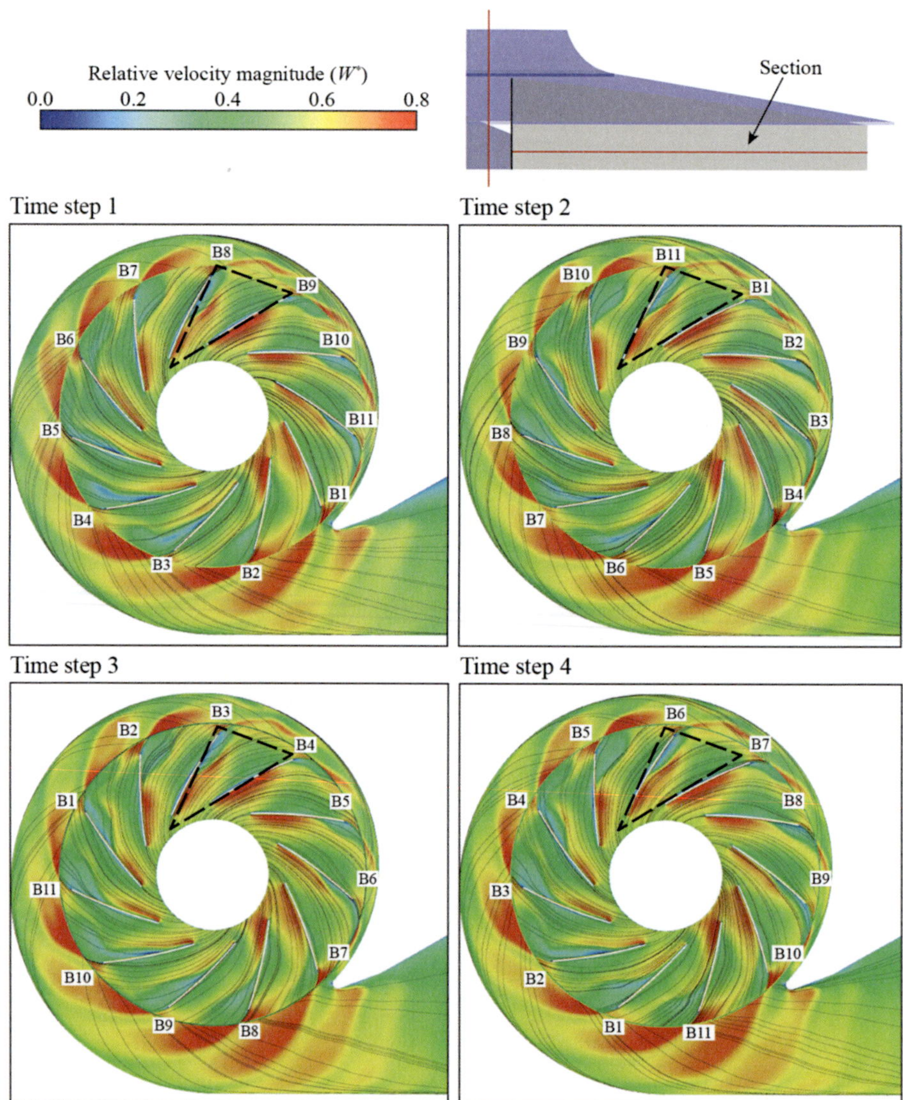

FIGURE 12.9. Contours of relative velocity magnitude (w^+) and relative velocity streamlines at an axial cross-section (see insert) at four time intervals corresponding to 90 degree impeller revolution.

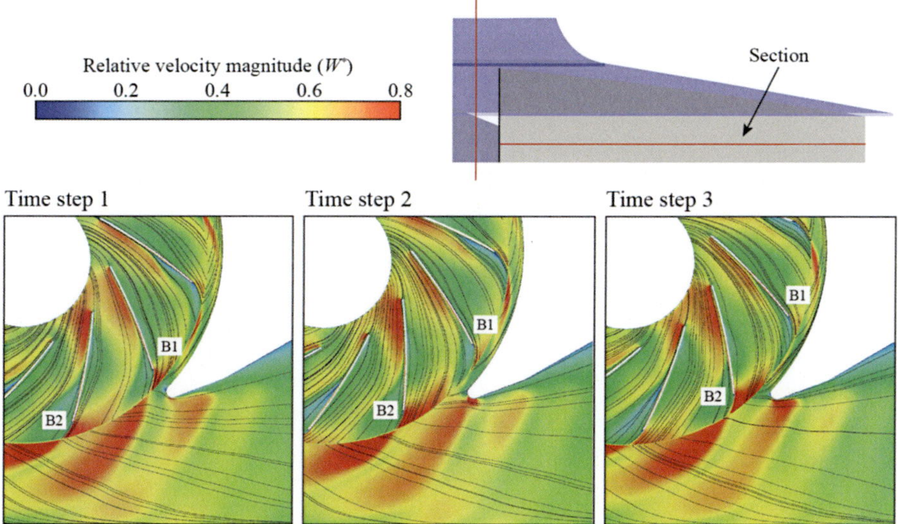

FIGURE 12.10. Contours of relative velocity magnitude (w^+) and relative velocity streamlines at an axial cross-section (see insert) at three time intervals corresponding to 10 degree of impeller revolution.

velocity is both periodic and multi-modal in nature, with a recognisable change in phase with changing immersion. Correctly tracking particles through the impeller's blade-to-blade passages requires a computational approach that can characterise the unsteady nature of the flow-field at all locations within the fan flow-field.

We may illustrate the full complexity of the blade-to-blade flow-field by considering streamlines through the blade-to-blade passages, Figure 12.12. Although the streamlines are complex, it is possible to identify some flow features. Most blades have a separation at the leading edge suction surface, and a series of secondary flow features within each blade-to-blade passage. We may associate the separation at the blade leading edge with the blade's non-optimal alignment with the incoming flow that in turn results in a secondary flow feature developing within each blade-to-blade passage.

We may provide additional insight into the complexity of the blade-to-blade flow-field by considering velocity vectors when plotted along streamlines, Figure 12.13. We plotted the velocity vectors at three time steps over 32 degrees of the impeller's rotation, corresponding to one blade passage. Thus, if we showed the fourth time step, it would be identical to the first. The change in velocity vectors from one time step to the next illustrates the evolution of the three-dimensional leading edge separation. The velocity vectors illustrate that the angle of attack of incoming flow onto the blade leading edge changes as the impeller rotates. This change in angle of attack results in a variation in size of the three-dimensional separation zone, black circle, Figure 12.13, that results in a time varying blockage of each blade-to-blade passage.

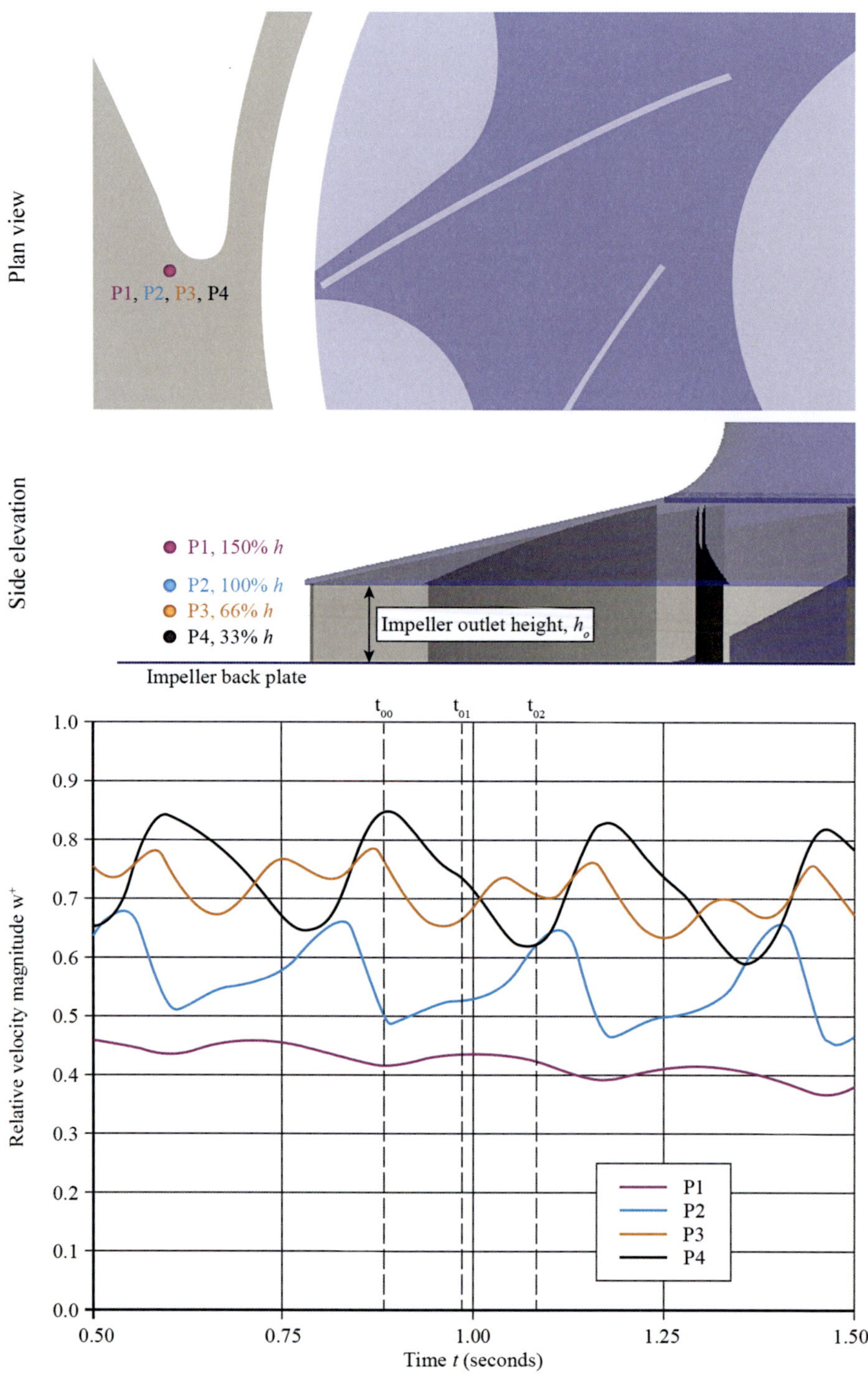

FIGURE 12.11. Relative velocity magnitude (w^+) predicted using an unsteady Reynolds-averaged Navier–Stokes numerical simulation. We present the velocity's predicted variation for four points near the volute's tongue at different axial immersions.

FIGURE 12.12. Three dimensional relative velocity streamlines inside the impeller predicted using an unsteady Reynolds-averaged Navier–Stokes simulation.

The main blade-to-blade passage flow comprises a core of high-velocity fluid (black arrow, Figure 12.13) that enters below the separation. This main passage flow is able to pass through the blade-to-blade passage without swirling. In contrast, secondary flow associated with the leading edge separation is driven across to the blade-to-blade passage towards the adjacent blade's pressure surface. As the secondary flow migrates across the blade-to-blade passage, it rolls up into a vortex forming a time varying secondary flow feature. This incoming flow angle of attack results in a variation in the size of the three-dimensional separation and consequently, the secondary flow features as it develops through the blade-to-blade passage. It is this time-varying change in secondary flow features that has a primary impact on particle trajectory.

Particle transport and erosion prediction

Corsini *et al.* (2012, 2013a) used their developed particle cloud tracking method to predict particle cloud trajectories through an axial fan's blade-to-blade passage. They calculated trajectories using as input a steady Reynolds-averaged Navier–Stokes prediction of the blade-to-blade flow-field. In the research reported in this chapter, we used an unsteady Reynolds-averaged Navier–Stokes approach to

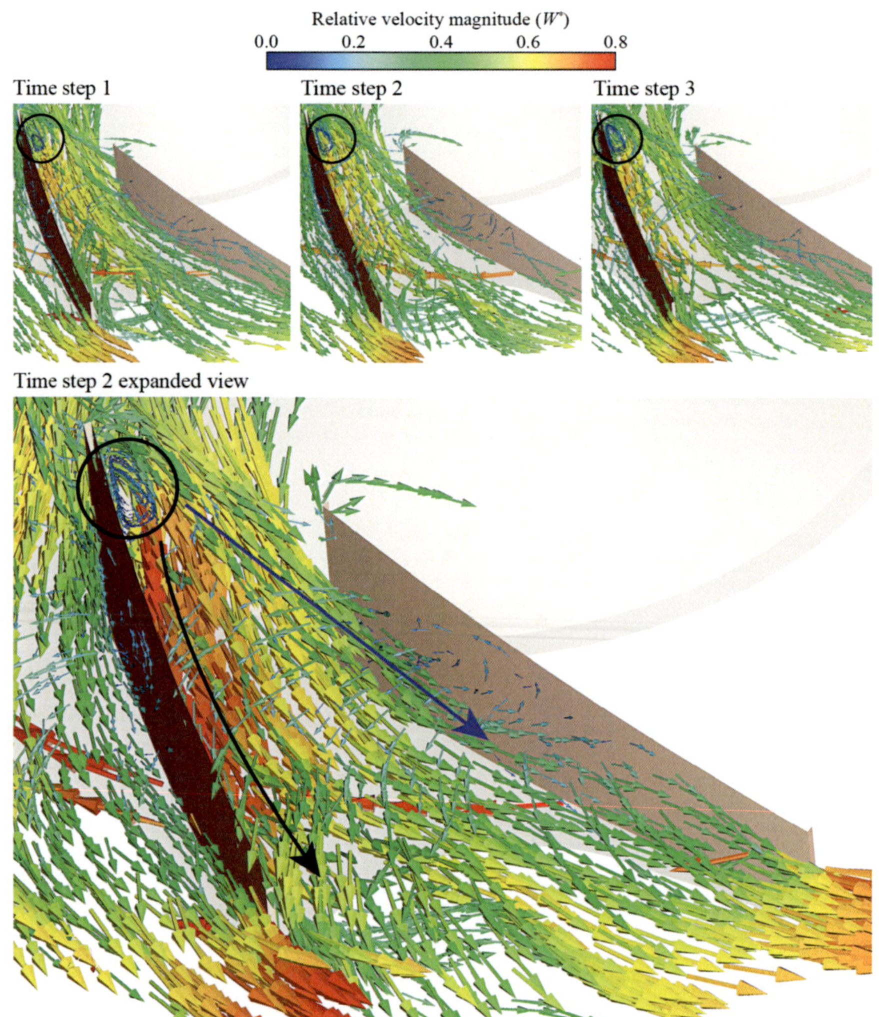

FIGURE 12.13. Relative velocity magnitude (w^+) vectors plotted along streamlines at three time steps corresponding to 32 degrees of impeller rotation that is equivalent to one blade passage, top, and a close-up view of the streamlines corresponding to the second time step, bottom. The black circle identifies the location of a blade leading edge recirculation. The black arrow indicates the direction of the main flow and the blue arrow indicates the direction of secondary flow.

predict the blade-to-blade flow-field. Although periodic, the blade-to-blade flow-field varied as each blade-to-blade passage rotated. We accounted for this variation by dividing one impeller revolution into 72 time steps, each corresponding to five degrees of rotation. By dividing a single revolution into time steps, we were able to generate the instantaneous flow-field for a complete impeller revolution, with an instantaneous flow-field prediction every five degrees of rotation. We then used these flow-field predictions as input to Corsini *et al.*'s (2012, 2013a) particle cloud tracking method. We used the flow-field predictions sequentially, assuming a steady flow-field during the time between predictions. Using this approximation we were able to account for the once-per-revolution periodic nature of the blade-to-blade flow-field and predict particle cloud trajectories without coupling the flow-field simulation to the particle cloud tracking algorithm.

Particle transport

We validated the particle cloud tracking model for turbulent particle dispersion using Wang and Stock's (1993) study. This study evaluates the dispersion of heavy particles that occurs as a consequence of turbulence. Wang and Stock performed a direct numerical simulation computing the Eulerian and Lagrangian integral time scale, T_{mE} and T_L respectively. They found that one may express the fluid integral time scale (T) seen by a particle, which is a function of the particle Stokes number (Stk). It is characterised as:

$$\frac{T(Stk)}{T_{mE}} = 1 - \frac{(1 - T_L / T_{mE})}{(1 + Stk)^{0.4(1+0.01Stk)}}. \tag{9}$$

In the research reported in this chapter, we predicted T_{mE} and T_L, for a range of Stokes numbers. This enabled us to predict the ratio T/T_{mE}. Since T_{mE} and T_L change at each time instant, we averaged the ratio over one complete impeller revolution. We then compared our predicted ratio with the ratio we calculated using Equation (9), Wang and Stock (1993) and with Sato and Yamamoto's (1987) experimentally measured ratio, Figure 12.14. Our predicted ratio is in reasonable agreement with Wang and Stock's method and Sato and Yamamoto's (1987) measurements. We may attribute the differences to differences in turbulence. Therefore, we concluded that the particle cloud tracking model, using as input the 72 flow-field predictions, resulted in a fair representation of particle dispersion when subjected to turbulent flow.

Erosion

Having verified the accuracy of the particle cloud tracking model, we may now use it to study cloud trajectories through the impeller, Figure 12.15. The cloud trajectories indicate that at the blade leading edge particle trajectories are influenced by the flow turning through 90 degrees as it turns from the inlet plenum into the blade-to-blade passage. Particles either migrate from the blade leading edge suction

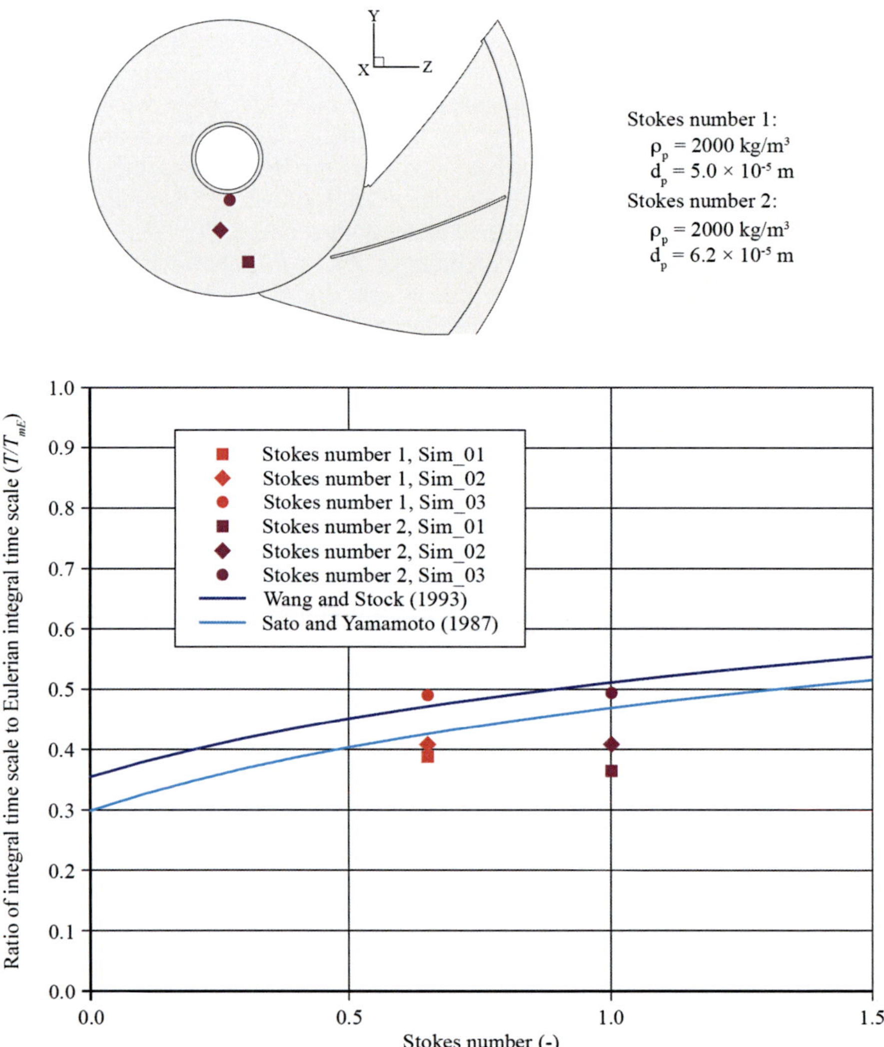

FIGURE 12.14. The ratio between the fluid integral time scale (T) seen by a particle and the Eulerian integral time scale (T_{mE}). The authors computed the ratio using the cloud particle tracking model at three starting positions, Sim_01, Sim_02 and Sim_03 and two stokes numbers, 0.62 and 1.01. The agreement was reasonable between the ratio predicted using the cloud particle tracking model and when predicted using Wang and Stock's (1993) method and Sato and Yamamoto's (1987) measurement.

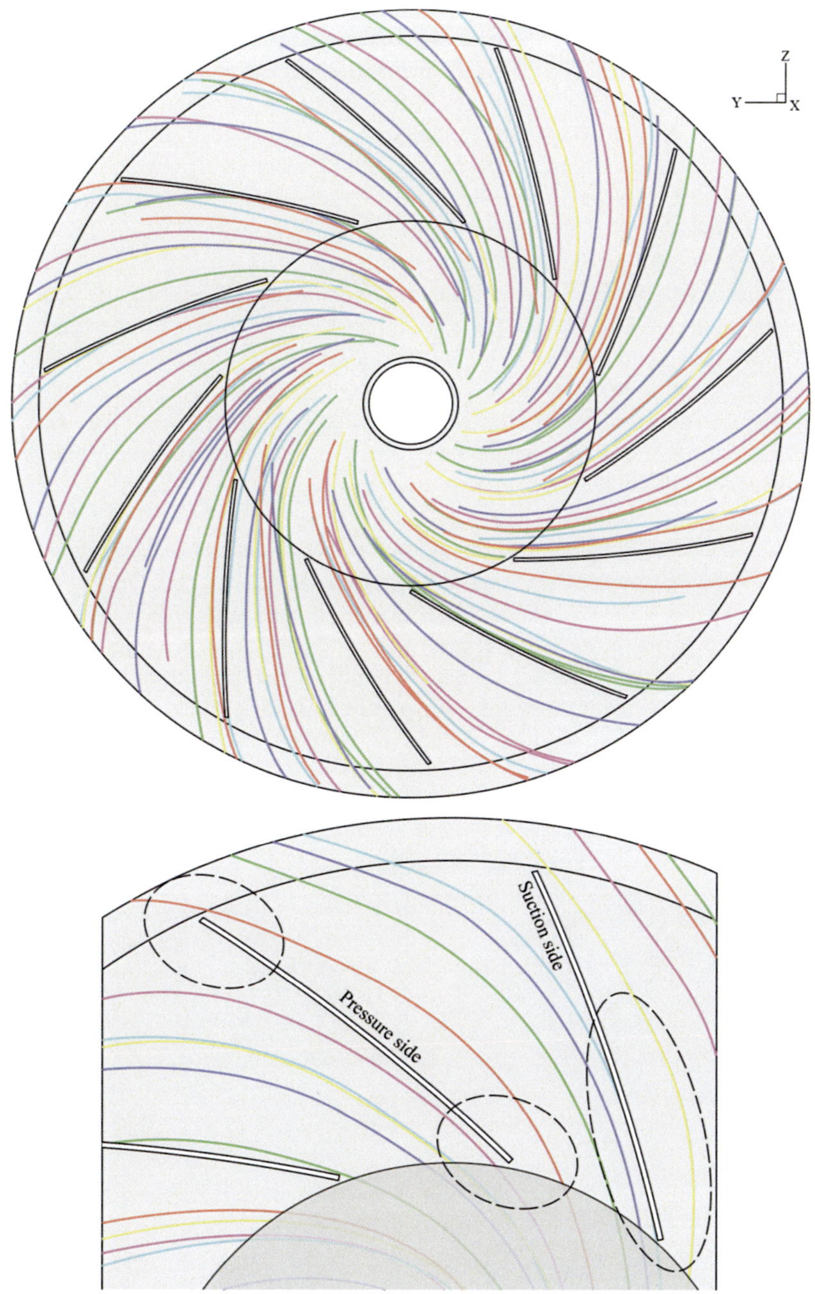

FIGURE 12.15. Cloud trajectories used in the numerical simulation (top) and blade passage details (bottom). The ellipses indicate the zones where trajectories are closer to blade leading and trailing edges.

surface to the blade trailing edge pressure surface or impact on the impeller hub disc. Close to the blade trailing edge, a combination of centrifugal force and forces induced on the particles as a consequence of the secondary flow features within the blade-to-blade passage drive the particles. The secondary flow feature forces are responsible for deflecting particle trajectories in regions of high curvature. A result of this deflection is that particles impact on the blade pressure surface in the trailing edge region, and then go on to slide along the pressure surface until they reach the blade trailing edge.

We may used a prediction of particle trajectory to predict the impact frequency of particles over the impeller shroud and hub discs, Figure 12.16. The impact frequency patterns indicate that particle trajectories are driven by the impingement of the inlet flow onto the hub disc, following which they rebound onto the shroud disc. One may facilitate further insight into the impact frequency patterns by considering an isometric view of the impeller, Figure 12.17. We see clear evidence of impact regions at the blade leading edge and over the blade pressure surface in the trailing edge region, confirming that it is the secondary flow features that drive particle trajectory through the blade-to-blade passage.

The predicted impact frequency patterns are useful as they facilitate an understanding of the mechanisms driving particle trajectories. However, impact frequency is not erosion rate and therefore we computed erosion rate over the impeller shroud and hub discs, Figure 12.18. Erosion occurs primarily over the conical surface on the hub disc opposite the inlet plenum and in the transition region between the hub disc and the blades. It is the transition between the hub disc and the blades that affects the local fluid velocity, with erosion rate patterns asymmetric as a

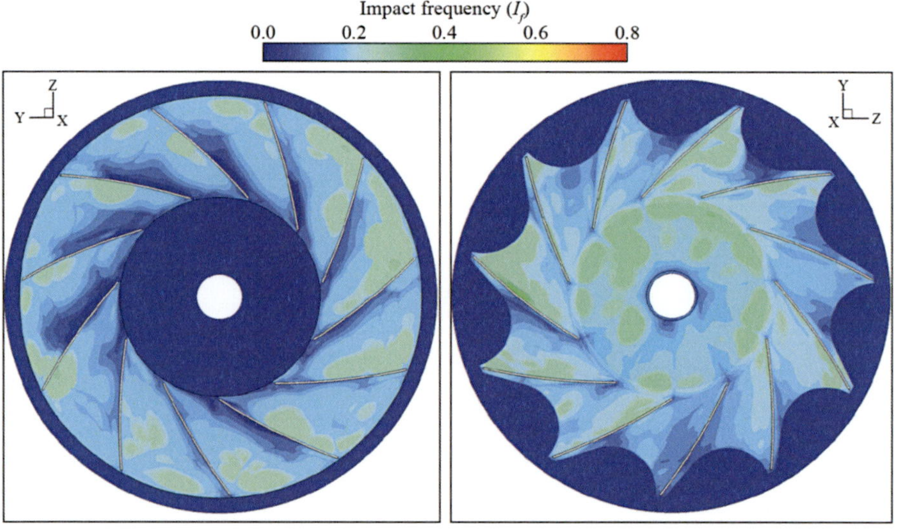

FIGURE 12.16. Normalised impact frequency I_f over the impeller's top, left and bottom of the impeller, right.

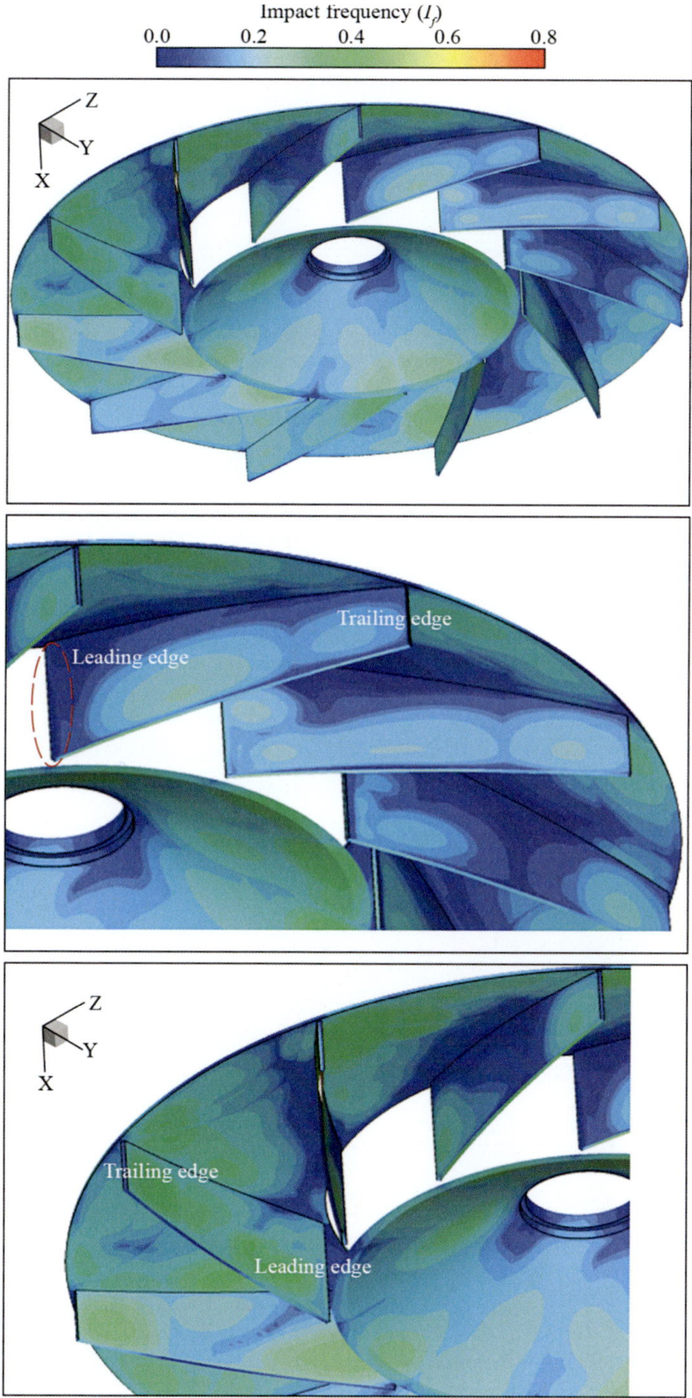

FIGURE 12.17. Normalised impact frequency I_f over the impeller blades, top with close up views of the suction surface of three blades, centre and the pressure surface of two blades, bottom.

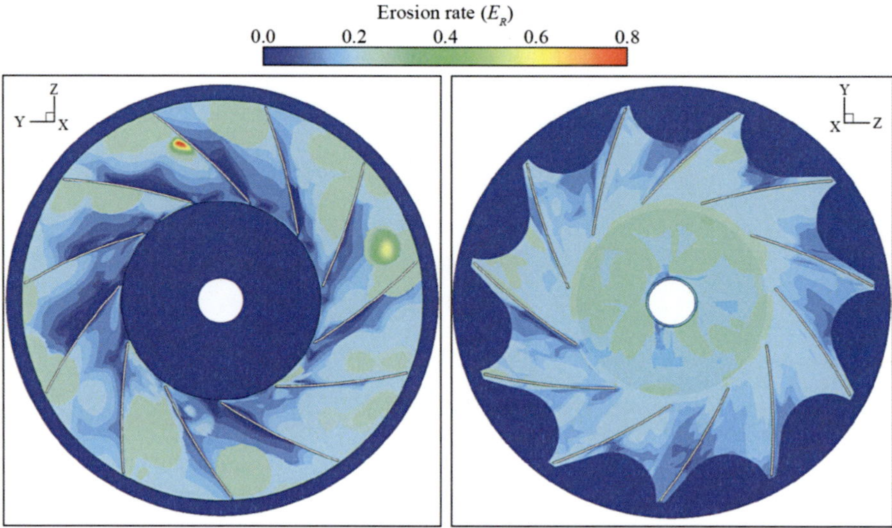

FIGURE 12.18. Normalised erosion rate E_R over the top of the impeller, left and bottom of the impeller, right.

consequence of the unsteady flow-field through the blade-to-blade passage. We may facilitate further insight into the erosion rate patterns by considering an isometric view of the impeller, Figure 12.19. The predicted erosion rate is clearly circumferentially non-uniform, confirming that the erosion rate is higher on the blade suction surface and in the middle of the blade pressure surface's leading and trailing edge regions.

The predicted erosion rate patterns were in reasonable agreement with those that we observed over an impeller removed from service in an erosive applications. Although the required computational effort was large, the resultant erosion rate predictions proved accurate enough to form an input into the fan design process. First, designers could use the predicted erosion rate for new combinations of particle sizes to predict the change in erosion rate and therefore fan service life. Second, designers were able to establish the erosive consequences of a change in blade geometry, and therefore optimise impeller design to produce an inherently more erosion tolerant design.

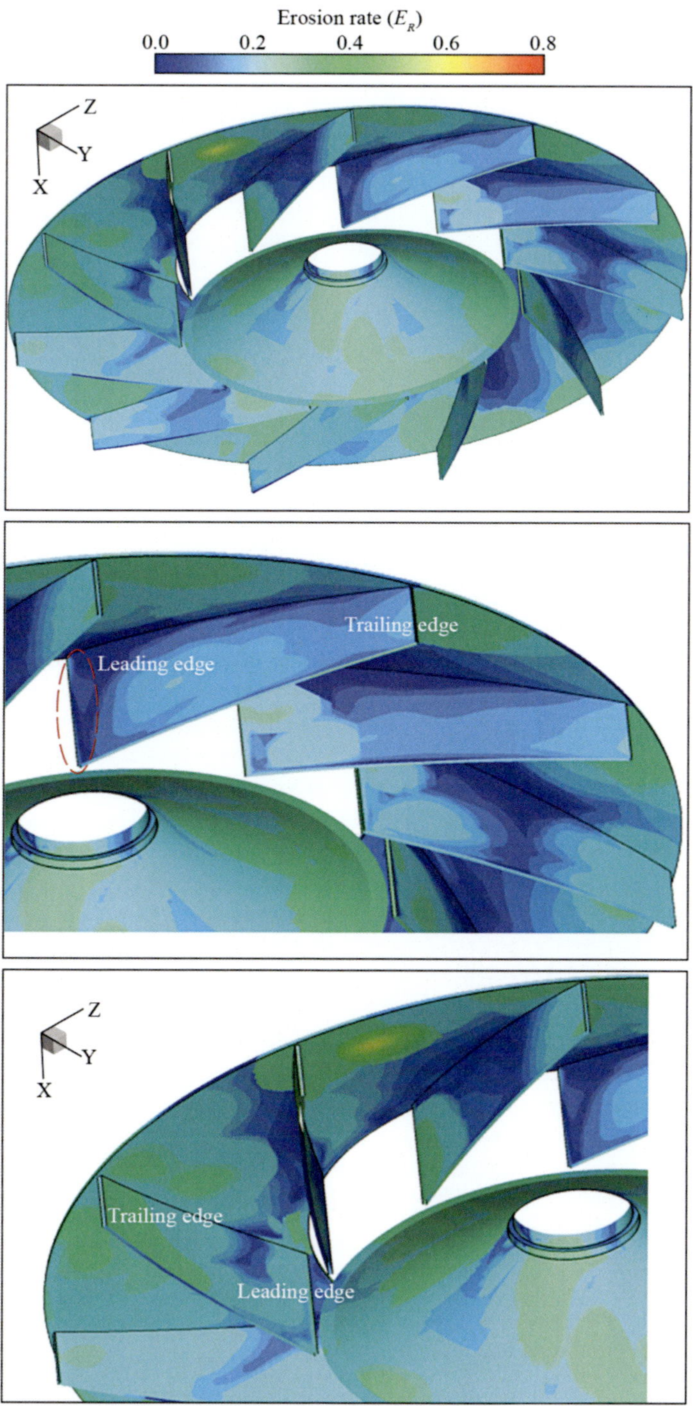

FIGURE 12.19. Normalised erosion rate E_R over the impeller blades, top with close up views of the suction surface of three blades, centre and the pressure surface of two blades, bottom.

CONCLUSIONS

The research reported in this chapter focused on predicting the flow-field through and erosion of a centrifugal fan. The studied fans were originally developed for use in power, steel and cement applications that occur with the presence of erosive particles as a consequence of the application. We predicted the flow-field through the fan by solving the unsteady Reynolds-averaged Navier–Stokes equations using the open-source code OpenFOAM. We predicted the trajectory of particles through the fan using a particle cloud tracking algorithm.

We predicted the flow-field through the fan inlet, the blade-to-blade passage and discharge volume by solving the unsteady Reynolds-averaged Navier–Stokes equations. This unsteady Reynolds-averaged Navier–Stokes prediction required approximately an order of magnitude more computational effort than a comparable steady Reynolds-averaged Navier–Stokes prediction. We utilised an unsteady prediction as the blade-to-blade flow-field was strongly influenced by the unsteady interaction of the impeller and the discharge volute. Accounting for the unsteady interaction enabled us to predict the trajectory of particle clouds as they transitioned from the stationary to rotating frame of reference. Additionally, accounting for the unsteady interaction resulted in a flow-field prediction of sufficient accuracy to enable us to predict the trajectory of particle clouds passing through the blade-to-blade passage.

The reported research highlights a series of unsteady flow features that are responsible for the initiation of secondary flow features that develop as they pass through the blade-to-blade passage. We conclude that the location of each blade-to-blade passage as the impeller rotates drives these unsteady periodic flow features. As a result, the secondary flow features are periodic, driving the dispersion of particles as they pass through a blade-to-blade passage as it rotates. The resulting prediction of particle trajectory facilitated a prediction of erosion rate patterns over the impeller shroud and hub discs and blade pressure and suction surfaces. The predictions of erosion rate with 5 μm particles were in reasonable agreement with those we observed on impellers removed from service in erosive applications where the average particle size was 5 μm.

The agreement between predicted erosion rate and the erosion we observed on impellers removed from service is good enough to enable fan designers to use the erosion prediction method to predict the effect of changing particle size on impeller erosion rate, and therefore, fan service life. Additionally, fan designers may use the method to establish the erosive consequences of a change in blade geometry. Knowing the effect of a change of geometry on erosion rate provides an input into an impeller optimisation process. The optimised impeller geometry is inherently more resistant to the effects of erosion.

REFERENCES

ISO 5801 (2007), *Industrial Fans – Performance Testing Using Standardized Airways.*

Baxter, L.L. (1989), 'Turbulent Transport of Particles', PhD thesis, Brigham Young University, Provo, UT, USA.

Baxter, L.L. and Smith, P.J. (1993), 'Turbulent Dispersion of Particles: the STP Model', *Energy and Fuels*, vol. 7(6), pp. 852–859.

Borello, D., Venturini, P., Rispoli, F. and Saavedra G.Z.R. (2013), 'Prediction of Multiphase Combustion and Ash Deposition within a Biomass Furnace', *Applied Energy*, vol. 101, pp. 413-422.

Cardillo, L., Corsini, A., Delibra, G., Rispoli, F. and Sheard, A.G. (2014), 'Predicting the Performance of an Industrial Centrifugal Fan Incorporating Cambered Plate Impeller Blades', *Periodica Polytechnica, Mechanical Engineering*, vol. 58(1), pp. 15–25.

Corsini, A., Marchegiani, A., Rispoli, F., Sheard, A.G. and Venturini, P. (2012), 'Predicting Blade Leading Edge Erosion in an Axial Induced Draft Fan', *Transactions of the ASME, Journal of Engineering for Gas Turbines & Power*, vol. 134, paper no. 042601, pp. 1–9.

Corsini, A., Rispoli, F., Sheard, A.G. and Venturini, P. (2013a), 'Numerical Simulation of Coal Fly-ash Erosion in an Induced Draft Fan', *Transactions of the ASME, Journal of Fluids Engineering*, vol. 135, paper no. 081303, pp. 1–12.

Corsini, A., Delibra, G. and Sheard, A.G. (2013b), 'A Critical Review of Computational Methods and Their Application in Industrial Fan Design', *International Scholarly Research Network, Mechanical Engineering*, vol. 2013, article ID 625175, pp 1-20.

Farrell, P.E. and Maddison, J.R. (2011), 'Conservative Interpolation between Volume Meshes by Local Galerkin Projection', *Computer Methods in Applied Mechanics and Engineering*, vol. 200, pp. 89–100.

Hu, G.R., Li, Y.M. and Jia, X.N. (2008), 'Simulation of Gas-solid Two-phase Flow of Centrifugal Fan', *Coal Mine Machinery*, vol. 2008(3), pp. 37–39.

Jain, S. (1995), 'Three-dimensional Simulation of Turbulent Particle Dispersion', PhD thesis, University of Utah, Salt Lake City, UT, USA.

Kær, S.K. (2001), 'Numerical Investigation of Ash Deposition in Straw-fired Furnaces', PhD thesis, Aalborg University, Aalborg, Denmark.

Launder, B.E. and Sharma, B.R. (1974) 'Application of the Energy-dissipation Model of Turbulence to the Calculation of Flow Near a Spinning Disc', *Letters in Heat and Mass Transfer*, vol. 1(2), pp. 131–137.

Litchford, R.J. and Jeng, S.M. (1991), 'Efficient Statistical Transport Model for Turbulent Particle Dispersion in Sprays', *AIAA Journal*, vol. 29, pp. 1443–1451.

Mei, D. and Xing, F.T. (2007), 'Numerical Simulation Research on Erosive Problem of Gas-solid Two Phase Flow Fan', *Fluid Machinery*, vol. 2007(10), pp. 25–28.

Rugang, C. (2000), 'Experimental Study on Erosion Characteristic for Different Blade Contours in the Centrifugal Can Exposed to Solid Particulate Environment', *Fluid Machinery*, vol. 2000(1), pp. 5–8.

Sato, Y. and Yamamoto, K. (1987), 'Lagrangian Measurement of Fluid-particle Motion in an Isotropic Turbulent Field', *Journal of Fluid Mechanics*, vol. 175, pp. 183–199.

Tabakoff, W., Kotwal R. and Hamed, A. (1979), 'Erosion Study of Different Materials Affected by Coal Ash Particles', *Wear*, vol. 52(1), pp. 161–173.

Venturini, P. (2010), 'Modelling of Particle Wall-deposition in Two Phase Gas-solid Flows', PhD thesis, Sapienza University of Rome, Rome, Italy.

Venturini, P., Borello, D., Iossa, C., Lentini, D. and Rispoli F. (2010), 'Modelling of Multiphase Combustion and Deposit Formation and Deposit Formation in a Biomass-fed Boiler', *Energy*, vol. 35(7), pp. 3008–3021.

Wang, L.P. (1990) 'On the Dispersion of Heavy Particles by Turbulent Motion', PhD thesis, Washington State University, Pullman, WA, USA.

Wang, L.P. and Stock, D.E. (1993), 'Dispersion of Heavy Particles by Turbulent Motion', *American Meteorological Society*, vol. 50(13), pp. 1897–1913.

Weller, H.G., Tabor, G., Jasak, H. and Fureby, C. (1998), 'A Tensorial Approach to Continuum Mechanics Using Object-Oriented Technique', *Computers in Physics*, vol. 12(6), pp 620–631.

Appendix 1

A Comparison of U.S. and European Approaches to Regulating Fan Efficiency

M.G. Ivanovich and N.M. Jones

ABSTRACT

The United States and European Union began regulating commercial and industrial fan efficiency around the same time in 2007; however, they took very different approaches. The European Union began with product regulations stemming from Ecodesign directives. The U.S. began with adopting fan efficiency requirements into model codes and standards for energy efficiency and green construction published by ASHRAE (Formerly the American Society of Heating, Refrigerating and Air Conditioning Engineers) and the International Code Council (ICC). A U.S. federal regulation began through the U.S. Department of Energy after the ASHRAE and International Code Council provisions had been initiated. This appendix describes and compares the U.S. and European approaches regarding fan efficiency regulation. For the U.S., we place emphasis on the use of model codes and standards including ASHRAE Standard 90.1 because the U.S. Department of Energy regulation will not become effective until 2020. For Europe, we provide an outline of Commission Regulation (EU) No. 327/2011.

This appendix is a revised and extended version of Ivanovich M.G. and Jones N.M. (2014), 'A Comparison of U.S. and European Approaches to Regulating Fan Efficiency', *Proceedings of the CIBSE ASHRAE Technical Symposium 03*, Dublin, Ireland, 3–4 April.

NOMENCLATURE

		Latin letters
D	[mm or inches]	Fan diameter
p_t	[Pa]	Total pressure
V_{max}	[m^3/s]	Air flow at the permissible window's high end
V_{min}	[m^3/s]	Air flow at the low end of the permissible window
V_{peak}	[m^3/s]	Air flow at the fan's peak total efficiency point
V	[m^3/s]	Volume flow rate
		Greek letters
η_M	[%]	Fan peak efficiency
η_t	[%]	Fan total efficiency

Acronyms

AMCA	Air Movement and Control Association
ANSI	American National Standard Institute
ASHRAE	American Society of Heating, Refrigerating and Air Conditioning Engineers
DOE	Department of Energy
EN	European Norm
EU	European Union
FEG	Fan Efficiency Grade
FMEG	Fan Motor Efficiency Grade
HVAC	Heating, ventilation and air conditioning
HVLS	High-volume, low-speed
ICC	International Code Council
IECC	International Energy Conservation Code
IES	Illuminating Engineering Society
IgCC	International Green Construction Code
ISO	International Standards Organization
MEP	Member of the European Parliament
MEPR	Minimum efficiency performance requirement
PRV	Powered wall/roof ventilators

INTRODUCTION

Fans applied in commercial heating, ventilation and air conditioning (HVAC) systems account for about 1.25 Quads per year for all commercial buildings in the U.S. (Westphalen and Koszalinski, 1999). This is a significant fraction of the U.S. energy budget. However, only recently have government agencies developed and incorporated rating standards for fan efficiency into codes, standards and regulatory policies in Europe and the U.S. (Hauer and Brooks, 2012; Cermak and Ivanovich, 2013).

In Europe, Commission Regulation (EU) No. 327/2011 (2011), the Regulation, sets minimum efficiency targets that building ventilation fans must meet. The Regulation targets fans that are driven by electric motors with input powers from 125 W to 500 kW. Compliance with Regulation Tier 1 targets became mandatory on 1 January 2013, and more stringent Tier 2 targets took effect on 1 January 2015. The Regulation is a product requirement, not a building code. Product requirements differ from building codes in that they place minimum efficiency restrictions directly on manufactured devices or equipment. Building codes place requirements on products that are applied in systems installed in facilities. These products may have applications, such as industrial processes, which are unaffected by building codes. Building code provisions may also have application requirements, such as size limits, that are not usually a component of product regulations.

In the U.S., the industry has addressed fan efficiency regulation in model codes and standards for energy efficiency and green construction. Model codes and standards place baseline (minimum) energy performance requirements on building components and systems. ASHRAE Standard 90.1 is a model energy standard; the International Energy Conservation Code is a model energy code. Green model construction codes and standards address buildings holistically by including provisions for the building site, envelope, lighting, plumbing, and HVAC systems that are more stringent than baseline codes and standards. ASHRAE Standard 189.1 is a model green construction standard. The International Green Construction Code is a model green construction code. Publishing authorities drive energy efficiency and green construction codes and standards to become progressively more stringent. For example, ASHRAE set a goal that the 2010 version of Standard 90.1 would be 30 per cent more stringent than the 2004 version of the Standard (Gowri, 2009). One may accomplish increasing stringency in two ways: increase the scope of coverage to include previously uncovered features or increase stringency of covered equipment or systems. Hence, fans, which a model energy code or standard did not previously cover, drew interest as a means for achieving increasing stringency (AMCA, 2009).

U.S. federal product efficiency standards have also not covered fans. However, in 2011 the U.S. Department of Energy (DOE) initiated a federal rating standard and minimum efficiency regulation for commercial and industrial fans (U.S. Department of Energy, 2013a). The rulemaking process will establish a product requirement, not a building code and should conclude in 2015 or 2016, with enforcement beginning in 2020. The Department of Energy rulemaking will include establishing a federal test standard for fans, which will serve for compliance verification and enforcement, and minimum energy performance standards for fans covered by the regulation (U.S. Department of Energy, 2013a).

Although they were started at approximately the same time, European and U.S. regulatory approaches are different in their approach. Europe and the U.S. have chosen different technical boundaries for regulated equipment and a different structure and rigour for market surveillance and enforcement. The significance of the difference is that many fan manufacturers are international. Having to comply with different fan efficiency standards in Europe and the U.S. imposes additional regulatory burden on affected companies. This creates market inefficiencies as it constitutes a diversion of time and effort. Perhaps most significantly it results in a fan that is acceptable in one geographic region not being acceptable in another. Therefore, manufacturers must upgrade fan sizing and selection software to account for where they sell a fan. This upgrading requires a research and development effort that deflects effort away from actually making the fan itself more efficient. U.S. and European stakeholders, including manufacturers, regulatory bodies, and non-governmental organisations, are discussing fan regulation harmonisation from an international perspective. However, there is no formal movement as yet to initiate a harmonisation effort. Although not covered in this appendix, stakeholders from Asia also are involved in these discussions because fan efficiency regulation is also taking place in Asia, and international business interests are similarly affected.

U.S. FAN CODES, STANDARDS, AND REGULATIONS

The basis of U.S. fan efficiency provisions in model codes and standards is the Air Movement and Control Association (AMCA) Standard 205 *Energy Efficiency Classification for Fans*. AMCA International is a not-for-profit manufacturers association with more than 320 member companies worldwide. First published in 2010, AMCA 205 was revised in 2012 and approved by the American National Standards Institute (ANSI/AMCA, 2012). AMCA 205 has two significant attributes. First, it rates fan efficiency using a non-dimensional index called a Fan Efficiency Grade (FEG), which one calculates from data obtained during a fan performance test. Fan Efficiency Grades are presented as a family of curves, Figure A1.1. Each Fan Efficiency Grade band represents a range of efficiencies that vary depending on fan impeller diameter. For example, FEG 67 includes fan efficiencies of 64, 65, 66 and 67 per cent at the plateau for fans with an impeller of 500 mm or greater in diameter, and efficiencies as low as 35 per cent for fans with an impeller of 125 mm.

The scope of AMCA 205 is limited to fans having an impeller diameter of 125 mm or greater, operating with a shaft power of 750 W. It also requires one to calculate fan efficiency in accordance with procedures specified in test standards for rating fan performance:

- ANSI/AMCA 210 (ANSI/ASHRAE 51) *Laboratory Methods of Testing Fans for Certified Aerodynamic Performance Rating;*
- ANSI/AMCA 230 *Laboratory Methods of Testing Air Circulating Fans for Rating and Certification;*
- AMCA 260 *Laboratory Methods of Testing Induced Flow Fans for Rating*; and
- ISO 5801 *Industrial Fans - Performance Testing using Standardised Airways.*

Fans that manufacturers cannot test to one of these standards are excluded from AMCA 205, which means they are subsequently excluded from codes, standards and regulations with fan efficiency provisions based on AMCA 205. Examples of excluded fan types include air curtains and jet fans. Compliance checking is further

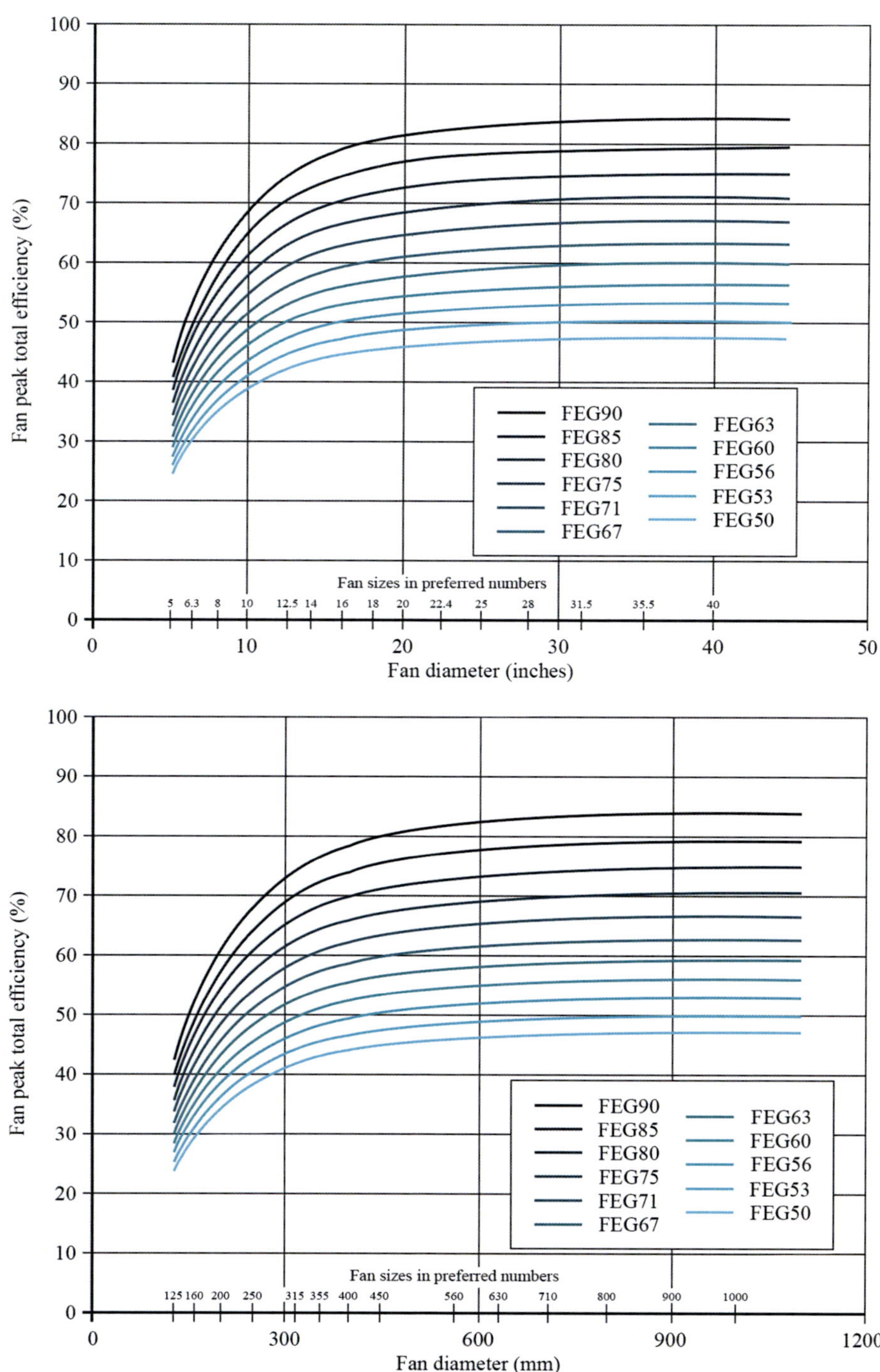

FIGURE A1.1. Fan Efficiency Grade curves (ANSI/AMCA, 2012).

complicated by the fact that, even if a fan type falls under the scope of a particular standard, the fan's size may exclude it. For example, the scope of AMCA 230 is limited to fans up to 1.8 m, which exempts the larger high-volume, low-speed (HVLS) ceiling fans.

A second attribute of AMCA 205 is that it goes beyond fan efficiency to recognise that fan energy consumption is highly dependent on sizing and selection practice. Thus, AMCA 205 prescribes that the '…fan operating efficiency at all intended operating point(s) shall not be less than 15 percentage points below the fan peak total efficiency', Figure A1.2.

The intent of the sizing and selection window is to encourage what AMCA refers to as 'right-sizing.' It is often the case that electronic sizing and selection software will provide multiple fan selections for a given airflow and pressure duty point. In most cases, the software's output will be a spectrum of fan types and diameters. The smaller diameter fans will typically be less efficient that the larger diameter fans, and therefore consume more energy than their larger-diameter counterparts.

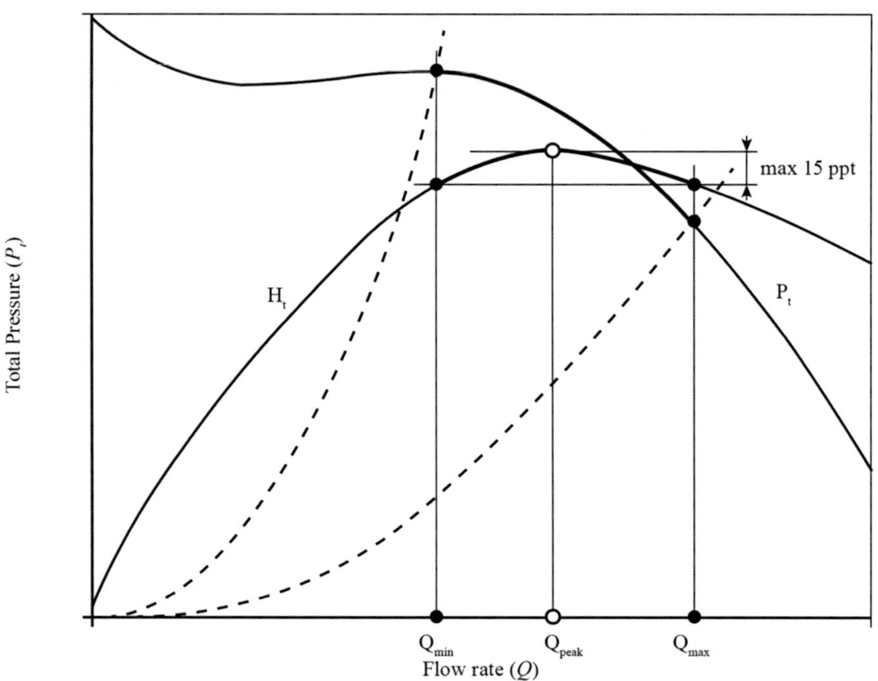

Legend of symbols for Figure A.2:
P_t = Fan total pressure line for the fan at varying flows and pressures
$_t$ = Fan total efficiency
Q_{min} = Airflow at the low end of the permissible window
Q_{peak}= Airflow at the fan's peak total efficiency point
Q_{max} = Airflow at permissible window's high end

FIGURE A1.2. Sizing and selection window (ANSI/AMCA, 2012).

Despite their inferior efficiency, smaller diameter fans have a lower first cost than larger diameter fans. There is a tendency for ventilation system designers to specify and purchase lowest-first-cost fans from the options available. A prescribed sizing and selection window will change a fan selection to a selection within 15 percentage points of the fan's rated peak total efficiency. This will effectively exclude the use of small fans, as their operating points are typically more than 15 percentage points of the fan's rated peak total efficiency. Consequently, AMCA 205 impacts manufacturers by defining an efficiency rating system for fans, and it impacts practitioners by prescribing a sizing and selection window that leads to larger, more efficient fans operating closer to their peak efficiency.

First-generation model codes and standards

When ASHRAE first considered fan efficiency provisions for ASHRAE Standard 90.1, they considered a 'straight line' 65 per cent efficiency requirement as a starting point (Cermak and Ivanovich, 2013). However, because smaller-diameter fans are inherently less efficient than larger diameter fans of the same type, setting a 65 per cent minimum fan efficiency would eliminate many fan types under 500 mm diameter, Figure A1.3.

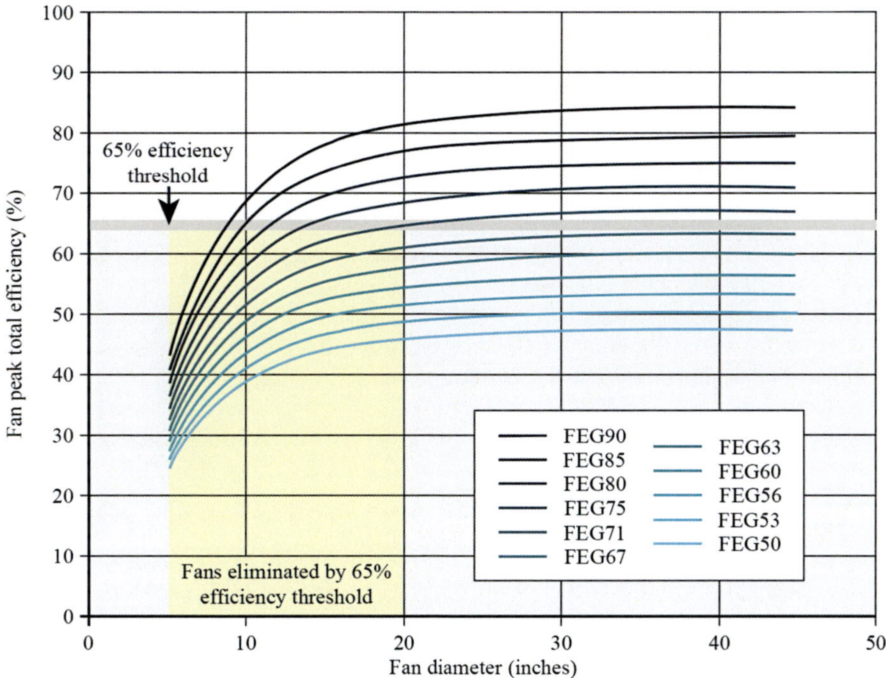

FIGURE A1.3. Fan Efficiency Grade (FEG) versus straight-line efficiency requirement (Cermak and Ivanovich, 2013).

About fan total pressure

Fan Efficiency Grades are based on fan total pressure for all fan ratings, and the sizing and selection window is also based on fan total pressure. The use of total pressure when sizing and selecting fans is counter to standard practice in the U.S., where the majority of design practice, manufacturers' data, and sizing and selection software programs are based on fan static pressure (Cermak and Murphy, 2011). This is especially true for fans that are installed in systems where the fan is not connected to an outlet duct. In such instances, there is little or no opportunity to recover the velocity component of total pressure.

Transitioning from static pressure to total pressure for a fan will require replacing a substantial amount of the manufacturers' hardcopy and electronic literature plus training of distributors and the design community. Since the publication of AMCA 205 and its adoption into codes and standards, manufacturers have started including fan total pressure data and Fan Efficiency Grade ratings in their software. Manufacturers are also participating in the AMCA Certified Ratings Program for Fan Efficiency Grades. The AMCA Certified Ratings Program is a voluntary program that assures buyers and specifiers that manufacturer's ratings are based on standard test methods and procedures, and are subject to review by AMCA International as an impartial authority. As of December 2013, 52 companies have certified more than 430 fan models (AMCA, 2013a).

Codes and standards based on AMCA 205

After AMCA 205 was published in 2010, work began in earnest to have U.S. model codes and standards adopt fan-efficiency provisions written around it. Table A1.1 lists the model codes and standards that have or are considering fan-efficiency provisions based on AMCA 205. The fan efficiency provisions referenced in Table A1.1 are written around AMCA 205, following its two-part fan-efficiency provisions. Part 1 sets a minimum fan efficiency rating using the Fan Efficiency Grade metric. Part 2 sets a sizing and selection window relative to the fan's rated peak total efficiency. Scope of coverage and exemptions are also part of these provisions, and these tailor the Fan Efficiency Grade and selection provisions to what today's state-of-the-art fan designs can practically attain. Table A1.2 summarises the efficiency provisions in the four model codes and standards. The Notes section lists the exemptions in ASHRAE 90.1, which are also explicitly listed in International Energy Conservation Code (IECC). The exemptions are adopted by reference in the proposals for ASHRAE 189.1-2014 and IgCC-2015.

Because these are first-generation requirements for fan efficiency, the provisions are not as stringent as some would expect and energy savings will most likely be negligible. The ASHRAE 90.1 Mechanical Subcommittee deemed this conservative approach essential, and their conservative nature carried through to the fan efficiency provisions in subsequent codes and standards. The rationale is that manufacturers and engineers need to become familiar with the Fan Efficiency Grade metric and accommodate wider use of fan total pressure before regulating bodies significantly increase minimum fan efficiency requirements.

Table A1.1. *Model codes and standards for energy efficiency and green construction.*

Publication	Type	Status
ASHRAE, ANSI/ASHRAE/IES 90.1, *Energy Standard for Buildings Except Low-Rise Residential Buildings*	Baseline energy standard	Exists in 2013 version
International Code Council, *International Energy Conservation Code (IECC)*	Baseline energy code	Exists in 2015 version
ANSI/ASHRAE/USGBC/IES 189.1, *Standard for Construction of High Performance Buildings Except Low-Rise Residential Building*	Green construction standard	Proposal undergoing public review for 2014 version
International Code Council, *International Green Construction Code (IgCC)*	Green construction code	Exists in 2012 version. Proposal for 2015 would harmonise with ASHRAE 189.1 and IECC

Energy codes/standards adoption and compliance

The amount of energy that the implementation of model codes and standards will save will depend upon the speed with which jurisdictions adopt the most recent versions. Jurisdictions exist at national, state, and municipal levels and adoption rates vary widely. At the time of writing only a few states have adopted ASHRAE 90-2010 or its equivalent, IECC-2012. ASHRAE 90.1-2013 has just been released and IECC-2015 has been finalised, each with their first fan efficiency provision. Energy savings from the fan efficiency provisions may not begin accruing until 2020 given the speed with which organisations are adoption energy-codes.

Two other factors that impact the energy savings of model codes and standards are compliance and enforcement. A meta-analysis of energy code compliance studies at the state level found that the local jurisdictional level generally enforces U.S. energy codes, even for state projects (Misuriello *et al.*, 2010). The study also found that compliance often suffers as a consequence of four issues:

- As-built conditions are sometimes different from plans. For example, contractors often have to change duct runs to go around obstacles not known during the design phase.
- There is little on-site review of construction to compare with plans. For example, construction features may be hidden from inspectors, such as installations within wall cavities.
- Substitution of non-complaint products is common, for example, replacing one model of a fan for another.
- Education and training is needed for many states.

Table A1.2. *Fan efficiency provisions in U.S. model codes and standards.*

Model code/standard	Basis	Scope of coverage	Fan size[1] [hp]	Minimum Fan Efficiency Grade	Sizing/selection window[2] [percentage points]	Certified FEG and energy label required	Exemptions
2012 IgCC	AMCA 205-10	For buildings <25,000 sq ft, stand-alone supply, return and exhaust fans	> 1	FEG 71	10	No	None
2013 ASHRAE 90.1	ANSI/ AMCA 205-12	Buildings other than low-rise residential buildings	> 5	FEG 67	15	No	Yes
2014 ASHRAE 189.1 (proposed)	ANSI/ AMCA 205-12	Buildings other than low-rise residential buildings	> 5	FEG 67	10	No	Yes[3]
2015 IECC	ANSI/ AMCA 205-12	Buildings other than low-rise residential buildings	> 5	FEG 67	15	Yes	Yes[3]
2015 IgCC (proposed)	ANSI/ AMCA 205-12	Buildings <25,000 sq ft	> 5	FEG 67	10	Yes	Yes[3]

Notes for Table A1.2:

[1]Applies to fan arrays with aggregated motor nameplate ratings.

[2]Expressed as fan operating point selected to be within X percentage points of the fan's maximum total pressure (this is a rated value provided by the manufacturer).

[3]List of exemptions below is in 90.1-2013 in IECC-2015. The list also is adopted by reference in change proposals for 189.1-2014 and IgCC-2015.

- Single fans with a motor nameplate horsepower of 5 hp (3.7 kw) or less.
- Multiple fans in series or parallel (e.g., fan arrays) that have a combined motor nameplate horsepower of 5 hp (3.7 kw) or less and are operated as a single fan's functional equivalent.
- Fans that are part of the equipment listed under Section 6.4.1.1 (note: in IECC, it is Section C403.2.3).
- Fans included in equipment bearing a third party-certified seal for the equipment package's air or energy performance.
- Powered wall/roof ventilators (PRV).
- Fans outside the scope of AMCA 205.
- Fans that are intended to operate only during emergency conditions.

With these challenges in mind, AMCA has developed four resources to assist interpreting fan efficiency codes and standards:

- Requirements for independently certified Fan Efficiency Grade ratings and an energy label in model codes (IECC and IgCC).
- Online fan model database that has been AMCA-certified for Fan Efficiency Grade and peak total efficiency (www.amca.org/feg/feg-finder.aspx).
- ANSI/AMCA 205-12 is available online at no cost (www.amca.org/feg/codes-and-standards.aspx).
- An online library of technical articles, white papers and PowerPoint presentations about Fan Efficiency Grades and relevant code and standard developments (www.amca.org/feg/best-practices.aspx).

U.S. Department of Energy Action

In 2011, the U.S. Department of Energy signaled its intent to regulate commercial and industrial fans. The Department of Energy published a 'Request for Information' that stated they had preliminarily ruled that the Department of Energy has the authority to regulate fan efficiency. The Request for Information also indicated a need for stakeholders such as manufacturers to provide the Department of Energy with information about fans and the fan market (U.S. Department of Energy, 2011). As part of the next stage of the regulatory process, the Department of Energy released a Framework Document for public review on February 1, 2013. The Framework Document described how the Department of Energy is considering regulating fan efficiency, and identified fan efficiency rating standards, fan market data, the European Union fan regulation, and other information the Department of Energy would consider. Following the completion of milestones, the Department of Energy will issue a final rule in 2015 or 2016, and enforcement should begin around 2020 (U.S. Department of Energy, 2013b).

Based on the Framework Document, the Department of Energy's vision for regulating fan efficiency is very different from the approach that the model codes and standards have taken. The Department of Energy's scope is much broader, including industrial process fans and a size range from 125 W to 500 kW. The Department of Energy will define multiple fan categories, each having an independently considered metric, test standard, minimum efficiency performance requirement and range of covered products. Also, the Department of Energy is considering regulating 'extended products', defined as fans sold with motors and fans sold with motors and drives, in addition to fans sold without motors or drives.

Because the Department of Energy appliance and equipment standards are restricted to products, the federal regulation cannot include the provisions in model codes and standards having a sizing and selection window. The significance of this is that operating efficiency is dependent on fan size. A fan efficiency requirement without a sizing and selection window might not yield energy savings. In fact, such a requirement could lead to *increasing* fan energy consumption if one selects

Table A1.3. *Schedule of model code and standard revisions.*

	1st generation year (year finalised)	2nd generation year (year finalised)	3rd generation year (year finalised)
ASHARAE 90.1	2013 (2013)	2016 (2016)	2019 (2019)
ASHRAE 189.1	2014 (2014)*	2017 (2017)	2020 (2020)
IECC	2015 (2013)	2018 (2016)	2021 (2019)
IgCC	2015 (2014)*	2018 (2017)	2021 (2020)

* Organisations are evaluating fan efficiency proposals during the current update cycle.

and purchases smaller fan sizes to recoup the higher purchasing costs of higher efficiency fans.

The Framework Document generated considerable commentary from industry stakeholders, including proposed redefinitions of the fan categories by AMCA International to align with current industry practice (AMCA, 2013b). Whilst the Department of Energy is developing the federal fan efficiency standard and regulation, ASHRAE and the International Code Council will have implemented or finalised second and third generations of fan efficiency provisions in model codes and standards, Table A1.3. As discussed earlier, first-generation provisions are already in place or proposed. Development of second-generation provisions will soon begin, possibly appearing first in ASHRAE 90.1-2016. Some of the early thinking on second-generation provisions is to incorporate knowledge gained by participating in the Department of Energy rulemaking process, such as defining fan categories and setting separate requirements for each. Third-generation provisions would harmonise model codes and standards with applicable Department of Energy requirements; hence, their development would begin after the Department of Energy publishes its final rule in 2015 or 2016.

Of particular significance to the U.S. federal regulation will be strict market surveillance. The Department of Energy's regulations for equipment include establishing a testing standard for each type of covered product. The test standard is for compliance testing, which manufacturers use to certify that their products meet federal minimum energy performance requirements. The Department of Energy will also use the same test standard for enforcement testing, whereby independent third party contractors conduct periodic check tests on product samples and compare results against manufacturers' data. Products that fail the check tests can result in fines and removal from the market.

EUROPEAN FAN REGULATIONS

The European Union (EU) has been intensively regulating energy efficiency and emissions relating to climate change under the European Ecodesign Directive. This Directive provides a framework for establishing requirements for energy-related products and the Directive's Implementing Regulations that will impact all member states. The EU has published more than a dozen 'Implementing Measures,' covering products ranging from television sets to light bulbs and fans. These regulations are for products only. They are not building code provisions.

To identify targets for Implementing Measures, the European Commission relies on a network of consultants to launch a series of studies with input from stakeholders. Impact assessments establish priorities based on estimated potential for energy saving. The Commission then prepares working documents based on the study outcomes and impact assessments. The department, known as a Directorate General, that is responsible for a particular Implementing Measure, will convene one or more Consultation Forum meetings to discuss the working document with industry stakeholders and member states' government representatives. Based on feedback, the Directorate General prepares a draft Implementing Measure to discuss with Council of Member state experts and the Commission in a Regulatory Committee. Once the Regulatory Committee has endorsed the draft Implementing Measure, the Council and European Parliament have typically from two to four months to consider the proposed legislation prior to its adoption as a legally binding European Regulation.

An impact assessment will evaluate the environmental aspects from design through manufacture to disposal of a product. However, the regulation of the efficiency of a typical energy-using product or an energy-related product is primarily focused on the amount of electricity consumed as a result of using or deploying the product. Hence, the Commission has chosen to regulate fan efficiency on the assumption that an electric motor drives the fan, whether it is fitted to the product at the point of sale or not. Consequently, the EU adopted Fan Motor Efficiency Grades (FMEG), as defined by the International Standards Organization (ISO) EN ISO 12759 *Fans — Efficiency Classification for Fans*, as an efficiency metric rather than Fan Efficiency Grades. The Fan Motor Efficiency Grades designation addresses the overall efficiency of an entire fan assembly, accounting for losses in, for example, a belt-drive system, electric motor, and variable speed drive and for losses from suboptimal combinations of those components, Figure A1.4.

The European Commission has used relevant information from EN ISO 12759, together with market benchmarking studies to determine the target Fan Motor Efficiency Grades for fans based on fan type in both ducted and non-ducted configurations for building ventilation applications (Radgen *et al.*, 2008). Ducted applications use fan total pressure as the basis for efficiency calculation; non-ducted fans use fan static pressure. Manufacturers declare the Fan Motor Efficiency Grade based on full fan speed and they list the chosen efficiency category as static or total. For most types of fans, manufacturers provide efficiencies for static and total pressure installations. However, the way one installs and uses many energy-using products determines efficiency and energy consumption. This is often beyond the control of the product manufacturer or supplier. Therefore, the European Commission decided that it would set minimum efficiency performance requirements (MEPRs) at optimum or peak efficiency operating points for electric motors and fans and then adopt an 'extended product approach' to regulate them.

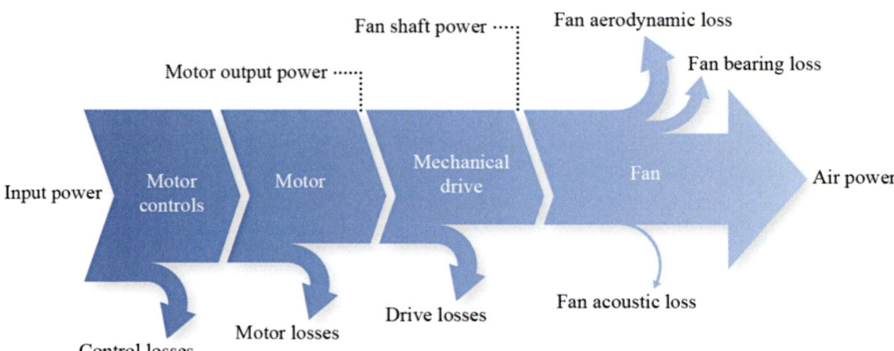

FIGURE A1.4. Power and losses through an entire fan assembly (Hauer and Brooks, 2012).

In the case of fans as an 'extended product,' this could mean that the motor that drives the fan must comply with a minimum efficiency performance requirement before a manufacturer can legally place it on the market. The fan manufacturer may not always be using the compliant motor at its optimum efficiency point, but the fan must comply with the minimum efficiency performance requirement that is required in the fan regulation. In continuation of this extended product approach, the EU is introducing and drafting more implementing measures, which will set minimum efficiency performance requirements for a range of equipment and products that incorporate both the fans and the motors. Note that the extended product approach also means that whereas fans used in 'fan units' such as air handling units, box fans and roof fans are within the scope of Regulation 327/2011, the 'fan units' themselves are not. These will be dealt with in a separate regulation. For fans and fan-using equipment installed in heating, ventilation and air conditioning applications in buildings, implementing the Energy Performance of Buildings Directive should further control the minimum efficiency of the systems that utilise them. Some member states have already introduced total specific fan power targets, based on the electricity usage to provide the required air volume flow, for the fans installed in building heating, ventilation and air conditioning systems.

Fans regulation

Commission Regulation (EU) No. 327/2011 sets minimum efficiency targets that fans driven by electric motors with input powers from 125 W to 500 kW must meet. Compliance with Tier 1 targets became mandatory from 1 January, 2013, and more stringent Tier 2 targets applied on 1 January, 2015. Determination of Fan Motor Efficiency Grades is based on applying two standards: EN ISO 12759 to determine target efficiency and Fan Motor Efficiency Grade and EN ISO 5801 *Industrial Fans — Performance Testing Using Standardised Airways* for setting up and measuring air performance and air power. Subsequently, the Commission has issued a mandate (M500) for drafting a specific Harmonised EN Standard that will pull together all the relevant information and specify the required methodologies and calculations to demonstrate compliance with Regulation 327/2011. This work is now progressing in the European Committee for Standardization CEN TC/156 WG/17, but meanwhile, this regulation is still, and will be, in force. Tabulated minimum efficiencies according to EU 327/2011 are available as a PDF and an Excel spreadsheet at www.amca.org/feg/fmeg.aspx. To guide the application and enforcement of Regulation 327/2011, the European Commission has published two, non-legally binding, frequently asked questions guidance documents:

Frequently Asked Questions on the Ecodesign Directive 2009/125/EEC. This document contains general information and pages dealing with specific issues from Implementing Measures, including Regulation 327/2011.

Guidelines. This document contains answers to frequently asked questions specific to Regulation 327/2011.

Both documents are downloadable from the European Commission website at http://ec.europa.eu/enterprise/policies/sustainable-business/documents/eco-design/guidance/index_en.htm. These, unlike Regulation 327/2011, are 'living documents' in so far as the Commission can update or add to them at any time.

Once the European Commission publishes an implementing measure, it is difficult to persuade it to consider any amendments. Even then, an amendment agreed by the Commission would have to go through the Council and Parliamentary procedures and take some considerable time. However, although the Tier 2 fan efficiency targets are already set and compliance has been mandatory from January 2015, the European Commission is committed to complete a review of Regulation 327/2011 in 2015. To meet this schedule, the review process commenced early in 2014 and has involved scrutiny, including impact assessments to date and stakeholder consultations. The review process is providing opportunity for introducing amendments if review participants find sufficient justification.

The energy and emissions savings projected to result from Regulation 327/2011 are significant. A European Commission Impact Assessment study identified a stock of 143 million fans in 2005, possibly rising to 227 million units in 2020. When assessed at the EU level, fan energy consumption was 390 TWh annually representing approximately 179 million tons of CO_2 in emissions for fossil fuel burning power stations in 2005, with an estimated rise to 630 TWh annually in 2020, a 60 per cent increase.

The Impact Assessment predicted that Regulation 327/2011 could achieve a reduction of 54 TWh annually by 2020. This equates to savings of 25 million tons of CO_2 and an associated 7 billion Euro reduction in the cost of electricity needed to drive the installed fan base. To give this a European perspective, this is approximately equal to the amount of electricity consumed in Greece in 2006. The Impact Assessment also identified that member states monitor the situation through market surveillance which is intended to ensure that manufacturers comply with the regulation requirements. However, the level of enactment of market surveillance is limited by the availability of resources and is fragmented across the EU.

Market surveillance and regulation in the EU

One of the mainstays of the EU single market is the free movement of goods. In this context, the proper functioning of the single market rests ultimately on consumer trust. Consumers and purchasers must have confidence that, irrespective of where a product is manufactured and in which member state they buy it, the product is both safe and compliant. This is the message from an explanatory statement in the draft report on the *Proposal for a Regulation on Market Surveillance of Products and Amendments to Existing Council Directives* that the 'European Parliamentary Committee on the Internal Market and Consumer Protection' published in June 2013.

Finnish Member of the European Parliament (MEP) Ms Sirpa Pietikäinen, rapporteur on market surveillance, explained that efficient and high-quality market surveillance should act as the ultimate guarantor. This is a stance that the European AMCA has encouraged and supported. Market surveillance should not only ensure that products will not endanger the consumers' health and safety or present a hazard in the workplace, but also guarantee that products are compliant with applicable EU legislation such as rules for minimum efficiency, noise levels and emission limits.

However, it is clear that member states' market surveillance is already failing to adequately address the growing number of imports to the EU and the increased number of products circulating within the single market. Recent figures from the European Commission suggest that authorities check only 0.3 per cent of all goods placed on the market in the European Union, including imported goods. Meanwhile, there is still an expectation that market surveillance authorities will be able to rise to new challenges, such as the growth in e-commerce.

The enactment of adequate market surveillance is made more difficult by a complex legal structure. In 2011, a European parliament resolution on the revision of the General Product Safety Directive and Market Surveillance stated that the '*current legislative framework for market surveillance does not provide enough coherence and should therefore be reviewed and further coordinated*'. The resulting proposal would establish a common European framework for market surveillance that will impact all products circulating within the internal market, including those imported to the EU.

Surveillance needed beyond health and safety

Traditionally, the fragmented market surveillance activity within the EU has largely focused on health and safety. EU member states now recognise that the Market Surveillance Regulation should not only safeguard health and safety, but also serve as a means to enforce EU legislation that seeks to uphold other public interests such as energy efficiency and the environment.

The implementing measures that the EU is introducing for various products falling within the scope of the Ecodesign Directive 2009/125/EC all contain rules for enacting market surveillance. Honest manufacturers put effort and resources into ensuring that they are compliant with EU safety, environment and other requirements. Market surveillance should ensure that players who do not abide by the rules do not gain advantage in the market over compliant operators. It is in the interests of trustworthy manufacturers, importers and distributors, as well as consumers that market surveillance authorities ensure that a member state's legally empowered regulatory body can identify unsafe, harmful and otherwise non-compliant products and keep or take them off the market.

The surveillance resources issue

The biggest challenge facing member states wishing to enact effective market surveillance is the lack of available resources and the growing volumes of trade. In its search for longer-term solutions, the European Commission has indicated an interest in exploring new, innovative market-based solutions that can complement market surveillance actions that the authorities currently solely undertake. Although there is an underlying suspicion of industry-sponsored certification and accreditation programmes, it is possible that the Commission may have to evaluate genuine contributions to effective market surveillance that responsible industry can provide, despite resistance to the idea on the part of the European Commission. Another suggested solution could involve third party auditing of operators' quality control systems, as well as of the end products.

COMPARING U.S. AND EUROPEAN FAN REGULATIONS

Although they were started at approximately the same time, European and U.S. regulatory approaches to fan efficiency regulation are quite different, Table A1.4. Differences include: the technical boundaries of the regulated equipment; fans alone verses fans plus motors plus drives, to the present structure and rigor of market surveillance; and enforcement. The significance of the difference is that many fan manufacturers have international scope, and having to comply with different fan efficiency standards imposes extra regulatory burden on affected companies. Moreover, regulatory differences also create market inefficiencies as they constitute a diversion of staff time, marketing effort, communications programmes and engineering literature.

Ideally, harmonisation will occur by 2020, which is when the U.S. Department of Energy regulation for fan efficiency should become effective. Some degree of harmonisation already exists in that EN ISO 12759 has a definition of Fan Efficiency Grades that is consistent with AMCA 205. AMCA 205 currently is developing a metric consistent with Fan Motor Efficiency Grades (Ivanovich, 2014), and the Department of Energy is considering adopting or developing such a metric for regulating extended products (U.S. Department of Energy, 2013b).

Table A1.4. *Comparison of U.S. and EU fan efficiency regulatory approaches.*

Parameter	United States	European Union
Promulgating bodies	U.S. Department of Energy (DOE)	European Commission
Effectiveness dates	2020 (estimated)	January 1, 2013, for Tier 1 January 1, 2015, for Tier 2
Fan efficiency rating standard	Draft Department of Energy test standard is expected in 2014; finalisation in 2015/2016; taking effect in 2019/2020	Draft of harmonised EN standard for fans — procedures and methods to determine the energy efficiency in preparation. 2015 target for introduction
Scope — application	Fans for commercial and industrial applications; not restricted to building ventilation systems	Fans, including those integrated into other energy-related products. Excludes fans for use in toxic, corrosive, flammable and abrasive environments, and at supply voltage >1,000V ac or >1,500V dc
Scope — size	125 W to 500 kW proposed in Framework Document	Electrical input power from 125 W to 500 kW
Product boundaries	Fan only, with possible extended product provisions, fans sold motors and fans sold with motors and drives	Extended product: fan + motor + drive (if specified)
Metric	Fan Efficiency Grade (FEG) for fan-only, as defined in ANSI/AMCA Standard 205. Metric for extended products is under development by AMCA	Fan Motor Efficiency Grade (FMEG), as defined in EN ISO Standard 12759
Pressure basis	To be determined	Fan total pressure for ducted fans. Fan static pressure for non-ducted fans
Market surveillance	Manufacturer declaration of compliance using federal test standard, with federally administered periodic check tests	Manufacturer documentation and declaration of compliance with FMEG target. Member state's targeted documentation check and possible independent test. European Union wide sharing of results
Projected annual energy savings	20 per cent of annual fan power for covered fans (U.S. Department of Energy, 2011)	54 TWh per year by 2020

CONCLUSION

Developing fan efficiency regulations for commercial and industrial fans is a relatively new phenomenon. Although started at approximately the same time in 2007, the U.S. and EU have taken very different approaches. The U.S. effort has primarily been through collaboration between AMCA International and ASHRAE and the International Code Council that publish model codes and standards for energy efficiency and green construction. Local code jurisdictions will eventually adopt and enforce these codes and standards. Adopting fan efficiency provisions began in 2012, but the provisions should not have significant market impact until 2019 or 2020 because of the slow uptake of revised model codes and standards by code jurisdictions.

The U.S. approach is characterised by having a single Fan Efficiency Grade (FEG) requirement for all applicable fans, and a sizing and selection window that impacts design practice such that fans are sized upwards in diameter to operate closer to peak total efficiency more of the time. The primary difference between versions of the codes and standards is that the provisions in model codes (those under the auspices of the International Code Council) include requirements for independently certified Fan Efficiency Grade ratings and an energy label, Table A1.2. Provisions in the model standards (those under the auspices of ASHRAE) do not have certification and labeling requirements. All code and standard provisions, both in place and proposed, have applicability to fans greater than 3.7 kW, with a list of exemptions to help ease implementing first-generation provisions.

Development of a federal fan efficiency test standard and minimum energy performance standard began in 2011 and should be completed in late 2015 or early 2016. The U.S. federal regulations, administered by the U.S. Department of Energy, should become effective in 2020. The Department of Energy regulation will establish a number of fan categories, each with its own minimum energy efficiency standard and scope of coverage. The federal requirement will have uniform applicability and enforcement throughout the U.S. and its territories. Enforcement of U.S. federal product efficiency standards is rigorous.

U.S. model codes and standards will eventually be updated to be consistent with federal requirements. Because of the three-to-five year lag between publishing final federal requirements and enforcing them, model codes and standards could be updated to reflect federal requirements ahead of their enforcement. This convergence will harmonise fan efficiency standards throughout the U.S.

The European approach began with a market study and proposal that much of the industry considered unrealistic. This prompted the rapid drafting of a reference ISO standard for fan efficiency, and led to a European Commission regulation that went into effect 1 January, 2013, with an automatic escalation of stringency taking effect 1 January, 2015. The scope of the European Commission regulation ranges from fans with a low electrical consumption of 125 W to fans requiring up to 500 kW. The European Commission approach uses the metric Fan Motor Efficiency Grade (FMEG) for fans, regardless of whether or not manufacturers actually sell them with the motor and specified ancillaries that will be required to enable them to function, and treats non-ducted and ducted fans differently.

If an extended product approach is included in the U.S. Department of Energy regulation and in second- and third-generation fan efficiency provisions in model codes and standards, these measures will bring U.S. and EU approaches closer together. However, it will not result in harmonising their approaches.

REFERENCES

AMCA (2009), *Fan Industry – Facing up to the Energy Challenges*, Air Movement and Control Association International, 13 May.

AMCA (2013a), *Certified Ratings Program*, Air Movement and Control Association International, www.amca.org/feg/feg-finder.aspx.

AMCA (2013b), *Response to U.S. Dept. of Energy Framework Document for Commercial and Industrial Fans*, docket no. EERE-2013-BT-STD-0006, 31 May.

ANSI/AMCA (2012), *Standard 205: Energy Efficiency Classification for Fans*, Air Movement and Control Association International.

Cermak, J. and Ivanovich, M.G. (2013), 'Fan Efficiency Requirements for Standard 90.1-2013', *ASHRAE Journal,* vol. 55(4), pp. 24–30.

Cermak, J. and Murphy, J. (2011), 'Select Fans using Fan Total Pressure to Save Energy', *ASHRAE Journal,* 53(7), pp. 44–47.

Commission Regulation (EU) No 327/2011 (2011), *Official Journal of the European Union.*

Gowri, K. (2009), 'What's New in ASHRAE 90.1-2010?', *Proceedings of the 17th National Conference on Building Commissioning*, Seattle, WA, USA, 3–5 June.

Hauer, A. and Brooks, J. (2012), 'Fan Motor Efficiency Grades in the European Market', *AMCA inmotion(2),* p. 14.

Ivanovich, M.G. (2014), 'Fan Efficiency Codes and Standards: The Pressure is On', *HPAC Engineering*, 6 January.

Misuriello, H., Penney, S., Eldridge, M. and Foster, B. (2010), 'Lessons Learned from Building Energy Code Compliance and Enforcement Evaluation Studies', *Proceedings of the American Council for an Energy-Efficient Economy Summer Study*, Pacific Grove, CA, USA, 15–20 August.

Radgen, P., Oberschmidt, J. and Cory, W.T.W. (2008), *EuP Lot 11: Fans for Ventilation in Non Residential Buildings: Final Report*, Fraunhofer Institute Systems and Innovation Research.

U.S. Department of Energy (2011), 'Notice of Proposed Determination', *Federal Register*, 76 FR 37678, 28 June.

U.S. Department of Energy (2013a), 'Energy Efficiency Program for Commercial and Industrial Equipment: Public Meeting and Availability of the Framework Document for Commercial and Industrial Fans and Blowers', *Federal Register,* 78/22: 7387, 1 February.

U.S. Department of Energy (2013b), 'Presentation Slides: *Energy Conservation Standards for Commercial & Industrial Fans and Blowers: Framework Public Meeting*', docket number EERE-2013-BT-STD-0006, 15 February.

Westphalen D. and Koszalinski, S. (1999), 'Energy Consumption Characteristics of Commercial Building HVAC Systems Volume II: Thermal Distribution, Auxiliary Equipment, and Ventilation', *U.S. Department of Energy*, contract no. DE-AC01-96CE23798.

Appendix

2

Variable Pitch Fans

A.G. Sheard and A. Rhoden

ABSTRACT

This appendix is adapted from the patent 'Variable pitch fans'. The adaption aims to present the invention in a more accessible way to the reader, without significantly altering the patent's content.

A need to prevent the premature failure of bearings used to support each blade of a variable pitch in motion fan inspired the research that resulted in the intellectual property that forms the basis of the patent. A variable pitch in motion fan comprises a hub with blades extending radially from the hub. The blades are each mounted on an associated bearing mounted within the hub. The bearing comprises a radially inner race fastened to the blade and a second radially outer race fastened to the hub. The balls in the bearing are separated, with each separator being cylindrical and having its cylindrical axis coaxial with the orbital axis of the bearing balls. The separators are adapted to rotate about the orbital axis, the side face having a part spherical recess adapted to receive a ball. The separators' rotation about the orbital axis entrains lubricant in the bearing and transports it from the outer race over the balls towards the inner race. By transporting lubricant over the balls the separator transports it into the ball-to-separator contact zone. In combination with the separators' cylindrical design, transporting lubricant into the ball-to-separator contact zone reduces the rate at which the radially inner race and balls wear, thus preventing premature bearing failure.

This appendix is a revised and extended version of Sheard, A.G. and Rhoden, A. (2012), 'Variable Pitch Fans'. Patent No. GB 2485634 B, 12 September 2012.

INTRODUCTION

The present invention relates to an axial air movement fan having a plurality of variable pitch in motion blades. Each blade extends radially from a hub that rotates about the fan's main axis. A bearing located within the hub supports each blade that enables it to pivot about its radial axis. The hub incorporates a mechanism to pivot the blades simultaneously, thus varying the pitch of all blades in unison.

Variable pitch in motion fans are classically applied into forced- and induced-draft application with coal-fired boilers. The variable calorific value of the coal burnt in the boiler results in a variable boiler pressure. The inertia of forced- and induced-draft fans is large, making it difficult to change fan speed fast enough to maintain a stable boiler pressure. However, one can change blade pitch rapidly and therefore a variable pitch in motion fan running at constant speed can react to changing boiler pressure quickly enough to maintain it within safe operating limits.

Engineers typically design variable pitch in motion fan blades to pivot through an angular range of movement that result in a blade pitch range between 10 and 80 degrees. Such arrangements have the disadvantage that, as the bearing does not fully rotate but pivots backwards and forwards there is a tendency, caused by centrifugal force, for lubricant in the bearing to move to the outer race. The bearing's limited degree of movement, which is usually a ball bearing, means that the lubricant does not transfer from the outer race to the inner race by the balls in the conventional manner since the balls do not rotate enough to entrain sufficient lubricant and move it to the inner race, which tends to run dry. This leads to premature wear and failure in the bearing and the present invention seeks to provide a solution to this problem.

According to the present invention there is an axial air movement fan with a plurality of variable pitch in motion blades. Each fan blade extends radially from a hub that rotates about a main axis and each blade further pivots about its radial axis to vary blade pitch. The blades are mounted on an associated bearing that then mounts within said hub, Figure A2.1.

The bearing comprises a radially inner race fastened to the blade via a counterweight and bolting arrangement and a second radially outer race fastened to the hub via a mushroom shaped bearing retainer, Figure A2.1. At its root, the fan blade has a recess which is a snug fit on the head of the mushroom shaped bearing retainer. The fan blade is secured to the inner race via a counterweight with bolts which, when tightened, clamp the fan blade firmly to the bearing and hence through the mushroom shaped mounting to the hub. In operation, as the fan rotates, the blades' centrifugal force tends to increase the load on the bearing through the inner race. The slight deformation caused by this tends to lift the blade from its clamped position on the mushroom shaped bearing retainer to facilitate its pivotal movement.

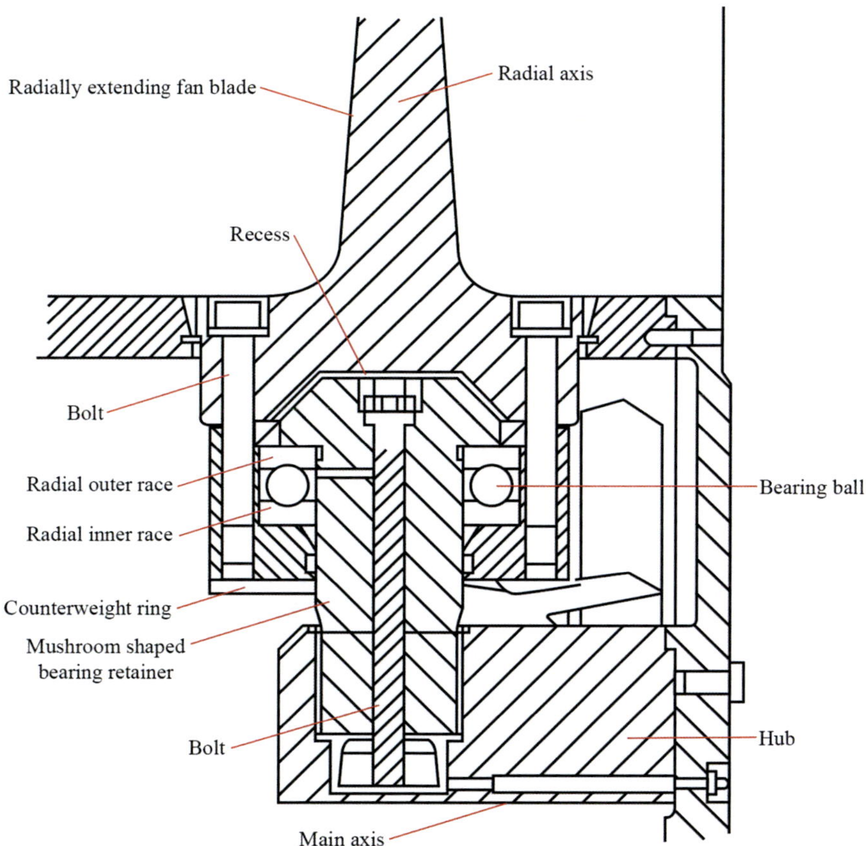

FIGURE A2.1. A sectional view through a hub and single fan blade of a variable pitch in motion fan illustrating the linkage between the blade and hub via a 'blade bearing' that facilitates blade pivoting whilst the fan rotates at constant speed.

BEARING BALL SEPARATOR CONCEPT

The bearing consists of an inner ball retaining ring and an outer ball retaining ring between which the bearing balls are located, Figure A2.2. Separators space apart adjacent balls, with the retaining rings holding the balls and the separators together as a sub-assembly prior to assembling the bearing. The separators have the additional advantage of assisting in retaining grease in the bearing during service operations. The separators retain grease when one assembles the bearing inner and outer ball retaining rings plus radially inner and outer raceways around the bearing balls and separators where there is only a very small gap between them, Figure A2.3.

Each separator consists of a generally cylindrical body having on its two side faces part spherical recesses which are adapted to receive and locate the ball

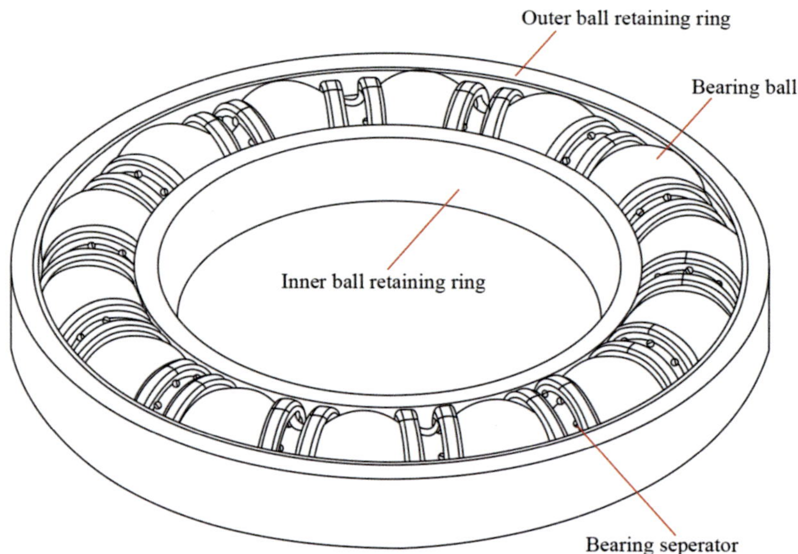

FIGURE A2.2. An isometric view of a blade bearing that facilitates blade pivoting whilst the fan rotates at constant speed.

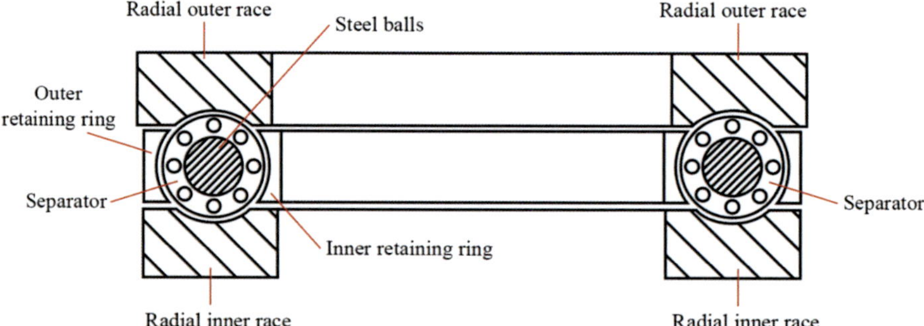

FIGURE A2.3. A sectional side view of a blade bearing illustrating the small gap between the bearing inner and outer ball retaining rings, radially inner and outer raceways that assists in retaining grease within the bearing.

bearings, Figure A2.4. The separators are dimensioned such that there is little or no free play between the balls and the separators. The recesses' radius is slightly larger than the balls' radius. In its circumferential periphery, the separator has a recess which forms an annular reservoir for lubricant, in the form of grease, for lubricating the bearing, Figure A2.5. The annular reservoir is sufficiently deep for walls to form between the reservoir and the part spherical recesses. A plurality of passages leads from the reservoir through the separator walls and into the part spherical recesses. These passages enable lubricant to pass from the reservoir to lubricate the balls.

Because the radius of the separator side face part spherical recesses is larger than that of the balls, a space forms between the balls and the passage openings on the faces of the part spherical recesses. This facilitates the passage of lubricant from the reservoir into the recesses. By having a larger radius for the recesses, the contact surface between the balls and separators reduces. This has the additional advantage that friction between the separators and the balls reduces as the material from which the separator is manufactured has a non-linear friction coefficient. A consequence of the non-linear friction coefficient is that a ball-to-ball force applied over a smaller separator area results in reduced ball-to-separator friction.

The passages extended through the walls in a direction parallel to the axis of rotation of the separator, Figure A2.5. The passages are inclined towards the axis of

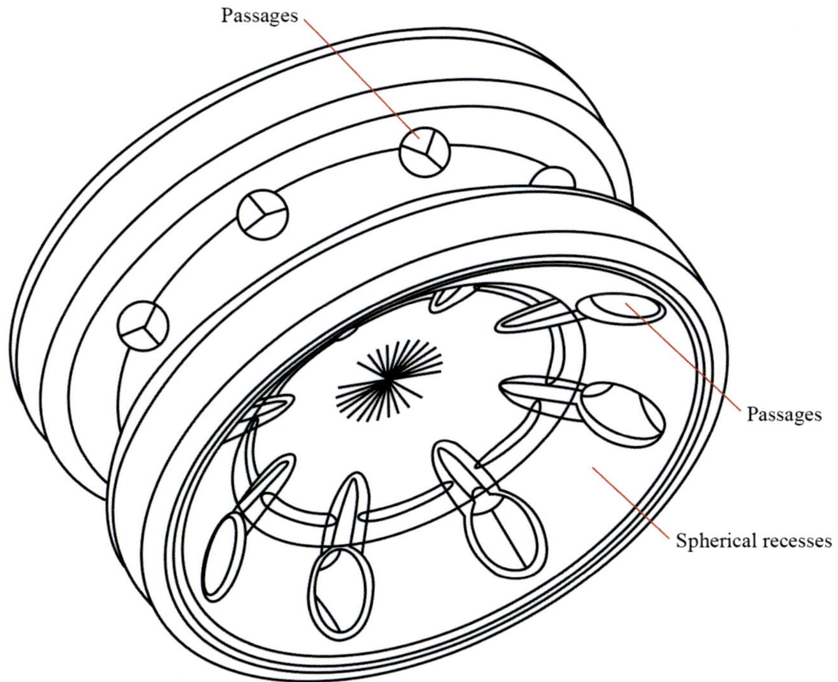

FIGURE A2.4. An isometric view of the separator used to divide the balls in each blade bearing.

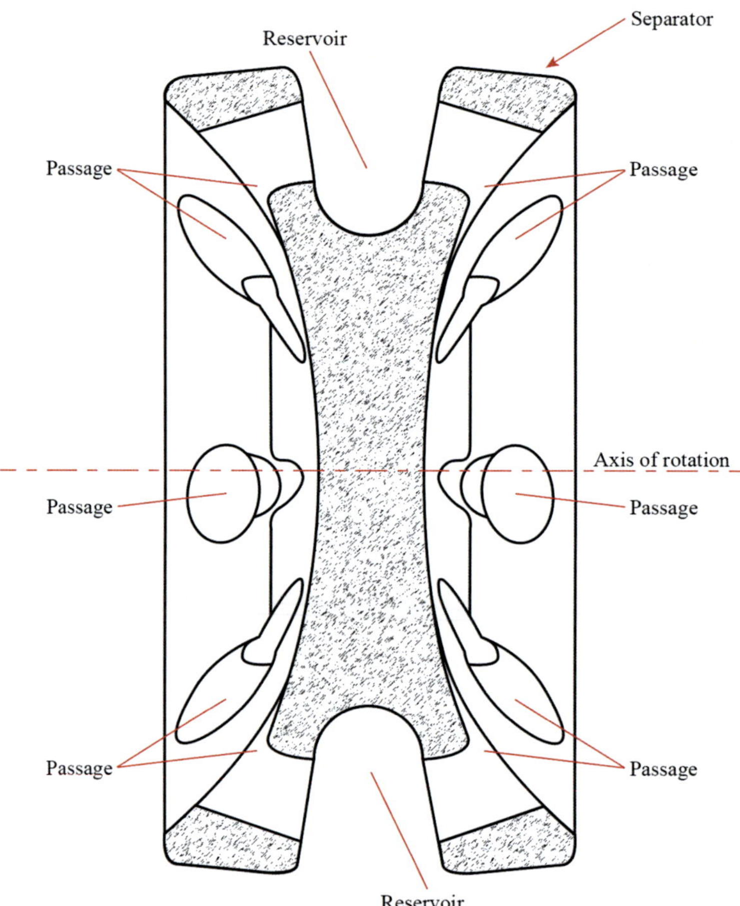

FIGURE A2.5. A cross sectional view through the separator illustrating the passages' location and orientation through which the reservoir transports grease from the radially outer to both the ball-separator contact area and the radially inner bearing race.

rotation of the separators in the direction from the reservoir towards the part-spherical faces. This inclination facilitates grease to pass from the reservoir into the recesses. Guides in the reservoir may assist in guiding the grease towards and into the passages. The provision of grooves extending from the openings radially inwardly, the depth and width of the grooves gradually reducing until they merge into the recesses' surface improves lubricant distribution across the surface of the part spherical recesses.

As the separators have a constant cross section in their axial plane that is symmetrical, the separators are able to rotate about their axis which is co-axial with the balls' orbital axis. In operation, when the fan blade and hence the bearing pivots, the balls rotate and move along a circular orbit determined by the diameter of the bearing raceways, Figure A2.6. As a result of circular orbit, each ball is biased to

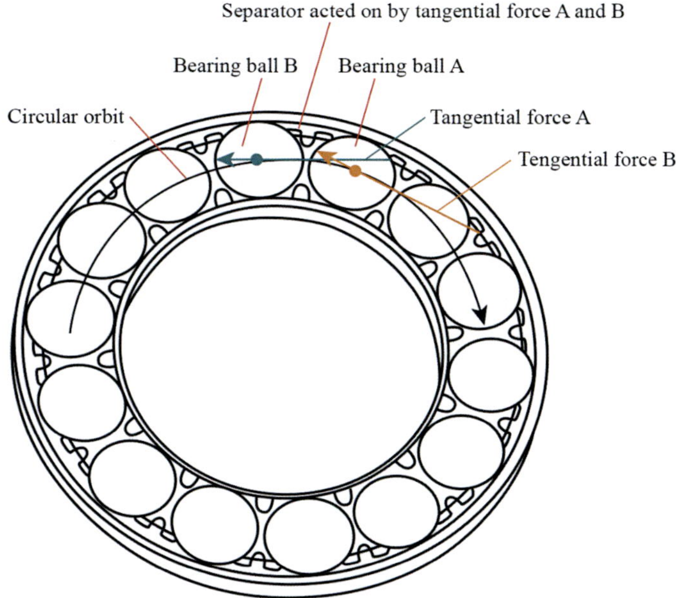

FIGURE A2.6. An illustration of the tangential forces imposed on each side of a separator by the bearing balls that induce the separators' rotation as the bearing pivots.

move in the tangential direction to the circular orbit, resulting in tangential forces acting on each separator, Figure A2.6. Two balls act upon each separator, one leading and one trailing labelled as bearing ball A and B in Figure A2.6. The tangential forces imposed on the separator by the two balls, tangential force A and B, are different since ball A and ball B are in different positions on the circular orbit. The difference between the point at which tangential force A and tangential force B act on each side of the separator results in the separator rotating about its axis, When the blade's pivotal movement reverses, the separators' rotation direction also reverses.

Centrifugal force tends to bias the lubricant towards the bearings radially outer race, from where it is urged through the separator passages successively as the separator rotates. In this way the lubricant spreads more efficiently over the balls' entire surface to improve the lubrication between the ball and the inner race. Using individual separators between the balls is required as it enables each separator to move independently. In traditional bearings in which the balls are located in an annular one piece bearing cage, ball-to-ball forces inevitably induce stresses in the one-piece bearing cage. The separators of the present invention permit slight bearing ball movement relative to one another which eliminates peak-stress regions that would otherwise occur in a one-piece bearing cage. Engineers have associated the peak-stress regions of a one-piece bearing cage with premature bearing failure in-service, and therefore eliminating them improves the bearing's service life.

CLAIMS

1. An axial air movement fan having a plurality of variable pitch in motion blades, each blade extending radially from a hub rotatable about a main axis with the ability of each blade to pivot about its radial axis to vary blade pitch. The blades are each mounted on an associated bearing that is mounted within the hub, with the bearing comprising a ball bearing with an outer race fastened to the hub and an inner race fastened to the blade. Separators divide the balls in the bearing, with each separator generally cylindrical and having its cylindrical axis coaxial with the orbital axis of the balls' orbit. The separator is adapted to rotate about the said orbital axis, as the separator's faces each have a spherical recess adapted to receive a ball. Forces acting on the separator induce the separator's rotation about the orbital axis. This results in entraining lubricant in the bearing as lubricant transports from the radially outer race over the balls towards the radially inner race.
2. An axial fan as claimed in Claim 1 wherein the part spherical recesses have a radius greater than that of the balls.
3. An axial fan as claimed in Claim 1 or 2, wherein the separators each have a circumferential recess in its exterior circumference forming an annular reservoir for lubricant, at least one passage extending from the reservoir into the part spherical recesses to enable lubricant to flow to and from the reservoir on to the recesses' surface.
4. An axial fan as claimed in Claim 3 wherein a plurality of the said passages are disposed about the annular reservoir to provide a plurality of pathways to the part spherical recesses.
5. An axial fan as claimed in Claim 3 or 4, wherein the passage openings into the part spherical recesses extend from the openings grooves in the recesses' surface to assist in distributing lubricant across the surface of the recesses and the balls.
6. An axial fan as claimed in Claim 5 wherein the grooves gradually reduce in depth and width away from the openings.
7. An axial fan as claimed in Claim 6 wherein the grooves extend substantially radially inwardly from the openings.
8. An axial fan as claimed in Claim 4 wherein the passages extend substantially parallel to the separator's axis of rotation.
9. An axial fan as claimed in Claim 4 wherein the passages extend at an angle inclined towards the separator's axis of rotation.

Bibliography

Page numbers in italics refer to figures or tables

AMCA, *Fan Industry — Facing up to the Energy Challenges*, Air Movement and Control Association International, 13 May, 2009 9

Angeli, D., 'OpenFOAM @ Mimesis and Numerical Simulation of the Mont Blanc Tunnel,' *HPC Enabling of OpenFOAM for CFD Applications,* CINECA, Casalecchio di Reno, Bologna, Italy, 26 November, 2012. 24

ANSI/AMCA, Standard 205: Energy Efficiency Classification for Fans, Air Movement and Control Association International, 2012 5

ANSI/ASHRAE Standard 149-2000, Laboratory Methods of Testing Fans Used to Exhaust Smoke in Smoke Management Systems, 2009 109

ANSYS, *CFX-Solver Theory Guide, ANSYS Inc., Canonsburg, PA, USA, 2009 165, 167*

Arakawa, C., Fleig, O., Makoto, I. and Masakazu, S., 'Numerical Approach for Noise Reduction of Wind Turbine Blade Tip with Earth Simulator,' *Journal of the Earth Simulator*, vol. 2, pp. 11–33, 2005 126

Archambeau, F., Mechotoua, N. and Sakiz, M., 'Code_Saturne: a Finite Volume Code for the Computation of Turbulent Incompressible Flows-Industrial Application, *IJFV International Journal of Finite Volumes,* vol. 1, pp.1-62, 2004 33

Balan, C. and Tabakoff, W., 'Axial Flow Compressor Performance Deterioration,' Proceedings of the AIAA.SAE/ASME 20th Joint Propulsion Conference, Cincinnati, OH, USA, 11–13 June, paper no. AIAA-84-1208, 1984 230, 256

Baxter, L.L., 'Turbulent Transport of Particles', *PhD thesis,* Brigham Young University, Provo, Utah, USA, 1989 26, 231, 235–7, 257, 262, 304, 305, 328, 330

Baxter, L.L. and Smith, P.J., 'Turbulent Dispersion of Particles: the STP Model,' *Energy and Fuels*, vol. 7(6), pp. 852–859, 1993 231, 262, 328

Bazilevs, Y., Hsu, M.-C., Akkerman, I., Wright, S., Takizawa, K., Henicke, B., Spielman, T. and Tezduyar, T.E. (2011), '3D Simulation of Wind Turbine Rotors at Full Scale. Part I: Geometry Modelling and Aerodynamics,' *International Journal for Numerical Methods in Fluids,* vol. 65, pp. 207–235, 2011 57

Beachkofski, B.K. and Grandhi, R.V.), 'Improved Distributed Hypercube Sampling,' *Proceedings of the 43rd AIAA/ASME/ASCE/AHS/ASC Structures, Structural Dynamics, and Materials Conference*, Denver, CO, 22–25 April, paper no. AIAA 2002-1274, 2002 163

Bellenoue, M., Moriniere, V. and Kageyama, T., 'Experimental 3-D Simulation of the Compression Wave, Due to Train-tunnel Entry,' *Journal of Fluids and Structures*, vol. 16, pp. 581–595, 2011 137

Betz, A., *Introduction to the Theory of Flow Machines* (D.G. Randall, trans.), Pergamon Press, Oxford, UK, 1966 24, 25

Bianchi, S., Borello, D., Corsini, A., Rispoli, F. and Sheard, A.G., 'Large-eddy Simulation of the Aerodynamic and Aeroacoustic Performance of a Ventilation Fan,' *Advances in Acoustics and Vibration,* vol. 2013, article ID 876973, pp. 1–10, 2013 109, 110, 138, 164

Bianchi, S., Corsini, A., Rispoli, F. and Sheard, A.G., 'Far-field Radiation of Tip Aerodynamic Sound Sources in Axial Fans Fitted with Passive Noise Control Features,' *Transactions of the ASME, Journal of Vibration & Acoustics,* vol. 133(5), paper no. 051001, pp. 1–11, 2011 96, 97

Bianchi, S., Corsini, A. and Paniagua, G., 'Amplification of the Force and Tonal Noise in Transonic HPT,' *Proceedings of the 9th European Conference on Turbomachinery, Fluid Dynamics & Thermodynamics,* Istanbul, Turkey, 21–25 March, pp. 1213 –1223, 2011 111, 115–16

Bianchi, S., Corsini, A. and Sheard, A.G., 'A Critical Review of Passive Noise Control Techniques in Industrial Fans,' *Transactions of the ASME, Journal of Engineering for Gas Turbines & Power,* vol. 136(4), paper no. 044001, pp. 1–9, 2014 46, 77, 330

Bianchi, S., Corsini, A. and Sheard, A.G., 'Demonstration of Stall Detection System for Induced Draft Fans,' *Proceedings of the Institution of Mechanical Engineers, Pat A: Journal of Power and Energy,* vol. 227, pp.272-284, 2013 4

Bianchi, S., Corsini, A. and Sheard, A.G., 'Installed Aeroacoustic Performance of Cooling Axial Fans Fitted with End-plates,' *Noise Control Engineering Journal*, vol. 60, pp. 519–527, 2012 181

Bianchi, S., Corsini, A. and Sheard, A.G., 'Sound Generation Mechanism in Low Speed Axial Fans,' *Journal of the Acoustical Society of America,* vol. 127, pp. 1837–1856, 2010 96, 97

Blake, W.K, *Mechanics of Flow-Induced Sound and Vibration, Volume 1: General Concepts and Elementary Sources,* Academic Press Inc., Orlando FL, USA, 1986 4

Bons, J.P., Taylor, R.J., McClain, S.T. and Rivir, R., 'The Many Manifestations of Turbine Surface Roughness,' *Transactions of the ASME, Journal of Turbomachinery*, vol. 123(4), pp. 739–748, 2001 256

Borello, D., Borrelli, P., Quagliata, E. and Rispoli, F., 'A Multi-grid Additive and Distributive Parallel Algorithm for FEM Turbomachinery CFD,' Proceedings of the European Congress on Computational Methods in Applied Sciences (ECCOMAS CFD 2001), Swansea, UK, 4–7 September. Corsini, A. and Rispoli, F. (2004), 'Using Sweep to Extend Stall-free Operational Range in Axial Fan Rotors,' *Proceedings of the IMechE Part A, Journal of Power and Energy*, vol. 218, pp. 129–139, 2001 238, 262

Borello, D., Corsini, A., Delibra, G., Fiorito, M. and Sheard, A.G., 'Large Eddy Simulation of a Tunnel Ventilation Fan,' *Transactions of the ASME, Journal of Fluids Engineering,* vol. 135(7), paper no. 071102, pp. 1-9, 2013 *30,* 111, 116, 143–4

Borello, D., Corsini, A., Delibra, G. and Sheard, A.G., 'Numerical Investigation of Detrimental Aerodynamic Effect of Pressure Pulses on a Metro Tunnel Fan', *Proceedings of the 10th European Turbomachinery Conference,* Lappeenranta, Finland, 15-19 April, pp. 573-582, 2013 30, *31, 32,* 135

Borello, D., Corsini, A., Minotti, S., Rispoli, F. and Sheard, A.G., 'U-RANS of a Large Industrial Fan Under Design and Off-Design Operations,' *Proceedings of the 9th European Conference on Turbomachinery, Fluid Dynamics & Thermodynamics,* Istanbul, Turkey, 21–25 March, pp. 1665–1676, 2011 91, 118, 123, *125*

Borello, D., Corsini, A., Rispoli, F. and Sheard, A.G., 'Investigation on the Unsteady Aerodynamics of an Industrial Fan,' *Proceedings of the Fan 2012 Conference,* Senlis, France, 18–20 April, 2012 203

Borello, D., Corsini, A. and Rispoli, F., 'A Finite Element Overlapping Scheme for Turbomachinery Flows on Parallel Platforms, *Computational Fluids*, vol. 32(7), pp. 1017–1047, 2003 231, 238, 257, 258

Borello, D., Delibra, G., Hanjalić , K. and Rispoli, F., 'Hybrid LES/RANS Study of Turbulent Flow in Linear Compressor Cascade with Moving Casing,' *Proceedings of the 55th American Society of Mechanical Engineers Gas Turbine and Aeroengine Congress,* Glasgow, Scotland, 14-18 June, paper no. GT2010-23755, 2010 29, 77, 138

Borello, D., Delibra, G., Hanjalić, K. and Rispoli, F., 'Large-eddy Simulations of Tip Leakage and Secondary Flows in Axial Compressor Cascade using a Near-wall Turbulence Model,' *Proceedings of the Institution of Mechanical Engineers Part A, Journal of Power and Energy,* vol.223 (6), pp. 645-655, 2009 29, 77, 111, 138

Borello, D., Hanjalić K. and Rispoli, F., 'Computation of Tip-leakage Flow in a Linear Compressor Cascade with a Second-moment Turbulence Closure,' *International Journal of Heat and Fluid Flow,* vol. 28(4), pp. 587-601, 2007 15, 77, 111, 138

Borello, D., Hanjalić K. and Rispoli, F., 'Prediction of Cascade Flows with Innovative Second-moment Closures,' *Transactions of the ASME, Journal of Fluids Engineering,* vol. 127(6), pp. 1059–1070, 2005 15, 77, 111, 138

Borello, D., Venturini, P., Rispoli, F. and Saavedra G.Z.R., 'Prediction of Multiphase Combustion and Ash Deposition within a Biomass Furnace,' *Applied Energy*, vol. 101, 2013 264, 304, 305, 328, 330

Brailko, I.A., Mileshin, V.I., Nyukhitkov, M.A. and Pankov, S.V., 'Computational and Experimental Investigation of Unsteady and Acoustic Characteristics of Counter-rotating Fans,' *Proceedings of the ASME 2004 Heat Transfer/Fluids Engineering Summer Conference,* Vol.2, Parts A and B, Charlotte, NC, USA, 11-15 July, paper no. HT-FED2004-56435, 2004 20

Brooks, A.N. and Hughes, T.J.R., 'Streamline Upwind / Petrov–Galerkin Formulations for Convection Dominated Flows with Particular Emphasis on the Incompressible Navier–Stokes Equations,' *Computer Methods in Applied Mechanics and Engineering,* vol. 32, pp. 199–259, 1982 48

Cardillo, L., Corsini, A., Delibra, G., Rispoli, F. and Sheard, A.G., 'A Numerical Investigation into the Aerodynamic Effect of Pressure Pulses on a Tunnel Ventilation Fan,' *Proceedings of the IMechE Part A, Journal of Power and Energy*, vol. 228(3), pp. 285 –299, 2014 161, 327, 330

Cardillo, L., Corsini, A., Delibra, G., Rispoli, F. and Sheard, A.G., 'A Numerical Investigation into the Aerodynamic Effect of Pressure Pulses on a Tunnel Ventilation Fan,' *Proceedings of the IMechE Part A, Journal of Power and Energy,* vol. 228(3), pp. 285 –299, 2014 330, *331,* 336

Cardillo, L., Corsini, A., Delibra, G., Rispoli, F. and Sheard, A.G., 'Predicting the Performance of an Industrial Centrifugal Fan Incorporating Cambered Plate Impeller Blades,' *Periodica Polytechnica, Mechanical Engineering*, vol. 58(1), pp. 15–25 2014 330

Carvel, C., 'The Effects of Ventilation on Fires in Tunnels,' *Proceedings of the International Tunnel Fire and Safety Conference,* Rotterdam, The Netherlands, 2–3 December, 1999 75, 109

Catabriga, L., Coutinho, A.L.G.A. and Tezduyar, T.E., 'Compressible Flow SUPG Parameters Computed from Element Matrices,' *Communications in Numerical Methods in Engineering,* vol. 21(9), pp. 465–476, 2005 48, 53

Chen, W.L., Lien, F.S. and Leschziner, M.A., 'Computational Prediction of Flow Around Highly Loaded Compressor-cascade Blades with Non-linear Eddy-viscosity Models,' *International Journal of Heat and Fluid Flow,* vol. 19(4), pp.307-319, 1998 15

Codina, R., 'Comparison of Some Finite Element Methods for Solving the Diffusion–convection-reaction Equation,' *Computer Methods in Applied Mechanics and Engineering,* vol. 156, pp. 185–210, 1998 47

Commission Regulation (EU) no. 327/2011, *Official Journal of the European Union,* 1 June, http://www.amca.org/UserFiles/file/COMMISSION%20REGULATION%20%28 EU%29%20No%20327-2011.pdf., 2011 4

Corsini, A., and Rispoli, F, 'Numerical Simulation of Three-dimensional Viscous Flow in an Isolated Axial Rotor,' *Polish Academy of Sciences, Archive of Mechanical Engineering (Archiwum budowy maszyn),* vol. XLVI (4), pp. 369–92, 1999 58

Corsini, A., Delibra, G., Rispoli, F., Sheard, A.G. and Venturini, P., 'Simulation of the Particle-laden Flows in a Large Centrifugal Fan', *Proceedings of the 58th American Society of Engineers Gas Turbine and Aeroengine Congress,* San Antonio, TX, USA, 3-7 June, paper no. GT2013-94982, 2013 9, *16,* 22, 23

Corsini, A., Delibra, G. and Sheard, A.G., 'A Critical Review of Computational Methods and Their Application in Industrial Fan Design,' *International Scholarly Research Network, Mechanical Engineering,* vol. 2013, article ID 625175, pp. 1–20, 2013 76, 162, 304, 327, 328

Corsini, A., Delibra, G.and Sheard, A.G., 'Leading Edge Bumps in Reversible Axial Fans,' *Proceedings of the 58th American Society of Mechanical Engineers Gas Turbine and Aeroengine Congress,* San Antonio, TX, USA, 3-7 June, paper no. GT2013-94853, 2013 9, *16,* 20

Corsini, A., Iossa, C., Rispoli, F. and Tezduyar, T.E., 'A DRD Finite Element Formulation for Computing Turbulent Reacting Flows in Gas Turbine Combustors,' *Computational Mechanics,* vol. 46(1), pp. 159–167, 2010 48

Corsini, A., Marchegiani, A., Minotti, S. and Sheard, A.G., 'Numerical Investigation on the Aerodynamic Influence of Eroded Leading-edge Geometry on Boiler Fan Performance,' *Proceedings of the 9th European Conference on Turbomachinery, Fluid Dynamics & Thermodynamics*, 21–25 March, Istanbul, Turkey, pp. 879–892, 2011 182, 270

Corsini, A., Marchegiani, A., Rispoli, F., Sheard, A.G. and Venturini, P., 'Predicting Blade Leading Edge Erosion in an Axial Induced Draft Fan,' Transactions of the ASME, Journal of Engineering for Gas Turbines & Power, vol. 134, paper no. 042601, pp. 1–9, 2012 182, 262, 266, 269, 305, 324, 328, 330, 341, 343

Corsini, A., Rispoli, F., and Santoriello, A., 'A High Order Petrov-Galerkin Stabilized Finite Element Method for Incompressible RANS in Presence of Strong Reaction Effects,' *Proceedings of CMFF-03, the Conference on Modelling Fluid Flow*, Budapest, Hungary, 3–6 September, 2003 238, 258

Corsini, A., Rispoli, F., Bencze, F., and Vad, J., 'Effects of Blade Sweep in a High Performance Axial Flow Rotor,' *Proceedings of the 4th European Conference on Turbomachinery,* Florence, Italy, 20–23 March, 2001 58

Corsini, A., Rispoli, F., Santoriello, A. and Tezduyar, T.E., 'Improved Discontinuity- capturing Finite Element Techniques for Reaction Effects in Turbulence Computation,' *Computational Mechanics,* vol. 38, pp. 356–364, 2006 48, 54, 231, 238, 258

Corsini, A., Rispoli, F., Sheard, A.G. and Kinghorn, I.R., 'Investigation of Improved Blade-Tip Concept for Axial Flow Fan,' *Proceedings of the 51st American Society of Mechanical Engineers Gas Turbine and Aeroengine Congress,* Barcelona, Spain, 8–11 May, paper no. GT2006-90592, 2006 59

Corsini, A., Rispoli, F., Sheard, A.G. and Tezduyar, T.E., 'Computational Analysis of Noise Reduction Devices in Axial Fans with Stabilized Finite Element Formulations,' *Computational Mechanics,* vol. 50, pp. 695–705, 2012 182, 258

Corsini, A., Rispoli, F., Sheard, A.G. and Venturini, P., 'Numerical Simulation of Coal Fly-ash Erosion in an Induced Draft Fan', *Transactions of the ASME, Journal of Fluids Engineering,* vol. 135, paper no. 081303, pp.1-12, 2013 26, *27,* 304, 305, 324, 328, 330, 341, 343

Corsini, A., Rispoli, F. and Santoriello, A., 'A New Stabilized Finite Element Method for Advection–diffusion-reaction Equations using Quadratic Elements,' in Vad, J., Lajos, T. and Schilling, R. (Eds), *Modelling Fluid Flow: The State of the Art,* Springer-Verlag, Berlin, Germany, 2004 57

Corsini, A., Rispoli, F. and Santoriello, A., 'A Variational Multiscale High-order Finite Element Formulation for Turbomachinery Flow Computations,' *Computer Methods in Applied Mechanics and Engineering,* vol. 194, pp. 4797–4823, 2005 48

Corsini, A., Rispoli, F. and Sheard, A.G., 'Aerodynamic Performance of Blade-tip End-plates Designed for Low-noise Operation in Axial-flow Fans,' *Transactions of the ASME, Journal of Fluids Engineering,* vol. 131, paper no. 081101, pp.1-13, 2009 9

Corsini, A., Rispoli, F. and Sheard, A.G., 'Development of Improved Blade-tip End-plate Concepts for Low-noise Operation in Industrial Fans,' *Proceedings of the IMechE Part A, Journal of Power and Energy,* vol.221, pp.669-681, 2007 9

Corsini, A., Rispoli, F. and Sheard, A.G., 'Shaping of Tip End-plate to Control Leakage-vortex Swirl in Axial-flow Fans', *Transactions of the ASME, Journal of Turbomachinery,* vol. 132, paper no. 031005, pp. 1-9, 2010 9, 60, 61

Corsini, A., Rispoli, F. and Tezduyar, T.E., 'Computer Modelling of Wave-energy Air Turbines with the SUPG/PSPG Formulation and Discontinuity-capturing Technique,' *Journal of Applied Mechanics,* vol. 79(1), paper no. 010910, pp. 1–8, 2012 48

Corsini, A., Rispoli, F. and Tezduyar, T.E., 'Stabilized Finite Element Computation of NOx Emission in Aero-engine Combustors,' *International Journal for Numerical Methods in Fluids,* vol. 65, pp. 254–270, 2011 48

Corsini, A., Santoriello, A., Tezduyar, T.E., Menichini, F. and Rispoli, F., 'A Multiscale Finite Element Formulation with Discontinuity Capturing for Turbulence Models with Dominant Reaction Like Terms,' *Transactions of the ASME, Journal of Applied Mechanics,* vol. 76(2), paper no. 021211, pp. 1–8, 2009 48, 258

Corsini, A. and Rispoli, F., 'Flow Analyses in a High-pressure Axial Ventilation Fan with a Non-linear Eddy-viscosity Closure,' *International Journal of Heat and Fluid Flow,* vol. 26(3), pp. 349-361, 2005 9, 15, 18, 19, 48, 77, 111, 138, 258

Corsini, A. and Rispoli, F., 'Using Sweep to Extend Stall-free Operational Range in Axial Fan Rotors,' *Proceedings of the Institute of Mechanical Engineering, Part A, Journal of Power and Energy,* vol. 218(3), pp.129-139, 2004 9, 46, 57, 241, 268

Corsini, A. and Sheard, A.G., ''Tip End-plate Concept Basedon Leakage Vortex Rotation Number Control', *Journal of Computational and Applied Mechanics,* vol. 8, pp. 21-37, 2007 47, 54, 58

Courant, R., Friedrichs, K. and Lewy, H., 'On the Partial Difference Equations of Mathematical Physics,' *Journal of Research and Development,* vol. 11(2), pp. 215-234, 1967 17, 92, 114

Craft, T.J., Launder, B.E. and Suga, K., 'Development and Application of a Cubic Eddy-viscosity Model of Turbulence,' *International Journal of Heat and Fluid Flow,* vol.17(2), pp. 108-115, 1996 14, 49, 50, 238, 258

Crowe, C.T., Troutt, T.R. and Chung, J.N., 'Numerical Models of Two-phase Turbulent Flows,' *Fluid Mechanics*, vol. 28, pp. 11–43, 1996 235

Cumpsty, N.A., *Compressor Aerodynamics,* Krieger Publishing Company, Malabar, Fl, USA, 2004 4

Cumpsty, N.A., 'Review — a Critical Review of Turbomachinery Noise,' *Transactions of the ASME, Journal of Fluids Engineering,* vol. 99(2), pp. 278–293, 1977 110

Curle, N., 'The Influence of Solid Boundaries upon Aerodynamic Sound,' *Proceedings of the Royal Society of London A,* vol. 231, pp. 505–514 1955 125

Daly, B.B., *Woods Practical Guide to Fan Engineering,* Woods of Colchester Ltd, Colchester, UK, 1978 297

Daly, B.B., *Woods Practical Guide to Fan Engineering,* Woods of Colchester Ltd, Colchester, UK, 1985 76, 109, 161, 162, 164, 174, 203, 207

Daneshkhah, K. and Ghaly, W.S., 'Aerodynamic Inverse Design for Viscous Flow in Turbomachinery Blading,' *AIAA Journal of Propulsion and Power,* vol. 23(4), pp. 814–820, 2007 162

Daneshkhah, K. and Sheard, A.G., 'A Parametric Study of Reversible Jet-fan Blades Aerodynamic Performance.' *Transactions of the ASME, Journal of Engineering for Gas Turbines and Power*, vol. 135(2), paper no. 024503, pp. 1–4, 2013 203

Davidson, L., 'Large Eddy Simulation: a Dynamic One-equation Subgrid Model for Three-dimensional Recirculating Flows,' *Proceedings of the 11th International Symposium on Turbulent Shear Flow,* Vol. 3, Grenable, 8-10 September, pp. 26.1-26.6, 1997 30, 77, 78, 79, 111, 138, 143

Davidson, P.A., *Turbulence, an Introduction for Scientists and Engineers,* Oxford University Press, Oxford, UK, 2004 6, 11, 13, 18

Delibra, G., Borello, D., Hanjalić, K. and Rispoli, F., 'A LES Insight into Convective Mechanism of Heat Transfer in a Wall-bounded Pin Matrix,' *Proceedings of the 14th International Heat Transfer Conference (IHTC),* Washington, DC, USA, 8–13 August, pp. 807–815, 2010 77, 79–80, 138

Delibra, G., Borello, D., Hanjalić , K. and RispoliF., 'U-RANS of Flow and Endwall Heat Transfer in a Pinned Passage Relevant to Gas-turbine Blade Cooling,' *International Journal of Heat and Fluid Flow,* vol. 30(3), pp. 549–560, 2009 77, 111, 138

Delibra, G., Hanjalić, K., Borello, D. and Rispoli, F., 'Vortex Structures and Heat Transfer in a Wall-bounded Pin Matrix: LES with a RANS Wall-treatment,' *International Journal of Heat and Fluid Flow,* vol. 31(5), pp. 740–753, 2010 77

Denton, J.D., *Extension of the Finite Volume Time Marching Method to Three Dimensions,* VKI lecture Series 1979-7, Rhode-St-Genèse, Belgium, 1979 18

Denton, J.D. and Dawes, W.N., 'Computational Fluid Dynamics for Turbomachinery Design,' *Proceedings of the Institution of Mechanical Engineers Part C, Journal of Mechanical Engineering Science,* vol. 213(2), pp. 107-124, 1999 18, 21, 300

Dubief, Y. and Delcayre, F., 'On Coherent-vortex Identification in Turbulence,' *Journal of Turbulence,* vol. 1, p. N11, 2000 82, *83, 94, 95, 96,* 118, *119*

Durbin, P.A., 'A Reynolds Stress Model for Near-wall Turbulence', *Journal of Fluid Mechanics*, vol. 249, pp. 465-498, 1993 15

Durbin, P.A., 'Separated Flow Computations with the k-epsilon-v-squared Model,' *AIAA Journal,* vol. 33(4), pp. 659-664, 1995 15

Durbin, P.A. and Pettersson Reif, B.A., *Statistical Theory and Modelling for Turbulent Flows,* John Wiley & Sons, Chichester, UK, 2001 6

Eck, B. (1973), Fans: Design and Operation of Centrifugal, Axial-flow, and Cross-flow Fans, Pergamon Press, Oxford, UK, 1973 297

Egorov, Y., Menter, F.R., Lechner, R. and Cokljat, D., 'The Scale-adaptive Simulation Method for Unsteady Turbulent Flow Predictions. Part 2: Application to Complex Flows,' *Flow, Turbulence and Combustion,* vol. 85(1), pp. 139–165, 2010 85

Eldred, M.S., Adams, B.M., Gay, D.M., Swiler, L.P., Haskell, K., Bohnhoff, W.J., Eddy, J.P., Hart, W.E., Watson, J.P., Hough, P.D. and Kolda, T.G. , 'DAKOTA, a Multilevel Parallel Object-oriented Framework for Design Optimization, Parameter Estimation, Uncertainty Quantification and Sensitivity Analysis,' *Sand Report SAND2006-4056,* Sandia National Laboratories, Livermore, CA, USA, 2006 33

Elfeki, S. and Tabakoff, W., 'Erosion Study of Radial Flow Compressor with Splitters,' Transactions of the ASME, Journal of Turbomachinery, vol. 109(1), pp. 62–69, 1987 230, 257

EN12101-3, Smoke and Heat Control Systems. Specification for Powered Smoke and Heat Exhaust Ventilators, 2002 109

Envia, E., Wilson, A.G. and Huff, D.L., 'Fan Noise: a Challenge to CAA,' *International Journal of Computational Fluid Dynamics,* vol. 18(6), pp. 471–480, 2004 27

Farassat, F., 'Linear Acoustic Formulas for Calculation of Rotating Blade Noise,' *AIAA Journal,* vol. 19(9), pp. 1122–1130, 1981 115

Farrell, P.E. and Maddison, J.R., 'Conservative Interpolation between Volume Meshes by Local Galerkin Projection,' *Computer Methods in Applied Mechanics and Engineering,* vol. 200, pp. 89–100, 2011 327

Fernández Oro, J.M., Díaz, K.M.A., Morros, C.S. and Marigorta, E.B., 'Unsteady Flow and Wake Transport in a Low-speed Axial Fan with Inlet Guide Vanes,' *Transactions of the ASME, Journal of Fluids Engineering,* vol. 129(8), pp. 1015–1029, 2007 20

Ferziger, J.H. and Peric, M. (2002), *Computational Methods for Fluid Dynamics*, Springer, London, UK, 2002 299, 300

Ffowcs Williams, J.E., 'Aeroacoustics,' *Annual Review of Fluid Mechanics,* vol. 9, pp. 447–468, 1977 47

Ffowcs Williams, J.E. and Hawkings, D.L., 'Sound Generation by Turbulence and Surfaces in Arbitrary Motion,' *Philosophical Transactions of the Royal Society of London A,* vol. 264, pp. 321–342, 1969 107, 111, 114, *125,* 126

Formaggia, L., Micheletti, S. and Perotto, S., 'Anisotropic Mesh Adaptation in Computational Fluid Dynamics: Application to the Advection–diffusion-reaction and the Stokes Problems,' *Applied Numerical Mathematics,* vol. 51(4), pp. 511–533, 2004 62

Franca, L.P. and Valentin, F., 'On an Improved Unusual Stabilized Finite Element Method for the Advective–reactive–diffusive Equation,' *Computer Methods in Applied Mechanics and Engineering,* vol. 190, pp. 1785–1800, 2000 47

Fukano, T. and Jang, C.M., 'Tip Clearance Noise of Axial Flow Fans Operating at Design and Off-design Condition,' *Journal of Sound and Vibration,* vol. 275, pp. 1027–1050, 2004 46

Furukawa, M., Inoue, M., Saiki, K. and Yamada, K., 'The Role of the Tip Leakage Vortex Breakdown in Compressor Rotor Aerodynamics,' *Transactions of the ASME, Journal of Turbomachinery,* vol. 121, pp. 469–480, 1999 120

Galpin, P.F. and Raithby, G.D., 'Numerical Solution of Problems in Incompressible Fluid Flow: Treatment of the Temperature-velocity Coupling,' *Journal of Numerical Heat Transfer*, vol. 10(2), pp. 105–129, 1986 187

Garg, A.K. and Leibovich, S., 'Spectral Characteristics of Vortex Breakdown Flowfields,' *Physics of Fluids,* vol. 22, pp. 2053–2064, 1979 56

Germano, M., Piomelli, U., Moin, P. and Cabot, H., 'A Dynamic Sub-grid-scale Eddy Viscosity Model,' *Physics of Fluids,* vol. 3, pp. 1760–1765, 1991 78, 138, 143

Ghenaiet, A., 'Numerical Simulations of Flow and Particle Dynamics within a Centrifugal Turbomachine,' *Proceedings of the Institution of Mechanical Engineers International Conference on Compressors and their Systems*, London, UK, 4–7 September, paper no. C639-652, pp. 497–506, 2005 230, 257

Ghenaiet, A., 'Numerical Study of Sand Ingestion through a Ventilating System,' *Proceedings of the World Congress on Engineering 2009,* Vol II, London, UK, 1–3 July, 2009 241, 257

Ghenaiet, A., 'Numerical Study of Sand Ingestion through a Ventilating System,' *Proceedings of the World Congress on Engineering 2009 Vol II,* London, UK, 1–3 July 2009 26–7

Ghenaiet, A., Tan, S.C. and Elder, R.L. 'Experimental Investigation of Axial Fan Erosion and Performance Degradation,' *Proceedings of the IMechE Part A, Journal of Power and Energy*, vol. 218(6), pp. 437–450, 2004 230, 256

Gho, S.L., 'AMCA Grows in Asia,' *AMCA inmotion(3),* p. 27, 2013 5

Giesen, B.J.M., Penders, S.H.A., Loomans, M.G.L.C., Rutten, P.G.S. and Hensen, J.L.M., 'Modelling and Simulation of a Jet Fan for Controlled Air Flow in Large Enclosures,' *Environmental Modelling and Software*, vol. 26(2), pp. 191–200, 2011 161

Godichon, A.F.E. and Sheard, A.G., 'Customer-oriented Design,' *International Cement Review*, October, pp. 95–99, 2009 297

Grant, G. and Tabakoff, W., 'Erosion Prediction in Turbomachinery Resulting from Environmental Solid Particles,' *Journal of Aircraft*, vol. 12(5), pp. 471–478, 1975 255, 256

Gravemeier, V. and Wall, W.A., 'A 'Divide-and-conquer' Spatial and Temporal Multiscale Method for Transient Convection–diffusion-reaction Equations' , *International Journal for Numerical Methods in Fluids,* vol. 54, pp. 779–804, 2007 48

Greenzweig, J.T., Henry, R. and Holm, T. , 'The Effect on Inlet Flow Distortion on the Performance of Centrifugal Fans for Utility Applications,' *Proceedings of the 56th American Society of Mechanical Engineers Turbine and Aeroengine Congress*, 6–10 June, Vancouver, Canada, paper no. GT2011-45422, 2011 182, 183, 302

Grundestam, O., Wallin, S. and Johansson, A.V., 'Direct Numerical Simulations of Rotating Turbulent Channel Flow,' *Journal of Fluid Mechanics,* vol. 598, pp. 177–199, 2008 80–1, *82*

Guédel, A., 'Prediction of the Noise Installation Effect Induced by a Bend at the Inlet of an Axial Flow Fan,' *Noise Control Engineering Journal*, vol. 54(1), pp. 21–26, 2006 181

Hah, C. and Shin, H.-W., 'Study of Near-stall Flow Behavior in a Modern Transonic Fan with Compound Sweep,' *Transactions of the ASME, Journal of Fluids Engineering,* vol. 134(7), paper no. 071101, pp. 1–7, 2012 77, 111

Hamed, A., Tabakoff, W. and Wenglarz, R., 'Erosion and Deposition in Turbomachinery,' *Journal of Propulsion and Power*, vol. 22, pp. 350–360, 2006 229, 230, 255, 256

Hanjalić, K., Kenjeres, S., Tummers, M.J. and Jonker, H.J.J., *Analysis and Modelling of Physical Transport Phenomena,* Delft Academic Press, Delft, The Netherlands, 2009 6, 15

Hanjalić, K., Popovac, M. and Hadziabdic, M., 'A Robust Near-wall Elliptic-relaxation Eddy-viscosity Turbulence Model for CFD,' *International Journal of Heat and Fluid Flow,* vol. 25(6), pp. 1047–1051, 2004 15

Hanjalić, K. and Jakirlić, S., 'A Model of Stress Dissipation in Second-moment Closures,' *Applied Scientific Research,* vol. 51, pp. 513–518, 1993 15

Hanjalić, K. and Jakirlić, S., 'Contribution Towards the Second-moment Closure Modelling of Separating Turbulent Flows,' *Computers and Fluids,* vol. 27(2), pp. 137–156, 1998 15

Hanjalić, K. and Launder, B., *Modelling Turbulence in Engineering and the Environment: Second-moment Routes to Closure,* Cambridge University Press, Cambridge, UK, 2011 15

Hauer, A. and Brooks, J., 'Fan Motor Efficiency Grades in the European Market,' *AMCA inmotion(2),* pp. 14–20, 2012 4

Hauke, G., 'A Simple Subgrid Scale Stabilized Method for the Advection–diffusion- reaction Equation,' *Computer Methods in Applied Mechanics and Engineering,* vol. 191, pp. 2925–2947, 2002 48

He, L. and Denton, J.D., 'Three-dimensional Time-marching Inviscid and Viscous Solutions for Unsteady Flows Around Vibrating Blades,' *Transactions of the ASME, Journal of Turbomachinery,* vol. 116(3), pp. 469–476, 1994 27

Hedefalk, J., Wahlstrom, B. and Rohlen, P., 'Lessons from the Baku Subway Fire,' *Proceedings of the 3rd International Conference on Safety in Road and Rail Tunnels,* Nice, France, 9–11 March, pp. 15–28, 1998 75, 109

Hirsch, C., *Numerical Computation of Internal and External Flows,* Butterworth- Heinemann, Oxford, UK, 2007 6, 11, 18, 19

Hoffman, J. and Johnson, C., 'Adaptive DNS/LES: a New Agenda in CFD,' in Franca, L.P., Tezduyar, T.E. and Masud, A. (Eds), *Finite Element Methods: 1970s and Beyond,* CIMNE, Barcelona, Spain, 2004 47, 62

Horlock, J.H. and Denton, J., 'A Review of Some Early Design Practice using Computational Fluid Dynamics and a Current Perspective,' *Transactions of the ASME, Journal of Turbomachinery,* vol. 127, pp. 5 –13, 2005 5, 18, 76, 110, 203

Hsu, M.-C., Akkerman, I. and Bazilevs, Y., 'Wind Turbine Aerodynamics using ALE- VMS: Validation and the Role of Weakly Enforced Boundary Conditions,' *Computational Mechanics,* vol. 50(4), pp. 499–511, 2012 57

Hsu, M.-C. and Bazilevs, Y., 'Fluid–structure Interaction Modelling of Wind Turbines: Simulating the Full Machine,' *Computational Mechanics,* vol. 50(6), pp. 821–833, 2012 57

Hu, G.R., Li, Y.M. and Jia, X.N., 'Simulation of Gas-solid Two-phase Flow of Centrifugal Fan,' *Coal Mine Machinery*, vol. 2008(3), pp. 37–39, 2008 324

Huang, L. and Wang, J. 'Acoustic Analysis of a Computer Cooling Fan,' *Journal of the Acoustical Society of America,* vol. 118, pp. 2190–2200, 2005 4, 137

Huang, Y.D. and Gao, W., 'A Numerical Study of the Train-induced Unsteady Airflow in a Subway Tunnel with Natural Ventilation Ducts Using the Dynamic Layering Method,' *Journal of Hydrodynamics,* vol. 22(2), pp. 164–172, 2010 137

Hughes, T.J.R., 'Multiscale Phenomena: Green's Functions, the Dirichlet-to-Neumann Formulation, Subgrid Scale Models, Bubbles and the Origins of Stabilized Methods', *Computer Methods in Applied Mechanics and Engineering,* vol. 127, pp. 387–401, 1995 48

Hussein, M.F. and Tabakoff, W., 'Computation and Plotting of Solid Particle Flow in Rotating Cascades,' *Computational Fluids,* vol. 2(1), pp. 1–15, 1974 230, 257

Inoue, M., Kuroumaru, M. and Furukawa, M., 'Behaviour of Tip Leakage Flow Behind an Axial Compressor Rotor,' *Transactions of the ASME, Journal of Engineering for Gas Turbines and Power,* vol. 108(1), pp. 7–14, 1986 56

ISO 5801, Industrial Fans — Performance Testing Using Standardized Airways, 2007 109, *114,* 214, 217, 242, 299, 306, 324, 336

ISO 10302, Fans for General Purposes. Methods of Noise Testing, Airborne Noise Emitted by Small Air-moving Devices.1996 (BS 848-2:2000) 174, 214, 217

ISO 13350, Industrial Fans — Performance Testing of Jet Fans, 1999 166, 172

ISO 21297-3. Smoke and Heat Control Systems — Part 3: Specification for Powered Smoke and Heat Exhaust Ventilators, 2006 75–6, 135

Ito, T., Suematsu, Y. and Hayase, T., 'On the Vortex Breakdown Phenomena in a Swirling Pipe-flow,' *Memoirs of the Faculty of Engineering, Nagoya University,* vol. 37, pp. 117–172 1985 56

Jagannathan, S., Schwänen, M. and Duggleby, A., 'Low Pressure Turbine Relaminarization Bubble Characterization using Massively-parallel Large Eddy Simulations,' *Transactions of the ASME, Journal of Fluids Engineering,* vol. 134(2), paper no. 021102, pp. 1–13, 2012 77, 111

Jain, S., 'Three-dimensional Simulation of Turbulent Particle Dispersion,' PhD thesis, University of Utah, Salt Lake City, UT, USA, 1995 231, 304, 328

Jang, C.M., Fukano, T. and Furukawa, M., 'Effects of the Tip Clearance on Vortical Flow and its Relation to Noise in an Axial Flow Fan,' *JSME International Journal, Series B, Fluids and Thermal Engineering,* vol. 46(3), pp. 356 –365, 2003 46

Jasak, H., 'OpenFOAM: a Year in Review,' *Proceedings of the Fifth OpenFOAM Workshop,* Gothenburg, Sweden, 21–24 June, 2010 33, 79, 111, 138, 298, 300

Jasnoch, U., Kress, H. and Rix, J., 'Towards a Virtual Prototyping Environment,' in Rix, J. (Ed.), *Virtual Prototyping: Virtual Environments and the Product Design Process,* Chapman and Hall, London, UK, pp. 173–183, 1994 110, 204

Jensen, C.E., 'Axial-flow Fan,' patent no. US 4,630,993, 23 December, 1986 47

Jones, W.P. and Launder, B.E., 'The Prediction of Laminarization with a Two-equation Model of Turbulence,' *International Journal of Heat and Mass Transfer,* vol. 15(2), pp. 301–314, 1972 9

Kær, S.K., 'Numerical Investigation of Ash Deposition in Straw-fired Furnaces,' PhD thesis, Aalborg University, Aalborg, Denmark, 2001 231, 235, 262, 304, 328

Karki, K.C. and Patankar, S.V., 'CFD Model for Jet Fan Ventilation Systems,' *Proceedings of the 10th International Symposium on Aerodynamics and Ventilation of Vehicle Tunnels Principles, Analysis and Design,* Boston, MA, USA, 1–3 November, 2000 161

Khelladi, S., Koudiri, S., Bakir, F. and Rey, R, 'Flow Study in the Impeller-diffuser Interface of a Vaned Centrifugal Fan,' *Transactions of the ASME, Journal of Fluids Engineering*, vol. 127(3), pp. 495–502, 2005 298

Kirk, B.S., Peterson, J.W., Stogner, R.H. and Carey, G.F., 'libMesh: A C++ Library for Parallel Adaptive Mesh Refinement/Coarsening Simulations,' *Engineering With Computers*, vol. 22, pp. 237–254, 2006 231, 238, 258

Klein, M., Sadiki, A. and Janicka, J., 'A Digital Filter Based Generation of Inflow Data for Spatially Developing Direct Numerical or Large Eddy Simulations,' *Journal of Computational Physics,* vol. 186(2), pp. 652–665, 2003 92

Kleis, I. and Kulu, P., *Solid Particle Erosion: Occurrence, Prediction and C*ontrol, Springer, London, UK, 2008 258

Kobayashi, T. and Yoda, M., 'Modified k-e Model for Turbulent Swirling Flow in a Straight Pipe,' *JSME International Journal,* vol. 30, pp. 66–71, 1987 15

Kundu, P.K., Cohen, I.M. and Dowling, D.R, *Fluid Mechanics,* 5th edition, Elsevier Academic Press, Oxford, UK, 2012 6, 11, 18

Kurz, R. and Brun, K., 'Degradation in Gas Turbine Systems,' *Transactions of the ASME, Journal of Engineering for Gas Turbines and Power*, vol. 123, pp. 70 –77, 2001 229, 255, 256

Labois, M. and Lakehal, D., 'Very-large Eddy Simulation (V-LES) of the Flow Across a Tube Bundle,' *Nuclear Engineering and Design,* vol. 241, pp. 2075–2085, 2011 78, 111, 138

Lacroix, D., 'New French Recommendations for Fire Ventilation in Road Tunnels,'

Proceedings of the 9th International Conference on Aerodynamics and Ventilation of Vehicle Tunnels: Developments for the 21st Century, Aosta Valley, Italy, 6–8 October, pp. 103–124, 1997 75, 109

Lakshminarayana, B. *Fluid Dynamics and Heat Transfer of Turbomachinery,* John Wiley & Sons, New York, NY, USA, 1996 4

Langtry, R.B. and Menter, F.R., 'Transition Modelling for General CFD Applications in Aeronautics,' *Proceedings of the 43rd AIAA Aerospace Sciences Meeting and Exhibit,* Reno, NV, USA, 10–13 January, paper no. AIAA 2005-522, 2005 17

Launder, B.E. and Sandham, N.D. (Eds), *Closure Strategies for Turbulent and Transitional Flows,* Cambridge University Press, Cambridge, UK, 2002 15

Launder, B.E. and Sharma, B.R., 'Application of the Energy-dissipation Model of Turbulence to the Calculation of Flow Near a Spinning Disc,' *Letters in Heat and Mass Transfer*, vol. 1(2), pp. 131–137, 1974 299, 300, 327

Lecrivain, G. and Hampel, U., 'Influence of the Lagrangian Integral Time Scale Estimation in the Near Wall Region on Particle Deposition,' Transactions of the ASME. Journal of Fluids Engineering, vol. 134(7), paper no. 074502, pp. 1–6, 2012 263

Lee, K.Y., Choi, Y.S., Kim, Y.L. and Yun, J.H., 'Design of Axial Fan using Inverse Design Method,' *Journal of Mechanical Science and Technology,* vol. 22, pp. 1883 –1888, 2008 22, 76, 110, 203

Lee, S., Heo, S. and Cheong, C., 'Prediction and Reduction of Internal Blade-passing Frequency Noise of the Centrifugal Fan in a Refrigerator,' *International Journal of Refrigeration,* vol. 33(6), pp. 1129–1141, 2010 22

Lee, Y.T., Ahuja, V., Hosagandi, A., Slipper, M.E., Mulvihill, L.P., Birkbeck, R. and Coleman, R., 'Impeller Design of a Centrifugal Fan with Blade Optimization,' *International Journal of Rotating Machi*nery, vol. 2011, paper no. 537824, pp. 1–16, 2011 298

Lee, Y.T., 'Impact of Fan Gap Flow on the Centrifugal Impeller Aerodynamics,' *Transactions of the ASME, Journal of Fluids Engineering*, vol. 132, pp. 1–7, paper no. 091103, 2010 288

Lian, Y. and Liou, M.S., 'Multi Objective Optimization using Coupled Response Surface Model and Evolutionary Algorithm,' *AIAA Journal*, vol. 43(6), pp. 1316–1325, 2005 162

Lieblein, S., 'Experimental Flow in Two-dimensional Cascades,' in NACA RME 56B03, reprinted in NASA SP-36, *Aerodynamic Design of the Axial-Flow Compressor,* pp. 183–226, 1956 120

Lien, F.S., Chen, W.L. and Leschziner, M.A., 'Low-Reynolds-number Eddy-viscosity Modelling Based on Non-linear Stress-strain/Vorticity Relations,' *Proceedings of the 3rd Symposium on Engineering Turbulence Modelling and Measurements,* Crete, Greece, 27–29 May, 1996 14, *16*

Lien, F.S. and Durbin, P.A. , 'Non Linear k- -v^2 Modelling with Application to High Lift,' *Proceedings of the 1996 Centre for Turbulence Research Summer Program,* Stanford University, CA, USA, 24 June–19 July, pp. 2–22, 1996 17

Lighthill, M.J., 'On Sound Generated Aerodynamically. I. General Theory,' *Proceedings of the Royal Society of London A,* vol. 211, pp. 564–587, 1952 115, 125

Lighthill, M.J., 'On Sound Generated Aerodynamically. II: Turbulence as a Source of Sound,' *Proceedings of the Royal Society of London A,* vol. 222, pp. 1–32, 1954 125

Litchford, R.J. and Jeng, S.M., 'Efficient Statistical Transport Model for Turbulent Particle Dispersion in Sprays,' *AIAA Journal*, vol. 29, pp. 1443–1451, 1991 231, 262, 304, 328

Longet, C.M.L., 'Axial Flow Fan with Noise Reducing Means,' patent no. US 2003/0123987 A1, 3 July, 2003 47

Longhouse, R.E., 'Control Tip-vortex Noise of Axial Flow Fans by Rotating Shrouds,' *Journal of Sound and Vibration,* vol. 58, pp. 201–214, 1978 46

Luo, X., Ji, B., Peng, X., Xu, H., Nishi, M. and Ji, B., 'Numerical Simulation of Cavity Shedding from a Three-dimensional Twisted Hydrofoil and Induced Pressure Fluctuation by Large-eddy Simulation,' *Transactions of the ASME, Journal of Fluids Engineering,* vol. 134(4), paper no. 041202, pp. 1–10, 2012 77, 111

Mansour, T. and Ghaly, W.S., 'An Implicit Geometric Representation of Turbine Blades using NURBS,' *Proceedings of the 11th CFD Conference of the Canadian Society of CFD,* Vancouver, Canada, 28–30 May, vol. 1, pp. 238–243, 2003 162

Marcinowski, H.,'Einfluss des Laufradspalts und der Luftfuehrung bei einem Kuehlgeblaese Axialer Bauart' *Motortechnische Zeitschrift,* vol. 14, pp. 259–262, 1953 46

Marshall, J.G. and Imregun, M., 'A Review of Aeroelasticity Methods with Emphasis on Turbomachinery Applications,' *Journal of Fluids and Structures*, vol. 10(3), pp. 237–267, 1996 27

Mattern, P., Sieber, S., Cantrak, D., Frohlig, F., Caglar, S . and Gabi, M. (2012), 'Investigations on the Swirl Flow Caused by Axial Fans — a Contribution to the Revision of ISO 5801,' *Proceedings of the Fan 2012 Conference*, Senlis, France, 18–20 April, 2012 214

Mei, D. and Xing, F.T., 'Numerical Simulation Research on Erosive Problem of Gas- solid Two Phase Flow Fan,' *Fluid Machinery*, vol. 2007(10), pp. 25–28, 2007 324

Mengistu, T., Ghaly, W.S. and Manour, T., 'Aerodynamic Shape Optimization of Turbine Blades using a Design-parameter-based Shape Representation,' *Proceedings of the 52nd American Society of Mechanical Engineers Gas Turbine and Aeroengine Congress*, Montreal, Canada, 14–17 May, paper no. GT2007-28041, 2007 162

Mengistu, T. and Ghaly, W.S., 'Aerodynamic Optimization of Turbomachinery Blades using Evolutionary Methods and ANN-based Surrogate Models,' *Journal of*, [*sic*] vol. 9(3), pp. 239–255, 2008 162

Menter, F. and Egorov, Y., 'A Scale-adaptive Simulation Model using Two-equation Models,' *Proceedings of the 43rd AIAA Aerospace Sciences Meeting and Exhibit,* Reno, NV, USA, 10–13 January, paper no. AIAA2005-1095, 2005 29

Menter, F.R., Langtry, R. and Völker, S., 'Transition Modelling for General Purpose CFD Codes,' *Flow, Turbulence and Combustion,* vol. 77, pp. 277–303, 2006 17

Menter, F.R., Langtry, R.B., Likki, S.R., Suzen, Y.B., Huang, P.G. and Völker, S., 'A Correlation-based Transition Model using Local Variables Part 1: Model Formulation,' *Transactions of the ASME, Journal of Turbomachinery,* vol. 128(3), pp. 1–10, 2004 17

Menter, F.R., 'Two-equation Eddy-viscosity Turbulence Models for Engineering Applications,' *AIAA Journal,* vol. 32(8), pp. 1598–1605, 1994 9

Menter, F.R., 'Zonal Two-equation k-w Turbulence Model for Aerodynamic Flows,' *Proceedings of the 24th Fluid Dynamics Conference,* Orlando, FL, USA, 6–9 July, paper no. AIAA1993-2906, 1993 9

Menter, F.R. and Egorov, Y., 'The Scale-adaptive Simulation Method for Unsteady Turbulent Flow Predictions. Part 1: Theory and Model Description,' *Flow, Turbulence and Combustion*, vol. 85(1), pp. 113–138, 2010 29

Menter, F.R. and Egorov, Y., 'The Scale-adaptive Simulation Method for Unsteady Turbulent Flow Predictions. Part 1: Theory and Model Description,' *Flow, Turbulence and Combustion,* vol. 85(1), pp. 113–138, 2010 85

Menter, F.R. and Egorov, Y., 'The Scale-adaptive Simulation Method for Unsteady Turbulent Flow Predictions. Part 2: Application to Complex Flows,' *Flow, Turbulence and Combustion*, vol. 85(1), pp. 139–165, 2010 29

Meyer, C.J. and Kröger, D.G., 'Numerical Simulation of the Flow Field in the Vicinity of an Axial Flow Fan,' *International Journal for Numerical Methods in Fluids,* vol. 36, pp. 947–969, 2001 24, 25

Miles, J.H., 'Procedure for Separating Noise Sources in Measurements of Turbofan Engine Core Noise,' *Proceedings of the 12th AIAA/CEAS Aeroacoustics Conference (27th AIAA Aeroacoustic Conference),* Cambridge Massachusetts, USA, paper no. AIAA 2006-2580, 2006 110

Mimura, M., 'Axial Flow Fan,' patent no. US 6,648,598 B2, 18 November, 2003 47

Mutama, K.R. and Hall, A.E., 'The Experimental Investigation of Jet Fan Aerodynamics using Wind Tunnel Modeling,' *Transactions of the ASME, Journal of Fluids Engineering*, vol. 118(2), pp. 322–328, 1996 161

Muthanna, C. and Devenport, W.J., 'Wake of a Compressor Cascade with Tip Gap, Part 1: Mean Flow and Turbulence Structure,' *AIAA Journal,* vol. 42(11), pp. 2320 –2331, 2004 83, *84, 85,* 86

Oka, Y.I., Ohnogi, H., Hosokawa, T. and Matsumura, M., 'The Impact Angle Dependence of Erosion Damage Caused by Solid Particle Impact,' *Wear*, vols 203–204, pp. 573–579, 1997 265

Okita, R., Zhang, Y., McLaury, B.S. and Shirazi, S.A., 'Experimental and Computational Investigations to Evaluate the Effects of Fluid Viscosity and Particle Size on Erosion Damage,' *Transactions of the ASME, Journal of Fluids Engineering*, vol. 134(6), pp. 1–13, paper no. 061301, 2012 256

Page, A.L., Elseewi, A.A. and Straughan, I.R., 'Physical and Chemical Properties of Fly Ash from Coal-fired Power Plants with Reference to Environmental Impacts,' *Residue Review*, vol. 71, pp. 83–120, 1979 230, 248, 288

Pandian, N.S., 'Fly Ash Characterization with Reference to Geotechnical Applications,' *Journal of the Indian Institute of Science*, vol. 84, pp. 189–216, 2004 257, 258, 265

Pascu, M., Miclea, M., Epple, P., Delgado A. and Durst, F., 'Analytical and Numerical Investigation of the Optimum Pressure Distribution along a Low-pressure Axial Fan Blade,' *Proceedings of the Institution of Mechanical Engineers, Part C: Journal of Mechanical Engineering Science*, vol. 223(3), pp. 643–657, 2009 193

Perkinson, R.H., Soule, M.C. and Holl, H.J., 'Separator for Bearing Assemblies with Cyclic Loads,' patent no. US 8,167,501 B2, 1 May, 2012 195

Petit, P., Page, M., Beaudoin, M. and Nilsson, H., 'The ERCOFTAC Centrifugal Pump OpenFOAM Case Study,' *Proceedings of the 3rd IAHR International Meeting of the Workgroup on Cavitation and Dynamic Problems in Hydraulic Machinery and Systems*, Brno, Czech Republic, 14–16 October, 2009 298, 300

Pinelli, M., Ferrari, C., Suman, A., Morini, M. and Rossini, M., 'Fluid Dynamic Design and Optimization of a Double Entry Fan Driven by Tractor Power Take Off for Mist Sprayer Applications,' *Proceedings of the Fan 2012 Conference,* Senlis, France, 18–20 April, 2012 9, *11*

Piotrowski, W., Elsner, W. and Drobniak, S., 'Transition Prediction on Turbine Blade Profile with Intermittency Transport Equation,' *Transactions of the ASME, Journal of Turbomachinery,* vol. 132(1), pp. 1–10, 2009 17, 18

Pope, S.B., *Turbulent Flows,* Cambridge University Press, Cambridge, UK, 2000 6, 7, 11, 17, 18

Popovac, M. and Hanjalić K., 'Compound Wall Treatment for RANS Computation of Complex Turbulent Flows and Heat Transfer,' *Flow, Turbulence and Combustion*, vol. 78(2), pp. 177–202, 2007 14

Powell, A., 'Mechanisms of Aerodynamic Sound Production,' *AGARD Report No. 466*, 1963 125, 126

Powell, A., 'The Theory of Vortex Sound,' Journal of the Acoustical Society of America, vol. 33, pp. 177–195, 1964 125, 126, 127

Pratt, M.J., 'Virtual Prototypes and Product Models in Mechanical Engineering,' In Rix, J. (Ed.), *Virtual Prototyping: Virtual Environments and the Product Design Process*, Chapman and Hall, London, UK, pp. 113–128, 1994 110, 204

Pucher, K., 'Fire in the Tauern Tunnel,' *Proceedings of the International Tunnel Fire & Safety Conference,* Rotterdam, The Netherlands, 2–3 December, 1999 75, 109

Pullan, G., Young, A.M., Day, I.J., Greitzer, E.M. and Spakovszky, Z.S., 'Origins and Structure of Spike-type Rotating Stall,' *Proceedings of the 57th American Society of Mechanical Engineers Gas Turbine and Aeroengine Congress,* Copenhagen, Denmark, 11–15 June, paper no. GT2012-68707, 2012 19

Quinlan, A.D. and Bent, P.H., 'High Frequency Noise Generation in Small Axial Flow Fans,' *Journal of Sound and Vibration,* vol. 218(2), pp. 177–204, 1998 46

Rábai, G. and Vad, J., 'Validation of a Computational Fluid Dynamics Method to be Applied to Linear Cascades of Twisted-swept Blades,' *Periodica Polytechnica, Mechanical Engineering,* vol. 49(2), pp. 163–180, 2005 7, *8*

Richardson, J.H., Sallee, G.P. and Smakula, F.K., 'Causes of High Pressure Compressor Deterioration in Service,' *Proceedings of the AIAA / SAE / ASME 15th Joint Propulsion Conference*, Las Vegas, NV, USA, 18–20 June, paper no. AIAA 79-1234, 1979 256, 262

Richardson, J.H., Sallee, G.P. and Smakula, F.K., 'Causes of High Pressure Compressor Deterioration in Service,' *Proceedings of the AIAA / SAE / ASME 15th Joint Propulsion Conference*, Las Vegas, NV, USA, 18–20 June, paper no. AIAA 79-1234, 1979 229

Rippl, A., 'Experimentelle Untersuchungen zuminstationaren Betriebsverhahen an der Stabilitarsgrenze eines mehrstufigen transsonischen Verdichters,' PhD thesis, Ruhr- Universität Bochum, Germany, 1995 136

Rispoli, F., Corsini, A. and Tezduyar, T.E., 'Finite Element Computation of Turbulent Flows with the Discontinuity-capturing Directional Dissipation (DCDD),' *Computer and Fluids,* vol. 36(1), pp. 121–126, 2007 48, 53

Roslyng, O., 'Installation Effect on Axial Flow Fan Caused by Swirl and Non-uniform Velocity Distribution,' *Proceedings of the Institution of Mechanical Engineers (IMechE) Conference 1984-4 on the Installation Effects in Ducted Fan Systems*, 1–2 May, London, UK, paper no. C114/84, pp. 21–28, 1984 181

Rugang, C., 'Experimental Study on Erosion Characteristic for Different Blade Contours in the Centrifugal Can Exposed to Solid Particulate Environment,' *Fluid Machinery*, vol. 2000(1), pp. 5–8, 2000 324

Saad, Y., *Iterative Methods for Sparse Linear Systems,* Society for Industrial and Applied Mathematics, Philadelphia, PA, USA, 1996 80, 114

Sallee, G.P., Kruckenburg, H.D. and Toomey, E.H. , 'Analysis of Turbofan Engine Performance Deterioration and Proposed Follow-on tests,' *Report NASA-CR-134769*, 1975 230, 256

Sander, G.F. and Lilley, D.G., 'The Performance of an Annular Vane Swirler,' *Proceedings of the AIAA/SAE/ASME 19th Joint Propulsion Conference,* Seattle, WA, USA, 27–29 June, paper no. AIAA-83-1326, 1983 15

Šaric , S., Kniesner, B., Mehdizadeh, A., Jakirlić, S., Hanjalić Ķ. and Tropea, C., 'Comparative Assessment of Hybrid LES/RANS Models in Turbulent Flows Separating from Smooth Surfaces,' in Peng, S.H. and Haase, W. (Eds), *Advances in Hybrid RANS- LES Modelling, Notes on Numerical Fluid Mechanics and Multidisciplinary Design,* vol. 97, pp. 142–151, 2008 7

Sato, Y. and Yamamoto, K., 'Lagrangian Measurement of Fluid-particle Motion in an Isotropic Turbulent Field,' *Journal of Fluid Mechanics*, vol. 175, pp. 183–199, 1987 343, 344

Schneider, H., von Terzi, D. and Bauer, H.-J., 'Large-eddy Simulations of Trailing- edge Cutback Film Cooling at Low Blowing Ratio,' *International Journal of Heat and Fluid Flow,* vol. 31, pp. 767–777, 2010 77, 111, 138

Seggiani, M., Bardi, A. and Vitolo, S., 'Prediction of Fly-ash Size Distribution: A Correlation between the Char Transition Radius and Coal Properties,' *Fuel*, vol. 79, pp. 999 –1002, 2000 257

Seo, S.J., Kim, K.Y. and Kang, S.H., 'Calculations of Three-dimensional Viscous Flow in a Multi-blade Centrifugal Fan by Modelling Blade Forces,' *Proceedings of the IMechE Part A, Journal of Power and Energy*, vol. 217, pp. 287–297, 2003 298

Sheard, A.G., Corsini, A., Minotti, S. and Sciulli, F., 'The Role of Computational Methods in the Development of an Aero-acoustic Design Methodology: Application in a Family of Large Industrial Fans,' *Proceedings of the 14th International Conference on Modelling Fluid Flow Technologies,* Budapest, Hungary, 9 –12 September, pp. 71–79, 2009 9, 18, 19, 76, *89,* 91, 110, 112, *113,* 118, 135, 136, 204, 206, 241, 268

Sheard, A.G., Corsini, A. and Rispoli, F., 'A Meridional Fan,' patent no. GB 2,452,104 B, 22 July, 2009 47

Sheard, A.G., Daneshkhah, K. and Corsini, A., 'Fan Conceptual Design as Applied to the Marmaray Tunnel Ventilation System,' *Proceedings of the 58th American Society of Mechanical Engineers Gas Turbine and Aeroengine Congress*, San Antonio, TX, USA, 3–7 June, paper no. GT2013-94548, 2013 135, 204, 210, 211, 213

Sheard, A.G. (Ed.), *Passive Noise Control in Industrial Fans,* Sigel Press, Cambridge, UK, 2014 47, 76

Sheard, A.G. (Ed.), *Tip Clearance Measurement in Aero and Industrial Turbomachinery,* Sigel Press, Cambridge, UK, 2012 76

Sheard, A.G. and Corsini, A., 'The Mechanical Impact of Aerodynamic Stall on Tunnel Ventilation Fans,' *International Journal of Rotating Machinery,* vol. 2012, paper no. 402763, pp. 1–12, 2012 76, 136, 137–8, 148–9, 202

Sheard, A.G. and Daneshkhah, K., 'The Conceptual Design of High Pressure Reversible Axial Tunnel Ventilation Fans,' *Advances in Acoustics and Vibration,* vol. 2012, article ID 562309, pp. 1–11, 2012 135–6

Sheard, A.G. and Jones, N.M., 'An Initial Assessment of the Changes that will Appear in a Forthcoming (2012) Revision of EN 12101-3,' *Proceedings of the Fan 2012 Conference*, Senlis, France, 18–20 April, 2012 221

Sheard, A.G. and Jones, N.M., 'Approval of High-temperature Emergency Tunnel- ventilation Fans: The Impact of ISO 21927-3,' *Proceedings of the ITA–AITES World Tunnel Congress and 34th General Assembly,* Agra, India, 19–25 September, pp. 1817–1826, 2008 76, 109, 135, 161

Sheard, A.G. and Jones, N.M., 'Emergency Ventilation for Vehicular, Rail and Metro Tunnels,' *Proceedings of the International Congress: Safety Innovation Criteria Inside*, 2005 109

Sheard, A.G. and Jones, N.M., 'High-temperature Certification of Large Fans for Emergency Ventilation,' *Proceedings of the 12th International Symposium on Aerodynamics and Ventilation of Vehicle Tun*nels, Portoroz, Slovenia, 11–13 July, pp. 123 –140, 2006 109

Sheard, A.G. and Jones, N.M., 'Powered Smoke and Heat Exhaust Ventilators: The Impact of EN 12101-3 and ISO 21927-3,' *Tunnelling and Underground Space Technology*, vol. 28, pp. 174–182, 2012 75, 109, 135, 161, 214, 221, 223

Sheard, A.G. and Jones, N.M., 'The Development of a Fan Range to Meet Increasingly Demanding Safety Criteria Inside Tunnels,' *Proceedings of the 14th Australasian Tunnelling Conference: Development of Underground Space*, Sky City, Auckland, New Zealand, 8–9 March, pp. 305–318, 2011 211, 223

Sheard, A.G. and Rhoden, A. , 'Variable Pitch Fans,' patent no. GB 2,485,634 B, 12 September, 2012 196, *197*, 198

Shipp, M., 'Investigating the Channel Tunnel Fire,' *Proceedings of the Seminar on Reducing Fire Risk in Trains,* Heathrow, England, 17 November, pp. 19 –29, 1997 75, 109

Smith, G.D.J. and Cumpsty, N.A., 'Flow Phenomena in Compressor Casing Treatment,' *Transactions of the ASME, Journal of Engineering for Gas Turbines and Power*, vol. 106(3), pp. 532–541, 1984 46, 288

Smith, P.J. , 'Three-dimensional Turbulent Particle Dispersion Submodel Development,' Department of Energy, Pittsburgh Energy Technology Center, Pittsburgh, PA, *Report DOE/PC/90094-T5*, 1993 264

Sommerfeld, M., van Wachem, B. and Oliemans, R., *Best Practice Guidelines for Computational Fluid Dynamics of Dispersed Multiphase Flows, European Research Community on Flow, Turbulence and Combustion* (ERCOFTAC), Brussels, Belgium, 2008 304

Son, P.N., Kim, J. and Ahn, E.Y., 'Effects of Bell Mouth Geometries on the Flow Rate of Centrifugal Blowers,' *Journal of Mechanical Science and Technology,* vol. 25(9), pp. 2267–2276, 2011 22

Sørensen, J.N. and Shen, W.Z., 'Numerical Modelling of Wind Turbine Wakes,' *Transactions of the ASME, Journal of Fluids Engineering,* vol. 124(2), pp. 393–399, 2002 25, 30

Spalart, P.R. and Allmaras, S.R., 'A One-equation Turbulence Model for Aerodynamic Flows,' *Proceedings of the 30th Aerospace Sciences Meeting and Exhibit,* Reno, NV, USA, 6 –9 January, paper no. AIAA-92-0439, 1992 7

Spall, R.E., Gatski, T.B. and Grosch, C.E. , 'A Criterion for Vortex Breakdown,' *Physics of Fluids,* vol. 30, pp. 3434–3440, 1987 56

Speziale, C.G., Sarkar, S. and Gatski, T.B., 'Modelling the Pressure-strain Correlation of Turbulence: an Invariant System Dynamic Approach,' *Journal of Fluid Mechanics,* vol. 227, pp. 245–272, 1991 17

Spisso, I., 'Parametric and Optimization Study: OpenFOAM and Dakota,' *HPC Enabling of OpenFOAM for CFD Applications,* CINECA, Casalecchio di Reno, Bologna, Italy, 26 November, 2012 33

Staiger, M. and Stetter, H., 'Aerodynamic Response of Axial Fan Bladings to Non- uniform Inlet Flow Fields,' *Institution of Mechanical Engineers Seminar on Installation Effects in Fan Systems*, 28 September, London, UK, 1993 181

Sugano, H., Yamaguchi, N. and Taguchi, S., 'A Study on the Ash Erosion of Axial Induced Draft Fans of Coal-fired Boilers,' *Mitsubishi Heavy Industries Technical Review*, vol. V19(N1), pp. 12-22, 1982 256

Suzuki, M. and Yamamoto, M. (2011), 'Numerical Simulation of Sand Erosion Phenomena in a Single-stage Axial Compressor,' *Journal of Fluid Science and Technology*, vol. 6, pp. 98 –113, 2011 257

Sweby, P.K., 'High Resolution Schemes using Flux Limiters for Hyperbolic Conservation Laws,' *SIAM Journal on Numerical Analysis,* vol. 21(5), pp. 995–1011, 1984 80, 112

Tabakoff, W., Kotwal, R. and Hamed, A., 'Erosion Study of Different Materials Affected by Coal Ash Particles,' *Wear,* vol. 52(1), pp. 161–173, 1979 26, 238, 244, 257, 265, 266, 269, 273, 305, 306, 329

Takata, H. and Tsukuda, Y., 'Stall Margin Improvement by Casing Treatment—its Mechanism and Effectiveness,' *Transactions of the ASME, Journal of Engineering for Gas Turbines and Power,* vol. 99(1), pp. 121–133, 1977 46

Takizawa, K., Henicke, B., Montes, D., Tezduyar, T.E., Hsu, M.-C. and Bazilevs, Y.,'Numerical-performance Studies for the Stabilized Space-time Computation of Wind-turbine Rotor Aerodynamics,' *Computational Mechanics,* vol. 48(6), pp. 647–657, 2011 48, 57

Takizawa, K., Henicke, B., Tezduyar, T.E., Hsu, M.-C. and Bazilevs, Y., 'Stabilized Space-time Computation of Wind-turbine Rotor Aerodynamics,' *Computational Mechanics,* vol. 48(3), pp. 333–344, 2011 48

TASCflow, *TASCflow Documentation, ANSYS Canada Ltd, Waterloo, Ontario, Canada, 2003 187*

Teixeira, J.A., Tourlidakis, A., Ivey, P.C., Sheard, A.G., Molin, J.A. and Kinghorn, I.R. , 'Computational Analysis of an Industrial Axial-flow Fan Including Installation Effects,' *Proceedings of the IMechE Advances of CFD in Fluid Machinery Design Seminar,* 18 November, London, UK, 2003 184

Tezduyar, T.E., 'Computation of Moving Boundaries and Interfaces and Stabilization Parameters,' *International Journal for Numerical Methods in Fluids,* vol. 43(5), pp. 555 –575, 2003 48, 53

Tezduyar, T.E., Mittal, S., Ray, S.E. and Shih, R., 'Incompressible Flow Computations with Stabilized Bilinear and Linear Equal-order-interpolation Velocity-pressure Elements,' *Computer Methods in Applied Mechanics and Engineering,* vol. 95(2), pp. 221–242, 1992 48

Tezduyar, T.E., Park, Y.J. and Deans, H.A., 'Finite Element Procedures for Time- dependent Convection–diffusion-reaction Systems,' *International Journal for Numerical Methods in Fluids,* vol. 7(10), pp. 1013–1033, 1987 47

Tezduyar, T.E., 'Stabilized Finite Element Formulations for Incompressible Flow Computations,' *Advances in Applied Mechanics,* vol. 28, pp. 1–44, 1992 48

Tezduyar, T.E. and Osawa, Y. , 'Finite Element Stabilization Parameters Computed from Element Matrices and Vectors,' *Computer Methods in Applied Mechanics and Engineering,* vol. 190, pp. 411–430, 2000 48, 53

Tezduyar, T.E. and Park, Y.J., 'Discontinuity-capturing Finite Element Formulations for Nonlinear Convection–diffusion-reaction Equations,' *Computer Methods in Applied Mechanics and Engineering,* vol. 59(3), pp. 307–325, 1986 47, 48, 52, 54

Thakur, S., Lin, W. and Wright, J. , 'Prediction of Flow in Centrifugal Blower using Quasi-steady Rotor-stator Models,' *Journal of Engineering Mechanics,* vol. 128, pp. 1039–1049, 2002 298

Thiart, G.D. and von Backstrom, T.W., 'Numerical Simulation of the Flow Field Near an Axial Flow Fan Operating under Distorted Inflow Conditions,' *Journal of Wind Engineering and Industrial Aerodynamics,* vol. 45(2), pp. 189–214, 1993 19

Thompson, D.W., King, P.I. and Rabe, D.C., 'Experimental Investigation of Stepped Tip Gap Effects on the Performance of a Transonic Axial-flow Compressor Rotor,' *Transactions of the ASME, Journal of Turbomachinery,* vol. 120(3), pp. 477– 486, 1998 46

Tropea, C., Yarin, A.L. and Foss, J.F. (Eds), *Springer Handbook of Experimental Fluid Mechanics,* Springer, Paris, France, 2007 4

Uchida, S., Nakamura, Y. and Ohsawa, M., 'Experiments on the Axisymmetric Vortex Breakdown in a Swirling Air Flow,' *Transactions of the Japan Society for Aeronautical Space Sciences,* vol. 27, pp. 206–216, 1985 56

US Department of Energy, 'Energy Conservation Standards Rulemaking Framework for Commercial and Industrial Fans and Blowers,' US Department of Energy, 1 February, 2013 4

Uselton, R.B., Cook, L.J. and Wright, T., 'Fan with Reduced Noise Generation,' patent no. US 7,351,041 B2, 1 April, 2008 47

Vad, J., Bencze, F., Corsini, A. and Rispoli, F., 'Non-free Vortex Flow Effects in an Axial Flow Rotor,' *Periodica Polytechnica, Mechanical Engineering,* vol. 45(2), pp. 201–216, 2001 76, 110, 203

Vad, J., 'Incorporation of Forward Blade Sweep in the Non-free Vortex Design Method of Axial Flow Turbomachinery Rotors,' *Periodica Polytechnica, Mechanical Engineering,* vol. 45(2), pp. 217–237, 2001 76, 110, 203

Vad, J. and Horváth, C., 'Study on the Effects of Axial Clearance Size on the Operation of an Axial Flow Electric Motor Cooling Fan,' *Proceedings of the 10th European Turbomachinery Conference,* Lappeenranta, Finland, 15–19 April, pp. 543–552, 2013 9

van der Spuy, S.J., von Backström, T.W. and Kröger, D.G., 'An Evaluation of Simplified Methods to Model the Performance of Axial Flow Fan Arrays,' *R & D Journal of the South African Institution of Mechanical Engineering,* vol. 26, pp. 12–20, 2010 25

Vanella, M., Piomelli, U. and Balaras, E., 'Effect of Grid Discontinuities on Large- eddy Simulation Statistics and Flow Fields,' *Journal of Turbulence,* vol. 9, paper ID 32, 2008 19

Vardy, A.E., 'Unsteady Airflows in Rapid Transit Systems Part 1: Measurements on the London Transport Victoria Line,' *Proceedings of the Institution of Mechanical Engineers,* vol. 194(1), pp. 341–348, 1980 137, *140,* 141, 142, 148

Vardy, A.E., 'Unsteady Airflows in Rapid Transit Systems Part 2: Theoretical Background and Design Parameters,' *Proceedings of the Institution of Mechanical Engineers*, vol. 194(1), pp. 349–356, 1980 137

Vasudeva Karanth, K. and Yagnesh Sharma, N., 'CFD Analysis on the Effect of Radial Gap on Impeller-diffuser Flow Interaction as Well as on the Flow Characteristics of a Centrifugal Fan,' *International Journal of Rotating Machinery*, vol. 2009, paper no. 293508, pp. 1–8, 2009 298

Venturini, P., Borello, D., Iossa, C., Lentini, D. and Rispoli F., 'Modelling of Multiphase Combustion and Deposit Formation and Deposit Formation in a Biomass-fed Boiler,' *Energy*, vol. 35(7), pp. 3008 –3021, 2010 26, 231

Venturini, P., Borello, D., Iossa, C., Lentini, D. and Rispoli F., 'Modelling of Multiphase Combustion and Deposit Formation and Deposit Formation in a Biomass-fed Boiler,' *Energy*, vol. 35(7), pp. 3008 –3021, 2010 237, 262, 264, 305, 328, 330

Venturini, P., 'Modelling of Particle-wall Deposition in Two-phase Gas-solid Flow,' PhD thesis, Sapienza Università di Roma, Rome, Italy, 2010 26, 330

Wadia, A.R., Szucs, P.N. and Crall, D.W., 'Inner Workings of Aerodynamic Sweep,' *Transactions of the ASME, Journal of Turbomachinery,* vol. 120(4), pp. 671–682, 1998 46

Wallis, R.A., *Axial Flow Fans: Design and Practice,* George Newnes Ltd, London, UK, 1961 76, 109, 203

Wang, L.P., 'On the Dispersion of Heavy Particles by Turbulent Motion,' PhD thesis, Washington State University, Pullman, WA, USA, 1990 231, 262, 304, 328

Wang, L.P. and Stock, D.E., 'Dispersion of Heavy Particles by Turbulent Motion,' *American Meteorological Society*, vol. 50(13), pp. 1897–1913, 1993 343, 344

Weller, H.G., Tabor, G., Jasak, H. and Fureby, C., 'A Tensorial Approach to Continuum Mechanics Using Object-Oriented Technique,' *Computers in Physics*, vol. 12(6), pp 620–631, 1998 298, 300, 327

Wilcox, D.C., 'Reassessment of the Scale-determining Equation for Advanced Turbulence Models,' *AIAA Journal,* vol. 26(11), pp. 1299–1310, 1988 9

Wilcox, D.C., 'Turbulence Modelling for CFD,' DCW Industries Inc., CA, USA, 1993 9

William-Louis, M. and Tournier, C., 'A Wave Signature Based Method for the Prediction of Pressure Transients in Railway Tunnels,' *Journal of Wind Engineering and Industrial Aerodynamics*, vol. 93(6), pp. 521–531, 2005 137

Wolfram, D. and Carolus, T.H., 'Experimental and Numerical Investigation of the Unsteady Flow Field and Tone Generation in an Isolated Centrifugal Fan Impeller,' *Journal of Sound and Vibration,* vol. 329(21), pp. 4380 — 4397, 2010 22, 29

Yang, Y., Lucius, A. and Brenner, G., '3D Unsteady CFD Simulation of the Pressure Fluctuation in a Centrifugal Fan,' *Proceedings of the Fan 2012 Conference,* Senlis, France, 18–20 April, 2012 17, 21, 22, 29

Zhang, M.J., Pomfret, M.J. and Wong, C.M., 'Three-dimensional Viscous Flow Simulation in a Backswept Centrifugal Impeller at the Design Point,' *Computers & Fluids, vol. 25(5), pp. 497–507, 1996 297–8*

Zhao, Y., Song, L., Wenqi, H., Weixiong, W., Dongtao, H. and Zhichi, Z., 'Numerical Simulation of Flow Field for a Whole Centrifugal Fan and Analysis of the Effects of Blade Inlet Angle and Impeller Gap,' *HVAC&R Research*, vol. 11(2), pp. 263–283, 2005 298

Author Index

Page numbers in italics refer to figures or tables

Subject Index

Page numbers in italics refer to figures or tables